Lecture Notes in Control and Information Sciences

Edited by M. Thoma and A. Wyner

Vol. 62: Analysis and Optimization of Systems
Proceedings of the Sixth International Conference on Analysis and Optimization of Systems
Nice, June 19–22, 1984
Edited by A. Bensoussan, J. L. Lions
XIX, 591 pages. 1984.

Vol. 63: Analysis and Optimization of Systems
Proceedings of the Sixth International Conference on Analysis and Optimization of Systems
Nice, June 19–22, 1984
Edited by A. Bensoussan, J. L. Lions
XIX, 700 pages. 1984.

Vol. 64: Arunabha Bagchi
Stackelberg Differential Games in Economic Models
VIII, 203 pages, 1984

Vol. 65: Yaakov Yavin
Numerical Studies in Nonlinear Filtering
VIII, 273 pages, 1985.

Vol. 66: Systems and Optimization
Proceedings of the Twente Workshop
Enschede, The Netherlands, April 16–18, 1984
Edited by A. Bagchi, H. Th. Jongen
X, 206 pages, 1985.

Vol. 67: Real Time Control of Large Scale Systems
Proceedings of the First European Workshop
University of Patras, Greece, Juli 9–12, 1984
Edited by G. Schmidt, M. Singh, A. Titli, S. Tzafestas
XI, 650 pages, 1985.

Vol. 68: T. Kaczorek
Two-Dimensional Linear Systems
IX, 397 pages, 1985.

Vol. 69: Stochastic Differential Systems – Filtering and Control
Proceedings of the IFIP-WG 7/1 Working Conference
Marseille-Luminy, France, March 12-17, 1984
Edited by M. Metivier, E. Pardoux
X, 310 pages, 1985.

Vol. 70: Uncertainty and Control
Proceedings of a DFVLR International Colloquium
Bonn, Germany, March, 1985
Edited by J. Ackermann
IV, 236 pages, 1985.

Vol. 71: N. Baba
New Topics in Learning Automata Theory and Applications
VII, 231 pages, 1985.

Vol. 72: A. Isidori
Nonlinear Control Systems: An Introduction
VI, 297 pages, 1985.

Vol. 73: J. Zarzycki
Nonlinear Prediction Ladder-Filters for Higher-Order Stochastic Sequences
V, 132 pages, 1985.

Vol. 74: K. Ichikawa
Control System Design based on Exact Model Matching Techniques
VII, 129 pages, 1985.

Vol. 75: Distributed Parameter Systems
Proceedings of the 2nd International Conference, Vorau, Austria 1984
Edited by F. Kappel, K. Kunisch, W. Schappacher
VIII, 460 pages, 1985.

Vol. 76: Stochastic Programming
Edited by F. Archetti, G. Di Pillo, M. Lucertini
V, 285 pages, 1986.

Vol. 77: Detection of Abrupt Changes in Signals and Dynamical Systems
Edited by M. Basseville, A. Benveniste
X, 373 pages, 1986.

Vol. 78: Stochastic Differential Systems
Proceedings of the 3rd Bad Honnef Conference, June 3–7, 1985
Edited by N. Christopeit, K. Helmes, M. Kohlmann
V, 372 pages, 1986.

Vol. 79: Signal Processing for Control
Edited by K. Godfrey, P. Jones
XVIII, 413 pages, 1986.

Vol. 80: Artificial Intelligence and Man-Machine Systems
Edited by H. Winter
IV, 211 pages, 1986.

For information about Vols. 1–61 please contact your bookseller or Springer-Verlag.

Lecture Notes in
Control and
Information Sciences

Edited by M. Thoma and A. Wyner

135

H. Nijmeijer
J. M. Schumacher (Eds.)

Three Decades of Mathematical System Theory

A Collection of Surveys at the Occasion of the 50th Birthday of Jan C. Willems

Springer-Verlag Berlin Heidelberg GmbH

Editors
Hendrik Nijmeijer
Department of Applied Mathematics
University of Twente
P. O. Box 217
7500 AE Enschede
The Netherlands

Johannes M. Schumacher
Centre for Mathematics and Computer Science
P. O. Box 4079
1009 AB Amsterdam
The Netherlands
and
Department of Economics
Tilburg University
P. O. Box 90153
5000 LE Tilburg
The Netherlands

ISBN 978-3-540-51605-7 ISBN 978-3-540-46709-0 (eBook)

DOI 10.1007/978-3-540-46709-0

Originally published by Springer-Verlag Berlin Heidelberg New York in 1989.

2161/3020-543210 Printed on acid-free paper.

Preface

The year of birth of a scientific discipline is not often clearly defined. However, in the case of mathematical system theory, the year 1959 is a strong candidate, with only 1958 and 1960 as serious opponents. One may take as evidence the testimony of George S. Axelby, the founding editor of both the *IEEE Transactions on Automatic Control* and the IFAC journal *Automatica*:

> The year 1959 was the prelude to drastic changes in the control field. (...) Then, the first IFAC Congress was held in Moscow, USSR, in June 1960. Three papers were presented that were to revolutionize the theory of automatic control and set the direction of research for years to come. They were the papers by Kalman, Bellman, and Pontryagin. It seemed that almost immediately after the IFAC Congress all papers were involved in modern control theory and the use of state variables with the theorem, lemma, proof format.

(Quoted from a speech held in Los Angeles on December 10, 1987, at the 24th IEEE Conference on Decision and Control; *IEEE Control Systems Magazine 8-2* (1988), p. 98.) The above words are about control, but, of course, system theory concerns more. Here's a description of what happened to filtering in the same period, following the early contributions of Wiener and Kolmogorov:

> As could be expected, there were many researchers who advanced, reexamined, reconsidered, generalized, etc., this Wiener-Kolmogorov theory and many applications of it were reported. In the opinions of some, these efforts had long reached the point of diminishing returns and the *IEEE Transactions on Information Theory* in fact felt it to be necessary to publish an editorial which essentially told these authors — albeit in nicer terms — to "get off it and get on with something else". This happened in 1958 at about the same time that Kalman and Bucy were putting together their papers which would result in one of the most rapid shifts of attention ever to be witnessed in a research field!

These lines come from the paper "Recursive Filtering", *Statistica Neerlandica 32* (1978), pp. 1-39 (the quote is on pp. 6-7). The author is Jan C. Willems, who himself turned twenty in 1959 and was largely innocent of the great turnover so aptly described by him. This was soon going to change. In 1968, Jan Willems received his doctorate in electrical engineering from MIT and started a career that would make him one of the great contributors to the emerging field. His early work is concerned with the stability analysis of nonlinear systems using Lyapunov techniques. Later on, he undertook a general study of dissipativity and the linear-quadratic optimal control problem, contributed to realization theory, and introduced the highly fruitful concept of 'almost invariance' in geometric control theory. In recent years, he has been systematically working out the consequences of his definition of a system as a 'behavior', a set of trajectories which may be described in various ways. A man of many ideas, Jan Willems is guided by a strong intuition and always looks for the basic issues. As a Professor at the University of Groningen, he has been responsible to a large extent for the flourishing of System Theory in the Netherlands; this has included the founding of the Dutch National Graduate School in System and Control Theory, of which he is Chairman. His leading position in the international system theory community is illustrated by the fact that he is Managing Editor of two prominent journals in the field, the *SIAM Journal on Control and Optimization* and *Systems and Control Letters.*

This year, Jan Willems will be fifty, and the field of mathematical system theory is three decades old. This volume has been compiled to let it serve as a surprise gift for Jan on his birthday, September 18, but also to look back on what system theory has achieved in thirty years, and to look ahead for the new challenges that are facing us. The contributions were written by invitation. All subjects covered in this book are related in some way to work done by Jan Willems, and all authors are related in some way to Jan Willems himself. In a volume on system theory, these are inactive constraints. The present book covers the wide area of mathematical system theory, and in particular indicates the great variety of methods that are being applied in this field. The contributors have been asked to write a survey-like paper, discussing past, present and future of a particular research field. We have also encouraged the authors, who are all responsible for leading contributions in their field of writing, to apply personal taste in selecting trends that they feel are important.

Jan Willems has many friends, and many among these are prominent system theorists. We, as the editors of this volume, have tried to keep the anniversary project manageable. We have not asked all of Jan's friends to contribute, and our selection is to some extent arbitrary.

We would like to thank the authors, for their enthousiastic response and for their fine papers. In particular, we are grateful to Jacques L. Willems, who happens to turn fifty on the same day as Jan, for his willingness to write a contribution with a more personal flavour than one usually finds in scientific articles. Furthermore, we thank the reviewers, who have helped the authors and us a lot to polish the contributions. Finally, our thanks go to Springer Verlag, for its willingness to publish the volume in the series *Lecture Notes in Control and Information Sciences* with its wide distribution.

Enschede/Amsterdam, June 1989

Henk Nijmeijer
Hans Schumacher

Contents

The Cascade Structure in System Theory

A. C. Antoulas

Department of Electrical and Computer Engineering
Rice University
Houston, Texas 77251-1892, U. S. A.
and
Mathematical System Theory
E. T. H. Zürich
CH-8092 Zürich, Switzerland

Abstract. An overview is presented of results which show the central role of the cascade structure in linear system theory. The first group of results is related to the recursive realization problem while the second group of results is related to passive network synthesis.

1. INTRODUCTION.

The cascade interconnection of two-pairs (systems having two inputs and two outputs) has played a major role in network theory. The impetus was given by the landmark result of Darlington's in 1939, who showed that every positive real (or bounded real) function can be synthesized as a *cascade* interconnection of lossless two-pairs terminated by a resistor. A quarter of a century later, Belevitch generalized this result to matrix-valued positive-real (or bounded-real) functions, whereby the *scalar* two-pairs are replaced by *matrix* two-pairs. This topic is also closely related to scattering matrix synthesis which has recently been investigated in detail in a series of papers by Dewilde and Dym (see e. g. Dewilde and Dym [1989]).

A few years ago Delsarte, Genin, and Kamp [1981] pointed out that the celebrated Nevanlinna-Pick recursive interpolation algorithm is closely related to the cascade structure. Furthermore, since stability tests for polynomials can be formulated equivalently as Nevanlinna-Pick problems, they are also related to the cascade structure. Actually, Vaidyanathan and Mitra [1987] showed that the classical Schur-Cohn test (for polynomials in the discrete transform variable z) as well as the Routh-Hurwitz test (for polynomials in the continuous transform variable s) can be interpreted in a unified way in terms of the cascade structure. Moreover, the parameters

Support was provided by N. S. F. through Grant ECS - 05293.

which define this decomposition can be used to determine the root distribution of these polynomials with respect to the unit circle and the imaginary axis, respectively.

Independently of the above developments, it has been recognized in the past decade, that the cascade structure plays a fundamental role in system theory; this is apparently unrelated with the passive network synthesis results. This came about by noticing that the cascade strucure is closely connected with the problem of recursive realization of a finite or infinite sequence of matrices or, equivalently, of a formal power series. Actually, Kalman [1979] noticed that the scalar version of the recursive realization problem can be represented as a *ladder* interconnection, which is equivalent to a continued fraction decomposition (well known in network theory). A few years later it was shown that the general recursive realization problem for matrix formal power series can be solved using the *cascade* interconnection of matrix two-pairs (Antoulas [1986]). It follows that in the scalar case the ladder interconnection is a special case of the cascade interconnection. This also shows that several results obtained in connection with the scalar recursive realization problem are actually connected with the cascade structure. Very recently (c.f. Willems and Antoulas [1989]) it has been recognized that the cascade structure is actually involved in the solution of the general (deterministic) modeling problem of arbitrary time series. These results are based on the powerful new framework of time series modeling introduced by Willems [1986-1987].

Besides the fact that the structure of the recursive realization problem is completely revealed (and the recursive Berlekamp-Massey algorithm shown to be connected to the cascade structure) there are further results of interest. Introducing the concepts of *jumps*, the structure of an arbitrary sequence of numbers (or formal power series) is revealed. From this the remarkable result follows that the Cauchy index of a rational function can be recursively computed from its cascade decomposition. Consequently, the classical test for the stability of a polynomial (obtained by decomposing it to even and odd part) has an interpretation in terms of the cascade structure. Finally, a connection between the cascade structure and geometric control theory can also be established (see Kalman [1979] and Antoulas and Bishop [1987]).

In the next section the cascade interconnection of two-pair systems is defined. In the first part of section 3, the consequences of a unimodularity assumption on the cascade interconnection are explored. This leads to the results related to recursive realization which are summarized in section 4; in section 4.1 results in connection with ladder interconnections are displayed. In the second part of section 3, lossless two-pair systems and related concepts are introduced; they are used in section 5, to give a brief overview of the relationship between the cascade structure, the Nevanlinna-Pick algorithm and various stability tests.

2. GENERAL DEFINITIONS AND NOTATION.

We will consider linear time-invariant systems, denoted by Σ, with two sets of inputs u, $\bar{u}$ and two sets of outputs y, $\bar{y}$, with dimensions

$$u, \bar{y} \in \mathbf{R}^m, y, \bar{u} \in \mathbf{R}^p.$$

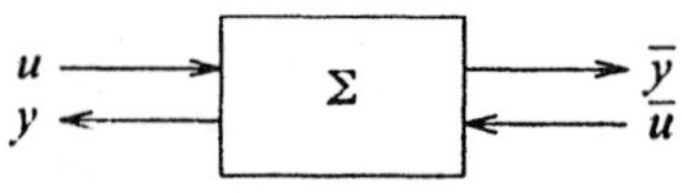

Such systems are known in network theory as *two-port* or *two-pair* systems. A *scalar* two-pair system is one for which $m = p = 1$. These systems can be described by means of the *chain parameters* or the *transfer parameters,* which are *rational* matrices in the variable σ. The former are:

$$\begin{aligned} &X(\sigma) \in \mathbf{R}^{p \times p}(\sigma), \quad L(\sigma) \in \mathbf{R}^{m \times p}(\sigma), \\ &Y(\sigma) \in \mathbf{R}^{p \times m}(\sigma), \quad M(\sigma) \in \mathbf{R}^{m \times m}(\sigma), \quad \det M(\sigma) \neq 0, \end{aligned} \tag{2.1a}$$

arranged in the *chain parameter matrix*

$$V(\sigma) := \begin{bmatrix} X(\sigma) & Y(\sigma) \\ L(\sigma) & M(\sigma) \end{bmatrix} \in \mathbf{R}^{(p+m)\times(p+m)}(\sigma), \tag{2.1b}$$

which satisfies the relationship

$$\begin{bmatrix} y(\sigma) \\ u(\sigma) \end{bmatrix} = V(\sigma) \begin{bmatrix} \bar{u}(\sigma) \\ \bar{y}(\sigma) \end{bmatrix}. \tag{2.1c}$$

The latter are:

$$\begin{aligned} &Z(\sigma) \in \mathbf{R}^{p \times m}(\sigma), \quad \Theta^{12}(\sigma) \in \mathbf{R}^{p \times p}(\sigma), \\ &\Theta^{21}(\sigma) \in \mathbf{R}^{m \times m}(\sigma), \quad \Theta^{22}(\sigma) \in \mathbf{R}^{m \times p}(\sigma), \quad \det \Theta^{21}(\sigma) \neq 0, \end{aligned} \tag{2.2a}$$

arranged in the *transfer parameter matrix*

$$\Theta(\sigma) := \begin{bmatrix} Z(\sigma) & \Theta^{12}(\sigma) \\ \Theta^{21}(\sigma) & \Theta^{22}(\sigma) \end{bmatrix} \in \mathbf{R}^{(p+m)\times(p+m)}(\sigma), \tag{2.2b}$$

which satisfies the relationship

$$\begin{bmatrix} y(\sigma) \\ \bar{y}(\sigma) \end{bmatrix} = \Theta(\sigma) \begin{bmatrix} u(\sigma) \\ \bar{u}(\sigma) \end{bmatrix}. \tag{2.2c}$$

Notice that the $(1, 1)$ element of $\Theta(\sigma)$ is singled out and denoted by Z instead of Θ^{11} as it will be of special significance in the sequel.

The chain and transfer parameters are related as follows:

$$\begin{aligned} &Z(\sigma) = Y(\sigma)M^{-1}(\sigma), \quad \Theta^{12}(\sigma) = X(\sigma) - Y(\sigma)M^{-1}(\sigma)L(\sigma), \\ &\Theta^{21}(\sigma) = M^{-1}(\sigma), \qquad \Theta^{22}(\sigma) = -M^{-1}(\sigma)L(\sigma), \end{aligned} \tag{2.3a}$$

and

$$\begin{aligned} &X(\sigma) = \Theta^{12}(\sigma) - Z(\sigma)(\Theta^{21}(\sigma))^{-1}\Theta^{22}(\sigma), \quad Y(\sigma) = Z(\sigma)(\Theta^{21}(\sigma))^{-1}, \\ &L(\sigma) = -(\Theta^{21}(\sigma))^{-1}\Theta^{22}(\sigma), \qquad M(\sigma) = (\Theta^{21}(\sigma))^{-1}. \end{aligned} \tag{2.3b}$$

In the sequel the dependence of the various quantities defined above on σ will be dropped to keep the notation simple.

Several two-pair systems Σ_k, $k = i, i+1, ..., j-1, j$, as defined above, can be interconnected by letting

$$\begin{bmatrix} \bar{u}_k \\ \bar{y}_k \end{bmatrix} = \begin{bmatrix} y_{k+1} \\ u_{k+1} \end{bmatrix}, \quad k = i, i+1, ..., j-1.$$

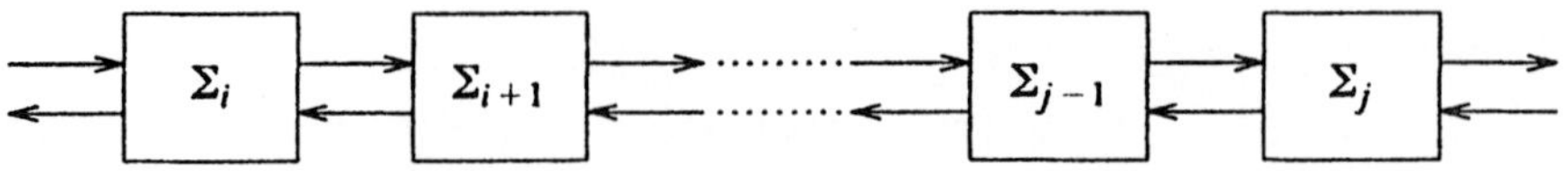

The overall system, denoted by $\Sigma_{i,j}$, will be referred to as the *cascade interconnection* of the subsystems Σ_k, $k = i, i+1, ..., j$. It readily follows that the chain parameter matrix $V_{i,j}$ of $\Sigma_{i,j}$ is

$$V_{i,j} = V_i V_{i+1} \cdots V_{j-1} V_j; \tag{2.4}$$

in other words

$$\begin{bmatrix} y_i \\ u_i \end{bmatrix} = V_{i,j} \begin{bmatrix} \bar{u}_j \\ \bar{y}_j \end{bmatrix}.$$

The (1, 1) entries Z_k of the Θ_k matrices are related in terms of *linear fractional transformations*, sometimes referred to as *homographic transformations*:

$$Z_{\alpha,\gamma} = (Y_{\alpha,\beta} + X_{\alpha,\beta} Z_{\beta+1,\gamma})(M_{\alpha,\beta} + L_{\alpha,\beta} Z_{\beta+1,\gamma})^{-1} =: V_{\alpha,\beta}[Z_{\beta+1,\gamma}], \tag{2.5}$$

for all $i \leqslant \alpha \leqslant \beta < \gamma \leqslant j$. This formula is sometimes interpreted as an *extraction* of the two-pair described by $\Theta_{\alpha,\beta}$ from the system described by $Z_{\alpha,\gamma}$ with *remainder* the system described by $Z_{\beta+1,\gamma}$.

3. SPECIAL PROPERTIES OF THE CHAIN PARAMETER MATRICES.

In this paper we will focus our attention on cascade interconnections where the chain parameter matrices are either *polynomial unimodular* or *lossless*. Some general properties of the corresponding cascade interconnections will be described next.

3.1 POLYNOMIAL UNIMODULAR CHAIN PARAMETER MATRICES.

If the chain parameter matrix V_k of Σ_k is polynomial unimodular, its inverse, denoted by W_k, is also polynomial unimodular; it will be partitioned in the same way as V_k:

$$W_k = \begin{bmatrix} T_k & -Q_k \\ -U_k & R_k \end{bmatrix}, \quad V_k W_k = I_{p+m}. \tag{3.1a}$$

From $\det M_k \neq 0$ follows $\det T_k \neq 0$. It will be assumed that

$$M_k^{-1} L_k, \; U_k T_k^{-1} \; \textit{are strictly proper rational.} \tag{3.1b}$$

Let the degree of the jth column of M_k, i^{th} row of T_k (i.e. i^{th} column of the transposed matrix T_k') be

$$\kappa_j(k) = \deg\,(M_k)_j,\ j \in \underline{m},\ \ \nu_i(k) = \deg\,(T_k')_i,\ i \in \underline{p}. \tag{3.2}$$

(For a positive integer r, $\underline{r} := \{1, 2, \cdots, r\}$.) We will assume without loss of generality, that M_k, T_k satisfy the following *normalization* conditions:

$$\sum_{j \in \underline{m}} \kappa_j(k) = \deg\det M_k,\ \ \sum_{i \in \underline{p}} \nu_i(k) = \deg\det T_k.$$

Such normalized matrices are called *column reduced, row reduced* respectively. In this case the indices $\kappa_j(k)$, $\nu_i(k)$ can be interpreted as the *Kronecker* or *reachability indices*, the *dual Kronecker* or *observability indices* of Z_k, respectively. As a consequence of the above assumptions

$$M_k^{-1},\ T_k^{-1}\ \textit{are proper rational.}$$

For details on reduced polynomial matrices and the Kronecker indices see Kailath [1980, ch. 6.3].

From the unimodularity of V_k it follows that the transfer parameters become:

$$\begin{aligned} Z_k &= Y_k M_k^{-1} = T_k^{-1} Q_k, & \Theta_k^{12} &= T_k^{-1}, \\ \Theta^{21} &= M_k^{-1}, & \Theta^{22} &= -\,M_k^{-1} L_k. \end{aligned} \tag{3.3}$$

The *McMillan degree* $\delta(Z)$ of a rational matrix Z is defined as the dimension of any minimal state space realization of Z. The unimodularity of V_k implies

$$\delta(\Theta_k) = \delta(Z_k). \tag{3.4a}$$

Furthermore, since the polynomial factorizations of Z_k given in (3.3) are coprime, we have

$$\delta(Z_k) = \deg\det M_k = \deg\det T_k. \tag{3.4b}$$

(3.4a) implies that Z_k *uniquely* determines the two-pair Σ_k and consequently the corresponding chain parameter matrix V_k as well. For details, and a proof of (3.4a), see Antoulas [1986].

Consider the cascade interconnection of the two two-pair systems Σ_α, $\Sigma_{\alpha+1}$. If the quantity $L_\alpha Z_{\alpha+1} M_\alpha^{-1}$ is strictly proper rational, i.e.

$$L_\alpha Z_{\alpha+1} M_\alpha^{-1} = C_1\sigma^{-1} + C_2\sigma^{-2} + \cdots,\ \ C_t \in \mathbf{R}^{m \times m}, \tag{3.5}$$

it can be shown that (cf. Antoulas [1986])

$$\delta(Z_{\alpha,\alpha+1}) = \delta(Z_\alpha) + \delta(Z_{\alpha+1}). \tag{3.6a}$$

If the above relationship holds for $\alpha = i, i+1, \cdots, j-1$, then

$$\delta(Z_{i,j}) = \delta(Z_i) + \delta(Z_{i+1}) + \cdots + \delta(Z_{j-1}) + \delta(Z_j). \tag{3.6b}$$

Therefore, if the chain parameter matrices of a cascade interconnection are polynomial unimodular and (3.5) is satisfied, the complexity of the overall system is equal to the sum of the complexities of each one of the subsystems.

Under the assumption (3.5) a minimal state space realization of $Z_{\alpha,\alpha+1}$ can be written down in terms of minimal state space realizations of Θ_α and $Z_{\alpha+1}$, by inspection. Let

$$\Theta_\alpha(\sigma) \; = \; \begin{bmatrix} H_\alpha \\ \tilde{H}_\alpha \end{bmatrix} (\sigma I - F_\alpha)^{-1} (G_\alpha \;\; \tilde{G}_\alpha) + \begin{bmatrix} 0 & J_\alpha \\ \tilde{J}_\alpha & 0 \end{bmatrix}, \tag{3.7a}$$

$$Z_{\alpha+1}(\sigma) \; = \; H_{\alpha+1}(\sigma I - F_{\alpha+1})^{-1} G_{\alpha+1}, \tag{3.7b}$$

be minimal state space realizations. Then

$$F_{\alpha,\alpha+1} \; := \; \begin{bmatrix} F_\alpha & \tilde{G}_\alpha H_{\alpha+1} \\ G_{\alpha+1}\tilde{H}_\alpha & F_{\alpha+1} \end{bmatrix}, \;\; G_{\alpha,\alpha+1} \; := \; \begin{bmatrix} G_\alpha \\ G_{\alpha+1}\tilde{J}_\alpha \end{bmatrix}, \tag{3.7c}$$

$$H_{\alpha,\alpha+1} \; := \; (H_\alpha \;\; J_\alpha H_{\alpha+1}),$$

is a minimal state space realization of

$$Z_{\alpha,\alpha+1} \; = \; V_\alpha[Z_{\alpha+1}]. \tag{3.7d}$$

These formulae can be extended to the cascade interconnection of several subsystems Σ_α, $\alpha = i, i+1, ..., j-1, j$. If these subsystems are such that the corresponding transfer parameter matrices are strictly proper rational (condition which is always satisfied for *scalar* two-pairs) the following triple

$$F_{i,j} = \begin{bmatrix} F_i & \tilde{G}_i H_{i+1} & & & & & \\ G_{i+1}\tilde{H}_i & F_{i+1} & \tilde{G}_{i+1} H_{i+2} & & \mathbf{0} & & \\ & G_{i+2}\tilde{H}_{i+1} & F_{i+2} & \cdot & & & \\ & & & \cdot & & & \\ & & & & \cdot & & \\ & & \mathbf{0} & & F_{j-2} & \tilde{G}_{j-2} H_{j-1} & \\ & & & & G_{j-1}\tilde{H}_{j-2} & F_{j-1} & \tilde{G}_{j-1} H_j \\ & & & & & G_j \tilde{H}_{j-1} & F_j \end{bmatrix},$$

$$G_{i,j} \; = \; (G_i' \;\; 0 \;\; 0 \; \ldots \; 0)', \;\; H_{i,j} \; = \; (H_i \;\; 0 \;\; 0 \; \ldots \; 0),$$

is a minimal state space realization of $Z_{i,j}$. The above formulae constitute the *cascade canonical form*, which shows explicitly how each F_α is *nested* in the overall $F_{i,j}$. For details see Antoulas and Bishop [1987].

As a consequence of the fact that (3.7a) is a minimal realization of $Z_{\alpha,\alpha+1}$, the corresponding state spaces, namely X_α, $X_{\alpha+1}$, $X_{\alpha,\alpha+1}$, satisfy

$$X_{\alpha,\alpha+1} \; = \; X_\alpha \oplus X_{\alpha+1}. \tag{3.8}$$

The isomorphism which appears in the above formula can be converted into equality by using polynomial spaces as introduced by Fuhrmann [1976]. Given a non-singular $n \times n$ polynomial matrix $A(\sigma)$, the set

$$X_A := \{x \in \mathbf{R}^n[\sigma]: A^{-1}x \textit{ is strictly proper rational}\},$$

is a finite dimensional vector space, containing polynomial vectors as elements, and

$$\dim X_A = \deg \det A.$$

With this notation, it turns out that we can define

$$X_\alpha := X_{T_\alpha},\quad X_{\alpha+1} := X_{T_{\alpha+1}},\quad X_{\alpha,\alpha+1} := X_{T_{\alpha,\alpha+1}}.$$

Isomorphism (3.8) then becomes

$$X_{\alpha,\alpha+1} = T_{\alpha+1}X_{T_\alpha} + X_{T_{\alpha+1}}. \tag{3.9}$$

The above formulae can easily be extended to the cascade interconnection of any number of sub-systems (see Antoulas and Bishop [1987]).

3.2 LOSSLESS CHAIN PARAMETER MATRICES.

The concepts and results presented in this sub-section are valid for discrete-time systems (i. e. in this section $\sigma = z$, the discrete-time transform variable). With appropriate modifications however, they are valid for continuous-time systems as well.

In the sequel D will denote the open unit disc in the complex plane $\mathbf{C}$, $\overline{D}$ its closure, $\overline{D}_c$ the complement of $\overline{D}$, and ∂D the unit circle. Superscript $*$ denotes complex conjugation of a scalar or a matrix. Given a rational matrix $A(\sigma)$,

$$A(\sigma)_* := A'(\sigma^{-1}),$$

where prime denotes complex conjugation followed by transposition. A two-pair system Σ is called *all-pass* if its transfer parameter matrix Θ satisfies

$$\Theta(\sigma)_*\Theta(\sigma) = I, \text{ for all } \sigma.$$

Σ is *bounded real (BR)* iff Θ is a bounded real matrix, i. e. all its poles are in D (is stable) and

$$\Theta'(\sigma)\Theta(\sigma) \leq I,\quad \sigma \in \overline{D}_c.$$

Σ is *lossless bounded real (LBR)* iff in addition Θ is unitary on ∂D, i. e.

$$\Theta'(\sigma)\Theta(\sigma) = I,\quad \sigma \in \partial D.$$

Notice that Σ is lossless bounded real iff it is stable all-pass.

In section 5 we will consider *elementary* scalar two-pair systems with chain parameter matrices of the following form:

$$V(\sigma) = \begin{bmatrix} 1 & \xi \\ \xi^* & 1 \end{bmatrix} \begin{bmatrix} 1 & 0 \\ 0 & \pi(\sigma) \end{bmatrix}, \tag{3.10a}$$

where $|\xi| \neq 1$ and $\pi(\sigma)$ is the following elementary scalar all-pass function:

$$\pi(\sigma) = \frac{1 - \sigma\zeta^*}{\sigma - \zeta}.$$

The corresponding transfer parameter matrix is:

$$\Theta = \begin{bmatrix} \xi & 1 - \xi\xi^* \\ 1 & -\xi^* \end{bmatrix} \begin{bmatrix} 1 & 0 \\ 0 & \pi(\sigma) \end{bmatrix}. \tag{3.10b}$$

If $1 - \xi\xi^* > 0$, we can normalize the above two-pair as follows:

$$\overline{V}(\sigma) = \frac{1}{(1 - \xi\xi^*)^{1/2}} V(\sigma). \tag{3.11a}$$

The resulting normalized transfer parameter matrix

$$\overline{\Theta}(\sigma) = \begin{bmatrix} \xi & (1 - \xi\xi^*)^{1/2} \\ (1 - \xi\xi^*)^{1/2} & -\xi^* \end{bmatrix} \begin{bmatrix} 1 & 0 \\ 0 & \pi(\sigma) \end{bmatrix}, \tag{3.11b}$$

is *all-pass*, i. e.

$$\overline{\Theta}(\sigma)_{*}\overline{\Theta}(\sigma) = I. \tag{3.11c}$$

If in addition, $|\zeta| < 1$, the normalized two-pair is LBR. We will refer to Σ with chain parameter matrix defined by (3.10a) as an *elementary unnormalized lossless* two-pair, and to $\overline{\Sigma}$ with chain parameter matrix defined by (3.11a) as an *elementary lossless* two-pair.

4. RECURSIVE REALIZATION AND RELATED TOPICS.

Consider the sequence of $p \times m$ constant matrices

$$S_{1,N-1} = (A_1, A_2, ..., A_{N-1}), \tag{4.1a}$$

which will also be referred to as a sequence of *Markov parameters*. The *realization problem* consists in obtaining a parametrization of *all* $p \times m$ rational matrices $Z_{1,N-1}(\sigma)$ such that

$$Z_{1,N-1}(\sigma) = A_1\sigma^{-1} + A_2\sigma^{-2} + \cdots + A_{N-1}\sigma^{-N+1} + \cdots, \tag{4.1b}$$

i. e. the coefficients of the first $N-1$ terms of the formal power series expansion (or equivalently, the Laurent expansion in the neighborhood of infinity) of the solution match the given $N-1$ Markov parameters. The complexity $\delta(Z_{1,N-1})$, which is defined as the McMillan degree (see sec. 3.1), is to be the parameter. Of particular interest are the minimal complexity realizations.

The basic problem of *existence* and *uniqueness* of realizations was first studied by Kalman (see Kalman, Falb, and Arbib [1968]). In the sequel we will address a deeper question, namely, how $Z_{1,N}$ and $\delta(Z_{1,N})$ depend on N. In other words, we are looking for the function ϕ such that

$$Z_{1,N} = \phi(Z_{1,N-1}). \tag{4.2}$$

This question is closely related to the *recursive* realization problem: given $Z_{1,N-1}$ and the new Markov parameter A_N, compute the updated solution $Z_{1,N}$ as a function of $Z_{1,N-1}$. *As it turns out, the* **cascade** *structure provides the key to obtaining explicit answers to the above questions.*

Let $Z_{1,N-1}$ be a minimal realization of the sequence $S_{1,N-1}$. Consider the coprime polynomial factorizations

$$Z_{1,N-1} = Y_{1,N-1}M_{1,N-1}^{-1} = T_{1,N-1}^{-1}Q_{1,N-1} = \tag{4.3}$$

$$= A_1\sigma^{-1} + \cdots + A_{N-1}\sigma^{-N+1} + \bar{A}_N\sigma^{-N} + \cdots .$$

Because of coprimeness, there exist polynomial matrices $X_{1,N-1}$, $L_{1,N-1}$, $U_{1,N-1}$, $R_{1,N-1}$, such that conditions (3.1a,b) are satisfied. It will be assumed that $M_{1,N-1}$, $T_{1,N-1}$, are column, row reduced (cf. (3.2a,b)). Thus to the original system, characterized by the transfer matrix $Z_{1,N-1}$, we attach a *two-pair* system $\Sigma_{1,N-1}$ whose chain parameter matrix is

$$V_{1,N-1} = \begin{bmatrix} X_{1,N-1} & Y_{1,N-1} \\ L_{1,N-1} & M_{1,N-1} \end{bmatrix}.$$

From (3.4) it follows that

$$\delta(\Theta_{1,N-1}) = \delta(Z_{1,N-1}).$$

Recall the definition of a homographic (or linear fractional) transformation given in (2.5). We can state the following basic

(4.4) **Theorem.** *Given* $\bar{Z} \in \mathbf{R}^{p\times m}(\sigma)$,

$$\hat{Z} := V_{1,N-1}[\bar{Z}] = (Y_{1,N-1} + X_{1,N-1}\bar{Z})(M_{1,N-1} + L_{1,N-1}\bar{Z})^{-1}, \tag{4.5}$$

is a realization of $S_{1,N-1}$ *if and only if*

$$L_{1,N-1}\bar{Z}M_{1,N-1}^{-1} \tag{4.6}$$

is strictly proper rational.

This result shows that given the fact that $\Sigma_{1,N-1}$ encodes the information contained in $S_{1,N-1}$, the parameter $\bar{Z}$ is in one-to-one correspondence with the continuations of $S_{1,N-1}$, namely

$$\bar{Z} \overset{1-1}{\leftrightarrow} (A_N, A_{N+1}, \cdots). \tag{4.7}$$

Consequently, if any of the Markov parameters A_t, $t \geqslant N$, is changed, *only* $\bar{Z}$ will be thereby affected, *not* $\Sigma_{1,N-1}$. The cascade interpretation of this formula is

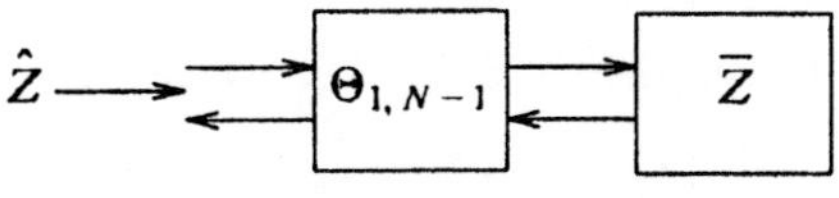

From (3.6a) follows that

$$\delta(\hat{Z}) = \delta(Z_{1,N-1}) + \delta(\overline{Z}); \tag{4.8}$$

this implies that $\hat{Z}$ is minimal if and only if $\overline{Z}$ is minimal.

The above result establishes the fundamental connection between the cascade structure and the realization problem. It shows that the function ϕ in (4.2) is in essence the homographic or linear fractional transformation. The complete solution of the recursive realization problem can now be obtained, based on the above theorem. Given $S_{1,N-1}$ and a minimal realization $Z_{1,N-1}$ thereof, our goal is obtain a parametrization of all realizations of the updated sequence

$$S_{1,N} = (A_1, A_2, \cdots, A_{N-1}, A_N),$$

and in particular, of all minimal ones.

Recall (4.3) as well as the fact that $M_{1,N-1}$ and $T_{1,N-1}$ are column, row reduced with column, row indices $\kappa_j(1,N-1), j \in \underline{m}, \nu_i(1,N-1), i \in \underline{p}$. Let

$$\Lambda_\kappa := \text{diag}\,(\sigma^{\kappa_1(1,N-1)}, \cdots, \sigma^{\kappa_m(1,N-1)}), \tag{4.9a}$$

$$\Lambda_\nu := \text{diag}\,(\sigma^{\nu_1(1,N-1)}, \cdots, \sigma^{\nu_p(1,N-1)}), \tag{4.9b}$$

while

$$M_{hc} \in \mathbf{R}^{m\times m},\ T_{hr} \in \mathbf{R}^{p\times p}, \tag{4.9c}$$

are the constant matrices made out of the highest column, row coefficients of $M_{1,N-1}, T_{1,N-1}$.

According to Theorem (4.4), given A_N, there exists a rational matrix $\overline{Z}$, denoted in this case by Z_N, such that

$$Z_{1,N} = V_{1,N-1}[Z_N], \tag{4.10a}$$

is a realization of $S_{1,N}$ and

$$\delta(Z_{1,N}) = \delta(Z_{1,N-1}) + \delta(Z_N). \tag{4.10b}$$

Moreover, all realizations of $S_{1,N}$ can be obtained in this way. There remains to show *how* Z_N can be determined. Let

$$E := T_{hr}(A_N - \overline{A}_N)T_{hc} =: (\epsilon_{ij}). \tag{4.11a}$$

The sequence

$$S_N := (E_1, \cdots, E_\rho),\ \rho := \max_{i,j}\,\{N - \kappa_j(1,N-1) - \nu_i(1,N-1)\} \tag{4.11b}$$

is defined as follows:

$$(E_t)_{i,j} := \begin{cases} \epsilon_{i,j} \ \textit{iff}\ t = N - \kappa_j(1,N-1) - \nu_i(1,N-1) > 0, \\ 0 \ \textit{iff}\ t = 1, 2, \cdots, N - \kappa_j(1,N-1) - \nu_i(1,N-1) > 0, \\ ? \ \textit{otherwise}, \end{cases} \tag{4.11c}$$

for $j \in \underline{m}$, $i \in \underline{p}$, where "?" stands for an element which can be chosen freely. In the scalar case $p = m = 1$, according to whether $\rho = N - 2n$, $n := \delta(Z_{1,N-1}) = \deg M_{1,N-1}$, is positive or not, $S_N = (0, \cdots, 0, \epsilon)$ or $S_N = (?)$; in the latter case the sequence is completely undetermined. We can now state the following fundamental

(4.12) **Theorem.** *(a)* $Z_{1,N}$ *is a (minimal) realization of* $S_{1,N}$ *if and only if* Z_N *is a (minimal) realization of* S_N.

(b) One minimal realization of the sequence S_N *is given by the formula*

$$Z_N^* := \Lambda_\nu(E\sigma^{-N})\Lambda_\kappa. \tag{4.13}$$

Notice that S_N depends on N, the Kronecker, the dual Kronecker indices, and the difference between the desired and the actual N^{th} Markov parameters. Thus by using the cascade structure, the computational complexity of updating a minimal realization is the same as the computational complexity of determining the realization of a single-term-sequence. The second part of the theorem provides a closed formula expression for one minimal update.

The above theorem has interesting consequences. The jth Kronecker index of the updated system $Z_{1,N}$ depends on the Kronecker and the dual Kronecker indices of $Z_{1,N-1}$ as follows: one of the three relationships given below will occur

$$\kappa_j(1,N) = \kappa_j(1,N-1); \tag{4.14a}$$

$$\kappa_j(1,N) = \kappa_i(1,N-1) > \kappa_j(1,N-1); \tag{4.14b}$$

$$\kappa_j(1,N) = N - \nu_i(1,N-1) > \kappa_j(1,N-1). \tag{4.14c}$$

A similar result holds for the dual Kronecker indices.

The above relationships show that the Kronecker (and the dual Kronecker) indices of the overall minimal realization are non-decreasing functions of N. They also show how the fine structure of the complexity builds up as successive Markov parameters A_t are supplied. Thus $Z_{1,N-1}$ already determines *all* possible fine structures of the updated system $Z_{1,N}$. Which one actually occurs depends on the value of the particular A_N.

The Kronecker indices of $Z_{1,N}$, $Z_{1,N-1}$ satisfy a further relationship, namely

$$\kappa_j(1,N) = \kappa_j(1,N-1) + \kappa_j(N), \tag{4.15}$$

where $\kappa_j(N)$ is the jth Kronecker index of Z_N^*. This shows that for minimal realizations, besides the McMillan degrees, the fine structure of the complexity (the Kronecker indices) is additive as well. Again, (4.15) holds also for the dual Kronecker indices.

We have shown how given $S_{1,N-1}$, a minimal realization $Z_{1,N-1}$ thereof, and the updated sequence $S_{1,N}$, we can construct a minimal Z_N such that

$$Z_{1,N} = V_{1,N-1}[Z_N],$$

or equivalently,

$$V_{1,N} = V_{1,N-1}V_N,$$

represents a minimal realization of $S_{1,N}$. Repeating the procedure, we can take care of further updatings $A_{N+1}, A_{N+2}, \cdots$, using the cascade structure. For details and proofs, see Antoulas

[1986].

The above recursive realization for the scalar case $m = p = 1$, and in particular formula (4.13), is known as the *Berlekamp-Massey algorithm*. The matrix case presented above was solved by Antoulas [1986]; Remark (5.19i) of this reference gives a short history of the problem.

We can now summarize the connection between recursive realization and the cascade structure in the following

(4.16) **Theorem.** *Consider the sequence of* $p \times m$ *constant matrices*

$$S = (A_1, A_2, \cdots, A_N, \cdots),$$

and the associated formal power series

$$Z(\sigma) = A_1\sigma^{-1} + A_2\sigma^{-2} + \cdots + A_N\sigma^{-N} + \cdots .$$

There is a bijective correspondence between S, Z *and a (possibly infinite) sequence of two-pair systems* Σ_i, *described by the corresponding chain parameter matrices* $V_i(\sigma)$. *These systems are interconnected in cascade. Furthermore, for every* $t > 0$ *there exists a positive integer* i_t, $t > 0$, *such that*

$$Z_{1,t} = Y_{1,t}M_{1,t}^{-1},$$

where

$$V_{1,t} = V_1 V_2 \cdots V_t = \begin{bmatrix} X_{1,t} & Y_{1,t} \\ L_{1,t} & M_{1,t} \end{bmatrix},$$

is a minimal realization of the following subsequence of S:

$$S_{1,t} = (A_1, A_2, \cdots, A_{i_t}).$$

The Kronecker indices, the dual Kronecker indices and consequently the McMillan degrees of the subsystems add up to the corresponding quantities for any finite part of the overall system. Rationality of $Z(\sigma)$ *is equivalent to the finiteness of the cascade structure.*

4.1 CASCADE AND LADDER STRUCTURES.

Whenever the Markov parameters are scalar quantities, i.e. $m = p = 1$, the cascade structure can be simplified. First we notice that the chain parameter matrix of the ith subsystem in the cascade decomposition has the form

$$V_i(\sigma) = \begin{bmatrix} 0 & 1 \\ -1 & m_i(\sigma) \end{bmatrix},$$

where m_i is a polynomial of degree δ_i. Consequently, the corresponding transfer parameter matrix is

$$\Theta_i(\sigma) = \frac{1}{m_i(\sigma)} \begin{bmatrix} 1 & 1 \\ 1 & 1 \end{bmatrix}.$$

It readily follows that the formal power series

$$Z(\sigma) = a_1\sigma^{-1} + a_2\sigma^{-2} + \cdots$$

can be expressed as a *continued fraction* in terms of the polynomials m_i:

$$Z(\sigma) = \cfrac{1}{m_1(\sigma) + \cfrac{1}{m_2(\sigma) + \cfrac{1}{\ddots}}}.$$

As stated in Theorem (4.16), rationality of Z is equivalent to the finiteness of the cascade decomposition, which in turn is equivalent to the finiteness of the above continued fraction decomposition.

Thus the cascade (or continued fraction) decomposition in the scalar case becomes:

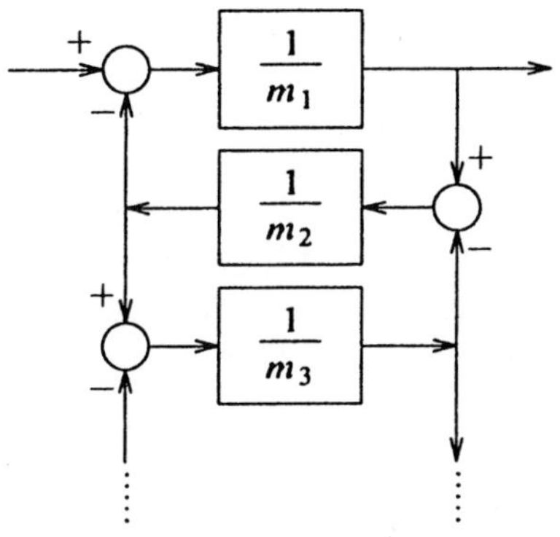

In network theory the above structure is recognized as a *ladder realization* (see e. g. Balabanian and Bickart [1981, ch. 12]).

Clearly, both the cascade and the ladder structures are related to the *Euclidean algorithm*. Let Z be rational:

$$Z(\sigma) = y(\sigma)/m(\sigma), \ \deg m(\sigma) > \deg y(\sigma).$$

Applying the Euclidean algorithm to the (coprime) pair m, y, the successive quotients turn out to be the polynomials m_i:

$$q_{i-1} = q_i m_i + q_{i+1}, \ i \in \underline{k},$$

where $q_0 := m$ and $q_1 := y$, and $q_k = 1$.

The equivalence of the cascade decomposition with continued fractions in the scalar case suggests a way of defining a continued fraction decomposition of rational matrices. Actually, as shown in Antoulas [1986], one can define the *matrix* continued fraction decomposition as the cascade decomposition. Consequently, the cascade decomposition provides a means of defining a matrix Euclidean algorithm (see Antoulas [1986]). For a different way of defining matrix continued fractions and a matrix Euclidean algorithm, see Fuhrmann [1983].

Next we will derive the consequences of relationships (4.14) for the scalar case. Let $\delta(1, t)$ denote the dimension of the minimal realizations of the sequence $(a_1, \cdots, a_t)$. Clearly, if the sequence is rational, for large enough t we have

$$\delta(1, t) = \delta_1 + \delta_2 + \cdots .$$

The positive integer N_i will be called the ith *jump point* of S iff

$$\delta(1, t) = \delta_1 + \delta_2 + \cdots + \delta_i, \text{ for all } N_i \leq t < N_{i+1},$$

i.e. iff the dimension of minimal realizations of all sequences $(a_1, \cdots, a_t)$ with t in the above interval, is equal to the sum of the first i δ's. Moreover, δ_i will be referred to as the ith *jump* of S. Clearly, $\delta(1, t)$ is a *staircase* function of t, with jumps of size δ_i taking place at the points $t = N_i$. Because of (4.14), for *any* sequence, there is a bijective correspondence between the set of jump sizes and the set of jump points. Let

$$\delta := (\delta_1 \ \delta_2 \ \cdots \ \delta_k)', \quad \mathbf{N} := (N_1 \ N_2 \ \cdots \ N_k)'.$$

It follows that

$$\mathbf{N} = A\delta,$$

where A is a lower triangular Töplitz matrix with first column equal to $(1 \ 2 \ 2 \ \cdots \ 2)'$; A^{-1} is also a lower triangular Töplitz matrix with first column equal to $(1 \ -2 \ 2 \ \cdots \ (-1)^k 2)$. These relationships imply in particular

$$\delta_{i+1} + \delta_i = N_{i+1} - N_i;$$

This means that *the larger the interval between jumps, the larger the new jump*. It is interesting to notice that for a generic sequence

$$\delta_i = 1, \text{ for all } i \Rightarrow N_i = 2i - 1, \text{ for all } i.$$

In conclusion, to any *arbitrary* sequence of numbers $S = (a_1, a_2, \cdots)$, one can associate two sets of positive numbers δ_i, N_i which satisfy the above relationship. This result for scalar sequences was first derived by Kalman [1979].

An issue of interest in applied mathematics, which is related to the cascade structure is that of the *Cauchy index* of a rational function

$$Z(\sigma) = p(\sigma)/q(\sigma) = a_1\sigma^{-1} + a_2\sigma^{-2} + \cdots ,$$

which is defined as the number of jumps of Z from $-\infty$ to $+\infty$ minus the number of jumps from $+\infty$ to $-\infty$. The Cauchy index is useful in investigating questions involving the distribution of roots of polynomials with respect to a given region.

A classical result asserts that the Cauchy index of Z is equal to the *signature* of the Hankel matrix associated with the sequence $(a_1, a_2, \cdots)$. (Recall that the signature of a symmetric matrix is equal to the number of positive minus the number of negative eigenvalues.)

It was first shown by Kalman [1979] that the cascade decomposition of Z provides a way of computing the Cauchy index or equivalently the signature of the associated Hankel matrix. Let

$$m_i(\sigma) = \mu^i \sigma^{\delta_i} + \textit{lower order terms.}$$

The distribution of the positive and the negative eigenvalues of the corresponding Hankel matrix can be recursively computed as a function of δ_i and μ^i:

(a) If δ_i is even, the signature remains unchanged, i.e. the number of positive and the number of negative eigenvalues increase by $\delta_i/2$.

(b) If δ_i is odd, the signature changes by one. If $\mu^i > 0$, the signature increases by one, i.e. the number of positive eigenvalues increases by $(\delta_i + 1)/2$, while the number of negative eigenvalues increases by $(\delta_i - 1)/2$. If $\mu^i < 0$, the signature decreases by one, i.e. the number of positive eigenvalues increases by $(\delta_i - 1)/2$ while that of negative eigenvalues increases by $(\delta_i + 1)/2$.

It is well known (see e.g. Gantmacher [1960]) that the above procedure applied to the rational function

$$Z(\sigma) = \frac{m_{even}(\sigma)}{m_{odd}(\sigma)} \text{ or } Z(\sigma) = \frac{m_{odd}(\sigma)}{m_{even}(\sigma)} = \alpha_1\sigma^{-1} + \alpha_2\sigma^{-2} + \cdots,$$

where m_{even}, m_{odd} are the even, odd parts of a given polynomial m, leads to a count of the number of roots of m in the left-, right-half of the complex plane. The polynomial m is *stable*, i.e. has all its roots in the left-half of the complex plane if and only if all the jumps of the sequence $(\alpha_1, \alpha_2, \cdots)$ are equal to one (i.e. the sequence is generic) and the leading coefficients μ^i of all subsystems of the cascade (ladder) decomposition of Z are positive.

By combining the above result with the general state-space representation of systems decomposed in cascade, a canonical form of (F, G, H) is derived; the elements of F are the Hurwitz determinants, i.e. stability checking is reduced to checking the positivity of the non-zero elements of F (see Kalman [1979]).

A final application of the cascade interconnection is in *geometric control* (see Wonham [1979]). Given a scalar triple (F, g, h) subspaces V of the state-space satisfying

$$FV \subset V + \text{im } g,$$

$$V \subset \ker h,$$

$$\dim V\text{: } \textit{maximal},$$

are of central importance. It is easy to show that there is a unique subspace, denoted by V^*, which satisfies the above conditions. Let

$$Z(\sigma) := h(\sigma I - F)^{-1}g = a_1\sigma^{-1} + a_2\sigma^{-2} + \cdots.$$

Consider the cascade decomposition of Z. It follows that V^* is equal to the state-space of the interconnection of all but the first subsystem in this decomposition (see Kalman [1979]). This result can be generalized to the matrix case (see Antoulas and Bishop [1987]). A similar result was shown by Fuhrmann [1983].

5. STABILITY ISSUES.

In this section we will show that the well-kown Nevanlinna-Pick recursive algorithm is closely related to the cascade structure. We will also show that various stability tests for polynomials have also an interpretation in terms of the cascade structure.

5.1 THE NEVANLINNA-PICK ALGORITHM.

Consider the array of pairs (σ_i, ϕ_i), $i \in \underline{N}$, $|\sigma_i| > 1$. For simplicity of exposition we will assume that $\sigma_i \neq \sigma_j$, $i \neq j$. The *Nevanlinna-Pick interpolation problem* consists in parametrizing all rational functions $\phi(\sigma)$ such that

$$\phi(\sigma_i) = \phi_i,\ i \in \underline{N}, \quad \text{and} \quad |\phi(\sigma)| \leqslant M,\ |\sigma| > 1,$$

where M is a given positive constant. With

$$\phi^k(\sigma) := n^k(\sigma)/d^k(\sigma),\ k = 0, 1, \cdots, N,$$

let

$$V_k(\sigma) := \begin{bmatrix} 1 & \dfrac{\phi_k^{k-1}}{M} \\ \dfrac{(\phi_k^{k-1})^*}{M} & 1 \end{bmatrix} \begin{bmatrix} 1 & 0 \\ 0 & \pi(\sigma) \end{bmatrix},$$

where $\pi(\sigma) = \dfrac{1 - \sigma\sigma_k{}^*}{\sigma - \sigma_k}$, for $k = 1, 2, \cdots, N$, and

$$\begin{bmatrix} n^{k-1}(\sigma) \\ Md^{k-1}(\sigma) \end{bmatrix} = V_k(\sigma) \begin{bmatrix} n^k(\sigma) \\ Md^k(\sigma) \end{bmatrix},$$

with $\phi_i^0 := \phi_i$, $i \in \underline{N}$. Using these relationships we can recursively construct the constants

$$\phi_m^k := \phi^k(\sigma_m), \text{ for } k = 1, 2, \cdots, N-1, \text{ and } m = k+1, \cdots, N.$$

Thus the matrices V_k, $k \in \underline{N}$, are completely determined. Recall notation (2.5). It follows that for a fixed, arbitrary ϕ^N, the function

$$\phi(\sigma) := \phi^0(\sigma) = V_1(\sigma)V_2(\sigma) \cdots V_N(\sigma)[\phi^N(\sigma)],$$

interpolates the given points. Furthermore

$$|\phi(\sigma)| \leqslant M \textit{ iff } |\phi_k^{k-1}| \leqslant M,\ k \in \underline{N}, \text{ and } |\phi^N(\sigma)| \leqslant M, \text{ for } |\sigma| > 1.$$

The above considerations show that $\phi(\sigma)$ can be synthesized, in system theoretic terms, as a cascade interconnection of two-pairs with chain parameter matrices V_i terminated by any bounded real system having transfer function $\phi^N(\sigma)$. A solution to the Nevanlinna-Pick interpolation problem exists if the parameters ϕ_k^{k-1} of each one of these subsystems have magnitude less than M.

If a solution exists, by normalizing the chain parameter matrices we can achieve that the two-pairs involved are all-pass systems (cf. sec. 3.2). The normalized chain parameter matrices are

$$\overline{V}_k(\sigma) := \frac{1}{(1 - |\phi_k^{k-1}/M|^2)^{1/2}} V_k(\sigma).$$

while the corresponding transfer parameter matrices are

$$\overline{\Theta}_k(\sigma) = \begin{bmatrix} \phi_k^{k-1}/M & (1 - |\phi_k^{k-1}/M|^2)^{1/2} \\ (1 - |\phi_k^{k-1}/M|^2)^{1/2} & -(\phi_k^{k-1}/M)^* \end{bmatrix} \begin{bmatrix} 1 & 0 \\ 0 & \pi(\sigma) \end{bmatrix},$$

The above considerations can be summarized in

(5.1) **Theorem.** *The Nevanlinna-Pick interpolation problem is solvable if and only if the interpolating function can be represented as a cascade interconnection of all-pass two-pairs, terminated with some BR transfer function.*

5.2 ON THE ROOT DISTRIBUTION OF POLYNOMIALS.

The method described in the previous subsection can also be used in order to determine the distribution of the roots of a given polynomial $d(\sigma)$ with respect to the unit circle. Let $\deg d = N$. We define the all-pass function

$$\phi(\sigma) := \frac{\sigma^N d(\sigma^{-1})}{d(\sigma)}.$$

Choose N points σ_i outside the unit disc, and define $\phi_i := \phi(\sigma_i)$. It readily follows that $d(\sigma)$ has *all* its roots inside the unit circle if and only if the Nevanlinna-Pick problem corresponding to the pairs (σ_i, ϕ_i), $i \in \underline{N}$, defined above, is solvable with $M = 1$.

If we make use of the unnormalized decomposition of subsection 3.2, it is actually possible to determine, iteratively, the distribution of the roots of an arbitrary polynomial $d(\sigma)$ with respect to the unit circle. For this we need the sequence of functions

$$\phi^k(\sigma) = n^k(\sigma)/d^k(\sigma),\ k \in \underline{N},$$

introduced in the previous sub-section. Let $\lambda_+(p)$ denote the number of roots of the polynomial $p(\sigma)$ inside the unit circle. The following rules, applied iteratively for $k = N, N-1, \cdots, 2, 1$, are simple consequences of Rouché's theorem:

$$\lambda_+(d^k) = \lambda_+(d^{k+1}),\ \text{if } |\phi_k^{k-1}|^2 < 1,$$

$$\lambda_+(d^k) = 1 + \lambda_+(n^{k+1}),\ \text{if } |\phi_k^{k-1}|^2 > 1.$$

Note that $\lambda_+(d^k) + \lambda_+(n^k) = k$, i.e. there are no roots on the unit circle. Suppose that only one parameter ϕ_k^{k-1} has magnitude greater than unity. It follows that the number of roots of $d(\sigma)$ outside the unit circle is $N - k + 1$; therefore if $k = 1$, *all* roots of $d(\sigma)$ lie outside the unit circle, while if $k = N$, only *one* root lies outside the unit disc. For a detailed exposition on this topic the reader is referred to Vaidyanathan and Mitra [1987], and Delsarte, Genin, and Kamp [1981].

5.3 DARLINGTON SYNTHESIS.

Finally, we would like to mention that the famous *Darlington synthesis* for bounded real functions, which is the cornerstone of passive network synthesis, is also closely related to the cascade structure.

(5.2) **Theorem.** *Every bounded real function can be expressed as the transfer function of a cascade interconnection of lossless bounded real functions which in addition are reciprocal as well, terminated by systems of unity transfer function.*

(5.3) **Remark.** The three results mentioned in this section, i.e. the Nevanlinna-Pick algorithm, checking the root distribution of polynomials, and the Darlington synthesis, have been generalized to the matrix case. Just as in the scalar case, there is a very close connection of the matrix version of these results with the cascade structure. For details, see e.g. Delsarte, Genin, and Kamp [1979], Vaidyanathan and Mitra [1987], and the references therein.

6. REFERENCES.

A.C. Antoulas [1986], *On recursiveness and related topics in linear systems*, IEEE-AC **31**: 1121-1135.

A.C. Antoulas and R.H. Bishop [1987], *Continued fraction decomposition of linear systems in the state space*, Systems and Control Letters, **9**: 43-53.

N. Balabanian and T. Bickart [1981], **Linear network theory**, Matrix Publishers.

Ph. Delsarte, Y. Genin, and Y. Kamp [1979], *The Nevanlinna-Pick problem for matrix-valued functions*, SIAM J. Appl. Math., **36**: 47-61.

Ph. Delsarte, Y. Genin, and Y. Kamp [1981], *On the role of the Nevanlinna-Pick problem in circuit and system theory*, Int. J. Circuit Theory Appl., **9**: 177-187.

P. Dewilde and H. Dym [1989], *Inverse scattering and networks*, short course given during the MTNS-89 Symposium, Amsterdam, June 19-23.

P.A. Fuhrmann [1983], *A matrix Euclidean algorithm and matrix continued fraction expansion*, Systems & Control Letters, **3**: 263-271.

F.R. Gantmacher [1960], **Matrix theory**, 2 volumes, Chelsea.

R.E. Kalman, P.L. Falb, and M.A. Arbib [1969], **Topics in mathematical system theory**, McGraw-Hill.

R.E. Kalman [1979], *On partial realizations, transfer functions, and canonical forms*, Acta Polytechnica Scandinavica, **Ma31**: 9-32.

T. Kailath [1980], **Linear systems**, Prentice Hall.

P.P. Vaidyanathan and S.K. Mitra [1987], *A unified structural interpretation of some well-known stability-test procedures for linear systems*, Proc. IEEE, **75**: 478-497.

J.C. Willems and A.C. Antoulas [1989], *Recursive time series modeling*, Technical Report, Department of Mathematics, E.T.H. Zurich.

J.C. Willems [1986-1987], *From time series to linear system*, Automatica; *Part I: Finite dimensional linear time invariant systems*, **22**: 561-580; *Part II: Exact modeling*, **22**: 675-694; *Part III: Approximate modeling*, **23**: 87-115.

Smooth Dynamical Systems which Realize Arithmetical and Logical Operations

R. W. Brockett

Division of Applied Sciences, Harvard University
Cambridge, MA 02138, U. S. A.

Dedicated to Jan Willems on the Occasion of his 50th Birthday

Abstract[1]

Although many biological and man-made systems combine aspects of digital and analog processing, until recently there has been very little theoretical work on models this type and many basic questions remain unresolved. In this paper we describe input-output systems governed by ordinary differential equations i) whose behavior is robust in the sense that certain well-defined qualitative aspects of the output depend only on certain well-defined qualitative aspects of the input and ii) are capable of generating behavior of the type one usually associates with digital systems. It is show that rather simple differential equation models can robustly execute arithmetical and logical operations; in particular, we show that continuous-time dynamical systems can simulate arbitrary finite automata.

[1]This work was supported in part by the U.S. Department of the Air Force under grant AFOSR-96-00197, in part by the U.S. Army Research Office under grant DAAL03-86-K-0171, and in part by the National Science Foundation under grant CDR-85-00108.

1. Introduction

Over the last 25 years we have learned a great deal about the properties of the differential equation models for systems with a given input-output behavior. Having its roots in passive network synthesis, the work of Kalman and Youla put the linear realization theory on its present course. The stochastic theory for second order statistics developed rapidly in the late 1960's and brought to the subject of stochastic differential equations some of the fundamental insights of Bode and Shannon. A little later developments in the theory of integral representations for input-output maps via Volterra series, resulted in a mathematical theory for nonlinear realization having sufficient specificity so as to bring Wiener's conceptual ideas on nonlinear input-output descriptions within reach of practically-minded people. Aspects of these developments have made their way into textbooks and can now be regarded as fundamental results in the subject of mathematical modeling.

On the other hand, there are numerous input-output synthesis problems arising in electronics, especially digital electronics, calling for a different formalism. The realization of analog counters and quantizers are examples of the kind of "signal-to-symbol" transduction problems that we have in mind. In this paper we develop a formalism targeted at such questions. We claim to show that it is possible to combine analog and digital computation in a theory which is both computationally powerful and aesthetically pleasing.

Ordinary analog computing is completely non-robust in that the values of the computed quantities depend on the values of the input and the system parameters. Here we want to develop models which are robust in the sense that small changes in the input produce little or no change in the output, and any changes which do occur do not propagate through successive stages of the signal processing. This suggests that one might want to attempt to find systems whose output behavior is governed more by some "topological" aspect of the input (e.g. a winding number) rather than by the details of its time history. For example, if the space of admissible input values is not simply connected, then one can attempt to find systems having the property that qualitative features of their output depend only on the homotopy type of the input curve. The performance of such systems would be robust in a very strong sense and would, at the same time, provide a mathematically precise link between aspects of analog and digital computing.

In order to make this somewhat more concrete, consider figure 1. We show there a curve which is parametrized by t and which lies in an annulus in $(u, \dot{u})$-space. Trajectories in this space are subject to certain constraints by virtue of the fact that when du/dt is positive, u must be increasing, and when du/dt is negative, u must be decreasing. This is true regardless of the details about how t parametrizes the curve. The problem of counting the number of zero crossings of a function of time is, for unrestricted functions, non-robust;

the function may come arbitrarily close to zero without crossing zero, or it may have a narrow spike which crosses zero for an extremely brief period of time but is otherwise safely removed from zero. However, functions which are constrained to the annulus of figure 1 have well-defined zero crossings. the exclusion of the inner disk means that when the value of the function is near $u = 0$, the velocity is large enough so that it will pass through zero in an unambiguous way and the exclusion of the outer region prohibits high frequency spikes. Within this annulus the idea of a zero crossing can justifiably be called robust. In a later section we will construct a differential equation which counts the zero crossings of a curve u provided it satisfies a restriction of this type. In our treatment of the realization question for automata we will use a similar but slightly more complicated input space. The counter and the automata provide examples of solving a realization problem in which the output of the dynamical system is sensitive only to qualitative aspects of the input.

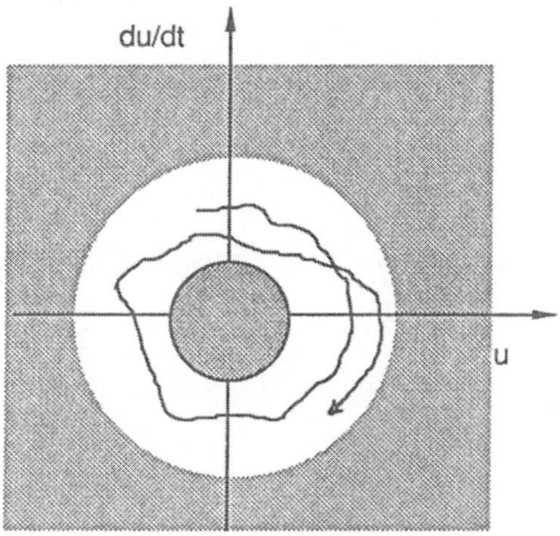

Figure 1: The input space for a system which counts zero crossings.

The differential equations we will use to realize these qualitatively defined input-output maps will typically have a large number of stable equilibria whose locations are insensitive to the value of the input. Roughly speaking we will set up a situation in which the input forces the state from one equilibrium to another, reminiscent of finite automata. From this point of view one could think of our construction as being a way to add flesh in the form of differential equations to the skeleton provided by a finite automaton. What distinguishes this work from standard digital electronics are the fact that we do not make a hard and fast distinction between the logical model and the circuit model and the fact that we require no explicit sampling or synchronizing clock.

2. Preliminary Results

We collect here a few ideas about the differential equations we will be working with in the later sections. We use $\|A\|$ to denote the square root of the sum the squares of the entries of A. Let $\text{Sym}\{\lambda_1, \lambda_2, \ldots, \lambda_n\}$ denote the set of all real n by n symmetric matrices with eigenvalues $\lambda_1, \lambda_2, \ldots, \lambda_n$ and let $[A, B] = AB - BA$. In a recent paper [1] we investigated the equation in the space of symmetric matrices,

$$\dot{H} = [H, [H, N]] \tag{1}$$

where $N = N^T$. Our attention here is devoted to some variations on this theme. Equation (1) has a number of remarkable properties which have to do with its origins as a steepest descent equation on the orthogonal group [2]. Some of these are explored in [1]. The more recent work of Bloch [3] and Bloch *et al.* [4] take the study of this equation in other directions. For our present purposes we want to consider the addition of a linear term of the form $[H, \Omega]$ so that we get

$$\dot{H} = [H, [H, N]] + [H, \Omega] \tag{2}$$

where, again, H and N are symmetric and Ω is restricted to be is skew symmetric.

Lemma 1: The eigenvalues of the solution of equation (2) do not change as time evolves. Considered as a differential equation on $\text{Sym}\{\lambda_1(H(0)), \lambda_2(H(0)), \ldots, \lambda_n(H(0))\}$, if N is diagonal and has no repeated entries and if $H(0)$ has no repeated eigenvalues, then there exists $\epsilon > 0$ such that for $\|\Omega\| < \epsilon$ equation (2) has $n!$ equilibrium points exactly one of which is asymptotically stable.

Proof: Expressing $\dot{H}$ as $\dot{H} = [H, [H, N] + \Omega] = [H, f(H)]$ we see that this is the standard isospectral form. If $\Omega = 0$ we know from [1] that the equilibrium points take the form $H = \text{diag}\ (\lambda_{\pi(1)}, \lambda_{\pi(2)}, \ldots, \lambda_{\pi(n)})$ with π being any permutation and that the linearization of $[H, [H, N]]$ at these points has eigenvalues of the form $(\lambda_i - \lambda_j)(n_{\pi(i)} - n_{\pi(j)}$. As discussed in [1], all $n!$ of the equilibria are nondegenerate and exactly one of the equilibria is asymptotically stable. For $\Omega \neq 0$ we see that

$$[H, [H, N]] + [\Omega, H] = 0$$

is satisfied if $[H, N] = \Omega$. The symmetric solutions of this equation take the form

$$H = D + \Delta$$

where D is diagonal and Δ is given by

$$\delta_{ij} = \omega_{ij}/(n_i - n_j)$$

To first order in ω_{ij} we obtain the correct eigenvalues for $D+\Delta$ by letting $D = \text{diag}\ (\lambda_{\pi(1)}, \lambda_{\pi(2)}, \ldots, \lambda_{\pi(n)})$. There are $n!$ such solutions, corresponding to the $n!$ choices of the permutation π. Because the zeros of $[H,[H,N]] = 0$ are nondegenerate and isolated there are no other solutions of $[H[H,N]] + [\Omega, H] = 0$ for $\|\Omega\|$ sufficiently small.

For reasons which will appear later, we consider in more detail a special case of equation (2).

Remark 1: Let H be two by two and suppose $H(0)$ has +1 and -1 as its eigenvalues. If N takes the form $N = \text{diag}\ (0, u)$, then for $h_{11} = \cos(\theta)$, equation (2) takes the form

$$\dot{\theta} = \omega + u \sin\theta$$

Proof: This is just a calculation. We have

$$\begin{aligned} \dot{h}_{11} &= -uh_{12}^2 + \omega h_{12} \\ \dot{h}_{12} &= u(h_{11} - h_{22})h_{12} - \omega h_{11} \\ \dot{h}_{22} &= uh_{12}^2 - \omega h_{12} \end{aligned}$$

Because $\dot{H} = [H,[H,N]] + [\Omega, H]$ evolves in such a way as to keep the eigenvalues of H constant we see that

$$h_{11} + h_{22} = 0$$

and

$$h_{11}h_{22} - h_{12}^2 = -1$$

Thus if we use $h_{11} = -h_{22} = \cos\theta$ and let $h_{12} = \sin\theta$, we get

$$\frac{d}{dt}\cos\theta = -\dot{\theta}\sin\theta = \omega\sin\theta + u\sin^2\theta$$

or

$$\dot{\theta} = \omega + u\sin\theta$$

This calculation allows us, in this case, to lift equation (2) from the circle to the real line. That is to say, we can regard θ as being real valued, not just circle valued. This will be important in the next section.

We now consider a simple class of systems which, unlike the ones considered above, have a large number of (locally) asymptotically stable equilibria. By diag H we understand the matrix whose diagonal entires are the same as those of H and whose off-diagonal terms are zero.

Lemma 2: Let H and L be n by n matrices. The solution of the coupled equations

$$\begin{aligned} \dot{H} &= [H,[H, \text{diag}\ L]] \\ \dot{L} &= [L,[L, \text{diag}\ H]] \end{aligned} \tag{3}$$

evolves in such a way as to leave the eigenvalues of H and L unchanged. If H and L are symmetric, tr(diag H)(diag L) satisfies

$$\begin{aligned}\frac{d}{dt}\mathrm{tr}(\mathrm{diag}\,(H)\mathrm{diag}\,(L)) &= \frac{d}{dt}\mathrm{tr}(L\,\mathrm{diag}\,(H)) \\ &= \frac{d}{dt}\mathrm{tr}(H\mathrm{diag}\,(L)) \\ &= \|[H,\mathrm{diag}\,(L)]\|^2 + \|[L,\mathrm{diag}\,(H)\|^2\end{aligned}$$

If $H(0)$ and $L(0)$ are symmetric with eigenvalues $\lambda_1 < \lambda_2 < \ldots < \lambda_n$, and $\mu_1 < \mu_2 < \ldots < \mu_n$ (unrepeated in each case), then as an equation on $\mathrm{Sym}\{\lambda_1, \lambda_2, \ldots, \lambda_n\} \times \mathrm{Sym}\{\mu_1, \mu_2, \ldots, \mu_n\}$, this pair of equations has $(n!)^2$ nondegenerate equilibrium points corresponding to each of the possible orderings of the eigenvalues along the diagonals. Exactly $n!$ of these equilibria are asymptotically stable. The remaining stationary points occur at points where diag H and/or diag L have repeated entries; none of these equilibria are asymptotically stable.

Proof: Isospectrality follows from the fact that any solution of $\dot{A} = [A, f(A)]$ is necessarily isospectral. The derivative computation is completely analogous to the one given in [1] and uses the fact that the Lie bracket of two symmetric matrices is necessarily skew symmetric. There are $n!$ different diagonal matrices having eigenvalues $\lambda_1, \lambda_2, \ldots, \lambda_n$ and $n!$ different diagonal matrices having eigenvalues $\mu_1, \mu_2, \ldots, \mu_n$. Thus there are $(n!)^2$ pairs of diagonal matrices which satisfy $[H, L] = 0$.

If diag(L) has unrepeated entries, then the only matrices which commute with it are diagonal. On the other hand, if diag(L) has repeated entries, then there are non-diagonal H's which commute with L and if diag(H) has repeated entries, there are non-diagonal L's which commute with diag(H), however the Liapunov function tr(diag Hdiag L) shows that none of these can be asymptotically stable.

Remark 2: (Compare with [1]) If $\{\lambda_1, \lambda_2, \ldots, \lambda_n\}$ is a set of n real numbers, we denote by $C\{\lambda_1, \lambda_2, \ldots, \lambda_n\}$ the convex hull of the $n!$ points in $\mathbf{R}^n$ of the form $(\lambda_{\pi(1)}, \lambda_{\pi(2)}, \ldots, \lambda_{\pi(n)})$, where π ranges over the set of all permutations. The Shur-Horn theorem asserts that the possible diagonals of elements of $\mathrm{Sym}\{\lambda_1, \lambda_2, \ldots, \lambda_n\}$ coincide with this set. It is not difficult to see that equation (3) evolves in such a way as to solve the following problem. Let $x \in C\{\lambda_1(H(0)), \lambda_2(H(0)), \ldots, \lambda_n(H(0))\}$ and let $y \in C\{\lambda_1(L(0)), \lambda_2(L(0)), \ldots, \lambda_n(L(0))\}$. Find x and y subject to these constraints such that the euclidean inner product $\langle x, y \rangle$ is as large as possible. Of course there are at least $n!$ optimizing values for x and y. Our result shows that if the $\lambda_i(H(0))$ are distinct and the $\lambda_i(L(0))$ are distinct, then there are exactly $n!$ points of local (and global) maxima.

3. Counting Zero Crossings

As an illustration of what we mean by a dynamical realization of an arithmetical operation, we consider the question of finding a dynamical system which counts. More precisely, we will define a dynamical system which generates a running total of the number of zero crossings of a scalar function u as it evolves in time. Because the output y will be a continuous function of time, we need to explain in what sense this, possibly noninteger-valued, output can be taken to be a count of anything. In fact, what we will show is that for a suitable restriction on u there is an inequality of the form

$$n-1 \leq y(t) \leq n+\epsilon$$

with n being the number of times u crosses zero on $[0,t)$. It will also happen that as the time since the last change in the sign of the input goes to infinity, the output $y(t)$ will approach a value which is, to within some assignable tolerance, the true count n.

The differential equation we will use for counting zero crossings is the real line version of the two by two form of equation (2) which we explored above, i.e. the equation

$$\dot{\theta} = \omega + u \sin\theta$$

The output $y(t)$ is just $\theta(t)/\pi$.

Our first result applies to the case where $u(\cdot)$ is piecewise constant.

Lemma 3: Let ω be a given real number $\omega \in (0,1)$. If $u(\cdot)$ is piecewise constant and if it takes on values in $|u(t)| > 1/\omega$ with the points of discontinuity being separated by at least π/ω units of time, then the solution of the input-output system

$$\dot{\theta}(t) = \omega + u(t)\sin\theta(t) \quad ; \quad \theta(0) = 0 \quad ; \quad y(t) = \theta(t)/\pi$$

is such that

$$(n(t)-1) < y(t) < n(t) + \sin^{-1}(\omega^2)$$

where $n(t)$ is the number of times u changes sign on $[0,t]$. (Our convention is to declare that a jump occurs at $t=0$.)

Proof: (See figure 2). We analyze the case where $u(t)$ is initially positive. the opposite case is similar. It is not difficult to show that if $\theta(0) = 0$ and if initially $u(t) > 1/\omega$, then because the time between jumps exceeds π/ω, at the first jump time $T > 0$,

$$\theta(T) = \int_0^T \omega + u(t)\sin\theta \, dt$$

is between π and $\pi + \sin^{-1}(\omega^2)$. At time T u jumps to a negative value which is less than $-1/\omega$. Because $u \sin\theta(T)$ will necessarily be positive, θ increases over the next π/ω units of time until it is to within $\sin^{-1}(\omega^2)$ units of 2π. Continuing in this way leads to the result claimed.

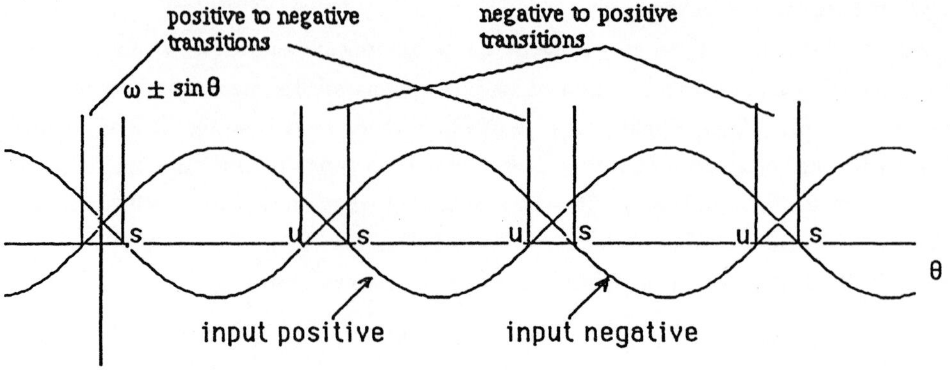

Figure 2: The function of $\omega + u(t)\sin\theta$ for $u(t) = \pm 1$ showing stable $(= s)$ and unstable $(= u)$ equilibrium values. The intervals on which $\omega + u(t)\sin\theta$ are positive for any admissible choices of $u(t)$ are the θ-values for which switching can occur.

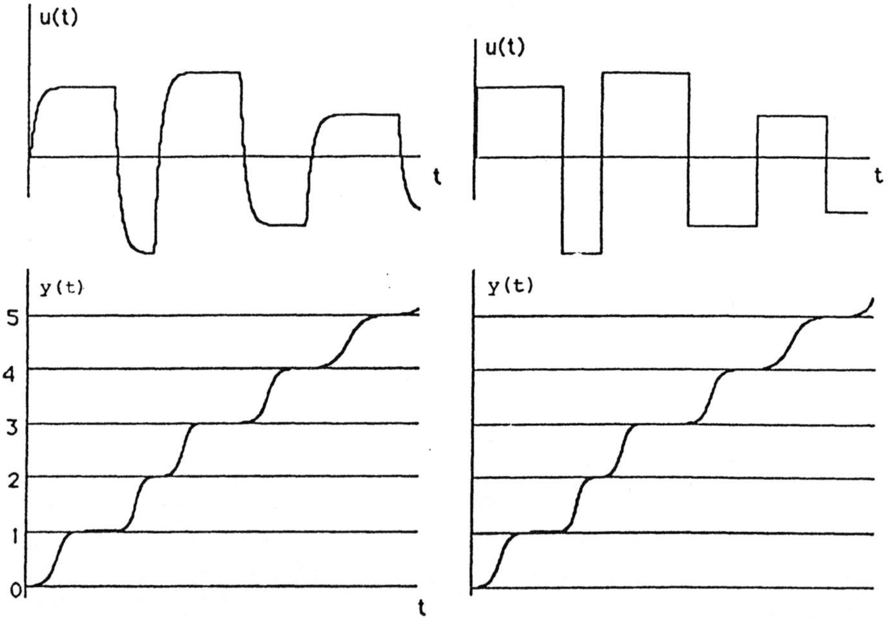

Figure 3: The response of $\dot{\theta} = \omega + u\sin\theta$; $y = \theta/\pi$ to smoothed and piecewise constant inputs.

This argument shows that at no time does $\theta(t)/\pi$ exceed the number of zero crossings by more than $\pi^{-1}\sin^{-1}(\omega^2)$ and at no time does it fall more than one behind. Clearly if u is constant, the solution approaches the appropriate value monotonically.

By an annular region in $(u, du/dt)$-space we understand a space which is formed by removing from a simply connected open set a second simply connected open set which is interior to the original. Using the fact that the solutions of this type of differential equation defines a continuous mapping from a function space $L_\infty(0,t)$ where u lies to the function space L_∞ where θ lies, we can assert the following theorem (whose proof is omitted).

Theorem 1: There exists an annular region A in $(u, \dot{u})$-space such that for u in A the equation $\dot{\theta} = \omega + u \sin\theta$ counts the zero crossings of u in the above sense.

In figure 3 we show the results of simulating this system for an ideal and a "generic" input.

4. Quantization without Sampling

Quantizers are ordinarily part of any system which uses digital computers to control analog systems. Usually quantizers operate by sampling the signal which is to be quantized and then operating on the sampled value. Subsequent stages of processing associated with the system are often synchronized with the sampling operation. Because there may not be any obvious synchronization signal present in some applications, the question arises as to the feasibility of quantization without sampling. The input space will be topologically "interesting" as suggested by figure 4.

Our realization will again use a form of equation (2). This time the input u enters as

$$\dot{H} = [H[H, N_0 + uN_1]] + [\Omega, H] \qquad ; \qquad y = \mathrm{tr}(N_2 H)$$

with N_0, N_1 and N_2 being diagonal matrices and Ω being a small skew symmetric matrix.

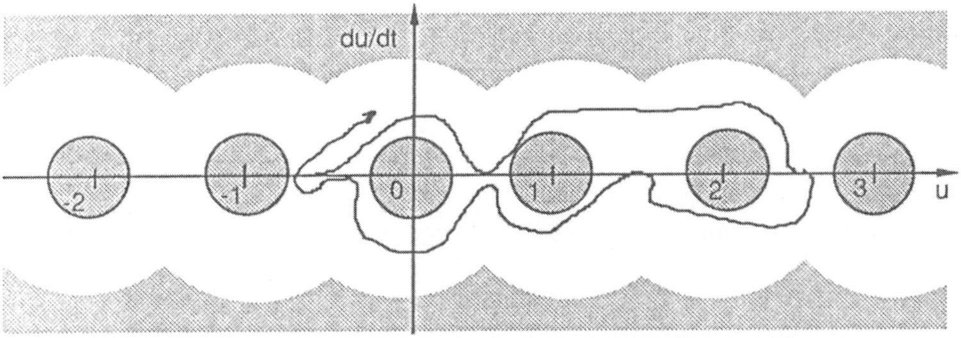

Figure 4: The qualitative form for the input space of a quantizer.

Lemma 4: Given $n-1$ real numbers $a_1 < a_2 < \ldots < a_{n-1}$ and n real numbers $b_1, b_2, \ldots, b_n$, there exists an $H(0)$ and a system of the form $\dot{H} = [H, [H, N_0 + uN_1]]$; $y = \mathrm{tr}(N_2 H)$ such that for each constant value of u which is not equal to any of the a's there is a unique equilibrium point $H(u)$ of $\dot{H} = [H, [H, N_0 + uN_1]]$ such that $H(u)$ is asymptotically stable, and for all $a_{i-j} < u < a_i$ $y = \text{tr}(N_2 H(u)) = b_i$. (For $u < a_i$; $\text{tr}(H(u)N_2) = b_1$ and for $a_{n-1} < u$; $\text{tr}(H(u)N_3) = b_n$.)

Proof: We construct the system in the following way. $N_0 = \text{diag}\,(a_1, a_2, \ldots, a_{n-1}, 0)$, $N_1 = \text{diag}\,(0, 0, \ldots, 0, 1)$ and $N_2 = N_1$. $H(0)$ has eigenvalues $(b_1, b_2, \ldots, b_n)$. The results then follow from theorem 4 of [1].

This lemma does not always provide a practical solution to the quantization problem because the solution H may spend too much time escaping from an unstable equilibrium point. Experience with simulations suggests that it is better to add a linear term so that a change in u affects $\dot{H}$ additively. We use

$$\dot{H} = [H, [H_1 N_0 + uN_1]] + [\Omega, H] \quad ; \quad y = \text{tr}(N_3 H)$$

This has an important implication. The value of y for a constant value of u is no longer exclusively dependent on the interval (a_i, a_{i+1}) to which u belongs. There is a dependence on the precise u value in this range. This sensitivity to u can be made as small as one likes by taking Ω to be small. This defect is, however, compensated for by an increased speed of response. A simulation of a seven level quantizer is shown in figure 5.

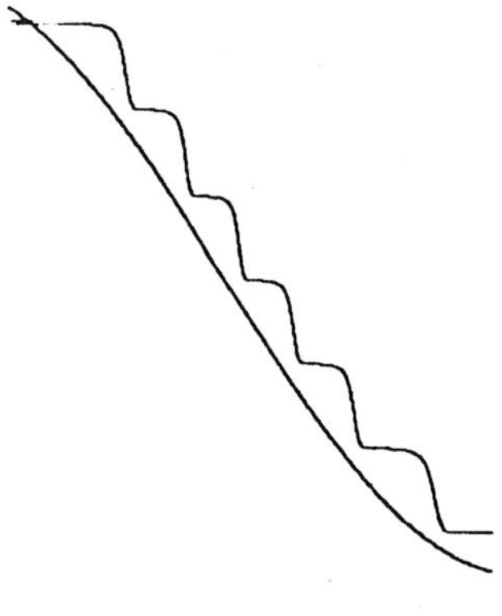

Figure 5: The response of a seven level quantizer to a $\sin t$ input.

5. Realizing Arbitrary Finite Automata

In this final section we show how to define continuous time systems which can perform any computation which a finite automaton can perform. This shows how one might use a continuous time (unsampled) system to perform any computation that a digital computer can perform.

Let U and X be finite sets; $U = \{u_1, u_2, \ldots, u_m\}$; $X = \{x_1, x_2, \ldots, x_n\}$. Suppose that we are given $\delta : X \times U \to X$ and wish to simulate the difference equation

$$x(k+1) = \delta(x(k), u(k))$$

We can think of this equation as defining a finite automaton with input alphabet U and state space X. We describe a system governed by ordinary differential equations which simulates this machine.

The differential equations we will use are a variation of equation (3) and take the form

$$\begin{aligned} \dot{H} &= [H, [H, \psi(u, L)]] + [\Omega_1, H] \\ \dot{L} &= [L, [L, \psi_0(u) \text{diag } H]] + [\Omega_2, L] \end{aligned}$$

with H and L being p by p symmetric matrices with eigenvalues $(1, 2, \ldots, p)$. The idea is this. Let p be an integer such that $p! > n$. Let f be a function $f : X \to \Pi$ which associates with each state x_i of the given automaton a permutation, $f(x_i) = \pi_i$ such that $\pi_i \neq \pi_j$ if $x_i \neq x_j$. Associate with each u_i the integer i. Let $\psi(u, L)$ be a function of u and L which takes on values in the space of diagonal matrices. We take $\psi(0, L) = 0$.

Code a generic input sequence $u_{i_1} u_{i_2} \ldots u_{i_s}$ as a time function with the zero values being used to mark the end of a letter; i.e. associate to the above word the time function

$$u(t) = \begin{cases} i_1 & ; \quad 0 \leq t < T_1 \\ 0 & ; \quad T_1 \leq t < P_1 \\ i_2 & ; \quad P_1 \leq t < T_2 \\ 0 & ; \quad T_2 \leq t < P_2 \\ \ldots & \quad\;\; \ldots \end{cases}$$

In this way, the inputs are coded as integer valued levels with 0 interspersed to mark time. We will assume that $P_i - T_i$ and $T_{i+1} - P_i$ are both greater than some fixed T. We take ψ_0 to be a smooth function such as $e^{-\alpha u^2}$ which is one when u is zero and very small when u is larger than, say, 1/2.

We are now ready to show how to choose ψ so as to make the system mimic the given automaton. Let L_i be the diagonal matrix with entries $(1, 2, 3, \ldots, n)$ permuted by the permutation $f(x_i)$. Define $\beta(i, j)$ so that

$$\pi_{\beta(i,j)} = f(\delta(x_i, u_j))$$

and let ψ be a smooth function such that $\psi(u_i, L_j) = \text{diag } (\pi_{\beta(i,j)}(1) \ , \ \pi_{\beta(i,j)}(2) \ , \ \ldots, \pi_{\beta(i,j)}(n))$. With this definition we can see that if $u = i \neq 0$, then $\dot{L}$ is effectively zero and the H equation will flow in such a way as to permute the diagonal of H so as to make it match the diagonal of $\psi(u_i, L_j)$. When u switches back to zero, $\dot{H}$ is zero and the L equation flows to match the diagonal of L with that of H. If u remains zero for a suitable time, then H and L are both nearly diagonal and their diagonals are matched. When a nonzero u appears, this "shuts down" the L equation and realigns the diagonals of H; when it returns to zero, this "shuts down" the H equation and updates the L equation so that it again matches H.

Conclusions

The question of finding differential equations which evolve in such a way as to robustly carry out a calculation has considerable scientific interest because of its connection with problems in biology and computing. In this paper we have considered a class of input-output models which have a large number of equilibria and have given a construction which shows that smooth dynamical systems of this type can simulate one general computational model, namely that of a finite automaton.

References

[1] R. W. Brockett, "Dynamical Systems that Sort Lists, Diagonalize Matrices and solve Linear Programming Problems," *Proceedings of the 1988 IEEE Conference on Decision and Control*, December 1988.

[2] R. W. Brockett, "Least Squares Matching Problems," *Journal of Linear Algebra and Its Applications*, to appear.

[3] A. M. Bloch, "Steepest Descent, Linear Programming and Hamiltonian Flows," submitted for publication.

[4] A. M. Bloch, R. W. Brockett, and T. Ratiu, "A New Formulation of the Generalized Toda Lattice Equations and their Fixed Point Analysis via the Moment Map," submitted for publication.

Pole Assignment by Output Feedback

C. I. Byrnes

Department of Electrical and Computer Engineering and
Department of Mathematics
Arizona State University
Tempe, AZ 85287, U. S. A.

1. Introduction. The problem of tuning the natural frequencies of a linear system using output measurement has long been recognized ([1]–[3]) as central in both classical and modern automatic control. Fundamental work by Nyquist in 1932 provided graphical stability and instability criteria for output feedback, while the graphical root-locus method of Evans also gives a clean set of necessary and sufficient conditions for a self conjugate set $\{s_1, \ldots, s_u\}$ of frequencies to be assignable as the poles of an nth degree transfer function $g(s)$; viz. $g(s_i) = g(s_j)$. For multivariable systems, this problem has been the object of study for many authors using an impressive variety of techniques from linear systems theory, combinatorics, complex function theory and, most recently, geometry and topology. Beginning in the early 1970's, methods from algebraic geometry were applied to several open problems in linear systems, in part because more standard techniques (see chapter 2) were not able to resolve some of the fundamental nonlinear features surprisingly underlying what were phrased and regarded as problems about linear systems. Among the most striking examples of this phenomenon are the essential role played by algebraic geometry and topology in the solution of the "parametrization" problem and the related questions concerning canonical forms for realization (see e.g.

*Dedicated, in homage, to my collaborator, friend and teacher, Jan Willems

**Research supported in part by grants from AFOSR and NSF.

[4]–[7]) and the applications of algebraic geometry to questions about output feedback stabilization and pole-assignment for multivariable linear systems, which we survey in this article. As early as 1972, the language of hypersurfaces, generic point, etc. was used (see e.g. [8]) to derive certain general position results concerning multiple roots of a closed-loop characteristic polynomial, while in [9] Anderson, Bose and Jury showed that the solvability of particular pole-assignment problems could be decided using polynomial criteria. Within just a few years, there were several quite sophisticated and novel applications of algebraic geometry to this problem.

Indeed, the most striking message of the work by A. G. J. MacFarlane et al. (see e.g. [10]) is the recognition that, for scalar gains and square multivariable systems, the behaviour of multivariable root-loci—which had previously been elusive—is a problem best stated and solved using the theory of rational functions defined on Riemann surfaces, allowing for a clear understanding of precisely how multivariable root-loci evolve with changes in the output gain. Simultaneously, in a series of seminal papers (see e.g. [11]–[13]) R. Hermann and C. F. Martin introduced a variety of useful tools and concepts from algebraic geometry. Perhaps the most widely appreciated of these techniques, at the time, was the application of the "dominant morphism theorem" proving that for generic systems, in the correct dimension range, pole-assignment is achievable using complex feedback (see 4.2). In retrospect, the Hermann-Martin curve [12] turned out to be an essential piece of this problem. However, the "dominant morphism" result coupled with a positive pole assignment result proved by Kimura [14] using combinatorial and geometric methods (§3), sparked renewed interest in the problem of arbitrary pole-assignment by real feedback. In a fundamental calculation [15], J. C. Willems and W. H. Hesselink showed by example that the positive results obtained over the complex field, failed to hold for real feedback systems in a serious way (see §4.3). Their results show that for 2×2 multivariable systems of degree 4, in general there will be two pole assigning feedback laws which are either both real or complex conjugates "with equal probability." The quadratic nature of this problem, underscoring some of the nonlinear features, also implies that linear formulae do not exist for pole-assigning output feedback laws, in contrast to the state feedback situation. This discovery considerably expanded the problem of pole-assignment by static or dynamic output feedback beyond the simple question of existence of real or even complex solutions to the pole-assignment equations, including questions such as the nature of possible formulae for pole assigning gains if these should exist and the behaviour of various nonlinear computational schemes for their numerical solution.

About a decade ago, inspired by a profound geometric construction due to Hermann and Martin (see §5.4) and by the questions raised by Willems and Hesselink, we developed a geometric framework for analyzing pole-assignment problems which incorporates the earlier algebraic geometric methods with the combinatorial methods pioneered by Hautus and by Kimura, making systematic use of classical enumerative geometry and its modern version, intersection theory. Using this framework, Brockett and Byrnes [16] were able to provide a multivariable Nyquist criterion for nonscalar gains and to explicitly enumerate the number of gains (possibly complex) which assign a given set of closed loop poles for the general $p \times m$ system of degree $mp = n$, viz.

$$d_{m,p} = \frac{1! \cdots (p-1)!(mp)!}{m! \cdots (m+p-1)!} \tag{1.1}$$

While $d_{2,2} = 2$, in harmony with Willems-Hesselink, $d_{2,3} = 5$ so that real pole-assigning gains exist for real data (see §5.1), a result which was unexpected at the time. One can also glean a great deal about the structure of the solution to the pole assignment equation from the analysis (see §5.2) underlying the determination of (1.1). For example, it is possible to compute the Galois groups for several generic problems (§6.3) thereby answering open questions about the kinds of formulae which might exist for pole-assigning gains. Moreover, this analysis also makes it possible to prove convergence for certain numerical schemes, e.g. homotopy continuation methods (see §5.3), for the solution of the pole-assignment equations. Finally, these techniques have been refined to study pole-assignability and stabilizability by real feedback when $mp \geq n$ (see §5.4). While, as we shall point out, we are still far from a complete solution to this important problem through the work of many people the results obtained using geometric methods contain, to the best of my knowledge, all existing results. On the other hand it is also fair to say that the major results of the geometric theory have still not been obtained by other methods.

In section 2.1, we define the pole assignment problem and develop some preliminaries. Section 2.2 contains a treatment of the rank one situation, which can still be treated by standard linear systems theory. Djaferis [17] made the observation that rank one methods could be used to study special pole-assignment problems using dynamic compensation and in section 2.3 we use this approach to give a simpler proof of the Brasch-Pearson theorem, also discussing related results due to Seraji [18], Stevens [19], Vidyasagar-Viswanadham [20] and Ghosh [21]. In chapter 3, we describe the combinatorial geometric interpretation of the problem, following Kimura [14] and Hautus [22], as well as some recent improvements due to Rosenthal [23]. After a brief introduction, Chapter 4 concentrates on applications of affine algebraic geometric techniques. Specifically, in section 4.2, following Hermann-Martin we compute the differential of the pole-assignment maps both for output and for state feedback obtaining necessary conditions for pole assignment which are generically sufficient over **C**, using the "dominant morphism theorem". Over **R**, with a little more work we obtain a new, and rather clean, proof of Heymann's Lemma. In 4.3, we derive the Willems-Hesselink results through an explicit, general elimination of the pole-assignment equation, using a multilinear algebraic approach due to Anderson, Morse and Wolovich [24]. Following Byrnes and Anderson [35] we then use elimination theoretic methods to show that this necessary condition for pole-assignment is also necessary for generic stabilizability. In 5.1 we cast the pole-assignment problem as a problem in enumerative geometry to which the classical methods of Schubert calculus formally apply. Putting the Schubert calculus on a firm foundation was, of course, Hilbert's 15th Problem, which was solved by van de Waerden, by Ehresmann, by Hodge and by Chern in its modern realization as intersection theory on Grassmannians. In section 5.2, we give an exposition of our geometric framework for pole-assignment which combines the earlier combinatorial approaches and algebraic geometric methods with the tools of the modern Schubert calculus. In particular, section 5.2 contains a sketch of the proofs of the main results announced in section 5.1. In section 5.3 we prove, using geometry, Galois theory and numerical methods that, in general, pole-assigning gains cannot be solved for in terms of rational expressions and the extraction of n-th roots. Nonetheless, the geometric analysis in section 5.2 does provide a convergence proof for the numerical solution of problems by the homotopy continuation method, as is illustrated by an example in 5.3. As we note in section 5.4, using Schubert calculus and the notion of Ljusternik-Šnirel'mann category from the calculus of variations, it is possible to obtain some significant refinements of the results derived for $mp = n$, amounting to what are currently the sharpest results known about pole-assignment by real output feedback.

It is a pleasure to thank many people for stimulating correspondence, conversations, preprints and suggestions: J. Ackermann, B. D. O. Anderson, I. Berstein, R. W. Brockett, T. Djaferis, B. K. Ghosh, J. Harris, M. L. J. Hautus, R. Hermann, W. Hesselink, H. Hiller, T. Kailath, H. Kimura, C. F. Martin, S. K. Mitter, A. S. Morse, D. Mumford, J. Rosenthal, J.-P. Serre, P. K. Stevens, R. E. Stong and X.-C. Wang. Finally, I want especially to acknowledge the profound influence Jan Willems has had on my way of thinking about this problem and about systems and control theory in general. Happy birthday, maestro!

2. Rank One Methods for Pole Assignment by Static or Dynamic Output Feedback

2.1 Preliminaries. We begin with a statement of the problem of pole-assignment by output feedback, first stated in the more elementary, but less natural, state-space form. Suppose $x \in \mathbf{R}^n$, $u \in \mathbf{R}^m$, $y \in \mathbf{R}^p$ and that the $n \times n$ matrix A, the $n \times m$ matrix B and the $p \times n$ matrix correspond to the linear control system

$$\dot{x} = Ax + Bu$$
$$y = Cx$$

which provided a minimal dimension realization

$$G(s) = C(sI - A)^{-1}B$$

of the transfer function

$$\hat{y}(s) = G(s)\hat{u}(s). \tag{2.1}$$

The problem of pole-assignment by output feedback formalizes the problem of arbitrarily tuning the natural frequencies of a linear system using only memoryless sensor feedback: Given a self-conjugate set $\{s_1, \ldots, s_n\}$, of complex numbers, perhaps counted with multiplicity, find an output feedback law, $u = ky$, which places these frequencies, in the sense that the roots of the characteristic polynomial of $A + BKC$ coincide with $\{s_1, \ldots, s_n\}$. Given the input-output nature of the problem of tuning natural frequencies by output feedback it is, not surprisingly, more convenient to express an n dimensional $p \times m$ linear system in terms of its transfer function, (2.1) which we shall always assume is a matrix-valued rational function, vanishing at ∞. After constant output feedback

$$u = -Ky + v$$

the closed-loop transfer function $G_K(s)$ takes the familiar form

$$G_k(s) = G(s)(I + KG(s))^{-1} = (I + G(s)K)^{-1}G(s) \tag{2.2}$$

leading to a "characterization" of the closed-loop poles as the zeros of the return difference determinant

$$\det(I + KG(s)) = 0 \tag{2.3}$$

Of course (2.3) is a little unsatisfactory with respect to open-loop poles, which correspond for example to the law $K = 0$. This can be rectified by remembering that the left-hand side of (2.3) is a rational function with poles at the open loop poles and zeroes at the closed-loop poles, some of which may however cancel. While we shall typically work with (2.3) keeping this caveat in mind, it is also easy to remedy this lacuna using the method of coprime factorization (see for example [26]). Explicitly, there exist polynomial matrices $N(s), D(s)$ satisfying

$$G(s) = N(s)D(s)^{-1} \tag{2.4}$$

which are coprime in the sense that there exist polynomial solutions to the equation

$$U(s)N(s) + V(s)D(s) = I$$

Moreover the poles of $G(s)$ are defined as the roots of the equation

$$\det D(s) = 0 \tag{2.5}$$

From (2.2), one may derive the representation

$$G_K(s) = N(s)(D(s) + KN(s))^{-1} \tag{2.6}$$

which is also a coprime factorization and hence we have a formula for the closed-loop poles

$$\det(D(s) + KN(s)) = 0 \tag{2.7}$$

which gives the correct answer if $K = 0$, suggesting that the correct form for (2.3) should be

$$\det(I + KG(s)) \det D(s) = 0 \tag{2.3$'$}$$

which incorporates pole-zero cancellations.

We note that, in the scalar input-scalar output case, (2.7) gives a simple characterization of those self-conjugate sets $\{s_1, \ldots, s_n\}$ of frequencies which can be assigned by output feedback, viz.

$$g(s_i) = g(s_j), \quad \forall i, j.$$

For multivariable systems the heart of the pole assignment problem lies in the fact that for fixed $s_0 \in \mathbf{C}$ the solution set

$$\det(I + KG(s_0)) = 0 \tag{2.8}$$

is in general very difficult to describe. In geometric terms, (2.8) defines a hypersurface but there is of course a class of hypersurfaces which are easy to analyze; viz. hyperplanes. To put it algebraically, if for example $G(s_0)$ has rank 1 then (2.8) takes the vastly more simple, affine form

$$\det(I + KG(s_0)) = 1 - tr(KG(s_0))$$

In particular if $G(s)$ has rank ≤ 1, (2.3)$'$ takes the form (see e.g. [27]):

$$\det D(s) - tr(K \operatorname{adj} G(s)) = 0 \tag{2.9}$$

so that the closed-loop characteristic polynomial is affine in the gain, generalizing the scalar input-output case. In this chapter, we exploit this simple observation to develop simple proofs or improvements of several fundamental theorems.

2.2 Pole-assignment by Output Feedback in the Case $\min(m,p) = 1$.

If $\min(m,p) = 1$, then $G(s)$ is either a row or a column vector and hence the affine formula for the return difference determinants applies. We shall assume $m = 1$, the situation being symmetric in m and p. In this case, the coprime factorization (2.4) takes the form

$$G(s)^T = [n_1(s)/d(s), \dots, n_p(s)/d(s)] \tag{2.9}$$

and the closed loop polynomial associated to the output feedback gain

$$K = [K_1, \dots, K_p]$$

is simply

$$d(s) - \langle K, n(s) \rangle = d(s) - \sum_{i=1}^{p} K_i n_i(s) = 0 \tag{2.10}$$

In particular, to place the closed-loop poles at $\{s_1, \dots, s_n\}$ is to solve the affine equation

$$d(s) - d_r(s) = \sum_{i=1}^{p} K_i n_i(s) \tag{2.11}$$

where

$$d_r(s) = \prod_{i=1}^{n} (s - s_i)$$

Since $d_r(s)$ is arbitrary, one obtains the following necessary and sufficient conditions for pole assignment.

Theorem 2.1. *Suppose* $\min(m,p) = 1$ *and set* $r = \dim$ *span* $\{n_1(s), \dots, n_{\max(m,p)}(s)\}$.

(a) *If* $\max(m,p) \geq n$, *a necessary and sufficient condition for pole assignment is* $r = n$. *In particular, if* $\max(m,p) = n$, *then it is necessary and sufficient that the entries of a numerator coprime factor be linearly independent.*

(b) *If* $\max(m,p) < n$, *then arbitrary pole-assignment is not possible. However, one can place any self-conjugate set* $\{s_1, \dots, s_r\}$ *of closed-loop poles by a real output feedback law.*

Proof. Only the last claim in (b) has not been discussed. From (2.11) we see that to place a pole at s_i is to define an affine hyperplane $H(s_i)$ in the linear space of gains K. To say that r of the $n_j(s)$ are linearly independent is to say that $H(s_1), \dots, H(s_r)$ are not parallel and in fact intersect nontrivially.

2.3 Pole Assignment by Dynamic Compensation

Suppose (A, B, C) is a minimal triple in a state-space realization of a $p \times m$ matrix valued rational function $G(s)$ having degree n, so that

$$G(s) = C(sI - A)^{-1}B$$

We denote by $\kappa_i, i = 1, \ldots, m$ and ν_j, $j = 1, \ldots, p$ the controllability and observability indices of (A, B) and (A, C), respectively. Using some fundamental work by Forney [42], it is possible to compute these partitions of the integer n directly from the transfer function $G(s)$, but it is more traditional to interpret these output feedback invariants in terms of a state-space realization. In this section we will present pole-placement criteria due independently to Seraji [18] and Stevens [19] generalizing the well-known result in terms of the controllability and observability indices, due to Brasch-Pearson [28].

First, we recall the nontrivial fact that for any $G(s)$ of degree n there exist constant gain output feedback laws K such that the closed-loop transfer function $G_K(s)$ has n distinct (simple) poles (see also [28]). In the scalar case, this follows directly from a root-locus plot which shows also that this in fact holds for all but finitely many gains. In the $p \times m$ case, supposing $p > m$, for almost all $m \times p$ matrices F, the square transfer function $FG(s)$ has degree n, reducing the rectangular case to the square case by "squaring down". In the square case, we can again appeal to a root-locus argument using the beautiful multivariable root-locus theory developed by Postlethwaite and MacFarlane [10].

Our first formal result generalizes this observation, yielding a variant of Heymann's Lemma.

Lemma 2.2. *For any $p \times m$ transfer function $G(s)$ with degree $\delta(G_k(s))$, there exists an output feedback law K and an m vector v such that*

$$\delta(G(s)) = \delta(G_K(s) \cdot v)$$

Proof. Choosing K so that $G_K(s)$ has $n = \delta(G_K(s))$ simple poles $s_1, \ldots, s_n$, we can write

$$G_K(s) = \sum_{i=1}^{n} \frac{R_i}{s - s_i}$$

where $R_i = \bar{R}_j$ whenever $s_i = \bar{s}_j$. To say $\delta(G_K(s)) = n$ is of course to say rank $(R_i) = 1$, $i = 1, \ldots, n$. For any $v \in \mathbb{R}^m$ we have

$$G_K(s)v = \sum_{i=1}^{n} \frac{R_i v}{s - s_i}$$

and therefore $\delta(G_K(s)v) < n$ if, and only if, $R_i v = 0$ for some i. Therefore, the choice of any v not in the proper subset $\bigcup_{i=1}^{n}$ ker (R_i) of $\mathbb{R}^m$ proves the assertion.

Remark 2.3. (i) Our proof and remarks actually show that there is an open dense set of K and an open dense set of v for which the lemma holds.

(ii) Denoting $A - BKC$ by A_K we have

$$G_K(s)v = C(sI - A_K)^{-1}Bv$$

In particular, the observability indices of $G_K(s)v$ coincide with the observability indices of $G(s)$.

(iii) The same results hold, of course, for $w^t G_K(s)$ with observability indices replaced by controllability indices.

We can now give an elementary proof of the following classical result.

Theorem 2.4 (Brasch-Pearson [28]). *Arbitrary pole assignment for $G(s)$ can be achieved using dynamic compensation of order $q = \min(\kappa_{\max}, \nu_{\max}) - 1$.*

Proof [29]. Suppose a self-conjugate set of desired poles, $\{s_1, \ldots, s_{n+q}\}$ is given. Choosing K and v as in Lemma 2.2, we seek a $1 \times p$ compensator $K(s)$ placing these poles. If $G_K(s)v = N(s)D(s)^{-1}$ and $K(s) = Q(s)^{-1}P(s)$ are coprime factorizations, then the return difference determinant reduces to the constraint

$$Q(s)D(s) + P(s)N(s) = \prod_{i=1}^{n+q} (s - s_i) \tag{2.12}$$

which is linear in the polynomial pair $(Q(s), P(s))$ thought of as an element of a $(q+1)(p+1)$ dimensional vector space. In particular, arbitrary pole assignment by a compensator $vK(s)$ is possible if, and only if, the linear (Sylvester) map

$$S_q : \mathbb{R}^{(q+1)(p+1)} \to \mathbb{R}^{(n+q+1)},$$

obtained by equating coefficients in (2.12), is surjective. On the other hand, Bitmead et al. [30] have given an elegant formula for the rank of S_q in terms of the observability indices of $G_K(s)v$, which of course coincide with those of $G(s)$:

$$\operatorname{rank} S_q = (p+1)(q+1) - \sum_{\nu_i < q+1} (q + 1 - \nu_i)$$

Therefore, $G(s)$ can be pole-assigned by a compensator of order q, where q satisfies

$$(q+1)p - \sum_{\nu_i < q+1} (q + 1 - \nu_i) \geq n \tag{2.13}$$

However, the left-hand side of (2.13) is a nondecreasing function of q, achieving its maximum value, viz. n', when $q = \nu_{\max} - 1$. Since, by duality, the same results hold mutatis mutandis for $q = \kappa_{\max} - 1$, the assertion follows.

Since $\min(\kappa_{\max}, \nu_{\max}) \leq n$, one very easy corollary of the Brasch-Pearson Theorem is the well-known result that one can always pole-assign by a dynamic compensator of order $q \leq n-1$. In fact for systems having rank $C = 1$ and rank $B = 1$, $\kappa_{\max} = \nu_{\max} = n$ so that we must take $q = n - 1$. These rank conditions are however quite restrictive, or "nongeneric", and for generic systems one can obtain a much less conservative estimate for $\min(\kappa_{\max}, \nu_{\max})$. Recall that a property P of points in an affine space $\mathbf{R}^N$ or $\mathbf{C}^N$ is called generic provided the set of points which do not enjoy P lie in a proper subset X of the form

$$X = \{x : f_j(x) = 0 \quad , \quad j = 1, \ldots, \ell\} \tag{2.14}$$

where $f_j(x)$ are polynomials. A set of the form (2.14) is called an algebraic subset (of either $\mathbf{R}^n$ or $\mathbf{C}^n$). For example the set of all $n \times m$ matrices B having rank less than or equal to 1 is an algebraic subset of $\mathbf{R}^{nm}$. Also, those pairs (A, B) corresponding to a four-dimensional state space system having two inputs, which have controllability indices $(4, 0)$ or $(3, 1)$ are nongeneric while those pairs which have controllability indices $(2, 2)$ are generic. More generally, for systems having the generic choice of controllability indices it follows that

$$q \cdot m \geq n \quad \Longrightarrow \quad \kappa_{\max} \geq q \tag{2.15}$$

By duality we also have

$$q \cdot p \geq n \quad \Longrightarrow \quad \nu_{\max} \geq q \tag{2.15$'$}$$

leading to a generic form of the Brasch-Pearson Theorem.

Corollary 2.5. *Arbitrary pole assignment holds for the generic degree n, $p \times m$ linear systems using q-th order dynamic output compensation, where q is any natural number satisfying*

$$\max(m, p)(q + 1) \geq n \tag{2.16}$$

Of course, if $\min(m, p) = 1$ then Corollary 2.5 also implies that the generic system can be pole assigned using constant gain output feedback, provided $\max(m, p) \geq n$. More generally, by combining Corollary 2.5 with the technique used in the proof of the Brasch-Pearson result, we obtain a stronger form of (2.16). First, consider the problem of placing $n + q$ distinct real poles: $s_1, \ldots, s_{n+q}$. If $p \leq m$, the algorithm proceeds as follows

(i) Place the subset $\{s_1, \ldots, s_{p-1}\}$ by output feedback K, as in Corollary 2.5;

(ii) If R_i is the residue of $G_K(s)$ at $s = s_i$, choose a p-vector w so that $w^T R_i = 0$, $i = 1, \ldots, p - 1$.

For generic $G(s)$, $w^T G(s)$ has the generic set of controllability indices and McMillan degree $n - p + 1$. According to Corollary 2.5 we can place the remaining poles $s_p, \ldots, s_{n+q}$ by a compensator of order q, where q satisfies

$$m(q + 1) \geq n - p + 1$$

Recalling that $p = \min(m,p)$ and $m = \max(m,p)$, we can state some generic stabilization results.

Theorem 2.6 (Stevens). *For the generic $p \times m$ system of degree n any set of $n+q$ distinct real poles can be assigned by a real dynamic compensator of order q for any q satisfying*

$$(q+1)\max(m,p) + \min(m,p) - 1 \geq n \tag{2.16}$$

In particular, the generic $p \times m$ system of degree n can be stabilized by a q-th compensator, for any q satisfying (2.16).

The only difference arising when considering a self-conjugate set of complex poles, is that if $\min(m,p)$ is even then the subset of $\min(m,p) - 1$ poles selected in (i) must contain a real pole. This of course is not an obstruction if $n + q$ is odd.

Theorem 2.7 (Stevens). *The generic $p \times m$ system of degree n can be generically pole assigned by a compensator of order q if*

$$(q+1)\max(m,p) + \min(m,p) - 1 \geq n + \epsilon \tag{2.17}$$

where $\epsilon \in \{0,1\}$ with $\epsilon = 0$ unless both $\min(m,p)$ and $n+q$ are even.

This calculation also gives a proof of a related theorem due to Seraji [18].

Theorem 2.8 (Seraji). *If (2.16) holds then using a q-th order compensator closed-loop poles can be generically assigned at any set of $n+q$ frequencies containing a self-conjugate set of $\min(m,p) - 1$ frequencies.*

Remark 2.9. In [31], Youla, Bongiorno and Lu treat the interesting problem of stabilizing a scalar input-scalar output system using stable dynamic compensation, discovering the hard constraint that between every pair of real, nonnegative open loop zeroes there should exist an even number of real open-loop poles. This should, of course, underscore some of the differences between dynamic output compensation and observer based stabilization schemes. Saeks and Murray [32] discovered that this result was equivalent to a result on the existence of a single dynamic compensator simultaneously stabilizing two given plants. Because of its use as a method for designing reliable, or fault-tolerant, stabilization schemes, the simultaneous stabilization problem was formulated and researched in the general setting of stabilizing r given $p \times m$ systems having McMillan degrees n_i, $i = 1, \ldots, r$, by a single (nonswitching) compensator. In particular, Vidyasagar and Viswanadham [20] showed, using the "YBJ parameterization" of all stabilizing compensators for a given plant, that when $r = 2$ and

$\max(m,p) > 1$ the generic pair of systems is always simultaneously stabilizable, in sharp contrast to the case $\max(m,p) = 1$ studied in [32]. In [33], Ghosh and Byrnes proved that $\max(m,p) \geq r$ is sufficient for generic simultaneous stabilizability, generalizing the results of [20]. Moreover, in his thesis ([21]), Ghosh obtained a generalization of Theorems 2.6–2.8, yielding an explicit estimate for the order of the simultaneously stabilizing compensator

$$q(\max(m,p) + 1 - r) + \max(m,p) \geq \sum_{i=1}^{r} n_i \tag{2.17}$$

Finally, in the cases $r < \max(m,p)$ one can use transcendental methods, related to what we now call H^∞-methods, to obtain simultaneous stabilizability criteria for fixed open sets of r-tuples of systems (see [34]).

3. Combinatorial Geometric Methods.

In 1975, H. Kimura published a pole-assignment result which essentially generalizes both the rank 1 case (Corollary 2.5) and the classical results on eigenvalue assignment by state feedback [14]. Kimura's motivation was the problem of stabilizing mechanical systems by output feedback which proved intractible, for very good reason. Such a system typically has $n = 2m$ degrees of freedom, m inputs and m outputs but more generally, one might hope the condition, $m + p \geq n$, would suffice for pole assignability, for the generic system. Actually, a fundamental computation due to Willems and Hesselink ([15], see also section 4.3) shows that this fails in the first nontrivial case, $m = p = 2$ and $n = 4$. Nonetheless, Kimura was able to show that something close to the intuition gleaned from mechanics does hold (see also Davison-Wang [35], for an independent derivation).

Theorem 3.1 (Kimura). *If $m + p - 1 \geq n$, the generic system can be assigned any self-conjugate set of distinct poles, different from the open-loop poles.*

Remark 3.2. Actually one can sharpen this result to obtain arbitrary pole-assignment for the generic system.

The original proof of Theorem 3.1 given by Kimura contains an interesting combination of the geometry of linear subspaces and combinatorics involving dimension counting for subspaces with various incidence properties. There are of course some antecedents for such methods in the literature on eigenvalue assignment by state feedback, especially by M.L.J. Hautus [22]. In [22], Hautus is able to express the eigenvalue assignment as the problem of choosing n linearly independent $(n + m)$ vectors, one each from a set of n preassigned n-dimensional subspaces of $(n + m)$-space. Hautus' criterion for the existence of such a selection is, quite remarkably, a linear algebra analogue of a result of Rado's [36], generalizing Ph. Hall's famous solution of the "marriage problem", and is equivalent to controllability of the underlying linear system. Kimura's situation is similar, but of course more complicated since he requires output feedback.

More explicitly, from the return difference determinant we see that to say s_i, $i = 1, \ldots, n$ is a closed-loop pole for the output feedback gain $-K$ is to say

$$\det \begin{bmatrix} I & G(s_i) \\ K & I \end{bmatrix} = 0 \tag{3.1}$$

Equivalently, we have

$$\text{col. sp.} \begin{bmatrix} I \\ K \end{bmatrix} \cap \text{col. sp.} \begin{bmatrix} G(s_i) \\ I \end{bmatrix} \neq (0) \tag{3.2}$$

when s_i is not a pole of $G(s)$ and in general

$$\text{col. sp.} \begin{bmatrix} I \\ K \end{bmatrix} \cap \text{col. sp.} \begin{bmatrix} N(s_i) \\ D(s_i) \end{bmatrix} \neq (0)$$

where $G(s) = N(s)D(s)^{-1}$ is a coprime factorization. Regarding K and $G(s_i)$ as linear maps

$$K : \mathbf{C}^p \to \mathbf{C}^m \quad , \quad G(s_i) : \mathbf{C}^m \to \mathbf{C}^p$$

which define subspaces of $\mathbf{C}^{p+m}$ via

$$\begin{aligned} \text{gr } (K) &= \{(y,u) : u = Ky\} \\ \text{gr } G(s_i) &= \{(y,u) : y = G(s_i)u\} \end{aligned}$$

the pole-assignment condition (3.1) may be expressed as

$$\text{gr } (-K) \cap \text{ gr } G(s_i) \neq (0) \quad , \quad i = 1, \ldots, n. \tag{3.3}$$

In this language, the pole-assignment problem may be stated as follows:

Given the n m-planes gr $G(s_i)$ in $\mathbf{C}^{p+m}$ find a p plane W, having the form W = gr (K), such that

$$\dim (W \cap \text{ gr } G(s_i)) \geq 1 \quad , \quad i = 1, \ldots, n \tag{3.4}$$

For state feedback, $G(s) = (sI - A)^{-1}B$ and the approach taken by Hautus [22] was to construct W by choosing n linearly independent vectors, $w_i \in$ gr $G(s_i)$, setting $W = \text{sp } \{w_1, \ldots, w_n\}$. For output feedback and real poles, an alternate construction of W is given in the following result, due to Rosenthal.

Lemma 3.3 [23]. *Given $(m+p-1)$ m-planes V_i in $\mathbf{R}^{m+p}$ there exists a p-plane W intersecting each V_i nontrivially.*

Proof. Suppose $m \leq p$. If $\hat{W}$ is a $(p-m+1)$-plane intersecting $V_1, \ldots, V_{p-m+1}$ nontrivially, for any two planes V_i, V_j one finds a vector v_i so that $\hat{W} \oplus \text{sp}\,\{v_i\} \cap V_i \neq (0)$, $\hat{W} \oplus \text{sp}\,\{v_i\} \cap V_j \neq (0)$. By adding $m-1$ vectors, one therefore intersects $(p-m+1)+2(m-1)$ planes.

As Rosenthal shows, it is possible to improve on this result by choosing $\hat{W}$ more carefully. For example [23], if $V_1, \ldots, V_5 \subset \mathbf{R}^5$ are 2-planes the $V_1+V_2+V_3$ cannot be a direct sum; i.e. there is a 2-plane $\hat{W}$ intersecting V_1, V_2, V_3 nontrivially. Because $\hat{W}+V_4+V_5$ is not a direct sum, there is a 3 plane intersecting each V_i nontrivially. We need a little notation in order to formalize this conclusion: For $a, b \in \mathbf{Z}$, $b \neq 0$, the Gauss bracket $[a/b]$ denotes the greatest integer less than or equal to a/b.

Theorem 3.4 (Rosenthal). *Assume $p \geq m$ and set $k = \left[\frac{p}{m}\right]$. For the generic system, the condition*

$$p + \left[\frac{m}{1}\right] + \ldots + \left[\frac{m}{k}\right] - 1 \geq n$$

is sufficient for pole placement of real, distinct poles.

A rigorous proof of Theorem 3.4 requires a proof that if a p-plane W can be constructed, meeting each $V_i = \text{gr}\; G(s_i)$, then there exists such a p-plane having the form, $W = \text{gr}\;(K)$. This "general position" argument will be discussed in sections 5.2, 5.4.

4. Affine Algebraic Methods for Pole Assignment

4.1 Introduction. When rank $G(s) \geq 1$, equations (2.3)–(2.3)$'$ are no longer affine. This, of course, is also the case for eigenvalue assignment by state feedback and while we will examine some analogies with this problem in section 4.3 the differences between pole assignment by output and by state feedback should become clear from a variety of viewpoints. One of the possible differences was anticipated in [9], where several questions concerning the finer structure of the problem are raised, concerning for example what kinds of formulae would exist for pole-assigning feedback gains, assuming such feedback laws would exist. For example, Anderson, Bose and Jury [9] ask whether, as in the state feedback case, there would exist affine straight-line formulae for K i.e. universal formulae for K that are affine in the closed-loop poles with coefficients which can be "preprocessed" rationally in terms of the plant parameters. As we have seen, this does hold in the case $\min(m,p) = 1$. One of the corollaries of the now famous calculation by Willems and Hesselink [15] is that the answer to this question is negative; indeed, in the general 2×2 case one requires straight-line formulae using square roots (see Remark 4.10.) Furthermore, Byrnes and Stevens [37] showed in the 2×3 case that it is impossible to find solutions using straight-line formulae with radicals (see section 5.3). We now turn to the technical details, beginning with the fundamental analysis by Hermann and Martin of pole-assignment over $\mathbf{C}$.

4.2 Infinitesimal Analysis.

In terms of a minimal state-space realization

$$G(s) = C(sI - A)^{-1}B \tag{4.1}$$

of $G(s)$ with A an $n \times n$ matrix, B an $n \times m$ matrix and C a $p \times n$ matrix, the pole assignment equations (2.3)–(2.3)′ take the form

$$\det(sI - A - BKC) = s^n + c_1(K)s^{n-1} + \ldots + c_n(K) \tag{4.2}$$

Equating coefficients (4.2) defines a polynomial map

$$\chi : k^{mp} \to k^n, \tag{4.3}$$

defined for $k = \mathbf{R}$ or $\mathbf{C}$ via

$$\chi(K) = (c_i(K))_{i=1}^n \tag*{(4.3)$'$}$$

There is a nontrivial consequence of the formulation of the pole-assignment problem as a problem concerning a polynomial mapping (4.3), apparently first noticed by Willems and Hesselink [15]. Namely, the condition

$$mp \geq n \tag{4.4}$$

is necessary for arbitrary pole-assignment. While one might appeal to a "dimension count" in this case, such arguments are usually far more subtle than just counting a few numbers. A careful proof would even show that (4.4) is necessary for image (χ) to contain an open set. We denote by $d\chi|_K$ the Jacobian of χ at a point K. A point $y \in k^n$ is then called regular if either $\chi^{-1}(y)$ is empty or if rank $d\chi|_K$, for all $K \in \chi^{-1}(y)$, has the maximum possible value. Supposing that image (χ) contains an open set, Sard's theorem asserts that there exists a regular value in the image of χ, from which (4.4) then follows using the implicit function theorem.

Proposition 4.1 (Willems-Hesselink). *The condition $mp \geq n$ is necessary for pole-assignment by output feedback.*

Turning to sufficient conditions, any proposition asserting that

$$\text{rank } d\chi|_{K_0} = n \tag{4.5}$$

for some K_0 will have lots of corollaries over both **R** and **C**. As we noted, over either **R** or **C**, from the implicit theorem it follows that (4.5) implies image (χ) contains an open neighborhood of $\chi(K_0)$; i.e. that one can assign all poles sufficiently close to the roots of

$$p_{\chi_0}(s) = s^n + c_1(K_0)s^{n-1} + \ldots + c_n(K_0)$$

Since (4.3) is a polynomial map, over **C** a more remarkable fact follows from what is called either "the fundamental openness theorem" [38] or the "dominant morphism theorem" [39]; viz. (4.5) implies that image (χ) contains an open dense set of monic complex polynomials. In other words, over **C** the existence of one K_0 satisfying (4.5) implies that one can place a generic, or almost any, set of desired poles. The generic existence of such K_0 follows from a beautiful calculation given in [11].

Theorem 4.2 (Hermann-Martin). *For any $p \times m$ strictly proper transfer function of degree n,*

$$rank\ d\chi|_0 = \ dim\ span\ \{CB, CAB, \ldots, CA^{n-1}B\} \tag{4.6}$$

Proof. From (2.3) we compute the directional derivative of χ at 0, in the direction K, as

$$\begin{aligned} d\chi|_0(K) &= \det\ D(s) \lim_{\epsilon \to 0} \frac{\det\ (I + \epsilon KG(s) - \det\ (I)}{\epsilon} \\ &= \det\ D(s) \lim_{\epsilon \to 0} -\frac{tr(\epsilon KG(s))}{\epsilon} = -\ \det\ D(s) tr(KG(s)) \\ &= \det\ D(s) \langle -K, G(s) \rangle \end{aligned}$$

where $\langle K, M \rangle = tr(KM)$ is the canonical representation of the $p \times m$ matrix K as a linear functional on the space of $m \times p$ matrices M. Expanding $G(s)$ in a Laurent series

$$G(s) = \sum_{i=1}^{\infty} H_i s^{-i}$$

we find that rank $d\chi|_0$ is given by the dimension of space of n-vectors

$$(\langle -K, H_1 \rangle, \ldots, \langle -K, H_n \rangle)$$

with K arbitrary. Since $H_i = CA^{i-1}B$, (4.6) follows.

Of course, if $mp \geq n$ the property that the right hand side of (4.6) is n-dimensional is generic.

Corollary 4.3 (Willems-Hesselink). *For the generic real system, the condition $mp \geq n$ is necessary and sufficient to be able to assign an open neighborhood of the open-loop characteristic polynomial via real output feedback.*

Corollary 4.4 (Hermann-Martin). *For the generic real or complex system, the condition $mp \geq n$ is necessary and sufficient to be able to assign an open, dense set of complex closed-loop characteristic polynomials via complex output feedback.*

Remark 4.5. Results such as Corollary 4.4 are often described as providing generic pole-assignment for the generic system. In fact, it is known [43] that if $mp \geq n$ for the generic system the polynomial map χ is proper, i.e. $\chi^{-1}(F)$ is compact for compact sets F. Therefore, χ maps closed sets to closed sets and, in particular, image χ is also closed (cf. section 5). In summary, any result on generic pole-assignment for generic systems can be sharpened to arbitrary pole-assignment for generic systems. Moreover, in many cases the particular generic property of such systems is explicitly known, just as in the case of state feedback and reachable systems.

The geometric analysis of eigenvalue assignability via state feedback is similar but, naturally, much simpler. In fact, using infinitesimal methods and one of the techniques presented in chapter 2 one can give an elementary, clean proof of Heymann's Lemma.

Theorem 4.6 (Heymann). *The pair (A, B) is reachable if, and only if, there exists a state feedback law K and a vector $v \in \mathbb{R}^m$ such that the pair $(A+BF, Bv)$ is reachable.*

Since eigenvalue assignment, via state feedback, for scalar reachable systems is quite easy to prove, the principal application of Heymann's Lemma is Wonham's celebrated result asserting that reachability is equivalent to arbitrary eigenvalue assignability. To prove Heymann's Lemma we first differentiate a rational map, as in the proof of Theorem 4.2, again following Martin and Hermann [12]. We need some notation:

$$\begin{aligned} Gl_n(\mathbb{R}) &= \{\text{real } n \times n \text{ matrices } T : \det T \neq 0\} \\ M_{m,n} &= \{\text{real } m \times n \text{ matrices}\} \\ [S_1, S_2] &= S_1 S_2 - S_2 S_1 \quad , \quad S_i \in M_{n,n} \end{aligned}$$

Consider the rational map

$$\Phi : G\ell_n \times M_{m,n} \to M_{n,n}$$

defined via

$$\Phi(T, F) = T(A + BF)T^{-1} \tag{4.7}$$

A straightforward, but enjoyable, calculation shows that the directional derivative of the map

$$A \mapsto TAT^{-1}$$

in the direction $S \in M_{n,n}$ is just $[S, A]$. More generally we have

$$d\Phi|_{(I,0)}(S, F) = [S, A] + BF$$

To say

$$d\Phi|_{(I,0)} : M_{n,n} \times M_{m,n} \to M_{n,n}$$

is onto is to say

$$(d\Phi)^* : M^*_{n,n} \to M^*_{n,n} \times M^*_{m,n}$$

is one-to-one, but if $L \in M^*_{n,n} \simeq M_{n,n}$ we have

$$d\Phi|^*_{(I,0)}(L)(S,F) = \langle [S,A] + BF, L\rangle = tr([S,A]L + BFL) \tag{4.8}$$

If $d\Phi^*_{(I,0)}(L)$ is zero, then (4.8) vanishes for all (S,F). In particular

$$\mathrm{tr}([S,A],L) = \mathrm{tr}([A,L]S) = 0 \qquad \forall S$$

and

$$\mathrm{tr}(BFL) = \mathrm{tr}(LBF) = 0 \qquad \forall F$$

In other words, we must have

$$[A,L] = 0 \qquad \text{and} \qquad LB = 0$$

so that

$$L(B, AB, \ldots, A^{n-1}B) = 0.$$

Lemma 4.7 (Hermann-Martin). *For Φ defined in (4.7)*

$$n^2 = rank\ d\Phi|_{(I,0)} \Leftrightarrow dim\ col.\ span\ (B, AB, \ldots, A^{n-1}B) = n \tag{4.9}$$

Hermann and Martin used (4.9) to prove an analogue of Corollary (4.4) using the dominant morphism theorem, viz. reachability implies almost arbitrary eigenvalue assignability via complex state feedback. Over $\mathbb{R}$, for (A,B) reachable, the implicit function theorem shows that image (Φ) contains an open set U of $n \times n$ matrices. Since the set X of matrices with nondistinct eigenvalues is a proper, algebraic subset of $M_{n,n}$, X has no interior. In particular, $U - U \cap X \neq \phi$ so that if (A,B) is reachable there exists F such that $(A+BF)$ has distinct eigenvalues. As in Lemma 2.2, we note that reachability of this pair is equivalent to

$$G_F(s) = (sI - A - BF)^{-1}B = \sum_{i=1}^{n} \frac{R_i}{s - s_i}$$

with s_i distinct, rank $R_i = 1$ and $R_i = \bar{R}_j$ whenever $s_i = \bar{s}_j$. Therefore any v satisfying

$$R_i v \neq 0 \qquad i = 1, \ldots, n$$

will also render $(A + BF, Bv)$ reachable, proving Heymann's Lemma and implying Wonham's Theorem.

We conclude this section with a recent extension [40]–[41] of these kinds of results to a more general class of additive inverse eigenvalue problems. In a state-space representation, we consider the problem of making the spectrum of

$$A + L \quad , \quad L \in \mathcal{L} \tag{4.7}$$

arbitrary where $\mathcal{L}$ is a subspace of the vector space M_n of complex $n \times n$ matrices. For pole-assignment by output feedback, we have

$$\mathcal{L}_0 = \{BKC : \beta, C \text{ fixed}, K \text{ arbitrary}\}$$

In general, defining

$$\chi_A : \mathcal{L} \to \mathbf{C}^n \tag{4.8}$$

via

$$\chi_A(L) = \det\,(sI - A - L)$$

an extension of the proof of Proposition 4.1 shows that the condition

$$\text{rank } \mathcal{L} \geq n$$

is necessary for arbitrary eigenvalue assignment, for any fixed A. This of course is not sufficient, even for generic A, as the example

$$s\ell_n = \{L : tr\ L = 0\}$$

shows in a dramatic way, since dim $s\ell_n = n^2 - 1$. A remarkable result, derived in [40], shows that these two considerations are in fact the essence of this problem.

Theorem 4.8 (Wang). *Necessary and sufficient conditions for image χ_A to contain an open dense set of polynomials for the generic A,* are

(i) dim $\mathcal{L} \geq n$

(ii) $\mathcal{L} \not\subset s\ell_n$.

We note that $\mathcal{L}_0 \subset s\ell_n$ if, and only if, $CB = 0$. Moreover, for generic B, C, we compute dim $\mathcal{L}_0 = mp$. In [41], Wang also gives explicit description of those A for which χ_A is "almost onto". The method of proof for Theorem 4.8 begins similarly, with an application of the "dominant morphism theorem," from which one derives equivalent conditions for χ_A to be "almost onto" for generic A: There exists an $L_0 \in \mathcal{L}$ such that

(a) rank $d\chi_A(L_0) = n$;

(b) The linear map $\phi : \mathcal{L} \to \mathbf{C}^n$, $\phi(L) = (tr(L), tr(A+L_0)L, \ldots, tr(A+L_0)^{n-1}L)$ is onto;

(c) Denoting by Q the Lie algebra of matrices generated by $\{I, A+L_0, \ldots (A+L_0)^{n-1}\}$ consider the restriction of the moment map $\Phi : \mathcal{L} \to Q^*$, $\Phi(L)M = tr(LM)$, then $\dim Q = n$ and $\Phi(\mathcal{L}) = Q^*$;

(d) $\mathcal{L} + Im\ ad_{(A+L_0)} = M_n$, where

$$ad_{A+L_0}(M) = [A+L_0, M] = (A+L_0)M - M(A+L_0);$$

(e) $\mathcal{L}^\perp \cap$ Ker $ad_{(A+L_0)^t} = \{0\}$, where

$$\mathcal{L}^\perp = \{M \in M_n : tr(M^t L) = 0 \ \forall\, L \in \mathcal{L}\}.$$

When $\mathcal{L}_1 = \{BK : B \text{ fixed}, K \text{ arbitrary}\}$, Hermann and Martin [11] proved that condition (e) is equivalent to controllability of (A, B). For $\mathcal{L}_0$ as defined as above, Willems and Hesselink [15] proved that (a), (b) and (e) are equivalent if $A + L_0$ has a cyclic vector. Wang's proof [41] of the general result proceeds from (c), using the Lie theoretic interpretation of the moment map from Hamiltonian mechanics and the conjugacy of Cartan subalgebras of M_n to eliminate the dependence of criteria (a)–(e) on L_0.

4.3 Elimination Theoretic Techniques.

In the aftermath of the introduction of infinitesimal (complex) algebraic geometric methods, some rather serious questions arose concerning pole-assignment using real output feedback. In the light of rank one methods and Kimura's theorem, the first nontrivial case to analyze is pole assignment for 2×2 systems having McMillan degree 4. Here, the system (4.2) of algebraic equations consists of 4 equations, of degrees 4, 3, 2 and 1 in 4 unknowns. General wisdom (gleaned from Bezout's Theorem) would then have it that generically there should be $4! = 24$ solutions. In contrast, eliminating all but one of the variables, Willems and Hesselink [15] found that generically the resulting equation, or "eliminant," had degree 2, so that there are $2 << 24$ solutions. This shows that the pole-assignment problem has a highly nongeneric special structure, which is probably most easily derived from the return difference determinant using the Binet-Cauchy expansion of an $(m+p) \times (m+p)$ determinant as an inner product of complementary m-th order and p-th order minors:

$$\det \begin{pmatrix} I & G(s_i) \\ K & I \end{pmatrix} = \left\langle m_\alpha \begin{pmatrix} I \\ K \end{pmatrix}, m_{\alpha'} \begin{pmatrix} G(s_i) \\ I \end{pmatrix} \right\rangle \tag{4.9}$$

In this case, there are six 2×2 minors of $\binom{I}{K}$, of which one is constant, four are linear and one quadratic. Thus, assuming s_i distinct (4.9) yields, for $i = 1, \ldots, 4$, four equations in K which contain the same quadratic term; i.e. we encounter four linear equations subject to a quadratic constraint.

Remark 4.9. In general, the pole assignment equations consist of n linear equations

$$\left\langle v, m_{\alpha'}\begin{pmatrix} G(s_i) \\ I \end{pmatrix}\right\rangle = 0 \quad , \quad i = 1, \ldots, n \tag{4.9}$$

subject to quadratic equations (i.e. Plücker relations) reflecting the fact that v is a vector consisting of the p-th order minors of a $(p+m) \times p$ matrix. This was first discovered in the general case by Brockett and Byrnes ([16], see also [43]) using both projective algebraic geometric methods (see chapter 5) and the Binet-Cauchy formula, and also independently by Morse, Anderson and Wolovich [24] using multilinear algebra. This observation has also been the starting point of a recent assault on the pole-assignment problem by Karcanias et al. [44]–[45].

In order to find an explicit form of the eliminant in the case $m = p = 2$, $n = 4$, we follow the derivation in [24], with the obvious notation

$$\begin{aligned} G(s) &= \frac{1}{\Delta(s)}\begin{bmatrix} P_1(s) & P_2(s) \\ P_3(s) & P_4(s) \end{bmatrix} \quad , \quad P_i(s) \text{ polynomial} \\ P_5(s) &= \Delta(s) \det G(s) \\ K &= \begin{bmatrix} K_1 & K_2 \\ K_3 & K_4 \end{bmatrix} \end{aligned}$$

and

$$K_5 = \det K = K_1 K_4 - K_2 K_3 \tag{4.10}$$

In particular, (4.2) may be expressed as

$$\Delta_c(s) = \prod_{i=1}^{4}(s - s_i) = \Delta(s) + \sum_{j=1}^{5} p_j(s) K_j.$$

In terms of the expansion

$$p_j = \sum_i p_{ji} s^i$$

(4.2) reduces to

$$c_i = \Delta_i + \sum_j p_{ji} K_j \tag{4.11}$$

together with the quadratic relation (4.10) defining K_5. We denote a general solution of (4.11) by $K_i = K_i^0 + \gamma e_i$ where (e_i) is a basis element of the (generically) one-dimensional kernel of the linear map defined in (4.11). The quadratic relation (4.10) thus becomes

$$\alpha\gamma^2 - \beta\gamma + \sigma = 0$$

where

$$\begin{aligned}
\alpha &= e_1e_4 - e_2e_3 \\
\beta &= K_2^0e_3 + K_3^0e_2 - e_1K_4^0 - e_4K_1^0 + e_5 \\
\sigma &= K_1^0K_4^0 - K_2^0K_3^0 - K_5^0
\end{aligned}$$

In particular, the explicit form of the solution, assuming $\alpha \neq 0$ is

$$K_i = K_i^0 + \frac{\beta \pm \sqrt{\beta^2 - 4\alpha\sigma}}{2\alpha} e_i \tag{4.12}$$

Remark 4.10. The condition $\alpha \neq 0$ is precisely the "nondegeneracy" condition used in [16], [43] for $mp \leq n$. This generic condition implies that image (χ) will be closed; e.g. that sequences of solutions don't "go off to infinity" for convergent sequences of data (see also Remark 4.5). Also, the generic condition $\sigma \neq 0$ then implies that χ will be 2 to 1 and that any expression for pole-assigning gains will require the use of radicals, answering the questions raised by Anderson, Bose and Jury [9] in the negative (cf. section 4.1).

Since examples for which $\alpha \neq 0$ and $\sigma \neq 0$, exist in great profusion, we have several corollaries of the calculation in [15] which hold for the generic system.

Theorem 4.11 (Willems-Hesselink). *For the generic 2×2 system having McMillan degree 4, there are two output feedback gains, counted with multiplicity, which assign a given set of closed-loop poles. Moreover, for the generic real system:*

(i) *There is an open set, of infinite Lebesgue measure, of real characteristic polynomials which cannot be assigned using real output feedback;*

(ii) *The set of assignable, real, closed-loop characteristic polynomials is a closed set, containing an open subset having infinite Lebesgue measure, and*

(iii) *There is no straight-line, linear formula for pole assigning gains using just rational preprocessing of the system parameters and the desired characteristic coefficients.*

As one might expect, the situation becomes far more complicated for larger m, p, n and straightforward elimination soon becomes prohibitive. Before turning to the general case, it is therefore important to understand what elimination theory can and cannot imply. We will also illustrate this with some quite different applications of elimination methods to the pole-assignment problem.

Put geometrically, the main problem of elimination theory can be described as follows. Suppose $Z \subset k^N \times k^M$ is an algebraic set and consider the projection map

$$p_1 : Z \to K^N$$

defined via

$$p_1(x, y) = x$$

The main problem is to describe the subset

$$p_1(Z) = \{x : \exists\, y \text{ such that } (x, y) \in Z\}$$

as explicitly as possible. In other words $x \in p_1(Z)$ if, and only if, for some y the point (x, y) satisfies the polynomial equations defining Z. Therefore, if $p_1(Z)$ were itself described by polynomial equations we would have succeeded in eliminating y from the equations defining Z. This is not always the case, since projections are not closed maps–take, for example,

$$Z = \{(x, y) : xy - 1 = 0\} \subset \mathbf{C} \times \mathbf{C}.$$

There are nevertheless several results describing $p_1(Z)$ which are useful.

If $k = \mathbf{C}$, then Chevalley's Theorem [39] asserts that $p_1(Z)$ is "constructible"; i.e. $p_1(Z)$ can be described by polynomial equations

$$f_i(x) = 0 \qquad i = 1, \ldots, r \tag{4.13}$$

and polynomial inequations

$$g_j(x) \neq 0 \qquad \text{for some } j = 1, \ldots, s \tag{4.14}$$

If $p_1(Z)$ is closed, it is algebraic but this is not always the case. The main theorem of elimination theory, over $\mathbf{C}$, is that $p_1(Z)$ will be closed if the equations

$$h_k(x, y) = 0 \qquad k = 1, \ldots, t \tag{4.15}$$

defining Z are homogeneous in y (see [46]). Remarkably, for the pole-assignment problem this turns out to be the case for "nondegenerate" systems (see section 5).

If $k = \mathbf{R}$, then $p_1(Z)$ may not even be constructible, as the basic example

$$Z = \{(x, y) : y = x^2\} \subset \mathbf{R} \times \mathbf{R}$$

shows. The fundamental theorem of Tarski-Seidenberg [47] asserts, however, that $p_1(Z)$ will always be "semialgebraic"; i.e. $p_1(Z)$ is described by (4.13)–(4.14) together with polynomial inequalities

$$p_m(x,y) > 0 \qquad \text{some } m = 1,\dots,u \tag{4.16}$$

or

$$q_n(x,y) \geq 0 \qquad \text{some } n = 1,\dots,v \tag{4.17}$$

The main theorem of elimination theory also holds over $\mathbf{R}$, so that $p_1(Z)$ is closed if (4.15) is homogeneous in y. In this case, a recent refinement of Tarski-Seidenberg, by Delzell [48], asserts that $p_1(Z)$ can be described by (4.13) and (4.17), as in the second example given above.

Finally, there exist somewhat more symmetric versions of these results: Over $\mathbf{C}$, the image of a constructible set is constructible and, over $\mathbf{R}$, the image of a semialgebraic set is semialgebraic under a projection or, slightly more generally, a polynomial map. For example, since the open left-half complex plane is semialgebraic and hence so is its n-fold product, the set of n-th degree Hurwitz polynomials is semialgebraic, being the image of this product space under the polynomial map

$$(s_j) \mapsto c_i(s_j) \text{ where } \prod_j (s - s_j) = \sum_{i=0}^{n} c_i(s_j)s^i.$$

And, the Routh-Hurwitz conditions explicitly define this open semialgebraic set.

Remark 4.12. Semialgebraic sets are often called "decidable" since membership in such sets can be decided in a finite number of polynomial operations. For example, it is possible to decide for which (A,B,C) a given characteristic polynomial $p(s)$ can be assigned via output feedback, i.e. if

$$Z = \{((A,B,C),K) : \det(sI - A - BKC) = p(s)\}$$

and

$$p_1((A,B,C),K) = (A,B,C)$$

then $p_1(Z)$ is semialgebraic (see [9] for several refinements of such arguments).

As another illustration, we consider the problem of generic stabilizability [25]. Using a standard bilinear transformation, it is not hard to show that if stabilization by output feedback can be achieved for the generic $p \times m$ continuous-time system of degree n, then it is also possible for the generic $p \times m$ discrete-time system of degree n. A similar calculation also enables one to achieve generic stabilization with a pre-assigned margin of stability; i.e., for all $K \in \mathbf{Z}^+$ it must also follow that for the generic system one can assign poles somewhere in the disc about 0 of radius $1/K$. Using the fact that, generically, image (χ) is closed (cf. Remark 4.5) and the Baire Category Theorem, it follows that for a dense set of systems it must be possible to place the closed-loop poles all at 0. Moreover, an application of the Tarski-Seidenberg theorem then allows one to conclude that it must also be possible, for an open dense set of systems, to place the closed-loop poles all at 0. Finally, since the dimension of the algebraic set of $n \times n$ nilpotent matrices is $n^2 - n$ (see [49]), a dimension count as in the proof of Theorem 4.1 before yields the surprising result that (4.4) is in fact also necessary for generic stabilization.

Theorem 4.13 (Anderson-Byrnes). *The condition $mp \geq n$ is necessary for output feedback stabilization of the generic $p \times m$ system having degree n.*

We remark that all known examples suggest that, for given m, n and p, stabilizability may hold generically if, and only if, pole-assignability holds generically. For example, P. Molander (unpublished) has modified the Willems-Hesselink calculation to show that generic stabilizability does not hold for $m = p = 2$ and $n = 4$. This, of course, could also be seen from the formulae (4.12) and in [25] an explicit example of a (nondegenerate, i.e. $\alpha \neq 0$) system which cannot be stabilized is given. Using (4.12) and the Routh-Hurwitz inequalities, one concludes that small perturbations of this system also cannot be stabilized.

Proposition 4.14 (Molander, Anderson-Byrnes). *If $m = p = 2$, $n = 4$ there is an open set of systems which cannot be stabilized by output feedback. In particular, for each such system the set of characteristic polynomials which cannot be assigned contains a set having infinite Lebesgue measure.*

Remark 4.15. There is of course an open set of 2×2 systems having degree 4 which can be pole assigned and therefore stabilized. However, the existence of such open sets also follows from more classical methods; e.g. $m \times m$ minimum phase systems with invertible high frequency gain can always be stabilized using output feedback. This can be seen using a multivariable root-locus plot (see [10]) or using geometric linear systems theory by intepreting zeroes in terms of (A, B), or almost (A, B), invariant subspaces (see [50]).

In the next chapter we will turn to the general case $mp = n$, obtaining the explicit formula (1.1) for the degree of the eliminant, calculated by Brockett-Byrnes without using elimination theory. This implies several positive results. For example, if $m = 2$ and $p = 3$, the eliminant has degree 5, a result obtained independently by Anderson, Morse and Wolovich by explicit elimination. This result implies, or course, that the generic 2×3 system having degree ≤ 6 can be arbitrarily pole-assigned using real output feedback.

Remark 4.16. It is not surprising, yet nontrivial to prove (see [37]), that the Galois group of the generic 2×3, degree 6 problem is in fact the symmetric group on 5 letters, S_5, which of course is not a solvable group. In particular, there do not exist formulae for pole-assigning gains which use rational preprocessing and the extraction of r-th roots, underscoring the nonlinear nature of this problem.

5. Projective Algebraic Geometric Methods

5.1 Enumerative Geometry and the Schubert Calculus

Having analyzed the case $m = p = 2$, $n = 4$ in section 4.3 using an explicit elimination argument, the next nontrivial case would correspond to the choices

$$\min(m,p) = 2 \quad , \quad \max(m,p) = 3 \quad n = mp = 6 \tag{5.1}$$

We know both from the general formula (1.1) and an explicit elimination argument [24] in this case that the system (4.3) of 6 equations in 6 unknowns generically will have an eliminant having degree 5. This is remarkable since, in general, a system of six equations of degrees $1, 2, \ldots, 6$ will have an eliminant of degree $6! = 720$, implying that for the pole assignment equation 715 unexpected cancellations of a rather complicated nature occur, suggesting that correctly carrying out an elimination argument even for modest sizes of m, n and p will be extremely difficult. On the other hand, as the earlier work by Hautus and by Kimura suggests, the degree of the eliminant might possibly be interpreted in combinatorial geometric terms and therefore be computed without resorting to an explicit elimination. In fact, using methods of enumerative geometry, Brockett and Byrnes [16] were able to compute the number (counted with multiplicities) $d_{m,p}$ of possibly complex feedback laws placing a given set of poles of a (specific) generic class of $p \times m$ systems having degree $n = mp$. Explicitly, the degree $d_{m,p}$ of the eliminant is given by the formula

$$d_{m,p} = \frac{1! \cdots (p-1)!(mp)!}{m! \cdots (m+p-1)!} \tag{5.2}$$

We note that if $\min(m,p) = 1$, then $d_{m,p} = 1$, in harmony with the results of chapter 2. Moreover, $d_{2,2} = 2$, while $d_{2,3} = d_{3,2} = 5$. More generally, $d_{2,p}$ is the p-th Catalan number which is odd precisely when $p = 2^r - 1$. For example, if $\min(m,p) = 2$ and $\max(m,p) = 7$ there are 429 solutions of the pole-assignment equations and, in particular, at least one real feedback law for a real system and a self-conjugate set of desired poles.

Formula (5.2), and several very useful generalizations, can be derived in a formal way from the Schubert calculus of enumerative geometry, but it is known that there are several technical conditions which must be verified to ensure that such calculations will be correct. Indeed, Hilbert's 15th Problem, first solved by van der Waerden in 1929, asked for the rigorous justification of the Schubert calculus and of general methods in enumerative geometry. In this section we outline the basic framework, and necessary technical conditions, one obtains from Schubert's calculus. In 1886 Schubert [51] addressed the problem:

Given mp m-planes V_i in $\mathbf{C}^{m+p}$, how many p-planes W intersect each V_i nontrivially?

According to Schubert, if the planes V_i are in "general position" then this number is given by (5.2). Moreover, taking $V_i = gr(G(s_i))$, we see from (3.4) that, when it applies, (5.2) gives an upper bound on the number of possibly complex feedback laws K placing the closed-loop poles at $s = s_i$. Indeed, if each such p-plane W were of the form

$$W = \mathrm{gr}(K) \tag{5.3}$$

then (5.2) would yield the exact number of pole-assigning gains.

There are then two points which need clarification. First, in rigorous treatments of the Schubert calculus (see [52] and also the survey [53]), "general position" can be interpreted as the condition that only a finite number of p-planes W meet the planes V_i nontrivially. The second is to guarantee a priori that every such W has the form (5.3). Fortunately, these conditions are implied by a simple, system theoretic condition which involves some interesting geometric techniques (see sections 5.2–5.3). Not surprisingly, the condition $mp \leq n$ is necessary for the n m-planes gr $G(s_i)$ to be in "general position". Explicitly, "general position" will be implied by

$$\det \begin{bmatrix} K_1 & I \\ K_2 & G(s) \end{bmatrix} \equiv 0 \Longrightarrow rk \begin{bmatrix} K_1 \\ K_2 \end{bmatrix} < p \tag{5.4}$$

In [16], condition (5.4) is referred to as "nondegeneracy" of $G(s)$ and it is known that nondegenerate systems are generic if $mp \leq n$ (see [43]). Since nondegeneracy also implies image (χ) is closed, we can deduce

Theorem 5.1 (Brockett-Byrnes). *Suppose $mp = n$. For any nondegenerate $p \times m$ system having degree n and for any monic polynomial $p(s)$ of degree n there are, counting with multiplicity, $d_{m,p}$ output feedback laws assigning $p(s)$ as a closed-loop characteristic polynomial. In particular, if $\min(m,p) = 1$ or if $\min(m,p) = 2$ and $\max(m,p) = 2^r - 1$, for nondegenerate real transfer functions and real $p(s)$ there is always a real output feedback law assigning $p(s)$ as the closed-loop characteristic polynomial.*

Example 5.2. Consider the 3×2 system having degree 6

$$A = \begin{bmatrix} 1 & 2 & -1 & 2 & 1 & 1 \\ 1 & 3 & 1 & 2 & 3 & 1 \\ -1 & 1 & 2 & 3 & -1 & 1 \\ 3 & 2 & 1 & -3 & -1 & -2 \\ -1 & -3 & -2 & -1 & 1 & -3 \\ -2 & -1 & 1 & 3 & 2 & 1 \end{bmatrix}, \quad B = \begin{bmatrix} 1 & 0 \\ 0 & 1 \\ 0 & 0 \\ 0 & 0 \\ 0 & 0 \\ 0 & 0 \end{bmatrix} \qquad (5.5)$$

$$C = \begin{bmatrix} 1 & 0 & 0 & 0 & 0 & 0 \\ 0 & 1 & 0 & 0 & 0 & 0 \\ 0 & 0 & 1 & 0 & 0 & 0 \end{bmatrix}$$

In general, nondegeneracy is equivalent [43] to the non-existence of m linearly independent functionals $\phi_1, \ldots, \phi_m$ such that

$$\det\,[\phi_i(g_j(s))] \equiv 0$$

where $g_j(s)$ is the j-th column of $G(s)$. Using this criterion, in the case $m = 2$, it is tedious but straightforward (see [37]) to check that this system is nondegenerate. Moreover, there exists 5 feedback laws, three real and one complex conjugate pair, yielding a characteristic polynomial equal to the open-loop characteristic polynomial. Explicitly, one may compute, with maximal error ± 0.01, the solutions

$$K_1 = \begin{bmatrix} 0 & 0 & 0 \\ 0 & 0 & 0 \end{bmatrix}, \quad K_2 = \begin{bmatrix} -3.86 & -6.52 & 43.56 \\ 2.47 & 3.86 & 36.24 \end{bmatrix}$$

$$K_3 = \begin{bmatrix} 27.6 & -12.36 & -10.35 \\ 58.29 & -27.6 & -18.18 \end{bmatrix}$$

$$K_{4,5} = \begin{bmatrix} -.91 \pm j2.52 & 2.3 \pm j4.5 & -2.33 \pm j2.94 \\ -1.76 \pm j1.36 & .91 \pm j2.52 & -.65 \pm j6.7 \end{bmatrix}$$

We remark that the error bounds show that these are simple roots; i.e. roots with multiplicity one. We shall return to this example in section 5.3.

5.2 The General Position Lemma

In this section, we outline a rigorous justification for the calculation (5.2) of the number $d_{m,p}$ of roots to the pole-assignment equations for "nondegenerate" systems. In this program, we need to make a significant use of the concept, and properties, of a Grassmann variety (or Grassmannian) for two reasons. First, the modern formulation and justification of the Schubert calculus–in particular of formula (5.2)–is inextricably interwoven with the geometry of Grassmannians and, second, the only method we currently know to show that generically the m-planes gr $G(s_i)$ are in general position is through the Hermann-Martin map, a geometric interpretation of matrix transfer functions as curves in Grassmannians.

We use the notation

$$\text{Grass}\,(m, N) = \{V \subset \mathbf{C}^N : V \text{ a subspace of dimension } m\} \qquad (5.6)$$

The special case $m = 1$, corresponding to the projective space of lines through 0 in $\mathbf{C}^N$, is typically denoted by $\mathbf{P}^{N-1}$. Of course, the notations Grass (n, N) and $\mathbf{P}^{N-1}$ make sense for any field of scalars and, when we need to restrict ourselves to real data and real objects we will use the notation $\text{Grass}_{\mathbf{R}}(m, N)$ and $\mathbf{RP}^{N-1}$.

One very useful point of view is that Grass (m, N) is a natural "compactification" of the space $M_{N-m,m}$ of $(N-m) \times m$ matrices. To see this, choose an m-dimensional space $U \subset \mathbf{C}^N$ and any complementary subspace $Y \subset \mathbf{C}^N$ so that $\mathbf{C}^N = U \oplus Y$. For any m-plane V there are two possibilities; either V is complementary to Y or V is contained in the set

$$\sigma(Y) = \{V : \dim V = m, \ \dim (V \cap Y) \geq 1\} \tag{5.7}$$

If V is complementary to Y then for some linear G

$$V = \text{gr } (G) \quad \text{when} \quad G : U \to Y.$$

Hence, there is a natural correspondence

$$\text{Grass } (m, N) - \sigma(Y) \longleftrightarrow M_{N-m,m} \tag{5.8}$$

As we vary Y, (5.8) defines a cover of Grass (m, N) by subsets in natural correspondence with $M_{N-m,m} \simeq \mathbf{C}^{m(N-m)}$. For example, if $m = 1$ and $N = 2$ one has

$$\mathbf{P}^1 = (\mathbf{P}^1 - \sigma(Y)) \cup \sigma(Y) \tag{5.9}$$

where $\sigma(Y)$ is a single point, while $\mathbf{P}^1 - \sigma(Y)$ is in natural correspondence with $\mathbf{C}$. Varying Y gives a covering of $\mathbf{P}^1$ which induces the natural structure of a compact manifold on $\mathbf{P}^1$. On the other hand (5.9) exhibits $\mathbf{P}^1$ as the one-point compactification of $\mathbf{C}$, i.e. $\mathbf{P}^1$ may be naturally identified with the 2-sphere S^2. Under the same correspondence, $\mathbf{RP}^1$ may be identified with the circle, S^1. In general (5.8) allows one to regard Grass (m, N) as a compact complex manifold of dimension $m(N-m)$ containing $M_{N-m,m}$ as an open dense submanifold.

Definition 5.3. For any $N - m$ plane Y, the subset $\sigma(Y)$ defined in (5.7) is referred to as a Schubert hypersurface.

Remark 5.4. When $m = 1$, a Schubert hypersurface $\sigma(Y)$ in $\mathbf{P}^{N-1}$ consists of the lines $\ell \subset Y$. Since dim $Y = N - 1$, we see that

$$\sigma(Y) \simeq \mathbf{P}^{N-2} \subset \mathbf{P}^{N-1}$$

In general, however, Schubert hypersurfaces are not smooth. If, for example, $m = 2$ and $N = 4$ then there are two kinds of points $V \in \sigma(Y)$ for any fixed Y: those V such that

$$\dim (V \cap Y) = 1$$

and $V = Y$, which is the unique singular point on $\sigma(Y)$. If Y_1, Y_2 are 2-planes in general position in $\mathbf{C}^4$, i.e. if $Y_1 \oplus Y_2 = \mathbf{C}^4$, then $\sigma(Y_1) \cap \sigma(Y_2)$ is nonsingular. In fact, if $V \in \sigma(Y_1) \cap \sigma(Y_2)$ then $V = V \cap Y_1 \oplus V \cap Y_2$, $\dim V \cap Y_i = 1$ so that

$$\sigma(Y_1) \cap \sigma(Y_2) \simeq \mathbf{P}^1 \times \mathbf{P}^1 \tag{5.10}$$

In particular, over $\mathbf{R}$ we have a 2-torus

$$\sigma_{\mathbf{R}}(Y_1) \cap \sigma_{\mathbf{R}}(Y_1) \simeq S^1 \times S^1 \simeq T^2 \tag{5.11}$$

Our interest is in Schubert hypersurfaces in Grass $(p, m+p)$ defined by

$$\sigma(s_i) = \sigma(\text{gr } G(s_i)) = \{V : \dim V \cap \text{gr } G(s_i) \geq 1\} \tag{5.12}$$

since, according to (3.4), to say K place the closed-loop poles at $s = s_i$, for $i = 1, \ldots, n$, is to say

$$\text{gr } (K) \in \bigcap_{i=1}^{n} \sigma(s_i) \tag{5.13}$$

Before analyzing the general case, it is interesting to interpret the cases $mp = n$ and $\min(m,p) = 1$ or $m = p = 2$, already understood in sections 2 and 4 by other methods. First suppose $p = 1$, $m = n$. As above any Schubert hypersurface $\sigma(s_i)$ is a linear hyperplane $\mathbf{P}^{n-1} \subset \mathbf{P}^n$; i.e. to $\sigma(s_i)$ corresponds an n-plane $V(s_i)$ in $\mathbf{C}^{n+1}$ such that

$$\sigma(s_i) = \{\text{lines } \ell \subset \mathbf{C}^{n+1} : \ell \subset V(s_i)\}$$

Since $V(s_i)$ and $V(s_j)$ are either equal or have a codimension 2 intersection, we see that imposing each (linear) condition, gr $(K) \in \sigma(s_i)$ amounts to a loss in dimension of at most one; i.e.

$$\dim \bigcap_{i=1}^{n} \sigma(s_i) \geq 0$$

As we have seen it is generically the case that

$$\dim \bigcap_{i=1}^{n} \sigma(s_i) = 0,$$

occurring precisely when the numerators are linearly independent, in which case $\bigcap_{i=1}^{n} \sigma(s_i) = \{V\}$. If V were not of the form $V = \text{gr } (K)$, this would impose another condition on V, viz. $V = Y$, which does not hold generically. Finally, if the set $\{s_1, \ldots, s_n\}$ is self-conjugate then we have

$$\{\bar{V}\} = \overline{\bigcap_{i=1}^{n} \sigma(s_i)} = \bigcap_{i=1}^{n} \sigma(s_i) = \{V\}$$

so that $V = \text{gr } (K)$ and hence K is real (compare Theorem 2.1).

The case $m = p = 2$, $n = 4$, is of course nonlinear. However, assuming the Schubert hypersurfaces $\sigma(s_i)$, $i = 1, \ldots, 4$ are in general position, this case can also be analyzed just from the definitions [23]. We consider the real situation, taking $s_i \in \mathbf{R}$. As above, the first two constraints on V

$$\sigma_{\mathbf{R}}(s_1) \cap \sigma_{\mathbf{R}}(s_2) \simeq \mathbf{RP}' \times \mathbf{RP}'$$

are equivalent to the fact that the 2-plane V lie on a 2-torus. Moreover, as in section 4.3 the additional constraint

$$V \in \sigma_{\mathbf{R}}(s_3) \cap (\sigma_{\mathbf{R}}(s_1) \cap \sigma_{\mathbf{R}}(s_2))$$

is linear; i.e. this intersection is the graph of a linear map

$$\phi_3 : \mathbf{RP}^1 \to \mathbf{RP}^1 \ , \ \text{gr } (\phi_3) \subset \mathbf{RP}^1 \times \mathbf{RP}^1$$

Explicitly,

$$\sigma_{\mathbf{R}}(s_i) \cap (\sigma_{\mathbf{R}}(s_1) \cap \sigma_{\mathbf{R}}(s_2)) = \text{gr } (\phi_i), \quad i = 3, 4.$$

Thinking of the 2-torus $T^2 = S^1 \times S^1 = \mathbf{RP}^1 \times \mathbf{RP}^1$ as a rectangle with opposite side identified we have one of two pictures

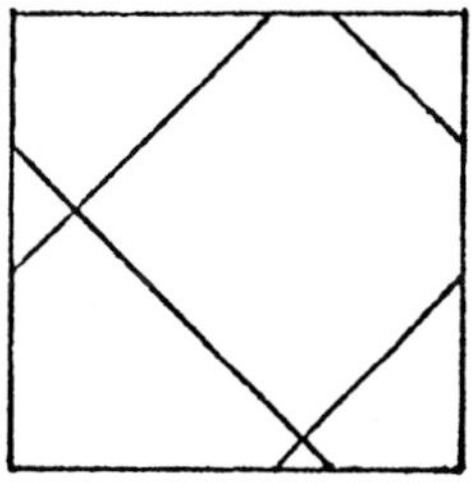

$\mathbf{RP}^1 \times \mathbf{RP}^1$

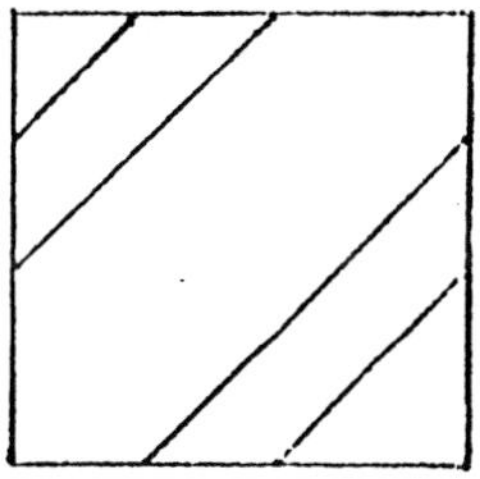

$\mathbf{RP}^1 \times \mathbf{RP}^1$

so that gr (ϕ_3) intersects gr (ϕ_4) either in 2 real points or not at all (compare Theorem 4.11).

There are several basic facts about Schubert hypersurfaces which are quite useful analyzing the intersection (5.13).

Theorem 5.5. *Consider n $(N-p)$-planes Y_i in $\mathbf{C}^N$.*

(1) *If* $\dim \bigcap_{i=1}^{r} \sigma(Y_i) = d$, *then*

$$\dim \bigcap_{i=1}^{r+1} \sigma(Y_i) \geq d - 1$$

(2) *If* $n = p(N-p)$, *then*

$$\dim \bigcap_{i=1}^{n} \sigma(Y_i) = 0$$

if, and only if,

$$\bigcap_{i=1}^{n} \sigma(Y_i)$$

is finite, in which case

$$\# \bigcap_{i=1}^{n} \sigma(Y_i) = \frac{1! \cdots (p-1)!(p(N-p))!}{(N-p)! \cdots (N-1)!}$$

counted with multiplicity.

Remark 5.6. Assertion (i), that the dimension of an intersection with a Schubert hypersurface can go down by at most one, is not a general fact about hypersurfaces. This follows from an analysis of the Plücker imbedding

$$\mathcal{P} : \text{Grass } (p, N) \to \mathbf{P}^{\binom{N}{p}-1}$$

exhibiting $\mathcal{P}(\text{Grass } (p,N))$ as a (projective) algebraic set defined by homogeneous quadratic (Plücker) equations. Under the Plücker imbedding any $\sigma(Y)$ can be realized as the intersection of $\mathcal{P}(\text{Grass } (p,N))$ with a projective hyperplane. It is in this sense that one may interpret the precise meaning of dimension, as well as claim (1), see [39]. Assertion (2) follows from an interpretation of $\cap\sigma(Y_i)$ in terms of any of the now standard intersection theories: homology, cohomology, algebraic intersection rings, see [53].

Now consider a $p \times m$ transfer function $G(s)$, having degree $n = mp$. We shall denote $\mathbf{C}^m$ by U and $\mathbf{C}^P$ by Y and consider the Schubert hypersurfaces $\sigma(s_i) \subset \text{Grass } (p, m+p)$ as defined in (5.12), for $s_i \in \mathbf{C}$ distinct. We note that, if $s = \infty$, the notation $\sigma(\infty)$ still makes sense, in fact since $G(\infty) = 0$ we have

$$U = \text{gr } G(\infty) \subset Y \oplus U$$

In particular, $\sigma(\infty) = \sigma(U)$ and therefore to say $V \in \sigma(\infty)$ is to say

$$\dim (V \cap U) \geq 1 \quad , \text{i.e.} \quad V \neq \text{gr } (K)$$

General Position Lemma 5.7. *Suppose* $mp = n$. *If* $G(s)$ *is nondegenerate, then for any choice of distinct* s_i, $i = 1, \ldots, n$

$$\# \bigcap_{i=1}^{n} \sigma(s_i) < \infty \tag{5.14}$$

Furthermore, the generic system is nondegenerate.

Proof of 5.17. Consider the rational equation in K_i

$$\det \begin{bmatrix} K_1 & G(s) \\ K_2 & I \end{bmatrix} = 0 \tag{5.15}$$

or, equivalently, the polynomial equation

$$\det \begin{bmatrix} K_1 & N(s) \\ K_2 & D(s) \end{bmatrix} = 0 \tag{5.15$'$}$$

where $G(s) = N(s)D(s)^{-1}$ is a coprime factorization and rank $\left[\begin{smallmatrix} K_1 \\ K_2 \end{smallmatrix}\right] = p$. For $G(s)$ fixed, if $K_1 = I$, $(5.15)'$ has roots at the closed-loop poles corresponding to K_2. Therefore, if rank $K_1 = p$, $(5.15)'$ is a polynomial equation of degree n. By continuity, for any K, the degree of $(5.15)'$ is at most n. The nondegeneracy condition (5.4) implies that $(5.15)'$ cannot be the zero polynomial and can therefore have at most n roots for any p-plane

$$V = \text{col. span} \begin{bmatrix} K_1 \\ K_2 \end{bmatrix}. \tag{5.16}$$

In particular, ∞ cannot be a root of (5.15). Put geometrically, this asserts

$$\bigcap_{i=1}^{n} \sigma(s_i) \cap \sigma(\infty) = \phi \tag{5.17}$$

since for any V in this intersection (5.15) would have the roots s_i, $i = 1, \ldots, n$ and ∞. By Theorem 5.5(1)

$$\dim \bigcap_{i=1}^{n} \sigma(s_i) \geq 0$$

If this dimension were positive, (5.17) would have nonnegative dimension, by Theorem 5.5(1). Therefore,

$$\dim \bigcap_{i=1}^{n} \sigma(s_i) = 0$$

and Theorem 5.5(2) applies, implying the theorem of Brockett and Byrnes for nondegenerate systems.

It remains, however, to show that nondegeneracy is generic; i.e. for m, n and p given there exists a (real) nondegenerate system. Before demonstrating this fact, we turn to another interesting corollary of the General Position Lemma.

Corollary 5.8 [43]. *For a nondegenerate system, image $(\chi) \subset \mathbf{C}^n$ is a closed subset. In particular, for a nondegenerate real system the set of assignable real monic polynomials is closed.*

Proof. We noted that the degree of $(5.15)'$ is at most n, the degree being less than n reflecting the fact some of the closed-loop poles have gone off to infinity in the high gain limit, det $K_1 \to 0$. This can be analyzed by homogenizing (5.15); i.e. replacing s by s/t and multiplying by t^r, r the highest power of t^{-1}, yielding a map

$$\begin{bmatrix} K_1 \\ \\ K_2 \end{bmatrix} \mapsto \Phi(s,t) \tag{5.18}$$

where $\Phi(s,t)$ is homogeneous of degree n, never zero if the system is nondegenerate. As in (5.16), this induces an extension of χ

$$\bar{\chi} : V \mapsto \text{span } \{\Phi(s,t)\},$$

where span $\{\Phi\}$ denotes the line through Φ in the $n+1$-dimensional space of n-th degree homogeneous polynomials; i.e.

$$\bar{\chi} : \text{Grass } (p, m+p) \to \mathbf{P}^n$$

Restricting $\bar{\chi}$ to the complement of $\sigma(\infty)$ we recover

$$\chi : \mathbf{C}^{mp} \to \mathbf{C}^n$$

as before. We claim

$$\text{image } (\chi) = \text{ image } (\bar{\chi}) \cap \mathbf{C}^n \tag{5.19}$$

and since image $(\bar{\chi})$ is always closed, the Corollary follows. But (5.19) is implied by the assertion that if a p-plane V assigns a set of n finite poles, counted with multiplicity, then $V = \text{gr } (K)$. If $V \neq \text{gr } (K)$ then $V \in \sigma(\infty)$ implying that (5.15) has n finite roots and one infinite root, counted with multiplicity, contradicting nondegeneracy.

We conclude this section with the construction of a nondegenerate system whenever $mp = n$. For this we need the Herman Martin map

$$\tilde{G} : \mathbf{P}^1 \to \text{Grass } (m, m+p)$$

defined for $s \in \mathbf{C}$ by

$$\tilde{G}(s) = \text{col. span} \begin{bmatrix} N(s) \\ \\ D(s) \end{bmatrix}$$

while

$$\tilde{G}(\infty) = \text{col. sp.} \begin{bmatrix} 0 \\ \\ I \end{bmatrix} = Y.$$

Each p-plane V defines a Schubert hypersurface $\sigma(V)$ in Grass $(m, m+p)$ and, by definition,

$$\tilde{G}(s) \in \sigma(V) \iff V \in \sigma(s) \tag{5.19}$$

In particular, recalling that $Y = \text{gr }(0)$ we see [12]

$$\tilde{G}^{-1}(\tilde{G}(\mathbf{P}^1) \cap \sigma(Y)) = \{\text{Poles of } G(s)\}$$

so that, counted with multiplicity,

$$\#\tilde{G}^{-1}(\tilde{G}(\mathbf{P}^1) \cap \sigma(Y)) = n$$

In other words, $\tilde{G}(\mathbf{P}^1)$ is a (rational) curve of degree n in Grass $(m, m+p)$, the Herman-Martin curve [12]. Moreover [12], every rational curve of degree n satisfying

$$\tilde{G}(\infty) = U$$

corresponds to a $p \times m$ transfer function of degree n. This correspondence allows for the application of both constructions and results from classical algebraic geometry to problems of linear systems. For example [43], nondegeneracy originally was formulated in this context.

In the light of (5.19), to say $G(s)$ is nondegenerate is to say that $\tilde{G}(\mathbf{P}^1)$ is not contained in any Schubert hypersurface. In particular, if $m = 1$ and $p = n$, $\tilde{G}(s)$ is a rational curve in $\mathbf{P}^n$ not contained in any hyperplane, the classical algebraic geometric definition of nondegeneracy. The classical example of such a curve

$$\tilde{G}(s) = \text{col. sp.} \begin{bmatrix} 1 \\ s \\ \vdots \\ s^n \end{bmatrix} \tag{5.20}$$

arises from the transfer function

$$G(s) = \begin{bmatrix} 1/s^n \\ 1/s^{n-1} \\ \vdots \\ 1/s \end{bmatrix} \tag{5.20$'$}$$

(5.20) is often referred to as the rational normal curve [53], but for $n = 3$ is called the twisted cubic, being an example of a space curve which lies in no plane [46]. In general, the rational normal curve is nondegenerate; i.e.

$$\sum_{i=0}^{n} a_i s^i = 0 \Longrightarrow a_i = 0$$

corresponding to linear independence of the numerators $n_i(s)$ as in Theorem 2.1. One can construct other nondegenerate curves, derived from the rational normal curve as follows. Consider the rational normal curve

$$\gamma : \mathbf{P}^1 \to \mathbf{P}^{p+1}$$

To each point $\gamma(s)$ on γ there corresponds a tangent line $\ell(s) \subset \mathbf{P}^{p+1}$. Being a line in $\mathbf{P}^{p+1}$ means of course that $\ell(s)$ corresponds to a 2-plane $V(s)$ in $\mathbf{C}^{p+2}$:

$$\ell(s) = \{\ell \subset \mathbf{C}^{p+2} : \ell \subset V(s)\}.$$

Therefore, we have a derived curve

$$\gamma^{(1)} : \mathbf{P}^1 \to \text{Grass } (2, 2+p)$$

which is also nondegenerate and has degree $2p$. More generally, to the rational normal curve

$$\gamma : \mathbf{P}^1 \to \mathbf{P}^{m+p-1}$$

we associate the derived curve

$$\gamma^{(m-1)} : \mathbf{P}^1 \to \text{Grass } (m, m+p)$$

where $\gamma^{(m-1)}(s)$ is the osculating $(m-1)$-plane to γ at $\gamma(s)$, see [53]. And, $\gamma^{(m-1)}$ is a nondegenerate curve in Grass $(m, m+p)$, having degree $n = mp$. This proves that the generic property, nondegeneracy, is not vacuous and hence holds generically.

5.3 Solution and Computation of Pole-Assignment Problems

In this section we shall focus on the possible explicit form of solutions to the pole-assignment problem in the case $mp = n$ and on the computation of feedback gains for explicit problems. As we have seen, the calculation by Willems-Hesselink in the case $m = p = 2, n = 4$ shows that, in general, linear formulae for pole assigning gains will not exist. Rather, any explicit formula will require the extraction of square roots. This is in fact the only case where the extraction of square roots suffices [25]. Furthermore, as we will demonstrate here, for $\min(m, p) = 2$, $\max(m, p) = 3$ and $n = 6$, the pole assignment equations are not even solvable by radicals [37]. Although there exist general formulae for the roots of an n-th degree polynomial in terms of theta functions, it is clear that most pole-assignment problems will need to be solved numerically.

Proposition 5.9 [37]. *If $mp = n$ and $G(s)$ is nondegenerate, the system of algebraic equations*

$$\chi(K) = c$$

can be solved numerically by the homotopy continuation method.

Proof. We begin with the observation that the proof of Corollary 5.8 shows more, viz. for a nondegenerate system and any compact set F of monic polynomials, the set $\chi^{-1}(F)$

is compact. As in [54] and [55], the homotopy continuation method–which allows one to deform a solution to a nominal problem, e.g. for (A_0, B_0, C_0) one takes the solution

$$\chi(0) = P_0(s) = \text{open-loop characteristic polynomial}$$

and continue it to a solution for (A_1, B_1, C_1) of the problem,

$$\chi(K) = p_1(s)$$

along paths from (A_0, B_0, C_0) to (A_1, B_1, C_1) and $p_0(s)$ to $p_1(s)$–will work, without a bifurcation analysis at the branch points, provided there is a path from (A_0, B_0, C_0) to (A_1, B_1, C_1) along which χ remains proper. Since over $\mathbf{C}$ the generic set of nondegenerate (A, B, C) is necessarily connected and since χ is always proper for (A, B, C) in this set, the homotopy continuation method applies.

Example 5.2 (bis). Consider the 3×2 nondegenerate system (5.5), having degree 6. Consider the following path linking the open loop characteristic polynomial to the polynomial s^6

$$s^6 - 5ts^5 + 4ts^4 + 12ts^3 - 87ts + 623ts - 246t \qquad 0 \leq t \leq 1$$

The solutions $K_i(t)$, $i = 1, \ldots, 5$ to the pole placement problem, as computed by the homotopy continuation method, can be schematically represented as follows:

$$\begin{array}{llccccc}
\text{Open-loop} & , t = 1 & K_1(t) & \neq K_2(t) & \neq K_3(t) & \neq K_4(t) \neq \bar{K}_4(t) \\
 & & \downarrow & \downarrow & \downarrow & \searrow\!\!\!\swarrow \\
s^6 & , t = 0 & K_1(t) & \neq K_2(t) & \neq K_3(t) & \neq K_4(t) \neq K_5(t)
\end{array}$$

with a unique branch point at $t_0 = 0.603 \pm 0.001$. The solution at the branch point t_0 takes the form

$$K_1(t_0) = \begin{pmatrix} 1.25 & 3.8 & 0.98 \\ -1.42 & 0.75 & -1.33 \end{pmatrix}$$

$$K_2(t_0) = \begin{pmatrix} -3.14 & -5.06 & 41.68 \\ 3.14 & 5.05 & 36.70 \end{pmatrix}$$

$$K_3(t_0) = \begin{pmatrix} 2.81 & 3.89 & -1.91 \\ 5.09 & -0.84 & -0.19 \end{pmatrix}$$

$$K_4(t_0) = K_5(t_0) = \begin{pmatrix} -1.24 & 1.34 & -0.81 \\ -2.40 & 3.24 & 3.37 \end{pmatrix}$$

with maximal error ± 0.01. The roots $K_4(t)$, $K_5(t)$ are real for $t \leq t_0$.

For general m, p, $mp = n$ one obtains the map

$$\chi : \mathbf{C}^{mp} \to \mathbf{C}^n$$

and if $E = \mathbf{C}(c_i)$ and $F = \mathbf{C}(k_{ij})$ are the fields of rational functions, then composition with χ gives a map

$$\chi^* : E \to F,$$
$$\chi^*(f) = f \circ \chi.$$

Since χ is surjective if (A, B, C) is nondegenerate by Theorem 5.1, χ^* is injective so one can regard

$$E \simeq \chi^* E \subset F$$

as an extension of fields. For example, to say that there exists rational formulae for the entries k_{ij} in terms of the c_i is to say $E = F$. From (5.2) it follows that this is the case if, and only if, $\min(m,p) = 1$. Indeed F is a vector space over the subfield E and from ([39], Theorems 6–7 on pp. 116–117) it follows that

$$\dim_E(F) = [F : E] = \deg_{\mathbf{C}}(\chi) = d_{m,p} \tag{5.20}$$

We first illustrate the use of (5.20) in conjunction with Galois theory to prove: The conditions $\min(m,p) = 1$ or $m = p = 2$ are necessary for the existence of formulae expressing pole assigning gains in terms of rational expressions and square roots [25]. For, by Galois theory, it is necessary that $[F : E]$ is a power of 2; i.e. $d_{m,p} = 2^r$ for some r. We claim that if $\min(m,p) \geq 2$ and $m + p \geq 5$, then $d_{m,p}$ is divisible by an odd prime. Indeed, by the strong form of Bertrand's postulate [56] there is a prime q, necessarily odd, such that

$$m + p - 1 < q < 2(m + p) - 4.$$

Clearly, q cannot divide the denominator of (5.2) but on the other hand if $\min(m,p) \geq 2$ then $mp > q$ so that q divides the numerator of $d_{m,p}$. Therefore, q divides $d_{m,p}$. We note, for example that if $m = 2$, $p = 3$ then $q = 5$ is the unique prime satisfying this inequality.

We shall now use these computations to calculate the Galois group of the pole-placement equations when $m = 2$, $p = 3$ and $n = 6$. Consider the fixed, but generic (indeed, a nondegenerate) system (A, B, C), where A is a 6×6, B is a 6×2, and C is a 3×6 real matrix. We shall prove that the pole-placement equations

$$\chi_{(A,B,C)}(K) = (c_1, \ldots, c_n) = c$$

cannot be solved by radicals.

Thus, if $m = 2$, $p = 3$ then (5.20) reduces to $[F : E] = 5$, therefore the minimal polynomial over $\mathbf{C}(c_i)$ of $k_{ij}(c)$, where $\chi(K(c)) = c$, has degree 5 for generic c ([39], pp. 116–117). And, since χ extends to a globally defined map $\bar{\chi}$ on $\mathrm{Grass}_{\mathbf{C}}(p, m+p)$, the minimal polynomial has its coefficients in $\mathbf{C}[c_i]$. Moreover, if $c_i \in \mathbf{R}$ then the coefficients of the minimal polynomial of $k_{ij}(c)$ are real polynomials in the c_i.

Theorem 5.10. *If* $\min(m,p) = 2$, $\max(m,p) = 3$ *and* $n = 6$, *then for generic* (A, B, C) *and for generic* $(c_i) \in \mathbf{R}^6$, *the equation*

$$\chi(K) = (c_i)$$

is not solvable by radicals.

Proof. To say that $\chi(K) = (c_i)$ is solvable by radicals, is to say that the minimal polynomial of $k_{ij}(c)$ is solvable by radicals. Since this is an equation of prime order defined over a subfield of $\mathbf{R}$, by Galois theory [57] one has a dichotomy provided the Galois group is in fact solvable: either

(i) all 5 roots $k_{ij}(c)$ are real; or

(ii) just 1 root $k_{ij}(c)$ is real.

In terms of the extended map, which is globally defined if (A, B, C) is nondegenerate,

$$\bar{\chi} : \mathrm{Grass}_{\mathbf{R}}(3,5) \to \mathbf{RP}^6$$

this is the assertion:

(i) $\bar{\chi}$ is 5 to 1 on an open subset, 1 to 1 on its complement, or

(ii) $\bar{\chi}$ is $1-1$ everywhere.

Lemma 5.11. *Suppose* $\min(m,p) = 2$, $\max(m,p) = 3$ *and* $n = 6$. *If for an open set of* (A, B, C) *the equation* $\chi(K) = (c_i)$ *is not solvable by radicals for an open set of* (c_i) *of* (A, B, C), *then this equation is not solvable by radicals for the generic choice of* (c_i) *and* (A, B, C).

Proof. Denote by $V \subset \mathbf{C}^{66}$ the open, dense subset of nondegenerate (A, B, C) and consider the map

$$\bar{\chi} : V \times \mathrm{Grass}_{\mathbf{C}}(3,5) \to V \times \mathbf{P}^6$$

defined by $\bar{\chi}((A, B, C), \Pi) = \bar{\chi}_{(A,B,C)}(\Pi)$ for a 3-plane $\Pi \subset \mathbf{C}^5$. If K_1 denotes the field of rational functions on $V \times \mathbf{CP}^6$ and K_2 denotes the field of rational functions on $V \times \mathrm{Grass}_{\mathbf{C}}(3,5)$ then $\bar{\chi}^* K_1 \subset K_2$ and it follows from the formula (5.3) that $\deg\,[K_2 : \chi^* K_1] = 5$. Moreover, the extension $K_2/\chi^* K_1$ is solvable if, and only if, the extension F/E defined in (5.20) is solvable for generic (A, B, C), by elementary Galois theory ([58], pp. 244–249). This, in turn, is solvable if, and only if, the extension field associated to the equation $\chi(K) = (c_i)$ is solvable for generic (c_i), again by Galois theory. From these statements, the assertion in the Lemma follows by taking contrapositives.

Turning to the proof of Theorem 5.10, one can see for purely topological reasons that (ii) can never occur for a nondegenerate system. That is, if $\bar{\chi}$ were $1-1$ then since

$\bar{\chi}$ is continuous and $\mathrm{Grass}_{\mathbf{R}}(3,5)$ is compact.

$$\bar{\chi} : \mathrm{Grass}_{\mathbf{R}}(3,5) \simeq \mathbf{RP}^6$$

would be a homeomorphism which is easily seen to be false by comparing higher homotopy groups. On the other hand, Example 5.1 shows that (i) is false for (5.5) and, since the roots K_i are simple, by a perturbation argument it follows that (i) is false for an open set of systems. In fact, we can prove more.

Theorem 5.12. *Let $n = 6$, $\max(m,p) = 3$ and $\min(m,p) = 2$. For generic (A,B,C) and generic (c_i), the Galois group of the equation*

$$\chi_i(K) = (c_i)$$

is the full symmetric group S_5 on 5 letters.

Proof. It follows from the above argument for the generic (A,B,C) and a generic choice of (c_i), that the equation

$$\chi(K) = (c_i)$$

is not solvable by radicals. Moreover, the minimal polynomial of the entries k_{ij} of K has degree 5 so that the Galois group G is a nonsolvable subgroup of S_5. It is a well known and straightforward proposition that the only such subgroups are A_5, the alternating subgroup, and S_5. Thus, we shall have $G = S_5$ if we can prove that G contains a simple transposition. Now, by elementary Galois theory ([58] pp. 244–249), it suffices to find a particular choice of nondegenerate (A,B,C) and c_i such that $G \simeq S_5$, and for this example we return to (5.5), leading to the map

$$\bar{\chi} : \mathrm{Grass}(3,5) \to \mathbf{P}^6.$$

By Lemma 5.11, the Galois group of the equation

$$\bar{\chi}(K) = (c_i) \in \mathbf{P}^6$$

is nonsolvable for generic (c_i). We prove that G contains a simple transposition by using two results due to Joe Harris [59]:

Lemma 5.13. *Let $\Pi : Y \to X$ be a holomorphic map of degree n. If there exists a point $p \in X$ such that the fiber of Y over p consists exactly of $n-1$ distinct points—i.e. $n-2$ simple points $q_1, \ldots, q_{n-2}$ and one double point q_{n-1}—and if Y is locally irreducible at q_{n-1}, then the monodromy group M of Π contains a simple transposition.*

Lemma 5.4. *If X, Y are irreducible algebraic varieties of the same dimension over the complex numbers $\mathbf{c}$, and $\Pi : Y \to X$ is a map of degree $d > 0$, the monodromy group equals the Galois group.*

On the other hand, we have already shown numerically that there exists (A, B, C) and a closed-loop characteristic polynomial for which there are three distinct solutions—three real simple roots K_1, K_2, K_3—and one real double solution K_4 to the pole-assignment equation. We have thus shown that the Galois group of the equation

$$\chi(K) = (c_i),$$

and thus of the extension field $\chi^* E \subset F$, is

$$\mathrm{Gal}(F/E) = S_5.$$

For generic (A, B, C) and generic (c_i) the Galois group G of the pole-placement equation is a subgroup

$$G \subset S_5,$$

while for fixed nondegenerate (A, B, C) and (c_i) the Galois group G' is a homomorphic image of G. In particular

$$G \to S_5$$

is surjective and therefore, by a counting argument, one has

$$G = S_5$$

for generic (A, B, C) and (c_i).

5.4 Topological Methods for Pole-Assignment and Stabilizability

There is a simple set–theoretic alternative to the interpretation of the pole-assignment problem as an intersection problem on Grass $(p, m+p)$. We shall work over the real field. Explicitly, given $G(s)$ and a set of desired, distinct real poles s_i, $i = 1, \ldots, n$, to say

$$\phi = \bigcap_{i=1}^{n} \sigma_{\mathbf{R}}(s_i) \subset \mathrm{Grass}_{\mathbf{R}}(p, m+p) \tag{5.21}$$

is to say the open sets $U_i = \mathrm{Grass}_{\mathbf{R}}(p, m+p) - \sigma_{\mathbf{R}}(s_i)$ cover $\mathrm{Grass}_{\mathbf{R}}(p, m+p)$

$$\bigcup_{i=1}^{n} U_i = \mathrm{Grass}_{\mathbf{R}}(p, m+p). \tag{5.21)'$$

As we have seen in section 5.2, the complement of a Schubert hypersurface is an open set, diffeomorphic with $M_{m,p}(\mathbf{R}) \simeq \mathbf{R}^{mp}$ and hence contractible. In particular, to say there is no real output feedback law placing the poles at s_i is to say that one can cover

$\mathrm{Grass}_{\mathbf{R}}(p, m+p)$ by n open subsets, contractible in $\mathrm{Grass}_{\mathbf{R}}(p, m+p)$. This simple observation is quite powerful in conjunction with general position arguments, such as Lemma 5.7.

In their study of global methods in the calculus of variations, Ljusternik and Šnirel'mann [60] discovered an important invariant of smooth manifolds:

Definition 5.15. Suppose X is a smooth n-manifold. The Ljusternik-Šnirel'mann category of X, denoted by L-S-cat (X) is the minimum cardinality of an open cover of X

$$\bigcup_{i=1}^{N} U_i = X$$

by open sets U_i which are contractible in X.

Using a generalization of the General Position Lemma to the case $mp \geq n$, one can prove [61] a generic pole-assignment result; placing poles at generic real frequencies. In particular, one has a generic stabilizability result.

Theorem 5.16 [61]. *Fix m, n and p. The generic system can be stabilized by real output feedback provided*

$$k_{m,p} = \textit{L-S-cat}\ (\mathrm{Grass}_{\mathbf{R}}(p, m+p)) \geq n+1. \tag{5.22}$$

Since $L - S -$ cat $(X) \leq$ dim $X + 1$, (5.22) reflects the necessary condition given in Theorem 4.13 for generic stabilizability. Indeed, one can always assert

$$m + p - 1 \leq k_{m,p} \leq mp$$

since, according to Eilenberg [62], L-S-cat (X) can be bounded from below topologically; in fact, by the maximum number of terms in a nontrivial product in any cohomology ring of X. Not surprisingly, for $X = \mathrm{Grass}_{\mathbf{R}}(p, m+p)$, this can be estimated in terms of the Schubert calculus, although over $\mathbf{R}$ the combinatorics becomes fairly involved (see [63]–[65]). Before turning to the problem of pole-assignment, we give several examples of what can be deduced from (5.22). First, define the integer s by

$$2^s < m + p \leq 2^{s+1} \tag{5.23}$$

Corollary 5.17 [61]. *If* $\min(m, p) = 2$, *then*

$$\max(m, p) + 2^s - 1 \geq n$$

implies generic stabilizability.

Corollary 5.18 [61]. *If* $\min(m, p) = 3$, *each of the following conditions imply generic stabilizability*

(i) $m + p = 2^{s+1} - 2^r + 1$ *and* $2^{s+2} - 3(2^{r-1}) - 4 \geq n$,

(ii) $m+p=2^{s+1}-2^r+2+t$ *where* $0\leq t\leq 2^{r-1}-2$ *and* $2^{s+2}-3(2^{r-1})-2+t\geq n$; *or*

(iii) $m+p=2^{s+1}$ *and* $2^{s+2}-5\geq n$

Corollary 5.19 [61]. *If* $\min(m,p)=4$, *each of the following conditions imply generic stabilizability*

(i) $m+p=2^s+1$ *and* $2^{s+1}+2^s-7\geq n$; *or*

(ii) $m+p=2^s+2^r+j+1$ *where* $s>r\geq 0, 0\leq j\leq 2^r-1$ *and* $2^{s+1}+2^s+s^{r+1}+j-7\geq n$.

There is a slightly more refined invariant which plays a similar role in the problem of pole assignment by real output feedback. If

$$f:X\to Y$$

is a continuous map, the category of f, denoted by cat (f), is defined as the minimum cardinality of an open cover (U_i) of X such that $f|_{U_i}$ is homotopic to a constant. If $f=i_X$, the identity map on X, then of course cat $(i_x)=$ L-S cat (x) and in general one has cat $(f)\leq$ L-S cat (X). An argument as before [66] shows that (5.21) implies

$$c_{m,p}=\text{ cat }(P_{m,p})\leq mp \tag{5.24}$$

where $P_{m,p}$ is the Plücker imbedding

$$P_{m,p}:\ \mathrm{Grass}_{\mathbf{R}}(p,m+p)\to \mathbf{RP}^{\binom{m+p}{p}-1}$$

Theorem 5.20 [66]. *Fix* m,n *and* p. $c_{m,p}\geq n$ *implies generic pole assignability by real output feedback.*

As before one can estimate $c_{m,p}$ from below using Schubert calculus; see [63]. Explicitly, with s as defined in (5.23) set

$$c'_{m,p}=\begin{cases}2^{s+1}-1 & \text{if } \min(m,p)=2,\ \max(m,p)\neq 2^s-1\\ 2^{s+1}-2 & \text{if } \min(m,p)=2,\ \max(m,p)=2^s-1\\ 2^{s+1}-1 & \text{if } \min(m,p)=3,\ m+p=2^K+1\\ 2^{k+1} & \text{otherwise}\end{cases}$$

Corollary 5.21 [66]. *Fix* m,n *and* p. *The condition*

$$c'_{m,p}\geq 2\,[n-1/2]+1$$

is sufficient for pole assignment for the generic $p\times m$ *system of degree* n.

Corollary 5.22 (Brockett-Byrnes). *If* $mp=n$, *then the conditions*

$$\min(m,p)=1 \text{ or } \min(m,p)=2 \text{ and } \max(m,p)=2^r-1$$

are sufficient for pole-assignment of the generic $p \times m$ system of degree n.

Corollary 5.23 (Kimura). *If $m + p - 1 \geq n$, then the generic $p \times m$ system of degree n can be arbitrarily pole assigned.*

Bibliography

[1] G. B. Airy, "On the regulator of the clock-work for effecting uniform movement of equatoreals," *Memoirs of the Royal Astronomical Society* **11** (1840) 249–267.

[2] G. B. Airy, "Supplement to the paper 'On the regulator of the clock-work for effecting uniform movement of equatoreals,'" *Memoirs of the Royal Astronomical Society* **20** (1851) 115–119.

[3] J. C. Maxwell, "On governors," *Proc. Royal Society London* **16** (1868) 270–283.

[4] R. E. Kalman, "Algebraic geometric description of the class of linear systems of constant dimension," *8th Annual Princeton Conference on Information Sciences and Systems*, Princeton, NJ, 1974.

[5] M. Hazewinkel and R. E. Kalman, "On invariants, canonical forms and moduli for linear, constant, finite-dimensional dynamical systems," in *Proc. of CNR-CISM Symp. on Alg. Sys. Th., Udine (1975)*, Lect. Notes in Econ. Math. Sys. Th., Vol. 131, Springer-Verlag, Heidelberg, 1976, 48–60.

[6] J. M. C. Clark, "The consistent selection of local coordinates in linear system identification," Proc. JACC, Purdue U., 1976.

[7] C. I. Byrnes, "On the moduli space for linear dynamical systems," in *Proc. of 1976 NASA-Ames Conference on Geometric Control Theory*, (C. F. Martin and R. Hermann, eds.) Math. Sci. Press, 1977, 229–276.

[8] E. J. Davison and S.-H. Wang, "Properties of linear time-invariant multivariable systems subject to arbitrary output and state feedback," *IEEE Trans. Aut. Control* **18** (1973) 24–32.

[9] B. D. O. Anderson, N. K. Bose and E. I. Jury, "Output feedback stabilization and related problems–solution via decision algebra methods," *IEEE Trans. Aut. Control*, **AC-20** (1975) 53–66.

[10] I. Postlethwaite and A. G. J. Mac Farlane, "A complex variable approach to the analysis of linear multivariable feedback systems," *Lecture Notes in Control and Inf. Sciences* **12** Springer-Verlag, New York 1979.

[11] R. Hermann and C. F. Martin, "Applications of algebraic geometry to system theory–Part I," *IEEE Trans. Aut. Control*, **AC-22** (1977) 19–25.

[12] C. F. Martin and R. Hermann, "Applications of algebraic geometry to system theory: The McMillan degree and Kronecker indices as topological and holomorphic invariants," *SIAM J. Control* **16** (1978) 743–755.

[13] R. Hermann and C. F. Martin, "Applications of algebraic geometry to system theory; Part II: Feedback and pole-placement for linear Hamiltonian systems," *Proc. of IEEE* **65** (1977) 841–848.

[14] H. Kimura, "Pole assignment by gain output feedback," *IEEE Trans. Aut. Control* **AC-20** (1975) 509–516.

[15] J. C. Willems and W. H. Hesselink, "Generic properties of the pole-placement problem," *Proc. of the 7th IFAC Congress* (1978) 1725–1729.

[16] R. W. Brockett and C. I. Byrnes, "Multivariable Nyquist criteria, root loci and pole placement: A geometric viewpoint," *IEEE Trans. Aut. Control* **AC-26** (1981) 271–284.

[17] T. E. Djaferis, "Generic pole assignment using dynamic output feedback," *Int. J. Control* **37** (1983) 127–144.

[18] H. Seraji, "Design of pole placement compensators for multivariable systems," *Automatica* **16** (1980) 335–338.

[19] P. K. Stevens, *Algebro-Geometric Methods for Linear Multivariable Feedback Systems*, Ph.D. Dissertation, Harvard University, 1982.

[20] M. Vidyasagar and N. Viswanadhan, "Algebraic design techniques for reliable stabilization," *IEEE Trans. Aut. Cont.* **AC-27** (1982) 1085–1095.

[21] B. K. Ghosh, "Simultaneous stabilization and pole-placement of a multimode linear dynamical system," Ph.D. dissertation, Harvard Univ., 1983.

[22] M. L. J. Hautus, "Stabilization, controllability, and observability of linear autonomous systems," *Proc. Kon. Nederl. Akadamie van Wetenschappen–Amsterdam*, Series A, **73** (1970) 448–455.

[23] J. Rosenthal, "Tuning natural frequencies by output feedback," *Computation and Control* (K. Bowers and J. Lund, eds.) Birkhäuser, Boston, 1989.

[24] A. S. Morse, W. A. Wolovich and B. D. O. Anderson, "Generic pole assignment: Preliminary results," *Proc. 20th IEEE Conf. on Decision and Control*, San Diego, 1981.

[25] C. I. Byrnes and B. D. O. Anderson, "Output feedback and generic stabilizability," *SIAM J. on Control* **22** (1984) 362–380.

[26] D. F. Delchamps, *State Space and Input-Output Linear Systems*, Springer-Verlag, New York, 1988.

[27] C. I. Byrnes, "On the control of certain infinite dimensional systems by algebro-geometric techniques," *Amer. J. Math.* **100** (1978) 1333–1381.

[28] F. M. Brasch and J. B. Pearson, "Pole placement using dynamic compensation," *IEEE Trans. Aut. Control* **AC-15** (1970) 34–43.

[29] C. I. Byrnes and P. K. Stevens, "Pole placement by static and dynamic output feedback," *Proc. of 21st IEEE Conf. on Dec. and Control* Orlando, 1982.

[30] R. R. Bitmead, S. Y. Kung, B. D. O. Anderson and T. Kailath, "Greatest common divisors via generalized Sylvester and Bezout matrices," *IEEE Trans. Aut. Contr.* **AC-23** (1978) 1043–1047.

[31] D. Youla, J. Bongiorno and C. Lu, "Single loop feedback stabilization of linear multivariable dynamic plants," *Automatica* **10** (1974) 159–173.

[32] R. Saeks and J. J. Murray, "Fractional representation, algebraic geometry, and the simultaneous stabilization problem," *IEEE Trans. Aut. Control*, **AC-27** (1982), 895–903.

[33] B. K. Ghosh and C. I. Byrnes, "Simultaneous stabilization and simultaneous pole-placement by non-switching dynamic compensation," *IEEE Trans. Aut. Control* **AC-28** (1983) 733–741.

[34] B. K. Ghosh, "Transcendental and interpolation methods in simultaneous stabilization and simultaneous partial pole placement problems," *SIAM J. Control and Opt.* **24** (1986) 1091–1109.

[35] E. J. Davison and S.-H. Wang, "On pole-assignment in linear multivariable systems using output feedback," *IEEE Trans. Aut. Contr.* **AC-20** (1975) 516–518.

[36] R. Rado, "A theorem on independence relations," *Quart J. Math* **13** (1962) 83–89.

[37] C. I. Byrnes and P. K. Stevens, "Global properties of the root-locus map," in *Feedback Control of Linear and Nonlinear Systems* (D. Hinrichsen and A. Isidori, eds.), *Springer-Verlag Lecture Notes in Control and Information Sciences* **39**, Berlin, 1982.

[38] D. Mumford, *Algebraic Geometry I: Complex Projective Varieties*, Springer-Verlag, NY, 1976.

[39] I. R. Shafarevich, *Basic Algebraic Geometry*, Springer-Verlag, NY, 1974.

[40] X.-C. Wang, "Geometric inverse eigenvalue problems," *Computation and Control*

(K. Bowers and J. Lund, eds.) Birkhäuser–Boston, 1989.

[41] X.-C. Wang, Ph.D. dissertation, Arizona State University, 1989.

[42] G. D. Forney, "Minimal bases of rational vector spaces with applications to multivariable linear systems," *SIAM J. Control* **13** (1975) 493–520.

[43] C. I. Byrnes, "Algebraic and geometric aspects of the analysis of feedback systems," in *Geometric Methods in Linear Systems Theory* (C. I. Byrnes and C. F. Martin, eds.), D. Reidel, Dordrecht 1980, 85–124.

[44] C. Giannakopoulis and N. Karcanias, "Pole assignment of strictly proper and proper linear systems by constant output feedback," *Int. J. Control* **42** (1985), 543–565.

[45] N. Karcanias and C. Giannakopoulis, "Grassmann matrices, decomposability of multivectors and the determinantal assignment problem," *Linear Circuits, Systems and Signal Processing: Theory and Applications* (C. I. Byrnes, C. F. Martin and R. Saeks, eds.), North-Holland, 1988, 307–312.

[46] D. Mumford, *Introduction to Algebraic Geometry*, Harvard Univ., 1964.

[47] N. Jacobson, *Lectures in Abstract Algebra* Vol. I, Van Nostrand, New York, 1953.

[48] C. Delzell, Ph.D. Dissertation, Stanford Univ., 1980.

[49] B. Kostant, "Lie group representations on polynomial rings," *Amer. J. Math.* **85** (1963), 327–404.

[50] J. M. Schumacher, "Almost stabilizability subspaces and high gain feedback," *IEEE Trans. Aut. Control* **AC-29** (1984), 620–629.

[51] H. Schubert, "Anzahlbestimmungen für lineare Räume beliebiger Dimension," *Acta Math* **8** (1886), 97–118.

[52] P. Griffiths and J. Harris, *Principles of Algebraic Geometry*, J. Wiley and Sons, New York, 1978.

[53] S. L. Kleiman, "Problem 15. Rigorous foundation of Schubert's enumerative calculus," *Proc. of Symp. in Pure Math.*, Vol. XXVIII, Amer. Math. Soc., Providence, 1976.

[54] F. J. Drexler, "A homotopy method for the calculation of all zeroes of zero dimensional polynomial ideals," in *Continuation Methods* (H. Wacker, ed.), Academic Press, NY, 1978, 69–93.

[55] *Continuation Methods* (H. Wacker, ed.), Academic Press, NY, 1978.

[56] G. H. Hardy and E. M. Wright, *An Introduction to the Theory of Numbers* 5th ed.,

Oxford, 1979.

[57] E. Artin, *Galois Theory*, Univ. of Notre Dame Press, Notre Dame, 1971.

[58] S. Lang, *Algebra*, Addison-Wesley, Reading, MA, 1971.

[59] J. Harris, "Galois Groups of Enumerative Problems," *Duke Math. J.* **46** (1979), 685–724.

[60] L. Ljusternik and L. Šnirel'mann, *Méthodes Topologiques dans les Problèmes Variationnels*, Hermann, Paris, 1934.

[61] C. I. Byrnes, "On the stabilizability of multivariable systems and the Ljusternik-Šnirel'mann category of real Grassmannians," *Systems and Control Letters* **3** (1983), 255–262.

[62] S. Eilenberg, "Sur un théorème topologique de M.L. Schnirelmann," *Mat. Sb 1* (1936), 557–559.

[63] H. I. Hiller, "On the height of the first Stiefel-Whitney class," *Proc. Amer. Math. Soc.* **79** (1980), 495–498.

[64] R. E. Stong, "Cup products in Grassmannians," *Topology and its Applications* **13** (1982), 103–113.

[65] I. Berstein, "On the Ljusternik-Šchnirel'mann category of Grassmannians," *Math. Proc. Camb. Phil. Soc.* **79** (1976), 129–134.

[66] C. I. Byrnes, "Control theory, inverse spectral problems, and real algebraic geometry," *Diff. Geom. Methods in Control Theory* (R. W. Brockett, R. S. Millman and H. J. Sussmann, eds.) Birkhäuser, Boston, 1982.

Zeros, Poles and Modules in Linear System Theory

G. Conte

Dip. Matematica, Università di Genova
Via L. B. Alberti 4, 16132 Genova, Italy

A. M. Perdon

Dip. Metodi Modelli e Matematici per le Scienze Applicate
Università di Padova
Via Belzoni 7, 35100 Padova, Italy

INTRODUCTION

The aim of this paper is to present an account of the module theoretic approach to the notions of pole and of zero of a linear transfer function and to show its naturalness and usefulness. The motivations for developing this approach arise from noticing that the classical definitions of pole and zero are not "rich" enough to allow a deep, satisfactory analysis of certain design problems. On the other hand, the notion of pole module and of zero module, which will be described in the paper, provide a powerful algebraic tool for investigating the behavior of poles and zeros in many interesting situations.

The definition of zeros of a multivariable systems is originally due to H. Rosenbrock ([20]), whose fundamental work stimulated in the past two decades a great number of researches on such concept (a review of notable publications on this topic in the period 1970-1987 can be found in [21]).The idea of defining a module which captures the concept of zero of a multivariable transfer function was introduced by B. Wyman and M. Sain in [27]. In that paper the zero module was employed to study the relationship between the zero structure of an invertible transfer function and the pole structure of its inverses. The result was a satisfactory description of such a relationship, which agrees with the basic intuition developed in the scalar case or in the square and invertible one. The original treatment considered only the finite zeros and poles and was extended to include poles and zeros at infinity by the authors (see [4]), by introducing the notions of infinite pole module and infinite zero module. Subsequently, various applications to transfer function equations ([5,9]) and to particular control and design problems ([6,7,8,28,29]) have been developed (see also [3,13,31,32]).

The paper is organized as follows. Section 1 contains some preliminaries and notations. The algebraic notions which will be used throughout the paper are very briefly recalled, emphasizing only the features we are interested in.

In Section 2, after a review of the classical definitions of multivariable finite pole and zero of a transfer function, we discuss two examples concerning the inversion problem and a more general transfer function equation. We show that a complete understanding of the relationship between the pole/zero structure of the data and the pole/zero structure of the solutions of the inversion problem or of the considered transfer function equation cannot be obtained , except in the scalar case or in the "trivial" nonsingular one, by means of the classical notions. Therefore we introduce in Section 3 two algebraic objects, namely the finite pole module X(T) and the finite zero module Z(T), associated with a transfer function T(z), which contain more information than the classical concepts of pole and zero. The notion of pole module can be easily described by means of the state module of the minimal realization of the considered transfer function. The notion of zero module, although quite natural, is less familiar and deserves to be investigated more deeply. We describe the way in which it is related to the transmission blocking properties and to the controlled invariant subspaces of the minimal realization. In particular, we have that the zero module Z(T) of a strictly proper transfer function T(z), with minimal realization (X,A,B,C), is naturally isomorphic to the quotient V^*/R^*, provided with the module structure over the ring of polynomials induced by (A+BF), where V^* is the the maximum controlled invariant subspace contained in Ker C, $R^*$is the maximum controllability subspace contained in Ker C and F is such that $(A+BF)V^* \subset V^*$.

In Section 4, the algebraic tools we have developed are applied to the study of the transfer function equation T(z) = H(z)G(z), where T(z) and G(z) are two given transfer functions with the same input space and H(z) is sought. This equation generalizes the examples discussed in Sec.2 and our aim is to clarify the relationship between the pole/zero structure of the involved transfer functions by employing the notions of pole module and of zero module. In particular, we want to know if there exists an "essential" pole structure which appears in any solution H(z). Moreover, in the affirmative case, we are interested in obtaining a description of it in terms of the pole/zero structure of T(z) and G(z) and in investigating the existence of solutions all of whose poles are essential. The relevance of the above questions in a design problem whose solution can be characterized as a solution H(z) of T(z) = H(z)G(z) (e.g. inversion problem, factorization through G(z), model matching) is clear, since the pole structure determines the dynamical properties of the minimal realization of a transfer function (e.g. its stability or dead-beat properties). The main result we obtain says that there exists a module P, described abstractly in terms of T(z) and G(z), which is naturally contained in the pole module of any solution H(z). The invariant factors of P characterize the essential part of the pole structure of any solution H(z). More concretely, P turns out to consist, in an algebraically meaningful sense, exactly of the poles of T(z) which are not poles of G(z) and of the zeros of G(z) which do not appear as zeros of $[T(z)^t\, G(z)^t]^t$. A simple procedure to compute the invariant factors of P in terms of coprime fractional representations of T(z) and G(z) is described. Then we show by construction that there exist essential solutions H(z), i.e. solutions all of whose poles are essential.

In Section 5, we briefly describe the extension of this approach to the case of poles and zeros at infinity and to the case of poles and zeros in a specific region of interest. This provides useful results for the design problems we have mentioned above and gives us the algebraic tools, represented by the infinite pole module and the infinite zero module, for obtaining a complete picture, including finite poles and zeros and poles and zeros at infinity, of the classical pole/zero structure of a linear transfer function.

Further constructions are however needed in order to get a complete understanding of the relations between the total number of poles and the total number of zeros of a transfer function. Therefore, following

[33], we introduce, in Section 6, new spaces associated with T(z) whose dimensions measure the difference, called defect, between the total number of poles and the total number of zeros when T(z) fails to be invertible. Such spaces, which are finite dimensional K-vector spaces without any natural or canonical module structure, can be interpreted as spaces of generic zeros, where generic means not located at any finite point nor at infinity, related to the kernel and cokernel of T(z). In this way, we can say that the total number of poles of any transfer function equals the total number of zeros, when everything is counted properly, that is when also the generic zeros are counted. Moreover, this result is stated in terms of existence of maps of algebraic objects and of exactness of a sequence, rather than in terms of equality of numbers. It therefore refines previous results which express the defect of T(z) in terms of Kronecker indices or Wedderburn numbers.

1. PRELIMINARIES AND NOTATIONS

Let K be a field and let K[z] and K(z) denote respectively the ring of polynomials and the field of rational functions in the indeterminate z over the field K. Given a K-vector space $V = K^n$, we denote by V(z) the K(z)-vector space $V \otimes_K K(z)$, whose elements are n-tuples with entries from K(z), and we denote by ΩV the K[z]-module $V \otimes_K K[z]$, whose elements are n-tuples with entries from K[z]. ΩV is, in an obvious way, a K[z]-submodule of V(z) and we denote by ΓV the quotient K[z]-module $V(z)/\Omega V$. Any element y(z) in ΓV can be uniquely represented as an n-tuple of strictly proper rational functions whose formal Laurent series expansion has the form $v(z) = \Sigma_{i=1\ldots\infty} v_t z^{-t}$, with $v_t \in V$. The K[z]-module structure is given, in this representation, by the usual product followed by the deletion of the polynomial part.

We assume that the reader is familiar with the theory of modules over a principal ideal domain, more specifically over K[z], and with the algebra of polynomial matrices (see, e.g. [1,14,18]). In particular, we will make use of the notions of Smith form of a polynomial matrix, torsion submodule of a K[z]-module, invariant factors of a finitely generated torsion K[z]-module, and short exact sequences of K[z]-modules and morphisms, which are briefly recalled below.

Smith form Given a pxm polynomial matrix M(z) (more generally a matrix with entries in a principal ideal domain), it is possible to factor M(z) as $M(z) = B_1(z)\begin{pmatrix} \mathrm{diag}\{\, p_1(z) \ldots\ldots p_r(z) \,\} & 0 \\ 0 & 0 \end{pmatrix} B_2(z)$, where $B_1(z)$, $B_2(z)$ are polynomial unimodular matrices (that is : polynomial nonsingular matrices with a polynomial inverse, or, equivalently, with constant nonzero determinant), r = rank M(z), the $p_i(z)$'s are monic polynomials such that $p_i(z) \mid p_{i+1}(z)$ ($p_i(z)$ divides $p_{i+1}(z)$) for i = 1,...,r . The $p_i(z)$'s are uniquely determined and the matrix blockdiag{ diag $\{p_i(z)\}$,0 } is called the Smith form of M(z).

Torsion submodule - Invariant factors If we consider the morphism $M(z) : \Omega U \rightarrow \Omega Y$ induced between ΩU and ΩY, for $U=K^m$ and $Y=K^p$, by the polynomial matrix M(z) with respect to the canonical bases, we have that the factorization of M(z) related to the Smith form induces a canonical isomorphism between the quotient module $\Omega Y/M\Omega U$ and the direct sum $\oplus_{i=1,r} K[z]/p_i(z)K[z] \oplus (K[z])^{p-r}$. The submodule of $\Omega Y/M\Omega U$ isomorphic to $\oplus_{i=1,r} K[z]/\, p_i(z)K[z]$ is the torsion submodule $t_{K[z]}(\Omega Y/M\Omega U)$ of $\Omega Y/M\Omega U$.

It is characterized by the fact that any element in it is annihilated by a suitable element of K[z] (in this case a factor of the product of the $p_i(z)$'s). The polynomials $p_i(z)$'s , which are the elements of the Smith form of M(z), are called the <u>invariant factors</u> of the module $t_{K[z]}(\Omega Y/M\Omega U)$. More generally, any finitely generated module M over K[z] has a unique direct sum decomposition of the form $\oplus_{i=1,q} K[z]/ p_i(z)K[z] \oplus (K[z])^s$, where the $p_i(z)$'s are monic polynomials such that $p_i(z) \mid p_{i+1}(z)$. Two modules are isomorphic iff their decompositions are equal. In particular, if M is a finitely generated torsion module, in particular if M consists of a finite dimensional K-vector space provided with a K[z]-module structure, we have s=0 and, as an abstract module, M is completely characterized by its invariant factors $p_i(z)$.

<u>Exact sequence</u> Given a sequence of K[z]-modules and morphisms

$$M_1 \xrightarrow{f} M_2 \xrightarrow{g} M_3$$

we say that the sequence is <u>exact at M_2</u> if Im f coincides with Ker g. In the following we will be interested in sequences of the form

$$0 \longrightarrow M_1 \xrightarrow{f} M_2 \xrightarrow{g} M_3 \longrightarrow 0$$

which are exact at M_1, at M_2 and at M_3. Such a sequence is called a <u>short exact sequence</u> . Exactness of the sequence at M_1 is equivalent to f being injective and exactness of the sequence at M_3 is equivalent to g being surjective. Together with exactness at M_2, this means that M_1 is isomorphic to Ker g and that M_3 is isomorphic to $M_2/f(M_1)$. Taking into account only the underlying structure of K-vector space of the modules M_1, M_2 and M_3 , the fact that the short sequence is exact implies that, as a K-vector space, M_2 is isomorphic to the direct sum $M_1 \oplus M_3 \cong M_1 \oplus M_2/f(M_1)$.

We will make use of the following result: given the diagram of modules and module morphisms

$$\begin{array}{ccccc} & i_1 & & p_1 & \\ A & \longrightarrow & B & \longrightarrow & B/A \\ \downarrow f & (1) & \downarrow g & & \\ A' & \longrightarrow & B' & \longrightarrow & B'/A' \\ & i_2 & & p_2 & \end{array}$$

where i_1, i_2 are inclusions and p_1, p_2 are canonical projections, if (1) is commutative, i.e. $g\, i_1 = i_2\, f$, then there exists an induced map $h: B/A \to B'/A'$ such that $h\, p_1 = p_2 g$.

<u>Transfer function - Fractional representations</u> Given two K-vector spaces $U = K^m$ and $Y = K^p$, by a <u>transfer function</u> we mean a K(z)-linear map $T(z) : U(z) \to Y(z)$ or, equivalently, the p x m matrix of rational functions which represents T(z) with respect to the canonical K(z)-basis of U(z) and Y(z).

By a left (resp. right) <u>coprime fractional representation</u> of T(z) we mean a representation of the form $T(z) = D^{-1}(z)N(z)$ (resp. $T(z) = N_1(z)D_1^{-1}(z)$), where D(z), N(z) (resp. $D_1(z)$,$N_1(z)$) are polynomial matrices whose common left (right) factors are unimodular and D(z) ($D_1(z)$) is nonsingular.

<u>Realization diagram</u> We assume that the reader is familiar also with the algebraic realization theory of T(z), or with the so-called polynomial model approach, described in [12,15,30]. In particular, we will frequently

make reference to the <u>realization diagram</u>, whose construction is described in detail in [2], induced by any coprime left fractional representation $T(z) = D^{-1}(z)N(z)$:

$$\begin{array}{ccccc}
 & & N(z) & & \\
U(z) & \xrightarrow{\quad T(z) \quad} & Y(z) & \xrightarrow{\quad D(z) \quad} & Y(z) \\
j\uparrow & & \downarrow \pi_- & & \downarrow \pi_- \\
\Omega U & \xrightarrow{\quad T^{\#} \quad} & \Gamma Y = Y(z)/\Omega Y & \xrightarrow{\quad D(z) \quad} & \Gamma Y \\
i\uparrow & \searrow \tilde{B} \qquad \tilde{C} \nearrow & \downarrow p & & \\
U & \xrightarrow[B]{} \; X = \mathrm{Ker}\, D \; \xrightarrow[C]{} & Y & & \\
 & A & & &
\end{array} \tag{1.1}$$

The notations are as follows:

- i and j are canonical inclusions, π_- is the canonical projection, p is defined by $p(\Sigma_{t=1\ldots\infty} y_t z^{-t}) = y_1$;

- $T^{\#}: \Omega U \rightarrow \Gamma Y$ is defined by $T^{\#} = \pi_- T j$, $D : \Gamma Y \rightarrow \Gamma Y$ is defined as the action of D(z) followed by the deletion of the polynomial part;

- Ker D, which consists of the strictly proper elements of Y(z) whose image by D(z) is polynomial, is a finitely generated torsion K[z]-module, hence it is a finite dimensional K-vector space ;

- $\tilde{C} : \mathrm{Ker}\, D \rightarrow \Gamma Y$ is the inclusion and $\tilde{B} : \Omega U \rightarrow \mathrm{Ker}\, D$ is uniquely determined by the condition $\tilde{C}\,\tilde{B} = T^{\#}$;

- forgetting the K[z]-module structure, we denote by X the underlying K-vector space (the equality X = Ker D in 1.1 is an abuse of notation, since it concerns only the K-vector space structure); $B : U \rightarrow X$ is defined by $B = \tilde{B}\, i$, $C : X \rightarrow Y$ is defined by $C = p\,\tilde{C}$, $A : X \rightarrow X$ is defined using the K[z]-structure of Ker D by $Ax = z\,x$.

It is well known that the dynamical system (X,A,B,C) arising from the realization diagram above, whose evolution is described by the equations

$$\begin{cases} \dot{x}(t) = A\,x(t) + B\,u(t) \\ y(t) = C\,x(t) \end{cases} \quad \text{or} \quad \begin{cases} x(t+1) = A\,x(t) + B\,u(t) \\ y(t) = C\,x(t) \end{cases},$$

(continous time) (discrete time)

is a minimal realization of the strictly proper part of T(z), which coincides with $C(zI-A)^{-1}B$. Remark that the K[z]-module structure of Ker D coincides with the one induced on the K-vector space X by z x = Ax. Denoting by (X,A) the K[z]-module consisting of X provided with the K[z]-structure induced by A, we therefore have that (X,A) is K[z]-isomorphic to Ker D. More generally, we have the following result:

1.2 PROPOSITION Let $T(z) = D^{-1}(z)N(z) = N_1(z)D_1^{-1}(z)$ be coprime fractional representations and let (X,A,B,C) be a minimal realization of T(z). Then, we have the following K[z]-isomorphisms: $(X,A) \cong \mathrm{Ker}\,D \cong \Omega Y/D\Omega Y \cong \Omega U/D_1\Omega U$.

Proof. The first two isomorphisms are proved in [2]. The last one follows, for instance, by [16] 6.5.

In dealing with matrices and morphisms induced by matrices, we use a superscript "t" to denote the transposition. Since no confusion can arise, we use a superscript "-1" to denote both the counterimage and, when the matrix or the morphism is invertible, the inverse.

2. *SOME EXAMPLES AND MOTIVATIONS*

In this section we want to show that the classical concepts of poles and zeros of a multivariable transfer function are not rich enough to allow a deep analysis of some design problems. This motivates the introduction, in the next section, of the more structured notions of pole module and zero module.

Let us start by recalling the classical definition of finite poles and zeros of the linear transfer function T(z) (see, for example, [16]). To this aim, we assume, without loss of generality, that the coefficient field K is algebraically closed, e.g. $K = \mathbb{C}$. If T(z) describes the input/output behaviour of a single-input/single-output system, it can be written as T(z) = p(z)/q(z), where p(z) and q(z) are coprime polynomials. Then, if $\alpha \in K$ is a root of q(z) of multiplicity ρ_α, we say that T(z) has a pole at α of order ρ_α and, if β is a root of p(z) of multiplicity ρ_β, we say that T(z) has a zero at β of order ρ_β. Roughly speaking, we can say that the poles of T(z) are the roots of its denominator and that the zeros of T(z) are the roots of its numerator. Actually, the notion of zero, or of pole, consists of a datum, expressed by an element of K, which determines the location of the zero, or of the pole, and of a multiplicity.

In the multivariable case, the definitions are given by means of the Smith-McMillan form of T(z). Recall that T(z) can be decomposed as

$$T(z) = B_1(z)\begin{pmatrix} \mathrm{diag}\,\{\,\varepsilon_1(z)/\psi_1(z).....\varepsilon_r(z)/\psi_r(z)\,\} & 0 \\ 0 & 0 \end{pmatrix} B_2(z)\,,$$

where $B_1(z)$ and $B_2(z)$ are polynomial unimodular matrices, r = rank T(z), $\varepsilon_i(z)$ and $\psi_i(z)$ are coprime monic polynomials such that $\varepsilon_i(z) \mid \varepsilon_{i+1}(z)$ and $\psi_{i+1}(z) \mid \psi_i(z)$ for i = 1,2,...,r. The elements $\varepsilon_i(z)$ and $\psi_i(z)$ of the Smith-McMillan form are uniquely determined. Then, if $\alpha \in K$ is a root of the $\psi_i(z)$'s of multiplicities $(\rho_1, \ldots, \rho_r)_\alpha$, we say that T(z) has a pole at α of total order $\Sigma\rho_i$ and we call the r-tuple $(-\rho_1, \ldots, -\rho_r)_\alpha$ the pole structure of T(z) at α (remark that $\rho_i \geq \rho_{i+1}$ and that some of the ρ_i's may be zero). If α is a root of the $\varepsilon_i(z)$'s of multiplicities $(\sigma_1, \ldots, \sigma_r)_\alpha$ we say that T(z) has a zero at α of total order $\Sigma\sigma_i$ and we call the r-tuple $(\sigma_1, \ldots, \sigma_r)_\alpha$ the zero structure of T(z) at α . Thus, the notion of multivariable zero, or pole, consists of a datum, expressed by an element of K, which determines the location of the zero, or of the pole, and of a string of multiplicities. Remark that, in the multivariable case, T(z) may have, at the same time, a pole and a zero at a given α. We speak in this case of numerically coincident pole and zero. In the multivariable case, a characterization in terms of numerator and denominator,in a suitable sense, of T(z) is expressed by the following

2.1 PROPOSITION Let $T(z) = D^{-1}(z)\, N(z) = N_1(z)\, D_1^{-1}(z)$ be coprime fractional representations. Then, the nontrivial elements in the Smith forms of D(z) and of $D_1(z)$ coincide with the $\psi_i(z)$'s and the non trivial elements in the Smith forms of N(z) and of $N_1(z)$ coincide with the $\varepsilon_i(z)$'s.

Proof. See [16] Sect. 6.5.

The dynamical interpretation of poles and zeros is well known (see [16] and also [34] and the references therein). Let us recall, in particular, that, assuming that T(z) is strictly proper and that (X,A,B,C) is its minimal realization, the poles of T(z) coincide with the eigenvalues of A. Therefore, they characterize the free dynamics of the system and, for example, its stability or dead-beat properties. In this sense, for K $= \mathbb{C}$, we will speak of unstable (i.e., in continous time, right half plane) poles and of stable ones. On the other hand, the zeros describe the transmission blocking properties of the system (X,A,B,C). If T(z) has a zero at α, in fact, the system matrix $\begin{pmatrix} zI-A & -B \\ C & 0 \end{pmatrix}$ loses rank at $z = \alpha$. Then, there exists a vector $\begin{pmatrix} x_o \\ u_o \end{pmatrix}$ such that $\begin{pmatrix} \alpha I-A & -B \\ C & 0 \end{pmatrix}\begin{pmatrix} x_o \\ u_o \end{pmatrix} = 0$. This means, in the discrete time model, that, taking the input sequence $u(t) = \alpha^t u_o$ for t=0,1,..., then the response corresponding to this input and to the initial state x_o is y(t)=0. In other terms, the state x_o blocks the transmission of the input u(t). An analogous result holds in the continous time situation. The vector $\begin{pmatrix} x_o \\ u_o \end{pmatrix}$ is called in [19] a zero direction associated to α.

Moreover, if we denote by V^* and by R^* respectively the maximum controlled invariant subspace of Ker C and the maximum controllability subspace of Ker C of the geometric theory (see [24]), we have that the zeros coincide with the eigenvalues of the automorphism of V^*/R^* induced by A + BF, where F: $X \to U$ is any feedback such that $(A + BF)\, V^* \subset V^*$ (such an F is called a "friend" of V^*).

The relevance of poles and zeros in the design of linear systems is illustrated in the following examples. They also show, by pointing out the differences between the scalar case and the multivariable case, the inadeguacy of the classical notions we recalled above in providing a clear description of the behaviour of multivariable poles and zeros in various interesting situations.

2.2 EXAMPLE - System inversion Many problems in control theory reduce ultimately to the inversion of a linear system. For an account of the literature on this, the reader is referred to [27]. Briefly, assuming that T(z) is a right or left invertible transfer function, one is interested in the solutions, respectively, of

2.3 $\qquad T(z)\, G(z) = I \qquad$ or of

2.4 $F(z)\,T(z) = I.$

Clearly, one would like to know from the analysis of T(z) which are the design limitation concerning G(z) or F(z), or, in other terms, which are the dynamical features of the inverses.

The analysis of the case in which T(z) is square and nonsingular, in particular when it is a scalar transfer function, presents no difficulty. In such a situation, in fact, there exists a unique inverse, whose pole structure coincides with the zero structure of T(z). This is obvious in the scalar case and follows easily in the multivariable case since, except for the ordering, the Smith-McMillan form of $T^{-1}(z)$ is the inverse of the Smith-McMillan form of T(z). In other terms, we have that when T(z) is square and nonsingular the dynamics of its inverse is completely described by the zero structure of T(z).

When T(z) is not square, we have many inverses with different pole structures. Consider, for instance, the left invertible transfer function $T(z) = [\, z/(z+1) \quad z/(z+2)\,]^t$, whose Smith-McMillan form is $[z/(z+1)(z+2) \quad 0\,]^t$. Both the transfer functions $F_1(z) = [\,(z+1)(z+2)/z \quad -(z+1)(z+2)/z\,]$ and $F_2(z) = [\,z+1)/(z-1) \quad -(z+2)/z(z-1)\,]$ are left inverses of T(z), and their Smith-McMillan forms are given respectively by $[\,(z+1)(z+2)/z \quad 0\,]$ and $[\,1/z(z-1) \quad 0\,]$. Hence, $F_1(z)$ has a pole structure consisting of a pole at 0 of order 1, which coincides with the zero structure of T(z), while $F_2(z)$ has also an unstable pole at 1 of order 1.

In this situation, it is therefore natural to consider the following questions:

i) is there an "essential" pole structure which appears in every inverse of T(z) ?

ii) given an affirmative answer to i), does there exist an inverse all of whose poles are essential ?

An answer to these questions would provide a characterization of the dynamical properties which are necessarily shared by all the inverses. In particular, it would allow one to investigate the existence of, e.g., stable inverses or dead-beat inverses. Clearly, what we expect is that, in accordance with the scalar case and, more generally, the multivariable square nonsingular case, an essential pole structure does exist and coincides with the zero structure of T(z). Roughly speaking, this is to say that the zeros of T(z) are, in some suitable sense, poles of any inverse. Unfortunately, mainly because of the presence of the unimodular matrices $B_1(z)$ and $B_2(z)$ which are not uniquely determined, it is difficult to obtain a satisfactory answer to i) and ii) using the classical definitions we have seen.

<u>2.5 EXAMPLE - Model matching and factorization problem</u> A class of problems more general than the system inversion is described by means of the equation

2.6 $T(z) = H(z)\,G(z)$ or, dually,

2.7 $T(z) = F(z)\,H(z)$

where T(z) and G(z) are two given transfer functions with the same input space (respectively, T(z) and F(z) are two given transfer functions with the same output space) and H(z) is sought. Such equations arise, for instance, in the model matching problem and in various factorization problems (see [5,9] and the references therein).

In the case in which the existence of solutions H(z) is assured, we are interested, in order to satisfy the design requirements, in their poles, and we can again formulate the same questions as in 2.2 i) and ii) :

i) is there an "essential" pole structure which appears in any solution H(z) of T(z) = H(z) G(z) (respectively T(z) = F(z) H(z)) ?

ii) given an affirmative answer to i), is there any solution H(z) all of whose poles are essential ?

Intuitively and roughly speaking, we expect that the essential part of the pole structure of any solution H(z) must supply the poles of T(z) which do not already appear in G(z), as well as the poles needed to cancel the zeros of G(z) which do not appear in $[\, T(z)^t \; G(z)^t \,]^t$ (note that it is necessary to compare the zeros of G(z) with those of $[\, T(z)^t \; G(z)^t \,]^t$, and not only with those of T(z), since some of the first may fail to appear as zeros of T(z) for non dynamical reasons, i.e. without being canceled by a pole of H(z), if T(z) is not full column rank).

In fact, similarly to 2.2, this is what happens clearly in the scalar case. If T(z) = p(z)/q(z) and G(z) = p'(z)/q'(z) are coprime representations, then the unique solution of 2.6 or 2.7 has a representation H(z) = p(z)q'(z)/ q(z)p'(z) , which is not necessarily reduced. However, the poles of H(z) consist of the roots of q(z), i.e. poles of T(z), which do not already appear as roots of q'(z), i.e. poles of G(z), together with the roots of p'(z), i.e. zeros of G(z), which do not appear as roots of p(z), i.e. zeros of T(z).

A straightforward generalization of these observations to the multivariable case, in which often the solution is not unique, is not possible. Numerator and denominator matrices, in fact, may not be invertible and, furthermore, numerically coincident poles and zeros do not necessarily cancel. Therefore, although some result on the location in K of the poles which are necessarily present in any solution H(z), called fixed poles, can be proved (see [25]), the information on the whole pole structure as well as a complete answer to i) and ii) seems difficult to obtain.

To close this section, we can conclude that, in the design problems mentioned in 2.2 and 2.5, it appears difficult, if we remain inside the framework characterized by the classical definitions, to extend beyond the scalar case the results suggested by the basic intuition about poles and zeros.

3. POLE MODULE AND ZERO MODULE

In this section we introduce the notions of pole module and of zero module. Overcoming the inadequacy of the classical definitions, they will allow us to develop, in the next section, a satisfactory algebraic treatment of the previously mentioned problems. Since the zero module is, in some sense, a concept less familiar than the pole module, we will spend more time in describing it and in comparing it with the objects of the geometric theory.

To begin with, let us remark that, given T(z) with a minimal realization of its strictly proper part (X,A,B,C), the whole information on the poles of T(z) can be given by means of the K[z]-module (X,A), consisting of the vector space X with the module structure induced by A. This , as we have recalled in 1.2, is isomorphic to $\Omega Y/D(z)\Omega Y$ and to $\Omega U/D_1(z)\Omega U$, where $T(z) = D^{-1}(z)\, N(z) = N_1(z)\, D_1{}^{-1}(z)$ are coprime fractional representations, and hence its invariant factors are, by 2.1, exactly the non trivial $\psi_i(z)$'s of the Smith-McMillan form of T(z). Let us clarify this point: the key fact is that the information contained in a set of polynomials, in particular in the $\psi_i(z)$'s, can be expressed by means of an abstract module having those polynomials as invariant factors. Hence, the poles can be computed directly from (X,A). In this sense we can speak of the pole module associated to T(z). Furthermore, since the problems we are dealing with are expressed at a transfer function level without employing the realizations, we choose, for the pole module, the representation given by the K[z]-module $X(T) = \Omega U / (T^{-1}(\Omega Y) \cap \Omega U)$. This is justified by the existence of an obvious isomorphism between (X,A) and X(T) which can be checked directly on 1.1. Formally, we have the following

3.1 DEFINITION The (finite) pole module of a transfer function $T(z) : U(z) \to Y(z)$ is the K[z]-module X(T) defined by $X(T) = \dfrac{\Omega U}{T^{-1}(\Omega Y) \cap \Omega U}$.

To introduce the notion of zero module we follow the original treatment of B. Wyman and M. Sain which appeared in [27]. It is helpful to start by considering the scalar case, representing T(z) as $T(z) = p(z)/q(z)$ with p(z) and q(z) coprime polynomials. Now, if the input $u(z) \in U(z)$ has a representation $u(z) = a(z)/b(z)$, with a(z) and b(z) coprime polynomials, and if b(z) and p(z) have nonunit factors in common, then the "modes" of u(z) represented by these factors fail to appear in the corresponding output $y(z) = T(z) u(z) = p(z)a(z)/q(z)b(z)$, because of cancellation. In other words, this means that the factors of the numerator, i.e. the zeros, can be revealed by looking at the modes of the inputs which fail to appear in the outputs. Henceforth, let us focus on the inputs which can produce no modes whatsoever in the response. In the scalar case we are considering, they are of the form $u(z) = a(z)q(z)/p(z)$, for a(z) in K[z], and produce responses $y(z) = a(z) \in K[z]$ having no modes. These excitations, whose modes describe the zeros of T(z), are characterized by being elements of $T^{-1}(K[z])$. Generalizing to the multivariable case, we will take into account the elements of $T^{-1}(\Omega Y)$. Now, remark that if u(z) is an element of K[z], no zero effect could be observed, since u(z) has no modes which can fail to appear in the output. At the same time, note that identically zero output is of little interest in the above discussion, since what is important is the failure of certain exciting modes to appear in the response. Although this last point has little significance in the scalar case, when there are no non-zero inputs which can produce zero responses, it is important in the multivariable generalization, where Ker T is not necessarily zero. Combining the above remarks, we have that, in the definition of an abstract module which captures the notion of zeros, the elements of Ker T and of ΩU, which generalizes K[z], can be safely neglected. This can be accomplished by forming an algebraic quotient modulo $\mathrm{Ker}\, T + \Omega U$. Thus, we are led to the following

3.2 DEFINITION The (finite) zero module of a transfer function $T(z) : U(z) \to Y(z)$ is the K[z]-module Z(T) defined by $Z(T) = \dfrac{T^{-1}(\Omega Y) + \Omega U}{\mathrm{Ker}\, T + \Omega U}$.

Remark that the addend ΩU in the numerator of the quotient is provided for consistency, so that the denominator is contained in the numerator.

Clearly we have now to show that Z(T) contains, in particular, the whole information on the numerator polynomials $\varepsilon_i(z)$'s of the Smith-McMillan form of T(z). This is proved by the following

3.3 PROPOSITION Let $T(z): U(z) \to Y(z)$ have the coprime fractional representation $T(z) = D^{-1}(z) N(z)$. Then, the zero module Z(T) is isomorphic to the torsion submodule $t_{K[z]}(\Omega Y / N\Omega U)$ of $\Omega Y / N\Omega U$.

The fundamental result stated by 3.3 says, by 2.1, that the invariant factors of Z(T) are exactly the nontrivial $\varepsilon_i(z)$'s of the Smith-McMillan form of T(z). Therefore, as for the pole module, we have an algebraic object described directly in terms of the transfer function which, in particular, contains the whole information on the zeros. Furthermore, the isomorphism with $t_{K[z]}$ (ΩY / $N\Omega U$), which is a finitely generated torsion K[z]-module, displays the structure of Z(T), showing that the latter is finitely generated over K with dimension as a K-vector space given by Σ_i deg ($\varepsilon_i(z)$) . The proof of 3.3 is given in [27] Th.1 and will not be repeated here. However, as an example of the algebraic techniques which will be used in the following, it is interesting to remark that the isomorphism between Z(T) and $t_{K[z]}$(ΩY / $N\Omega U$) can be defined by the map induced (dotted arrow) in the diagram below by the commutativity of (1) (see Sec. 1)

$$\begin{array}{ccccc} \mathrm{Ker}\, T + \Omega U & \longrightarrow & T^{-1}(\Omega Y) + \Omega U & \longrightarrow & Z(T) \\ N\downarrow & (1) & N\downarrow & & \vdots\downarrow \\ N\,\Omega U & \longrightarrow & \Omega Y & \longrightarrow & \Omega Y / N\Omega U \end{array}$$

3.4 REMARK Before going further, let us remark explicitly that Z(T) and X(T) contain more information than the sets { $\varepsilon_i(z)$ } and { $\psi_i(z)$ } , since they are defined not simply as abstract modules with a given set of invariant factors, but using a specific representation in terms of input and output spaces. However, it is important to note that these representations are not intended and are not suitable for computational purposes. In fact, it may be very difficult to compute the pole module or the zero module by means of their definitions, except in the scalar case (see 3.5 below). Thus, to determine explicitly the pole structure and the zero structure we have to use the coprime fractional representations and the isomorphism between X(T) and $\Omega Y/D\Omega Y$ and between Z(T) and $t_{K[z]}(\Omega Y/N\Omega U)$. On the other hand,the way in which the pole module and the zero module are defined and represented is an essential point in proving the results of the next Section.

3.5 EXAMPLE Let us compute the zero module Z(T) in the scalar case T(z) = p(z)/q(z). By definition, since Ker T = 0, we have Z(T) = $(T^{-1}(K[z]) + K[z]) / K[z]$. Now, $T^{-1}(K[z]) + K[z]$ = { a(z)q(z)/p(z) + b(z) , a(z) and b(z) $\in$ K[z] } = { (a(z)q(z) + b(z)p(z))/p(z) , a(z) and b(z) $\in$ K[z] } = 1/p(z) K[z]. The last step follows from the fact that p(z) and q(z) are relatively prime, so that any polynomial r(z) in K[z] can be written as r(z) = a(z)q(z) + b(z)p(z) for suitable a(z) and b(z) in K[z]. Therefore we have Z(T) = (1/p(z) K[z])/K[z] . Remark that there exists a K[z]-module isomorphism between (1/p(z)K[z])/ K[z] and K[z]/p(z)K[z] given by $[r(z)/p(z)]_{\mathrm{mod}\ K[z]} \to [r(z)]_{\mathrm{mod}\ p(z)K[z]}$. Thus we obtain the expected result, namely Z(T) $\cong$ K[z]/p(z)K[z].

In addition to 3.3, which states technically the relationship between Z(T) and the classical notion of zero, there is a natural connection between the concept of zero module and the transmission blocking properties of the minimal realization of T(z), when T(z) is strictly proper. To clarify this, let us write, for any u(z) $\in$ U(z), u(z) = $u_{pol} + u_{sp}$, where $u_{pol} \in \Omega U$ is the polynomial part of u(z) and u_{sp} is its strictly proper part. Now, any nontrivial element in Z(T) has the form [u(z)] ,where T(u(z)) is a nonzero element in ΩY. The fact that T(z) is strictly proper implies that y(z) = $T(u_{sp})$ is strictly proper and, since $T(u_{pol})$ = T(u(z)) - y(z), we have that u_{pol} is nonzero. The polynomial input u_{pol} sets up a state $x_o = \tilde{B}(u_{pol})$ of the minimal realization of T(z) (see 1.1), and the output corresponding to the initial condition x(0) = x_o and to the input sequence u_{sp} is given by $\tilde{C}(x_o) + \pi\_T(u_{sp}) = \pi\_T(u(z)) = 0$. In other words, the state x_o blocks the transmission of the input sequence u_{sp} and we can say that any non trivial element in Z(T) determines at least one pair (state,input sequence) with such a property. Comparing with the comments following 2.1, this shows that the zero module captures also the notion of zero direction described in [19].

By the previous discussion, it turns out in particular that the state $x_o = \tilde{B}(u_{pol})$ belongs to Ker C and has the property of being weakly unobservable, that is : there exists an input sequence, in this case u_{sp}, such that the corresponding output , with initial state $x(0) = x_o$, is zero. It is known that the set of weakly unobservable states coincides with the maximum controlled invariant subspace V^* contained in Ker C of the geometric theory [24]. Henceforth, we have a slightly different interpretation of the situation described above which point out a connection between Z(T) and V^*. This is made precise in the following :

3.6 PROPOSITION Let $T(z): U(z) \to Y(z)$ be a strictly proper transfer function with minimal realization (X,A,B,C). Denote respectively by V^* and R^* the maximum controlled invariant subspace of Ker C and the maximum controllability subspace contained in Ker C and let $F: U \to X$ be such that $(A + BF)\, V^* \subset V^*$. Then, there exists a natural K[z]-isomorphism between Z(T) and V^*/R^* provided with the K[z]-module structure induced by A + BF.

Two slightly different proofs of 3.6 are given in [31] Th.1 and in [3] Sect.4 (compare also with [23]). Here it is important to remark that 3.6 does not simply state the existence of an isomorphism between Z(T) and (V^*/R^*, A + BF) as abstract modules. This, on the other hand, is already known, since the invariant factors of the two modules are the same. The emphasis, in 3.6, is on the existence of a natural isomorphism, where "natural" means induced directly by the realization diagram 1.1. Actually, such an isomorphism is induced by the correspondence $[u(z)] = [u_{pol} + u_{sp}] \to \tilde{B}(u_{pol})$ we have already discussed. Alternatively, its inverse can be characterized as follows. For $v \in V^*$, let $a(z) \in \Omega U$ be such that $\tilde{B}(a(z)) = v$ and let $b(z) = \Sigma_{i=1...\infty} F(A + BF)^{i-1}(v)\, z^{-i}$, where F is such that $(A + BF)V^* \subset V^*$. It is possible to show that $(a(z) + b(z)) \in T^{-1}(\Omega Y)$ and that the correspondence $v \to [a(z) + b(z)] \in Z(T)$ between V^* and Z(T) is well defined and induces the isomorphism we are speaking of.

4. *FIXED POLES OF TRANSFER FUNCTION EQUATIONS*

We have now the algebraic tools needed to tackle the problems stated in 2.2 and 2.5 and to give an answer to i) and ii). In particular we focus on the equation 2.6, namely T(z) =H(z) G(z) where T(z) and G(z) are given, and , when a suitable necessary and sufficient condition for the existence of solutions H(z) is satisfied, we provide a description of the essential part of the pole structure of any solution. The equation 2.7 is treated in detail in [9]. The inversion problem of 2.2 is obviously a particular case of 2.6 and 2.7 and has been investigated in [27,31]. A somewhat less satisfactory treatment, if compared with that of [9], of 2.6 and 2.7 is contained in [5].

Before going further, let us recall that the equation 2.6, where $T(z) : U(z) \to Y(z)$ and $G(z) : U(z) \to W(z)$ are given, has solutions $H(z) : W(z) \to Y(z)$ if and only if

(A) $\qquad$ $\mathrm{Ker}\, G \subset \mathrm{Ker}\, T$.

Condition (A) simply means that, if an input produces zero response through $G(z)$, then it produces zero response also through $T(z)$. In the sequel we assume that the pair of transfer functions $T(z)$ and $G(z)$ we are dealing with, verifies the condition (A).

The approach we will follow in this Section can be summarized in the following way. First, we define abstractly, in terms of the data $T(z)$ and $G(z)$, a $K[z]$-module P which is shown to be contained in the pole module $X(H)$ of any solution $H(z)$. Therefore, P represents the essential pole structure of the solutions of 2.6. Then, we prove that P can be described in terms of poles and zeros of $T(z)$ and $G(z)$ in a way which agrees with the basic intuition explained in 2.5.

Given the equation 2.6 and assuming that (A) holds, let us consider the $K[z]$-module P defined by $P = \dfrac{G^{-1}(\Omega W)}{G^{-1}(\Omega W) \cap T^{-1}(\Omega Y)}$. The following proposition justifies the introduction of P.

<u>4.1 PROPOSITION</u> For any solution $H(z)$ of 2.6 there exists a natural inclusion $j : P \to X(H)$.

Proof. The map $j : P \to X(H)$ is the map induced (dotted arrow) in the diagram below by the commutativity of (1).

$$\begin{array}{ccccc} G^{-1}(\Omega W) \cap T^{-1}(\Omega Y) & \longrightarrow & G^{-1}(\Omega W) & \longrightarrow & P \\ \downarrow G & (1) & \downarrow G & & j \, \vdots\downarrow \\ \Omega W \cap H^{-1}(\Omega Y) & \longrightarrow & \Omega W & \longrightarrow & X(H) \end{array}$$

Assuming that $G(T^{-1}(\Omega Y))$ is contained in $H^{-1}(\Omega Y)$, we have used implicitly the fact that $H(z)$ is a solution of 2.6, i.e. $T(z) = H(z)\, G(z)$. It remains now to show that j is injective. For this purpose, let $[u(z)]$ be an element of P such that $j([u(z)] = 0$. By definition, this implies that $G(u(z))$ belongs to $\Omega W \cap H^{-1}(\Omega Y)$, hence $T(u(z)) = HG(u(z))$ belongs to ΩY. Thus, $u(z)$ belongs to $G^{-1}(\Omega W) \cap T^{-1}(\Omega Y)$ and, as a consequence, $[u(z)] = 0$.

By 4.1 we have immediately that the module P provides an answer to the question i) of 2.5. The invariant factors of P describe a pole structure which appears in any solution $H(z)$. Therefore, we have the following definition (compare with [25]):

<u>4.2 DEFINITION</u> The module $P = \dfrac{G^{-1}(\Omega W)}{G^{-1}(\Omega W) \cap T^{-1}(\Omega Y)}$ is called the <u>module of fixed poles</u> of the equation 2.6.

<u>4.3 EXAMPLE</u> Let $T(z) = I_2$, the 2 x 2 identity matrix, and let

$$G(z) = \begin{pmatrix} z/(z+1)^2 & 0 \\ 0 & (z^2+2z)/(z+1)^3 \\ z/(z+1)^2 & (z^2+2z)/(z+1)^3 \end{pmatrix}.$$ It is easily seen that the columns of $G(z)$ are independent, hence $\operatorname{Ker} G = 0$ and the condition (A) is trivially satisfied by $T(z)$ and $G(z)$. The equation 2.6 is, in this case, $I = H(z)\, G(z)$ and actually what we have is an inversion problem. In particular, $T^{-1}(\Omega Y) = \Omega Y$ and $P = \dfrac{G^{-1}(\Omega W)}{G^{-1}(\Omega W) \cap \Omega Y}$ and therefore, using a canonical isomorphism and the fact that $\operatorname{Ker} G = 0$, we have, in

accordance with [27], $P \cong \dfrac{G^{-1}(\Omega W) + \Omega Y}{\Omega Y} = Z(G)$. Using 3.3 and the coprime fractional representation

$$G(z) = D^{-1}(z)N(z) = \begin{pmatrix} (z+1)^2 & 0 & 0 \\ -(z^3+3z^2+3z) & 1 & z^3+3z^2+3z \\ 0 & 0 & (z+1)^3 \end{pmatrix}^{-1} \begin{pmatrix} z & 0 \\ 0 & z(z+2) \\ z(z+1) & z(z+2) \end{pmatrix},$$

since the Smith form of N(z) is $\begin{pmatrix} z & 0 \\ 0 & z(z+2) \\ 0 & 0 \end{pmatrix}$, we obtain $Z(G) \cong K[z] / zK[z] \oplus K[z]/ z(z+2)K[z]$. Therefore, any left inverse of G(z) must have a pole at 0 of total order 2 and structure (-1,-1) and a pole at -2 of total order 1 and structure (-1,0).

4.4 COMPUTATION OF P Before giving another example, let us show how to compute practically P, or better its invariant factors, in the general case. For this purpose, assume that the data T(z) and G(z) of 2.6 have coprime fractional representations $T(z) = N(z)D^{-1}(z)$, $G(z) = N_G(z)D_G^{-1}(z)$ and let $M(z) = D(z)A(z) = D_G(z)B(z)$ be the minimum common left multiple of D(z) and $D_G(z)$. The product by M(z) induces an isomorphism, described by $[u(z)] \to [M(uz)]$, between the module $\dfrac{(N_GB)^{-1}(\Omega W)}{(N_GB)^{-1}(\Omega W) \cap (NA)^{-1}(\Omega Y)}$ and P. Note that the multiplication by M(z) was the technique used in [5] in order to simplify the analysis of 2.6. Now, let S(z) be the maximum (nonsingular) common right divisor of $N_G(z)B(z)$ and $N(z)A(z)$, in particular let $N_G(z)B(z) = N'(z)S(z)$. Then, we have $(N_GB)^{-1}(\Omega W) \cap (NA)^{-1}(\Omega Y) = S^{-1}(\Omega U)$ and hence $P \cong (N'S)^{-1}(\Omega W)/ S^{-1}(\Omega U)$. As before, the multiplication by $N'(z)S(z)$ induces an isomorphism between P and $(\Omega W \cap \mathrm{Im}\, N') / N'(\Omega U)$. At this point it is easy to see that the last module is isomorphic to the torsion submodule $t_{K[z]}(\Omega W / N'(\Omega U))$ of $\Omega W / N'(\Omega U)$ and that, as a consequence, the invariant factors of P are the nontrivial elements in the Smith form of N'(z).

4.5 EXAMPLE Consider the transfer functions

$$T(z) = \begin{pmatrix} 1/(z+2) & -2/(z+2) \\ 1/(2z+2) & -1/(2z+2) \end{pmatrix} \text{ and } G(z) = \begin{pmatrix} 1/(z^2+z) & 1/(z+1) \\ 1/z & 1/(z+1) \\ 1/(z^2+z) & 1/(z+1) \end{pmatrix}.$$

It is easy to see that both T(z) and G(z) are full rank, hence the condition (A) is trivially satisfied since Ker G = 0 and Ker T = 0. We compute the invariant factors of P, the module of fixed poles of the equation T(z) = H(z)G(z), using the procedure explained in 4.4. Coprime fractional representations for T(z) and G(z) are the following:

$$T(z) = N(z)D^{-1}(z) \text{ with } N(z) = \begin{pmatrix} 2 & 1 \\ 1 & 1 \end{pmatrix}, D(z) = \begin{pmatrix} 2z & 3z+2 \\ -2 & z \end{pmatrix},$$

$$\text{and } G(z) = N_G(z)D_G^{-1}(z) \text{ with } N_G(z) = \begin{pmatrix} 1 & z \\ 0 & 1 \\ z+1 & 0 \end{pmatrix}, D_G(z) = \begin{pmatrix} z^2+z & z^2+z \\ 0 & -z-1 \end{pmatrix}.$$

The minimum common left multiple of D(z) and $D_G(z)$ is given by

$$M(z)= D(z)A(z)= D_G(z)B(z) = \begin{pmatrix} z^3+3z^2+2z & 2z^2+2z \\ 0 & -2(z+1) \end{pmatrix},$$

with $A(z) = \begin{pmatrix} z^2/2 & z+1 \\ z & 0 \end{pmatrix}$ and $B(z) = \begin{pmatrix} z+2 & 0 \\ 0 & 2 \end{pmatrix}$.

Then, since the matrices

$$N(z)A(z) = \begin{pmatrix} z^2+z & 2(z+1) \\ (z^2+2z)/2 & z+1 \end{pmatrix} \text{ and } N_G(z)B(z) = \begin{pmatrix} z+2 & 0 \\ z^2+3z+2 & 2z \\ z+2 & 0 \end{pmatrix}$$

turn out to be right coprime, their maximum common right divisor S(z) is unimodular. Hence $P \cong (N_G B)^{-1}(\Omega W)/\Omega U \cong t_{K[z]}(\,\Omega W/N_G B(\Omega U)\,)$ and the invariant factors of P are the nontrivial elements in the Smith form of $N_G(z)B(z)$, which is given by $\begin{pmatrix} 1 & 0 \\ 0 & z(z+2) \\ 0 & 0 \end{pmatrix}$. This means, in particular, that any solution H(z) of 2.6 has a pole of total order 1 at 0 whose structure is (-1,0) and a pole of total order 1 at -2, whose structure is (-1,0). That agrees with the fact that T(z) has a pole of total order 1 at -2 which does not appear as pole of G(z) , and G(z) has a zero of total order 1 at 0 which is not a zero of $\binom{T}{G}$.

Let us show that the abstract module P has a more concrete description in terms of zeros and poles of the data T(z) and G(z). Since our aim is to extend the intuitive interpretation we have in the scalar case, we need first to clarify what we mean by "the poles of T(z) which do not appear in G(z)" and by "the zeros of G(z) which do not appear in $[\,T(z)^t\ \ G(z)^t\,]^t$". This can be done quite naturally in the framework we have built up. First, since (A) holds, the zero module of the transfer function $[\,T(z)^t\ \ G(z)^t\,]^t$: $U(z) \to (Y\oplus W)(z)$ can be represented as

$$Z\binom{T}{G} = \frac{\binom{T}{G}^{-1}\Omega(Y\oplus W) + \Omega U}{\operatorname{Ker}\binom{T}{G} + \Omega U} =$$

$$\frac{T^{-1}(\Omega Y) \cap G^{-1}(\Omega W) + \Omega U}{\operatorname{Ker} T \cap \operatorname{Ker} G + \Omega U} = \frac{T^{-1}(\Omega Y) \cap G^{-1}(\Omega W) + \Omega U}{\operatorname{Ker} G + \Omega U}.$$

Therefore, as $Z(G) = \dfrac{G^{-1}(\Omega W) + \Omega U}{\operatorname{Ker} G + \Omega U}$, we have a natural inclusion $i : Z\binom{T}{G} \to Z(G)$ induced by the obvious inclusion of the numerator modules. On the other hand, we have a natural projection $p : X\binom{T}{G} \to X(G)$ between the pole modules

$$X\binom{T}{G} = \frac{\Omega U}{T^{-1}(\Omega Y) \cap G^{-1}(\Omega W) \cap \Omega U} \text{ and } X(G) = \frac{\Omega U}{G^{-1}(\Omega W) \cap \Omega U},$$

induced by the obvious inclusion of the numerator modules. Here the term "natural" means that the existence of the involved maps follows directly from the representations of the zero and pole modules that have been chosen.

4.7 NOTATIONS i) We denote by Z the K[z]-module defined, up to isomorphism, by the short exact sequence

$$0 \to Z\binom{T}{G} \xrightarrow{i} Z(G) \to Z \to 0$$

ii) We denote by X the K[z]-module defined, up to isomorphism, by the short exact sequence

$$0 \to X \to X\binom{T}{G} \xrightarrow{p} X(G) \to 0 .$$

The module Z, cokernel of i, is identified with the quotient $Z(G)/Z\binom{T}{G}$ and clearly it can be viewed as representing the zeros of G which are not zeros of $[\,T(z)^t\ G(z)^t\,]^t$. In particular we have that, as a K-vector space, Z(G) is isomorphic to $Z\binom{T}{G} \oplus_K Z$. In other terms Z represents exactly the zeros that one expects to be cancelled by the poles of any solution H(z) of 2.6. Analogously, since $X\binom{T}{G}$ describes the union of the poles of T(z) and of G(z), the kernel X of p can be viewed as representing the poles of T(z) which are not poles of G(z) (see also [5] 3.5). They are exactly the poles that one expects to see as poles of any solution H(z) of 2.6. More precisely, we can state the following

<u>4.8 PROPOSITION</u> There exists a natural inclusion $\phi: X \to P$ and a natural projection $\psi: P \to Z$ such that the following sequence

$$O \to X \xrightarrow{\phi} P \xrightarrow{\psi} Z \to O$$

is exact (i.e., in particular, $\phi(X) = \mathrm{Ker}\ \psi$).

Proof. See [9] 3.9.

From the above proposition it follows, in particular, that $P \cong X \oplus_K Z$ as a K-vector space. Then we can say, in a precise algebraic sense which extends and generalizes what we have seen in the scalar case, that the module of fixed poles of the equation 2.6 consists exactly of the zeros of G(z) which are not zeros of $[\,T(z)^t\ G(z)^t\,]^t$ and of the poles of T(z) which are not poles of G(z). It remains to give an answer to 2.5 ii). If we call <u>essential</u> a solution H(z) whose pole module X(H) contains only the fixed poles represented by P, i.e. for which the map j of 4.1 is an isomorphism, this amounts to investigating the existence of essential solutions. The result we have is contained in the following

<u>4.9 PROPOSITION</u> There exists a solution H(z) of 2.6 such that $j : P \to X(H)$ is an isomorphism.

Proof. For a complete proof, the reader is referred to [9] 4.4, here we simply show a possible construction of an essential H(z). First, let us choose a K[z]-basis $\{m_1,\ldots,m_r,v_{r+1},\ldots,v_q\}$ of ΩW, such that $\{m_1,\ldots,m_r\}$ is a K[z]-basis of $G(U(z)) \cap \Omega W$. Such a basis exists because both $G(U(z)) \cap \Omega W$ and $\dfrac{\Omega W}{G(U(z)) \cap \Omega W}$ are free K[z]-modules (see [27] and [5] 4.1 for details) and, moreover, it is also a K(z)-basis of W(z). Now, let $u_1(z),\ldots,u_r(z)$ be elements in U(z) such that $G(u_i(z)) = m_i$ for $i=1,\ldots,r$, and define $H(z) : W(z) \to Y(z)$ as follows

$$H(m_i) = T(u_i(z)) \qquad i = 1,\ldots, r$$
$$H(v_k) = 0 \qquad k = r+1,\ldots, q.$$

It turns out that H(z) is a solution of 2.6 such that $j : P \to X(H)$ is an isomorphism.

<u>4.10 REMARK</u> Different choices of the elements m_i's, v_k's, $u_i(z)$'s in the above construction lead to different essential solutions. However, not all the essential solutions can be obtained in this way. A complete description of the set of all the essential solutions is given in [9] 4.4, 4.5.

Summarizing the results of this Section, we can say, by 4.1 and 4.6, that the design problem represented by the equation 2.6 has solutions, if (A) holds, whose pole structure simply consists, in a meaningful way, of the poles of T(z) which do not appear in G(z) together with the zeros of G(z) which do not appear in $[\, T(z)^t \;\; G(z)^t \,]^t$. Moreover, this pole structure can be a priori computed by means of P, as we saw in 4.5, and a procedure to find these essential solutions can be given, as we mentioned in the proof of 4.9. The relevance of such results in solving the design problem, especially when solutions with particular properties as for instance, stability are sought, is obvious.

4.11 EXAMPLE Let us consider again the transfer functions of example 4.6. In this case we have, for instance , $G(U(z)) \cap \Omega W = \text{span}\,\{(1,z+1,1)^t,(0,1,0)^t\}$ and $G((\,z(z+1)\,,\,0)^t) = (1,z+1,1)^t$, $G((z+1,\,-(z+1)/z\,)^t) = (0,1,0)^t$. Adding the vector $(0,0,1)^t$, we obtain the basis for W(z) needed to define H(z) as

$$H((1,z+1,1)^t) = T((z(z+1),0)^t) = (\,z(z+1)/(z+2)\,,\,z/2)^t$$

$$H((0,1,0)^t) = T((z+1\,,-(z+1)/z)^t\,) = ((z^2-z-4)/(z(z+2)),(z^2+3z+4)/(2z(z+2)))^t$$

$$H((0,0,1)^t) = 0.$$

With respect to the canonical basis of W(z), we have

$$H(z) = \begin{pmatrix} (z+1)(z+4)/(z(z+2)) & (z^2-z-4)/(z(z+2)) & 0 \\ (-2z^2-7z-4)/(2z(z+2)) & (z^2+3z+4)/(2z(z+2)) & 0 \end{pmatrix}$$

and then it is not difficult to verify that H(z) is a solution of 2.6 whose pole module coincides with the fixed pole module P, computed as in 4.6. Hence, H(z) is an essential solution of the equation we are considering.

5. FURTHER DEVELOPMENTS AND GENERALIZATIONS

In the previous Sections we have considered only the finite poles and zeros, i.e. those which are located at a point α of K. In addition one can consider also the poles and zeros at infinity (see [16] for generalities). They play, in fact, a fundamental role in various design problems and the same questions concerning the finite poles and zeros we have seen can be formulated in the case of poles and zeros at infinity.

We will not develop here a complete treatement of the case of poles and zeros at infinity (see [4,6,7,9,10,26] for applications), but we will simply point out its general lines, assuming that the reader is familiar with the classical notions.

As in the finite case, the classical definitions, which are given in terms of the Smith-McMillan form at infinity, involve a string of multiplicities and, implicitly, a "location" at infinity. The same motivations we have described in 2.2 and 2.5 justify the attempt of developing a module theoretic approach. The key fact, in introducing the concept of module of poles at infinity (briefly: infinite pole module) and of module of zeros at infinity (infinite zero module), is that the ring of polynomials K[z] must be replaced, in our framework, by the ring O_∞ of proper rational functions, i.e. rational functions of the form p(z)/q(z) with $\deg p \leq \deg q$. In fact, in considering the structure at infinity, we have to take into account only the elements of K(z) which are "regular" (as meromorphic functions if, for instance, $K = \mathbb{C}$) at infinity, that is the proper rational functions, in the same way as, in considering the structure at every finite point of K, we had to take into account the elements of K(z) which are regular at every point of K, that is the polynomials. Note that O_∞ is, like K[z], a principal ideal domain, therefore all the results concerning the theory of modules and the algebra of matrices which have been used in the previous sections hold without changes. In fact, given the K-vector space $V = K^n$, denoting by $\Omega_\infty V$ the O_∞-module $V \otimes_K O_\infty$ (whose elements are n-tuples with entries from O_∞), we have the following definitions

5.1 DEFINITION The infinite pole module of T(z) is the O_∞-module $X_\infty(T)$ defined by $X_\infty(T) = \dfrac{\Omega_\infty U}{T^{-1}(\Omega_\infty Y) \cap \Omega_\infty U}$.

5.2 DEFINITION The infinite zero module of T(z) is the O_∞-module $Z_\infty(T)$ defined by $Z_\infty(T) = \dfrac{T^{-1}(\Omega_\infty Y) + \Omega_\infty U}{\text{Ker } T + \Omega_\infty U}$.

The analogy between 5.1 and 3.1 and between 5.2 and 3.2 is obvious. The invariant factors of $X_\infty(T)$ over O_∞ describe the structure of the poles at infinity and the invariant factors of $Z_\infty(T)$ describe the structure of the zeros at infinity . Moreover, if T(z) is strictly proper with minimal realization (X,A,B,C) and S^* denotes the minimal conditionally invariant subspace of X containing Im B, we have that $Z_\infty(T)$ is O_∞-isomorphic to S^*/R^* provided with the O_∞-structure given by the shift operator z^{-1}.

Considering again the equation 2.6, we can introduce the following :

5.3 DEFINITION The O_∞-module $P_\infty = \dfrac{G^{-1}(\Omega_\infty W)}{G^{-1}(\Omega_\infty W) \cap T^{-1}(\Omega_\infty Y)}$ is called the module of fixed poles at infinity of the equation 2.6.

Then we can prove, as in section 4:

5.4 PROPOSITION i) For any solution H(z) of 2.6 there exists a natural inclusion $j_\infty : P_\infty \rightarrow X_\infty(H)$.
ii) There exists a solution H(z) of 2.6 such that $j_\infty : P_\infty \rightarrow X_\infty(H)$ is an isomorphism.

The result of 5.4 can be applied when one is interested in the existence of a proper solution. A transfer function, in fact, is proper iff it has no poles at infinity, that is iff its infinite pole module is zero. Then, there exists a proper solution H(z) of 2.6 iff $P_\infty = 0$.

5.5 REMARK It is not possible to combine 4.9 (existence of essential solutions) together with 5.4 ii) in order to obtain an existence result for solutions H(z) having, at the same time, a pole module X(H) isomorphic to P and an infinite pole module $X_\infty(H)$ isomorphic to P_∞. The two propositions, in fact, concern two different frameworks, the polynomial one and the proper rational one, which, from the point of view of poles and zeros, are mutually exclusive. Furthermore, such solutions, in general, do not exist.

A further generalization, which leads to useful applications, can now be easily described (see [26]). Assume that we are interested in considering the pole/zero structure at every point of a proper subset S of $K \cup \{\infty\}$ which does not coincide with K (finite case) nor with $\{\infty\}$ (case at infinity). For instance, if $K = \mathbb{C}$ and we want to investigate the existence of proper and stable solutions, the set S we have to deal with consists of all the points with positive real part together with the point ∞. Noting that the subset of elements of K(z) which are regular at every point of S forms a ring O_S which is a principal ideal domain, it is not difficult to realize that one can develop an O_S-theory following the general lines along which the K[z]-theory and the O_∞-theory have been constructed. Therefore, using notations which are obvious, we have, in particular:

5.6 DEFINITION The S-pole module of T(z) is the O_S-module $X_S(T)$ defined by $X_S(T) = \dfrac{\Omega_S U}{T^{-1}(\Omega_S Y) \cap \Omega_S U}$.

5.7 DEFINITION The O_S-module $P_S = \dfrac{G^{-1}(\Omega_S W)}{G^{-1}(\Omega_S W) \cap T^{-1}(\Omega_S Y)}$ is called the module of fixed poles in S for 2.6

5.8 PROPOSITION i) For any solution H(z) of 2.6 there exists a natural inclusion $j_S : P_S \to X_S(H)$;
ii) There exists a solution H(z) of 2.6 such that $j_S : P_S \to X_S(H)$ is an isomorphism.

As a consequence of 5.8 we have, for instance, that there exist proper stable solutions H(z) of 2.6 iff $P_S = 0$ for $S = \{ \alpha \in \mathbb{C} \cup \{\infty\}, \mathrm{Re}\,\alpha > 0 \text{ or } \alpha = \infty \}$, or that there exists a dead beat solution H(z) of 2.6 iff $P_S = 0$ for $S = \{ \alpha \in K \cup \{\infty\}, \alpha \neq 0 \}$. The reader can easily list other examples. What is important to note is that, although the framework is different, the analysis of 2.6 with respect to a subset S of $K \cup \{\infty\}$ can be carried out, with the obvious modifications of notations, as we have seen in Section 4.

6. *GENERIC ZEROS AND THE WEDDERBURN-FORNEY CONSTRUCION*

Forgetting the module structures, the pole and zero modules, viewed as K-vector spaces, can be combined to form the global pole space $\mathbf{X}(T) = X(T) \oplus_K X_\infty(T)$ and the global zero space $\mathbf{Z}(T) = Z(T) \oplus_K Z_\infty(T)$. The dimensions of $\mathbf{X}(T)$ and of $\mathbf{Z}(T)$ are the total number of poles and the total number of zeros of T(z). In general, dim $\mathbf{X}(T)$ may be greater than dim $\mathbf{Z}(T)$ and the difference between the two is called the defect of T(z). This can be expressed in terms of Kronecker indices or Wedderburn numbers as in [11] and [16] Th. 6.5-11 and can be related to the failure of T(z) to be surjective or to be injective.

In the cascade composition of two transfer functions T(z)G(z), the presence of a nontrivial Kernel or Cokernel of one of the two may cause new zeros, that is zeros which are not in T(z) nor in G(z), to appear. This phenomenon, investigated in [8] and [17], suggests the idea of associating to Ker T and Cok T a notion of generic zero, not located at any point of K nor at infinity, of T(z). One can think that the Kronecker indices or the Wedderburn numbers we have mentioned above describe the number of such zeros. A precise algebraic formalization of these ideas again requires the use of the module theoretic approach. Following [33], given a K-vector space $V = K^n$, we consider the K-linear map $\pi_- : V(z) \to \Gamma V$ described in Section 1 which associates with any element of V(z) its strictly proper part, then, for any K(z)-subspace $S \subset V(z)$, we have the following

6.1 DEFINITION The Wedderburn-Forney space associated with $S \subset V(z)$ is the K-vector space W(S) defined by $W(S) = \dfrac{\pi_-(S)}{S \cap z^{-1}\Omega_\infty V}$.

For any $S \subset V(z)$, W(S) turns out to be a finite dimensional K-vector space. Therefore, for a given transfer function T(z), the construction of the Wedderburn-Forney spaces W(Ker T) and W(Im T) associated with Ker T and with Im T yields two objects of the same kind as X(T) and Z(T). We will think of the spaces W(Ker T) and W(Im T), which are trivial iff respectively T(z) is injective or surjective, as of spaces of generic zeros. Now, denoting by $\pi_+ : V(z) \to \Omega V$ the K-linear map which associates with any element of V(z) its polynomial part, we have that the mapping $u(z) \to (\pi_-u(z), \pi_+u(z))$ induces a K-linear map $h : W(\text{Ker } T) \to X(T)$. It turns out that h is injective, so that W(Ker T) can be viewed as a subspace of X(T) and we have the following

6.2 PROPOSITION For any transfer function T(z) there exists an exact sequence of K-vector spaces

$$0 \to Z(T) \xrightarrow{\;f\;} \frac{X(T)}{W(\text{Ker } T)} \xrightarrow{\;g\;} W(\text{Im } T) \to 0,$$

where $f : Z(T) \to \frac{X(T)}{W(\text{Ker } T)}$ is the map induced by $(u(z),v(z)) \to (\pi_+(u(z)+v(z)), \pi_-(u(z)+v(z)))$ for $u(z) \in T^{-1}(\Omega Y)$ and $v(z) \in T^{-1}(z^{-1}\Omega_\infty Y)$ and $g : \frac{X(T)}{W(\text{Ker } T)} \to W(\text{Im } T)$ is the map induced by $(u(z),v(z)) \to \pi_-T(z)(u(z)+v(z))$ for $u(z) \in \Omega U$ and $v(z) \in z^{-1}\Omega_\infty U$.

Proof. See [33] Main Th. 5.1.

As a direct consequence of 6.2 we have for any transfer function T(z) that dim X(T) = dim Z(T) + dim W(Ker T) + dim W(Im T). So, if we think of the dimensions over K of W(Ker T) and of W(Im T) as of the number of generic zeros of T(z), we have that the total number of poles equals the total number of zeros, including the generic ones. The above Proposition can therefore be viewed as a refinement, in structural terms, of the numerical results about the defect and the Kronecker indices or the Wedderburn numbers we mentioned at the beginning of the Section. Moreover, the above exact sequence can be viewed as a complete concise description of the relationship between zeros and poles of any transfer function.

REFERENCES

[1] M.F. Atiyah and I.G. MacDonald - Introduction to Commutative Algebra, Addison-Wesley, Reading (1969)

[2] G. Conte and A.M. Perdon - On polynomial matrices and finitely generated torsion K[z]-module, A.M.S. Lectures in Appl. Math., Vol.18 (1980)

[3] G. Conte and A.M. Perdon An algebraic notion of zero for system over rings, Lecture Notes in Control and Information Science, Springer-Verlag, 58 (1984)

[4] G. Conte and A.M. Perdon - Infinite zero module and infinite pole module, Lecture Notes in Control and Information Science, Springer-Verlag, 62 (1984)

[5] G. Conte and A.M. Perdon - Zero module and factorization problems, A.M.S. Series in Contemporary Mathematics, Vol. 47 (1985)

[6] G. Conte and A.M. Perdon - On the minimum delay problem, Systems and Control Letters, 5 (1985)

[7] G. Conte and A.M. Perdon - On the causal factorization problem, IEEE Trans. Aut. Control, AC-30 (1985)

[8] G. Conte and A.M. Perdon - Zeros of cascade composition, in Frequency Domain and State Space Methods for Linear Systems, C.Byrnes and A.Lindquist Eds., North-Holland,(1985)

[9] G. Conte, A.M. Perdon and B. Wyman - Fixed poles in transfer function equations, SIAM J. Control Opt., 26 (1988)

[10] G. Conte, A.M. Perdon and B. Wyman - Zero/pole structure of linear transfer functions, Proc. 24th IEEE CDC, Fort Lauderdale (1985)

[11] D.G. Forney - Minimal bases of rational vector spaces with applications to multivariable linear systems, SIAM J. Control, 13 (1975)

[12] P. Fuhrmann - Algebraic system theory : An analyst's point of view, J. Franklin Inst., 301 (1976)

[13] P. Fuhrmann and M. Hautus - On the zero module of rational matrix functions, Proc. 19th IEEE CDC , New York (1980)

[14] B. Hartley and T.O. Hawkles - Rings, Modules and Linear Algebra, Chapman and Hall, London (1970)

[15] M. Hautus and M. Heymann - Linear feedback. An algebraic approach, SIAM J. on Control and Opt., 16 (1978)

[16] T. Kailath - Linear Systems, Prentice Hall (1980)

[17] S. Kung and T. Kailath - Some notes on valuation theory in linear systems, Proc. 17th IEEE CDC, San Diego (1978)

[18] S. Lang - Algebra, Addison-Wesley, Reading (1965)

[19] A. MacFarlane and N.Karcanias - Poles and zeros of linear multivariable systems : a survey of the algebraic, geometric and complex-variable theory, Int J. Control, 24, (1976)

[20] H. Rosenbrock - State space and multivariable theory, John Wiley and Sons, N.Y., (1970)

[21] C. Schrader and M. Sain - Research on system zeros : a survey, Proc. 27th IEEE Conf. on Decision and Control, Austin, (1988)

[22] G. Verghese - Infinite-frequency behaviour in generalized dynamical systems, Ph. D. Thesis, Dept. Electrical Eng., Stanford Univ., (1978)

[23] J.C. Willems and P. Fuhrmann - A study of (A,B)-invariant subspaces via polynomial models, Int. J. Control, vol.31 (1980)

[24] W. M. Wonham - Linear multivariable control: A geometric approach, Springer-Verlag, (1979)

[25] W. Wolovich, P. Antsaklis and H. Elliott - On the stability of the solutions to minimal and non-minimal design problems, IEEE Trans. Autom. Control, AC-27 (1977)

[26] B. Wyman, G. Conte and A.M. Perdon - Local and global linear system theory, in Frequency Domain and State Space Methods for Linear Systems, C.Byrnes and A.Lindquist Eds., North-Holland,(1985)

[27] B. Wyman and M. Sain - The zero module and essential inverse systems IEEE Trans. Circuit and Systems, CAS-28 (1981)

[28] B. Wyman and M. Sain - Exact sequences for pole zero cancellation, Proc. MTNS 1981, Santa Monica ,(1981)

[29] B. Wyman, M. Sain et al. - The total synthesis problem of linear multivariable control,Proc. 20th Joint Autom. Control Conf., (1981)

[30] B. Wyman and M. Sain - Internal zeros and the system matrix, Proc. 20th Allerton Conf., Urbana (1982)

[31] B. Wyman and M. Sain - The zero module of a minimal realization, Linear Algebra and Appl., 50 (1983)

[32] B. Wyman and M. Sain - A unified pole-zero module for linear transfer functions, System and Control Letters, 5 (1984)

[33] B. Wyman, M. Sain, G. Conte and A.M. Perdon - On the zeros and poles of a transfer function, Linear Algebra and Its Application, to appear

[34] Outils et Modèles Mathematiques pour l'Automatique, l'Analyse des Systèmes et le Traitement du Signal, Ed. du C.N.R.S., Vol. 1 (1981)

Representations of Infinite-Dimensional Systems

R. F. Curtain

Mathematics Institute, University of Groningen
P. O. Box 800, 9700 AV Groningen, the Netherlands

This article is dedicated to Margherita Willems.

1. Introduction

The aim of this article is to clarify the relationships between the concept of an infinite-dimensional dynamical system as proposed in Kalman, Falb and Arbib [15], and the more recent well-posedness theories for controlled delay and partial differential equations in the literature. Despite the three decades of research in the area of infinite-dimensional sytems, it is only recently that this question has been completely resolved in an elegant manner in Salamon [28]-[29], Weiss [33]-[36], and Curtain and Weiss [7]. Earlier research tended to concentrate either exclusively on various differential state space formulations and questions of well-posedness of p.d.e.'s or delay equations (Lions [22], Curtain and Pritchard [3], Lasiecka [16], Lasiecka and Triggiani [17]-[21], Ho and Russell [14], Lions and Magenes [23], Washburn [31], Salamon [27] to mention a representative sample) or on abstract realization theories for transfer functions (Helton [12], Yamamoto [37], Fuhrman [11], Baras and Brockett [1]-[2]) and the twain took a long time to meet. The importance of a unified theory embracing both state space and frequency domain aspects of control synthesis and design for infinite-dimensional systems has been pointed out in Curtain [6], where special emphasis was laid on special classes of systems for which there already exists a substantial theory. Here we discuss a much larger class of infinite-dimensional linear systems.

There are many ways of representing linear infinite-dimensional systems, for example:

(*i*) directly in terms of p.d.e.'s or differential delay equations;

(*ii*) in terms of a quadruple (A,B,C,D) of abstract operators on a Banach (or Hilbert) space;

(*iii*) as a frequency domain relationship between inputs and outputs;

(*iv*) as a dynamical system in the sense of Kalman (an abstract linear system).

It is perhaps useful to think of (*i*) as a differential representation, of (*ii*) as a semigroup representation (since A is usually required to generate a C_0-semigroup on some Banach space) and of (*iv*) as an integral representation. (Following Weiss [33]-[36] we shall use the term "abstract linear system" for (*iv*).) These are all state space representations, whereas (*iii*) is a frequency domain representation. Each has its own mertis and uses and it is not the purpose of this article to survey, much less compare, all aspects of these different representations. Much research has been done on obtaining semigroup representations from differential representations (see Salamon [27]-[28] and the references therein) and on obtaining semigroup realizations from the transfer functions description (see Fuhrmann [11], Yamamoto [37] and the references therein). However, the question whether the differential representation (*i*) or the semigroup representation (*ii*) have a well-defined transfer function was not investigated systematically until fairly recently in Salamon [28]. Even more recently the link between the intergral representation (*iv*) and the representations (*ii*) and (*iii*) was made in the series of papers by Salamon [29], Weiss [33]-[36] and Curtain and Weiss [7], and it is this aspect we shall discuss in this article. Using the concept of an abstract linear system as the fundamental representation of an infinite-dimensional linear system, one obtains a very elegant theory which clarifies the relationships with the other three representations.

Let us illustrate the above ideas with the well-known finite-dimensional linear system whose differential representation is given by

$$x(t) = Ax(t) + Bu(t) \tag{1.1}$$

$$y(t) = Cx(t) + Du(t) \tag{1.2}$$

where $x(t) \in X = \mathbb{R}^n$, the state space, $u(t) \in U = \mathbb{R}^m$, the input space and $y(t) \in Y = \mathbb{R}^k$, the output space and (A,B,C,D) is a quadruple of matrices of matching dimensions, which coincides with the abstract operation representation (*ii*). An important property of the differential representation (*i*) is that $x(t)$ is the unique continuously differentiable solution of the differential equation (1.1) for any given initial state $x(0) = x_0 \in X$. The frequency domain representation (*iii*) is in terms of the transfer function

$$G(s) = C(sI - A)^{-1}B \tag{1.3}$$

The corrresponding abstract linear system representation is given by $\Sigma = (\mathbb{T}, \Phi, \mathbb{L}, \mathbb{F})$ defined by

$$\mathbb{T}_t = e^{At} \tag{1.4}$$

$$\Phi_t u = \int_0^t e^{A(t-\sigma)} Bu(\sigma)d\sigma \tag{1.5}$$

$$(\mathbb{L}_\tau x)(t) = \begin{cases} Ce^{At}x & ; t \in [0,\tau) \\ 0 & ; t \geq \tau \end{cases} \tag{1.6}$$

$$(\mathbb{F}_\tau u)(t) = \begin{cases} C\int_0^t e^{A(t-\sigma)}Bu(\sigma)d\sigma + Du(t) & ; t \in [0,\tau) \\ 0 & ; t \geq \tau \end{cases} \tag{1.7}$$

which explains why we described it as an integral representation. The relationship between representations (*i*) and (*ii*) and the implication of (*ii*) to (*iii*) are trivial and the implication of (*iii*) to (*ii*) is the subject of realization theory which does not concern us here. In order to explain the relationships between the abstract linear system representation and the other three we first need to define this concept. We follow Weiss [33]–[36].

An essential role is played by the composition properties under the *concatenation* of inputs, u and v, which is defined for any $\tau \geq 0$ by

$$(u \underset{\tau}{\diamond} v)(t) = \begin{cases} u(t) & \text{for } t \in [0,\tau) \\ v(t-\tau) & \text{for } t \geq \tau \end{cases} \tag{1.8}$$

for any $u,v \in L_2^{loc}(0,\infty;\, U)$.[1] We shall also use the *projection operators* P_τ from $L_2^{loc}(0,\infty;\, U)$ onto $L_2(0,\tau;\, U)$ which are defined for any $\tau \geq 0$ by

$$(P_\tau u)(t) = \begin{cases} u(t) & \text{for } t \in [0,\tau) \\ 0 & \text{for } t \geq \tau \end{cases} \tag{1.9}$$

Definition 1.1 $\Sigma = (\mathbb{T},\Phi,\mathbb{L},\mathbb{F})$ is a *abstract linear system* on the Hilbert spaces U, X and Y, if

(*i*) $\mathbb{T} = (\mathbb{T}_\tau)_{\tau\geq 0}$ is a strongly continuous semigroup of linear operators on X, i.e., $\mathbb{T}_t x$ is continuous in t for all $x \in X$,

$$\mathbb{T}_t \mathbb{T}_\tau = \mathbb{T}_{t+\tau} \tag{1.10}$$

for any $t,\tau \geq 0$ and $\mathbb{T}_0 = I$ (the identity).

(*ii*) $\Phi = (\Phi_\tau)_{\tau\geq 0}$ is a family of bounded linear operators from $L_2(0,\infty;\, U)$ to X such that

$$\Phi_{t+\tau}(u \underset{\tau}{\diamond} v) = \mathbb{T}_t \Phi_\tau u + \Phi_t v \tag{1.11}$$

for any $u,v \in L_2(0,\infty;\, U)$ and any τ, $t \geq 0$.

(*iii*) $\mathbb{L} = (\mathbb{L}_\tau)_{\tau\geq 0}$ is a family of bounded linear operators from X to $L_2(0,\infty;\, Y)$ such that

$$\mathbb{L}_{\tau+t} x = \mathbb{L}_\tau x \underset{\tau}{\diamond} \mathbb{L}_t \mathbb{T}_\tau x \tag{1.12}$$

for any $x \in X$ and any $t,\tau \geq 0$, and $\mathbb{L}_0 = 0$.

(*iv*) $\mathbb{F} = (\mathbb{F}_t)_{t\geq 0}$ is a family of bounded linear operators from $L_2(0,\infty;\, U)$ to $L_2(0,\infty;\, Y)$ such that

---

[1] $L_2^{loc}(0,\infty;\, U)$ denotes the space of functions f on $(0,\infty)$ such that for every $T>0$ the restriction of f to $(0,T)$ belongs to $L_2(0,T;\, U)$.

$$\mathbb{F}_{\tau+t}(u \underset{\tau}{\diamond} v) = \mathbb{F}_\tau u \underset{\tau}{\diamond} (\mathbb{L}_t \Phi_\tau u + \mathbb{F}_t v) \tag{1.13}$$

for any $u,v \in L_2(0,\infty; U)$ and any $t,\tau \geq 0$ and $\mathbb{F}_0 = 0$.

U is called the *input space*, X the *state space* and Y the *output space*. The operators Φ_t are called *reachability maps*, $\mathbb{L}_t$ are called *observability maps* and $\mathbb{F}_t$ are called *input–output maps*. The conditions (1.4), (1.5) and (1.6) are called *composition* properties.

Notice that this definition implies the *causality* of the reachability and input–output maps, i.e., for any $t \geq 0$

$$\Phi_t P_t = \Phi_t; \quad \mathbb{F}_t P_t = \mathbb{F}_t. \tag{1.14}$$

In addition there holds

$$P_t \mathbb{L}_\tau = \mathbb{L}_t; \quad P_t \mathbb{F}_\tau = \mathbb{F}_t \tag{1.15}$$

for any $0 \leq t \leq \tau$.

Both $\mathbb{L}_\tau$ and $\mathbb{F}_\tau$ have a limit as $\tau \to \infty$, where the convergence is in $L_2(0,\infty; Y)$ if $\mathbb{T}_t$ is exponentially stable or otherwise in the Fréchet space $L_2^{loc}(0,\infty; Y)$:

$$\mathbb{L}_\infty x = \lim_{\tau \to \infty} \mathbb{L}_\tau x; \qquad \mathbb{F}_\infty v = \lim_{\tau \to \infty} \mathbb{F}_\tau v \tag{1.16}$$

for any $x \in X$ and any $v \in L_2(0,\infty; U)$. In fact, $\mathbb{F}_\tau$ and $\mathbb{L}_\tau$ are defined uniquely by $\mathbb{F}_\infty$ and $\mathbb{L}_\infty$ respectively through (1.15), if we take $\tau = \infty$ there.

Returning to our finite-dimensional linear system we see that it is readily verified that (A,B,C,D) defines an abstract linear system $\Sigma = (\mathbb{T},\Phi,\mathbb{L},\mathbb{F})$ via (1.4)–(1.7) and conversely that Σ on the finite- dimensional spaces U, X and Y uniquely determines matrices $A \in \mathcal{L}(X)$, $B \in \mathcal{L}(U,X)$, $C \in \mathcal{L}(X,Y)$ and $D \in \mathcal{L}(U,Y)$ satisfying (1.4)–(1.7). In other words, the integral representation (*iv*) of an abstract linear system is equivalent with the abstract operator representation (*ii*) and this is equivalent with the differential representation (*i*).

Moreover, the transfer function given by (1.3) and the input–output map given by (1.7) are isomorphic using shifted Laplace transforms. This example has motivated much of the research into representations and well-posedness studies for infinite-dimensional linear systems, including the theme of this article, which uses the concept of an abstract linear system as the central representation.

We show that given an abstract linear system Σ on Hilbert spaces U, X and Y there exists a unique triple (A,B_a,C_a) of operators satisfying the equations (1.4)–(1.6) in a suitably generalized setting. Conversely we give sufficient conditions for a triple (A,B,C) of operators to generate a family of abstract linear systems. In general a triple (A,B,C) will not determine a unique transfer function (or input–output map), and although a generalization of the input–output formula (1.7) exists, it is rather complicated and depends on the transfer function. Similarly, a generalization of the differential representation (1.1)–(1.2) exists, but it also depends on the transfer function in general. Surprisingly, most abstract linear systems satisfy a regularity property which leads to great simplifications in the theory. These regular abstract linear systems allow for a natural, simple generalization of all the formulas (1.1)–(1.7). Moreover $x(t)$ has the interpretation as a strong continuous solution of differential equation of the form (1.1) and (1.2) holds. These results and much more can be found in Salamon [29], Curtain and Weiss [7] and Weiss [33]–[36]. Salamon [29] and Curtain and Weiss [7] consider the case where X, Y and U are Hilbert spaces and the input and output trajectories are locally L_2 and for simplicity this is the case treated here. In [33]–[36] Weiss extended this to the more general situation where X, Y and U are Banach spaces and the input and output spaces are locally L_p for arbitrary $p \in [1,\infty)$. Moreover the results of Weiss contain a much deeper analysis of the observation operator and related aspects of linear infinite-dimensional linear systems, which we shall not go into here.

Following this extended introduction, in section 2 we discuss the relationship between the usual well-posedness concepts for controlled p.d.e.'s and delay equations and the existence of a reachability map Φ_t satisfying (1.11). Since the concept of "well-posedness" is usually associated with p.d.e.'s and differential systems we prefer to use the term "admissible control operator" here which we show is compatible with the concept of an abstract linear system. In section 3 we examine the dual concept of an "admissible observation operator" and the existence of an observability map $\mathbb{L}_t$

satisfying (1.12). We show that the concept of an admissible observation operator is compatible with that of an abstract linear system. In addition, the key concept of the Lebesgue extension of the abstract observation operator is introduced. The representation of an abstract linear system in its entirety is then treated in section 4, where it is shown that a somewhat complicated generalization of the representation (1.1)–(1.7) is always possible, albeit using the transfer function of the system. Moreover, the relationships between the four different types of representations (differential, abstract operator, frequency domain and that of an abstract linear system) are clarified. Abstract linear systems define a unique transfer function and they correspond to a certain class of triples (A,B,C) which we call "input–output admissible" and a differential representation is always possible. A surprising and elegant result is that for regular abstract linear systems (and most are) a representation of the same simple form as (1.3)–(1.7) is possible in which the Lebesgue extension observation operator plays a crucial role. Moreover, the abstract state space equations (1.1)–(1.2), have a pointwise interpretation for almost all t with $x(t)$ as the unique strong continuous solution of (1.1). The concepts are illustrated by a detailed example in section 5.

2. Admissible control operators

The motivation for introducing the concept of admissibility of the control operator B stems from the need to make sense of the abstract differential equation

$$\dot{x}(t) = Ax(t)+Bu(t) \tag{2.1}$$

where $x(t)$ denotes the state of the system at time t, $x(t)\in X$, a Hilbert space (the state space), and A is the infinitesimal generator of the C_0–semigroup $\mathbb{T}=\{\mathbb{T}_\tau\}_{\tau\geq 0}$ on X, $u(t)$ is the input which takes values in the Hilbert space U (the input space) and B is a linear map on U. If B is a *bounded* operator, $B\in\mathcal{L}(U,X)$, then it is well-known, (Curtain and Pritchard [3]), that for any $u\in L_2(0,\tau;U)$ the following variation-of-parameters formula provides a suitable interpretation as the continuous solution of (2.1) on $[0,\tau]$

$$x(t) = \mathbf{T}_t x(0) + \int_0^t \mathbf{T}_{t-\sigma} B u(\sigma) d\sigma. \tag{2.2}$$

However, in many applications B does not map into X, but into a larger space, $V \supset X$, and (2.2) may not be well-defined. It is customary to call such control maps *unbounded*, although this does not agree with the meaning of unbounded in functional analysis. Unbounded control operators arise naturally in systems described by linear p.d.e.'s with boundary control action and in differential-delay systems with delayed control action. There exists a vast literature on the question as to the appropriate mathematical formulation of such systems and on sufficient conditions under which (2.2) will define a solution in some appropriate sense, for example Curtain and Pritchard [3], chapter 8, Curtain and Salamon [4], Desch, Lasiecka and Schappacher [8], Ho and Russell [14], Lasiecka [16], Lasiecka and Triggiani [17]–[20], Lions and Magenes [23], Russell [25], Salamon [27]–[28], Washburn [31]. The concept of a "solution" of an abstract differential equation (2.1) is an interesting topic in its own right and it depends very much on the mathematical formulation one chooses. Usually a p.d.e. formulation enables one to obtain the strongest regularity properties and so the strongest concept of solution (see for example Lasiecka and Triggiani [17]–]21]). Later on in this section we shall be able to say more about the interpretation of (2.2) as a solution to (2.1) in an abstract sense. In the natural formulation of p.d.e. systems and delay systems one is not presented with an (A,B) pair and Hilbert spaces X and V; a crucial step lies in formulating the (A,B) description for appropriate Hilbert spaces X and V. We shall not treat this aspect here, but refer the reader to Salamon [27] and [28] for one methodology of obtaining an abstract (A,B) formulation for suitable X, V spaces for delay and p.d.e. systems. The Salamon methodology may be seen as a synthesis and extension of existing techniques aimed at obtaining an abstract semigroup formulation for linear infinite-dimensional systems. In particular, this methodology coupled with the Carleson measure test in Ho and Russell [14] and Weiss [32] is very useful for establishing the well-posedness of p.d.e. systems generated by a spectral operator and having finite-rank inputs and outputs. We should at the same time point out that for many control applications it is not necessary to use an (A,B) or abstract semigroup formulation. However, in order to clarify the relationships between well-posed p.d.e. and delay systems and abstract linear systems it is a necessary first step.

At this stage we are only concerned with making sense of (2.1) for unbounded control operators. It is clearly necessary that the integral term in (2.2) have values in X for any input $u \in L_2(0,\tau; U)$, and this in turn requires that $\mathbb{T}_t$ have a continuous extension on V. This motivates the following defintion of an admissible control operator.

Definition 2.1 Suppose that A generates the C_0-semigroup $\mathbb{T} = (\mathbb{T}_\tau)_{\tau \geq 0}$ on the Hilbert space X and that $B \in \mathcal{L}(U,V)$ where U and V are Hilbert spaces. If X is dense in V and $\mathbb{T}$ has a continous extension to V, then we call B an *admissible control operator for* $\mathbb{T}$ *with respect to* V if the *reachability maps* $(\Phi_\tau)_{\tau \geq 0}$ are bounded from $L_2(0,\infty; U)$ to X for all finite $\tau \geq 0$, where

$$\Phi_t u := \int_0^t \mathbb{T}_{t-\sigma} B u(\sigma) d\sigma \tag{2.3}$$

The terminology "admissible control operator" comes from Weiss [33] and it is equivalent to the assumption S2 of Salamon [27], [28] and in some sense to the many other concepts of "well-posedness" which abound in the literature. Our concern is not their respective merits, but the type of (A,B)-pairs which lead to well-defined abstract linear systems in the sense of Definition 1.1. It is readily verified that if B is an admissible control operator for $\mathbb{T}$ with respect to V according to Definition 2.1, then the pair $(\mathbb{T},\Phi)$ satisfies the assumptions in Definition 1.1; we shall say it generates an abstract linear control system on X and U.

Definition 2.2 *An abstract linear control system* on the Hilbert spaces X and U is a pair $(\mathbb{T},\Phi)$, where $\mathbb{T} = (\mathbb{T}_\tau)_{\tau \geq 0}$ is a C_0-semigroup on X and $\Phi = (\Phi_\tau)_{\tau \geq 0}$ is a family of bounded operators from $L_2(0,\infty; U)$ to X such that the composition property (1.11) holds.

Clearly an abstract linear control system is a subset of Σ of Definition 1.1. Not only is $(\mathbb{T},\Phi)$ causal (see (1.14)), but $\Phi = (\Phi_\tau)_{\tau \geq 0}$ is a strongly continuous family of operators and the function $\Phi(t,u) := \Phi_t u$ is continuous on $[0,\infty) \times L_2(0,\infty; U)$, (Weiss, [33, Prop. 2.3]).

The question we are interested in here is whether an abstract linear control system determines an admissible control operator B. This was shown to be true in Salamon [29] for the $p=2$ and Hilbert space case and in Weiss [33] for the Banach space case and any $p\in[1,\infty)$. A related representation result for p.d.e. systems with boundary control also appeared in Desch, Lasiecka and Schappacher [8]. The main result in Weiss [33] is that any abstract linear control system has a representation $(\mathbb{T},\Phi)$ with Φ given by the form (2.3) for a unique admissible control operator $B_a\in\mathcal{L}(U,X_{-1})$, where X_{-1} is the completion of X induced by the following norm:

$$\|x\|_{-1} = \|(\beta I-A)^{-1}x\| \tag{2.4}$$

for a $\beta\in\rho(A)$.

Notice that (2.4) defines equivalent norms for different β's in $\rho(A)$ and so X_{-1} is independent of β and X is automatically dense in X_{-1}. Furthermore, $\mathbb{T}$ has a unique continuous extension to X_{-1} which coincides with its image via the isomorphism $(\beta I-A)$ from X to X_{-1}. In view of the complex, technical literature on well-posedness of controlled p.d.e. and delay equations, this is a surprisingly simple and elegant result.

Theorem 2.3 *Let X and U be Hilbert spaces and suppose that $(\mathbb{T},\Phi)$ is an abstract linear control system on X and U. Then there exists a unique operator $B_a\in\mathcal{L}(U,X_{-1})$ such that for any $t\geq 0$, Φ_t has the representation*

$$\Phi_t u = \int_0^t \mathbb{T}_{t-\sigma}\, B_a u(\sigma)d\sigma \tag{2.5}$$

for any $u\in L_2(0,\infty;U)$ and B_a is an admissible control operator for $\mathbb{T}$ with respect to X_{-1}.

Moreover, for any $x_0\in X$ and $u\in L_2(0,\infty;U)$ the function defined by

$$x(t) = \mathbb{T}_t x_0 + \Phi_t u \tag{2.6}$$

is the unique strong continuous solution of (2.1) with $B=B_a$ and the initial condition $x(0)=x_0$.

By the concept "strong continuous solution" is meant the following one.

Definition 2.4 We say that the function $x(\cdot)$ is a *strong continuous solution* of (2.1) on $[0,\tau]$ if $x(.)\in C(0,\tau;\,X)$ and for any $t\in[0,\tau]$ the following holds

$$x(t)-x(0) = \int_0^t [Ax(s)+Bu(s)]ds. \tag{2.7}$$

In Salamon [28], Lemma 2.5, it was shown that $x(t)$ given by (2.6) is the unique strong continuous solution of (2.1) with the original admissible control operator B. Although B and B_a generate the same reachability map Φ, they need not be equal. Sufficient conditions for $B=B_a$ are that B be an admissible control operator for $\mathbb{T}$ with respect to V as in Definition 2.1 and A have a continuous extension to an operator in $\mathcal{L}(X,V)$. For then $B\in\mathcal{L}(U,X_{-1})$ and since B_a is unique, $B=B_a$ (Weiss [33], Remark 3.14). An important special case is when B is a bounded operator. To avoid any confusion we introduce the following definition and notation.

Definition 2.5 Given the abstract linear control system $(\mathbb{T},\Phi)$ on the Hilbert spaces X and U we call the unique operator $B_a\in\mathcal{L}(U,X_{-1})$ of Theorem 2.3 the *abstract control operator*. It is defined by

$$B_a v = \lim_{\tau\to 0}\frac{1}{\tau}(\Phi_\tau v). \tag{2.8}$$

In the light of Theorem 2.4 it is natural to ask whether any $B\in\mathcal{L}(U,X_{-1})$ and C_0-semigroup $\mathbb{T}$ on X define an abstract control system $(\mathbb{T},\Phi)$ via (2.7). This conjecture is false. In fact if $B\in\mathcal{L}(U,X_{-1})$ does not define an abstract control system $(\mathbb{T},\Phi)$ (i.e. Φ_t defined by (2.6) is not in $\mathcal{L}(L_2(0,\infty;\,U),\,X)$ for all t), then (as shown in Weiss [33]) the state trajectory given by (2.6) can always be driven out of the state space X by some input $u\in L_2(0,\tau;\,U)\cap C^\infty([0,\tau),\,U)$ on an arbitrarily short time-interval. Consequently Theorem 2.3 has primarily theoretical value and does not relieve us of the often considerable burden of verifying that B is an admissible control operator for $\mathbb{T}$ with respect to V in p.d.e. and delay examples.

3. Admissible observation operators

The motivation for introducing the concept of admissible observation operators arises from the need to formulate the following type of observability map

$$y(t) = C\mathbb{T}_t x \qquad ; t \geq 0 \tag{3.1}$$

where $\mathbb{T} = (\mathbb{T}_\tau)_{\tau \geq 0}$ is a C_0-semigroup on the Hilbert space X (the state space), $x \in X$, and $C: W \to Y$ is a linear operator from a subspace W of X to another Hilbert space Y (the output space). If C is bounded on X then (3.1) is defined for all $x \in X$ and for all $t \geq 0$, but in many applications C is only defined on a proper subspace of X and so may be unbounded. Unbounded operators occur when one models boundary or point observations for systems described by linear p.d.e.'s or when one has delayed observations in delay equations. There is an extensive literature dealing with systems having unbounded observations operators, for example Curtain and Pritchard [3], Dolecki and Russell [9], Fuhrmann [11], Lasiecka and Triggiani [18], Pritchard and Wirth [24], Salamon [27], [28], Seidman [30], Yamamoto [37]. The usual interpretation of (3.1) is as a function from W to $L_2^{loc}(0,\infty; Y)$ and we follow the formulation used in Weiss [34].

Definition 3.1 Suppose $\mathbb{T} = (\mathbb{T}_\tau)_{\tau \geq 0}$ is a C_0-semigroup with infinitesimal generator A on the Hilbert space X and that W is a dense $\mathbb{T}$-invariant subspace of X. Then C is an *admissible observation operator for* $\mathbb{T}$ *with respect to* W if for some (and hence any) $\tau > 0$, the *observability map* $\mathbb{L}_\tau$ has a continuous extension to a bounded map from X to $L_2(0,\infty; Y)$, where $\mathbb{L}_\tau$ is defined for $x \in W$ by

$$(\mathbb{L}_\tau x)(t) = \begin{cases} C\mathbb{T}_t x; & t \in [0,\tau) \\ 0 & t \geq \tau. \end{cases} \tag{3.2}$$

Notice that the reachability and observability maps Φ_τ and $\mathbb{L}_\tau$ are duals of each other with respect to the pairing: $<u,y> = \int_0^T <u(T-t), y(t)>_{U,U^*} dt.$ Consequently B is an admissible control operator for $\mathbb{T}_t$ with respect to X_{-1} if and only if B^* is an admissible observation operator for $\mathbf{T}_t^*$ with respect to $(X^*)_1$, where X_{-1} is defined by (2.4) and $(X^*)_1$ is the domain of A^*, $D(A^*)$, with the graph norm. This duality was used extensively by Salamon [27], [28] to deduce results about admissible observation operators from dual admissible control operators, where he used the terminology "H1 and H2 hypotheses" and slightly different, but equivalent definitions. More general duality results can be found in Weiss [34]. First we examine the relationship of the concept of an admissible observation operator to that of an abstract linear observation system.

Definition 3.2 Let X and Y be Hilbert spaces. An ***abstract linear observation system*** on Y and X is a pair $(\mathbf{L},\mathbf{T})$ where $\mathbf{T}=(\mathbf{T}_\tau)_{\tau\geq 0}$ is a C_0-semigroup on X and $\mathbf{L}=(\mathbf{L}_\tau)_{\tau\geq 0}$ is a family of bounded maps from X to $L_2(0,\infty;Y)$ such that the composition property (1.12) holds and $\mathbf{L}_0=0$.

It is easy to verify that if C is an admissible observation operator for $\mathbf{T}_t$ with respect to W as in Definition 3.1, then it defines an abstract linear observation system with $\mathbf{L}_t$ defined by (3.2).

The following representation theorem for abstract linear observation systems is contained in Salamon [29] in a slightly different form; the given version follows Weiss [34] where he proved it for the more general situation in which X and Y are Banach spaces and the output functions are in $L_p^{loc}(0,\infty;Y)$, where $p\in[1,\infty]$. Let us denote by X_1 the space $D(A)$ with the graph norm.

Theorem 3.3 *Let X and Y be Hilbert spaces and suppose that $(\mathbf{L},\mathbf{T})$ is an abstract linear observation system on Y and X. Then there exists a unique $C_a\in\mathcal{L}(X_1,Y)$ such that for any $x\in X_1$ and any $t\geq 0$*

$$(\mathbf{L}_\infty x)(t)=C_a\mathbf{T}_t x. \tag{3.3}$$

Since C_a will in general be different from an original admissible observation operator C, we introduce the following definition.

Definition 3.4 Given the abstract linear observation system $(\mathbf{L},\mathbf{T})$ on the

Hilbert spaces X and Y, we call the unique operator $C_a \in \mathcal{L}(X_1,Y)$ of Theorem 3.3 the *abstract observation operator*. It is defined by

$$C_a x = (\mathbb{L}_\infty x)(0) \qquad \text{for } x \in D(A). \tag{3.4}$$

Clearly C_a is an admissible observation operator for $\mathbb{T}_t$ with respect to X_1 according to Definition 3.1.

So far the results on admissible observation operators and abstract linear observation systems parallel those for admissible control operators and abstract linear control systems and the duality between Theorems 2.3 and 3.3 is obvious.

The discussion which now follows has no counterpart for abstract linear control systems and it is a special feature of abstract linear observation systems, which was only recently discovered by Weiss [34]. It holds more generally for X and Y Banach spaces and with $L_p^{loc}(0,\infty; Y)$ as output trajectory space for $p \in [1,\infty]$.

Definition 3.5 Let X and Y be Hilbert spaces, $\mathbb{T}$ a C_0-semigroup on X and suppose that $C \in \mathcal{L}(X_1,Y)$. Then the *Lebesgue extension* of C (with respect to $\mathbb{T}$), C_L: $D(C_L) \to Y$ is defined by

$$C_L x = \lim_{T \to 0} C \frac{1}{T} \int_0^T \mathbb{T}_\sigma x d\sigma \tag{3.5}$$

$$D(C_L) = \{x \in X \mid \text{the limit in (3.5) exists}\}.$$

Weiss showed that C_L is an extension of C in the sense that

$$X_1 \hookrightarrow D(C_L) \hookrightarrow X. \tag{3.6}$$

Since C is typically not closed, $D(C_L)$ will not be complete under the graph norm, but it does become a Banach space under the following norm

$$\|x\|_C = \|x\| + \sup_{\tau\in(0,1]} \|C\frac{1}{\tau}\int_0^{\tau} \mathbb{T}_\sigma x d\sigma\|. \tag{3.7}$$

With respect to this norm $C_L \in \mathcal{L}(D(C_L), Y)$ and the embeddings in (3.6) are continuous.

The significance of the Lebesgue extension, C_L, is that it makes it possible to give a simple *pointwise* interpretation of the observation map (3.1) for *every* x *in the original state space*, X and almost all $t \geq 0$, whereas Theorem 3.3 only applies to x in the smaller space $D(A)$.

Theorem 3.6 *Under the same assumptions as in Theorem 3.3, if C_L is the Lebesgue extension of C_a, then for any $x \in X$, $t \geq 0$ we have that $\mathbb{T}_t x \in D(C_L)$ if and only if $\mathbb{L}_\infty x$ has a Lebesgue point in t. Furthermore,*

$$(\mathbb{L}_\infty x)(t) = C_L \mathbb{T}_t x \tag{3.8}$$

almost everywhere in $[0,\infty)$.[2]

The above is a surprisingly simple, elegant representation for the observability map which one could never have anticipated by examining the many examples of well-posed observation maps which abound in the literature. In the next section we shall use this Lebesgue extension to obtain a simple representation of the transfer function of the input-output map for regular abstract linear systems.

4. Input-output admissibility

In section 2 we considered admissible control operators and abstract linear *control* systems and in section 3 we considered admissible observation operators and abstract linear *observation* systems. Here we consider the full

---

[2]We recall that $y \in L_2^{loc}(0,\infty; Y)$ has a Lebesgue point in t if the limit of $\frac{1}{\tau}\int_t^{t+\tau} y(\sigma)d\sigma$ exists as $\tau \to 0$. Almost every $t \geq 0$ is a Lebesgue point for y and the limit equals $y(t)$ a.e.

abstract linear system $\Sigma=(\mathbb{T},\Phi,\mathbb{L},\mathbb{F})$ of Definition 1.1.

An obvious example of an abstract linear system on the Hilbert spaces U, X and Y can be obtained analogously to the finite-dimensional case (cf. (1.10)–(1.15)) by considering the following state space system:

$$\dot{x}(t) = Ax(t)+Bu(t) \tag{4.1}$$

$$y(t) = Cx(t)+Du(t) \tag{4.2}$$

where A generates the C_0-semigroup $\mathbb{T}$ on X and $B\in\mathcal{L}(U,X)$, $C\in\mathcal{L}(X,Y)$, $D\in\mathcal{L}(U,Y)$. This defines an abstract linear system $\Sigma=(\mathbb{T},\Phi,\mathbb{L},\mathbb{F})$, where the reachability and observability maps, Φ and $\mathbb{L}$, are given by (2.4) and (3.2) respectively and the input-output map $\mathbb{F}$ is given by

$$(\mathbb{F}_\tau u)(t) = \begin{cases} C\int_0^t \mathbb{T}_{t-\sigma}Bu(\sigma)d\sigma+Du(t) & ;\ 0\le t<\tau \\ 0 & ;\ t\ge\tau \end{cases} \tag{4.3}$$

When one tries to obtain an expression like (4.3) for the input-output map $\mathbb{F}$ for unbounded, but admissible control and observation operators B and C, one is confronted with the problem that $\int_0^t \mathbb{T}_{t-\sigma}Bu(\sigma)d\sigma$ will not be in $D(C)$ in general. In fact having admissible control and observation operators B and C is not sufficient to guarantee the existence of a bounded input-output map. This was realized in Salamon [28], where he imposed a time-domain condition for a "well-posed" system. This was replaced by a more easily verifiable frequency-domain condition in Curtain [5] and Salamon [29], and the relationship to the concept of an abstract linear system Σ was clarified in Curtain and Weiss [7], where they used the terminology "well-posedness of the triple (A,B,C) (in the sense of linear systems theory)". As remarked earlier in the introduction we feel that "well-posedness" is already used in connection with differential equations and so here we prefer to use the terminology "input-output admissibility of the triple (A,B,C)". The extra condition needed for input-output admissibility of a triple (A,B,C) is expressed in terms of its transfer functions.

Definition 4.1 Let U, X, Y, V and W be Hilbert spaces such that $W\subset X\subset V$ and let $B\in\mathcal{L}(U,V)$, $C\in\mathcal{L}(W,Y)$ and $\mathbb{T}=(\mathbb{T}_\tau)_{\tau\ge0}$ be a C_0-semigroup on X. Suppose that B is an admissible control operator for $\mathbb{T}_t$ with respect to V and that C is an admissible observation operator for $\mathbb{T}_t$ with respect to W. Then we define the *transfer functions* of the triple (A,B,C) to be the solutions, G: $\rho(A)\to\mathcal{L}(U,Y)$ of

$$\frac{G(s)-G(\beta)}{s-\beta} = -C(sI-A)^{-1}(\beta I-A)^{-1}B \tag{4.4}$$

for $s,\beta\in\rho(A)$, $s\neq\beta$.

We remark that since B is an admissible control operator with respect to $\mathbb{T}_t$, $(\beta I-A)^{-1}B$ is an $\mathcal{L}(U,X)$-valued analytic function and since C is an admissible observation operator with respect to $\mathbb{T}_t$, $C(sI-A)^{-1}$ is a $\mathcal{L}(X,Y)$-valued analytic function. Both $(\beta I-A)^{-1}B$ and $C(sI-A)^{-1}$ are analytic on some right half-plane $\mathbb{C}_\alpha=\{s\in\mathbb{C}: \operatorname{Re}s>0\}$. Consequently the transfer functions always exist as $\mathcal{L}(U,Y)$-valued functions which are analytic in some right half-plane, $\mathbb{C}_\alpha$. They differ only by an additive constant, $D\in\mathcal{L}(U,Y)$. The point is that they need not necessarily be bounded on any right half-plane, $\mathbb{C}_\alpha$. We impose this as an extra assumption on the triple (A,B,C) and call this input-output admissibility. Notice that as a consequence of Theorems 2.3 and 3.3, C and B in (4.4) may be replaced by the abstract observation and control operators C_a and B_a respectively.

Definition 4.2 Under the same assumptions as in Definition 4.1, we say that the triple (A,B,C) is *input-output admissible* if B is an admissible control operator for $\mathbb{T}$ with respect to the Hilbert space V, C is an admissible observation operator for $\mathbb{T}$ with respect to the Hilbert space W and its transfer functions are bounded on some half-plane $\mathbb{C}_\alpha$.

The main result in Curtain and Weiss [7] is that a triple (A,B,C) which is input-output admissible corresponds to the notion of an abstract linear system.

Theorem 4.3 *(i) An input-output admissible triple (A,B,C) determines a family of abstract linear systems $\Sigma=(\mathbb{T},\Phi,\mathbb{L},\mathbb{F})$ where $\mathbb{T}$ is the C_0-semigroup generated* *by A and Φ and $\mathbb{L}$ are the reachability and observability maps defined by* (2.3) *and* (3.2) *respectively. The family of input–output maps,* $\mathbb{F}$, *is defined by the equation*

$$(\mathbb{F}_\infty u)(t) = C_a[\int_0^t \mathbb{T}_{t-\sigma}B_a u(\sigma)d\sigma - (\beta I - A)^{-1}B_a u(t)] + G(\beta)u(t) \tag{4.5}$$

for $u \in W_{0,loc}^{1,2}(0,\infty;\, U)$[3] *and* $\mathbb{F}_t$ *is defined by* (1.8) *with* $\tau = \infty$.

(*ii*) *An abstract linear system* $\Sigma = (\mathbb{T},\Phi,\mathbb{L},\mathbb{F})$ *determines the unique input–output admissible triple* (A,B_a,C_a) *where* C_a *and* B_a *are the abstract observation and control operators associated with* $(\mathbb{T},\Phi)$ *and* $(\mathbb{L},\mathbb{T})$ *respectively.*

We emphasize that an input–output admissible triple (A,B,C) only determines G and hence $\mathbb{F}_\infty$ up to an arbitrary additive constant $D \in \mathcal{L}(U,Y)$. To a given transfer function G corresponds a unique $\mathbb{F}$ defined by (4.5) and conversely to a given $\mathbb{F}$ corresponds exactly one transfer function G. The key equation (4.5), was initially derived in Salamon [29] for smooth inputs and was extended to all $u \in L_2^{loc}(0,\infty;\, U)$ in Weiss [35], provided C_a is replaced by its Lebesgue extension, C_L. While B_a in (4.5) may be replaced by B, (4.5) may not hold for the original C operator (unless the domain of C includes $D(A)$). To understand the main idea of Theorem 4.3 let us consider the case where $\mathbb{T}$ is an exponentially stable C_0–semigroup. Then Theorem 4.3 says essentially that the input–output map $\mathbb{F}_\infty$ is bounded from $L_2(0,\infty;\, U)$ to $L_2(0,\infty;\, Y)$ if and only if its Laplace transform, the transfer function, G, is bounded on some half–plane. (An analagous but more complicated statement holds for the case that $\mathbb{T}$ is not stable.) This seemingly obvious fact is not so straightforward to prove as one might think. Two different proofs are given in Salamon [29] and Curtain and Weiss [7].

So Theorem 4.3 clarifies the relationship between an abstract linear

---

[3] $W_{0,loc}^{1,2}$ denotes the space of absolutely continuous functions whose derivatives are in $L_2^{loc}(0,\infty;\, U)$, and which are zero at $t=0$. It is a dense subspace of $L_2^{loc}(0,\infty;\, U)$.

system $\Sigma=(\mathbb{T},\Phi,\mathbb{L},\mathbb{F})$ on the Hilbert spaces U, X and Y and input-output admissible triples (A,B,C) and we have obtained the sought generalization of the finite-dimensional formulas (1.3) to (1.7). In other words, we have identified the abstract operator representation (*ii*) corresponding to the integral (abstract linear system) representation (*iv*) and at the same time the corresponding frequency domain representation (*iii*) via (4.5). It remains to give the appropriate differential representation (*i*) and the generalization of the finite- dimensional formulas (1.1) and (1.2).

Theorem 4.4 *Let $\Sigma=(\mathbb{T},\Phi,\mathbb{L},\mathbb{F})$ be an abstract linear system with input and output spaces U and Y respectively. Suppose that A is the infinitesimal generator of $\mathbb{T}$, B_a is the abstract control operator, C_a is the abstract observation operator and C_L is its Lebesgue extension. Then for any $\beta\in\rho(A)$, $x_0\in X$ and any $u\in L_2^{loc}(0,\infty;U)$, the functions x: $[0,\infty)\to X$ and $y\in L_2^{loc}(0,\infty;Y)$ defined by*

$$x(t) = \mathbb{T}_t x_0 + \Phi_t u \tag{4.6}$$

$$y = \mathbb{L}_\infty x_0 + \mathbb{F}_\infty u \tag{4.7}$$

satisfy the following equations a.e. in $t\geq 0$

$$\dot{x}(t) = Ax(t)+B_a u(t) \tag{4.8}$$

$$y(t) = C_L[x(t)-(\beta I-A)^{-1}B_a u(t)]+G(\beta)u(t). \tag{4.9}$$

Moreover, $x(t)$ is the unique strong continuous solution of (4.8) under the initial condition $x(0)=x_0$.

So (4.8) and (4.9) are the sought generalizations to the finite-dimensional formulas (1.1) and (1.2) and we have established the appropriate differential representation for an abstract linear system on the Hilbert spaces X, Y and U. Theorems 4.3 and 4.4 show that an input-output admissible triple (A,B,C) corresponds via Σ to the admissible triple (A,B_a,C_L) which satisfies the differential equations (4.8), (4.9) and this motivated the terminology "well-posed" used in Curtain and Weiss [7].

Although we have achieved our goal of clarifying the relationships between the four different types of representations for infinite- dimensional linear systems, the equations (4.5) and (4.9) are not as elegant as we might have wished for. Unlike the corresponding finite-dimensional formulas (1.7) and (1.2), they depend on the parameter β and the transfer function G. Moreover, both the transfer function G and the input-output map $\mathbb{F}$ are only defined up to an *arbitrary additive constant* by (A,B,C). In general this situation is a fact of life to be borne. However, in Weiss [35], a natural class of abstract linear systems was identified which has the property that Σ determines a unique feedthrough operator and for which (4.5) and (4.9) can be simplified considerably.

Definition 4.5 Let Σ be an abstract linear system with input space U and output space Y. We say that Σ is *regular* if for any $u \in U$, the corresponding *step response* has a Lebesgue point at 0. By the step response corresponding to $v \in U$ is meant the function

$$y_v = \mathbb{F}_\infty v \tag{4.10}$$

where in (4.10) v denotes the constant function on $[0,\infty)$ equal to v everywhere.

We remark that the step response defined by (4.10) is the natural generalization of the finite-dimensional concept, except that in the infinite-dimensional case y_v need not be continuous. An important consequence of Σ being regular is that the *feedthrough operator* $D \in \mathcal{L}(U,Y)$ is uniquely defined by the following limit

$$Dv = \lim_{T \downarrow 0} \frac{1}{T} \int_0^T y_v(\sigma) d\sigma. \tag{4.11}$$

Most systems arising in practice are regular and the assumption that Σ be regular eliminates the mathematically pathological cases. Alternative ways of defining regular abstract linear systems are given in the following proposition.

Proposition 4.6 *Suppose that $\Sigma=(\mathbb{T},\Phi,\mathbb{L},\mathbb{F})$ is an abstract linear system with input space U, state space X and output space Y. If A is the infinitesimal generator of $\mathbb{T}$, B_a and C_a are the abstract control and observation operators and C_L is the Lebesgue extension of C_a, then the following conditions are equivalent.*

(*i*) Σ *is regular;*

(*ii*) $C_L(sI-A)^{-1}B_a$ *is an analytic $\mathcal{L}(U,Y)$-valued function of s on $\rho(A)$;*

(*iii*) *for some $s\in\rho(A)$ and any $v\in U$, $(sI-A)^{-1}B_av\in D(C_L)$.*

We remark that Proposition 4.6 provides a way of testing whether an admissible triple (A,B,C) generates regular abstract linear systems using for example the results in Salamon [28]) and it motivates the following defintion.

Definition 4.7 Let (A,B,C) be an input-output admissible triple according to Definition 4.2. Then we call (A,B,C) a *regular admissible triple* if either of conditions (*ii*) or (*iii*) of Proposition 4.5 hold, where B_a and C_a are the abstract control and observation operators and C_L is the Lebesgue extension of C_a.

Regular abstract linear systems have a nice simple representation for the input-output map which is independent of the transfer function.

Theorem 4.8 *Suppose that Σ is a regular abstract linear system with input and output spaces U and Y respectively. If D is its feedthrough operator defined by* (4.11) *and $u\in L_2^{loc}(0,\infty;\,U)$, then for almost all $t\geq 0$*

$$\int_0^t \mathbb{T}_{t-\sigma}B_a u(\sigma)d\sigma\in D(C_L)$$

and

$$(\mathbb{F}_\infty u)(t) = C_L\int_0^t \mathbb{T}_{t-\sigma}B_a u(\sigma)d\sigma+Du(t) \tag{4.12}$$

Conversely, a regular admissible triple (A,B,C) together with a given

$D \in \mathcal{L}(U,Y)$ determines a regular abstract linear system $\Sigma = (\mathbb{T},\Phi,\mathbb{L},\mathbb{F})$ where Φ and $\mathbb{L}$ are defined by (2.3) and (3.3) (or (3.8)) as before and $\mathbb{F}$ is now defined by (4.12). This is now a nice generalization of the finite-dimensional case (1.7).

So here we see the crucial role played by the Lebesgue extension C_L of the abstract observation operator C_a; with C_a we can say no more than (4.5), in which we need the transfer function. Combining all the previous results we arrive at an elegant representation theorem for regular abstract linear systems, which generalizes theorem for systems with bounded operators given by (4.1), (4.2).

Theorem 4.8 *Let $\Sigma = (\mathbb{T},\Phi,\mathbb{L},\mathbb{F})$ be a regular abstract linear system with input and output spaces U and Y respectively. Suppose that A is the infinitesimal generator of $\mathbb{T}$, B_a is the abstract control operator, C_a is the abstract observation operator, C_L is its Lebesgue extension and D is the feedthrough operator of Σ. Then for any $x_0 \in X$ and any $u \in L_2^{loc}(0,\infty; U)$ the functions x: $[0,\infty) \to X$ and $y \in L_2^{loc}(0,\infty; Y)$ defined by*

$$x(t) = \mathbb{T}_t x_0 + \Phi_t u \tag{4.13}$$

$$y = \mathbb{L}_\infty x_0 + \mathbb{F}_\infty u \tag{4.14}$$

satisfy the following equations a.e. in $t \geq 0$

$$\dot{x}(t) = Ax(t) + B_a u(t) \tag{4.15}$$

$$y(t) = C_L x(t) + Du(t). \tag{4.16}$$

In particular, the function x is the unique strong continuous solution of (4.15) under the initial condition $x(0) = x_0$ and $x(t) \in D(C_L)$ a.e. in $t \geq 0$.

Notice that if (A,B,C) is an input-output admissible triple corresponding to Σ, then the trajectory $x(t)$ defined by (4.13) also satisfies (4.15) almost everywhere with B_a replaced by B (Salamon [28], Lemma 2.5). However, C_L in (4.16) may not be replaced by C in general. The correct interpretation of Theorem 4.8 is that regular abstract linear systems have a *representation* in terms of the regular admissible triple (A,B_a,C_a) and a unique feedthrough operator, D, and their trajectories $x(t)$ and $y(t)$ satisfy the abstract differential equations (4.15) and (4.16). In particular, if B and C are bounded, then $B=B_a$ and $C=C_a=C_L$.

A regular system also has a simple frequency domain representation.

Proposition 4.9 *With the notation of Theorem 4.8 if $u \in L_2(0,\infty; U)$, then, $\hat{y}$ the Laplace transform of y, exists and for $s \in \mathbb{C}$ with Re s sufficiently large*

$$\hat{y}(s) = C(sI-A)^{-1}x_0 + G(s)\hat{u}(s) \tag{4.17}$$

where the transfer function G is given by

$$G(s) = C_L(sI-A)^{-1}B + D \tag{4.18}$$

and moreover, for any $v \in U$

$$\lim_{\lambda\to\infty} G(\lambda)v = Dv \tag{4.19}$$

where λ is assumed to be real.

So we see that regular systems have differential and frequency domain representations which are reminiscent of the finite-dimensional ones, (1.1)–(1.7). It is a pleasing and rather unexpected result which took many years to be unravelled.

5. An example (see Curtain and Weiss [7] for details)

Let $X=\ell^2$, $U=Y=\mathbb{C}$, and define the following operators for sequences $x=(x_k) \in X$ and $u \in \mathbb{C}$

$$(Ax)_k = -kx_k; \qquad (Bu)_k = u \tag{5.1}$$

$$C_1 x = \sum_{k\in\mathbb{N}} x_k; \qquad C_2 x = \sum_{k\in\mathbb{N}} (-1)^k x_k \tag{5.2}$$

Then A generates the C_0-semigroup on ℓ^2 given by

$$(\mathbb{T}_t x)_k = e^{-kt} x_k; \qquad k \in \mathbb{N} \tag{5.3}$$

for any $x = (x_k) \in \ell^2$.

$$X_{-1} = \{(x_k) : (\frac{x_k}{k}) \in \ell_2\}; \quad X_1 = \{(x_k) : (k x_k) \in \ell_2\}$$

and $B \in \mathcal{L}(U, X_{-1})$, C_1 and $C_2 \in \mathcal{L}(X_1, Y)$. Using the Carleson measure criterion (see Ho and Russel [14] and Weiss [32]) it follows that B is an admissible control operator for $\mathbb{T}$ with respect to X_{-1} and C_1 and C_2 are admissible observation operators for $\mathbb{T}$ with respect to X_1. For input–output admissibility of the triples we need to evaluate the transfer function from (4.4). For (A,B,C_1) we obtain G_1 given by

$$G_1(s) = G_1(0) - \sum_{k \in \mathbb{N}} \frac{s}{k(s+k)} \tag{5.4}$$

and $G_1(s)$ is not bounded on any half-plane, so that (A,B,C_1) is not input–output admissible. On the other hand, (A,B,C_2) has the transfer function G_2 given by

$$G_2(s) = G_2(0) - \sum_{k \in \mathbb{N}} (-1)^k (\frac{1}{k} - \frac{1}{s+k}) \tag{5.5}$$

and this is bounded on $\mathbb{C}_0$, and so (A,B,C_2) is input–output admissible and generates abstract linear systems (depending on D). From Proposition 7.2 in Weiss [34] we have that the domain of Lebesgue extension C_L of C_2 contains the space

$$\{(x_k) : \sum_{k \in \mathbb{N}} (-1)^k x_k \text{ converges}\}$$

and for all x in this space we have

$$C_L x = \sum_{k \in \mathbb{N}} (-1)^k x_k$$

for $x \in D(C_L)$. It follows that

$$(sI-A)^{-1}Bv = (\frac{v}{s+k}) \in D(C_L)$$

for any $s \in \rho(A)$ and all $v \in \mathbb{C}$, so by Proposition 4.6, all systems Σ generated by (A,B,C_2) are regular and the input–output map has the representation (4.12), where D is the arbitrary feedthrough term. The transfer function has the representation

$$G_2(s) = C_L(sI-A)^{-1}B+D. \tag{5.6}$$

For another completely worked example (the one–dimensional heat equation with Dirichlet boundary control and point observation) see Curtain and Weiss [7]. Further examples can be found in Salamon [28] and Curtain [5] including some on retarded equations and hyperbolic p.d.e.'s.

Acknowledgements
The author is grateful to Dietmar Salamon, Hans Schumacher, George Weiss and Hans Zwart for their valuable comments and suggestions.

References

1. J.S. Baras, and R.W. Brockett: *H^2–Functions and Infinite Dimensional Realization Theory*, SIAM J. Control **13**, 1975, 221–224.
2. J.S. Baras, R.W. Brockett, and P.A. Fuhrmann: *State Space Models for Infinite Dimensional Systems*, in "Mathematical Systems Theory", G. Marchesini and S.K. Mitter, eds. pp. 204–225, LNEMS 131, Springer–Verlag, New York, 1976.
3. R.F. Curtain, A.J. Pritchard: *Infinite Dimensional Linear Systems Theory*, Lecture Notes in Information Sciences 8, Springer–Verlag, 1978.
4. R.F. Curtain, D. Salamon: *Finite Dimensional Compensators for Infinite Dimensional Systems with Unbounded Input Operators*, SIAM J. Control & Optim. **24**, 1986, pp. 797–816.
5. R.F. Curtain: *Well–Posedness of Infinite Dimensional Linear Systems in Time and Frequency Domain*, Research Report TW–287, University of Groningen, August 1988.

6. R.F. Curtain: *A Synthesis of Time and Frequency Domain Methods for the Control of Infinite-Dimensional Systems: A System Theoretic Approach* (to appear in the series SIAM Frontiers in Applied Mathematics).
7. R.F. Curtain, G. Weiss: *Well Posedness of Triples of Operators (in the Sense of Linear Systems Theory)*, Proceedings of the Conference on Distributed Parameter Systems, Vorau, Austria, July 1988, to appear.
8. W. Desch, I. Lasiecka, W. Schappacher: *Feedback Boundary Control Problems for Linear Semigroups*, Israel J. of Math. **51**, 1985, pp. 177–207.
9. S. Dolecki, D.L. Russell: *A General Theory of Observation and Control*, SIAM J. Control & Optim. **15**, 1977, pp. 185–220.
10. H.O. Fattorini: *Boundary Control Systems*, SIAM J. Control **6**, 1968, pp. 349–385.
11. P. Fuhrman: *Linear Systems and Operators in Hilbert Space*, McGraw-Hill, New York. 1981.
12. J.W. Helton: *Systems with Infinite-Dimensional State Space: The Hilbert Space Approach*, Proc. IEEE, **64**, 1976, pp. 145–160.
13. J.W. Helton: *Operator Theory, Analytic Functions, Matrices and Electrical Engineering*, Regional Conference Series in Mathematics, No. 68, AMS, Providence, RI, 1987.
14. L.F. Ho, D.L. Russell: *Admissible Input Elements for Systems in Hilbert Space and a Carleson Measure Criterion*, SIAM J. Control & Optim. **21**, 1983, pp. 614–640.
15. R.E. Kalman, P.L. Falb, M.A. Arbib: *Topics in Mathematical System Theory*, McGraw-Hill, New York, 1969.
16. I. Lasiecka: *Unified Theory for Abstract Parabolic Boundary Problems – A Semigroup Approach*, Appl. Math. Optim. **6**, 1980, pp. 287–333.
17. I. Lasiecka, R. Triggiani: *A Cosine Operator Approch to Modelling $L_2(0,T;\ L_2(\Gamma))$-Boundary Input Hyperbolic Equations*, Appl. Math. Optim. **7**, 1981, pp. 35–93.
18. I. Lasiecka, R. Triggiani: *Stabilization of Neumann Boundary Feedback of Parabolic Equations: The Case of Trace in the Feedback Loop*, Appl. Math. Optim. **10**, 1983, pp. 307–350.
19. I. Lasiecka, R. Triggiani: *Regularity of Hyperbolic Equations under $L_2(0,T;\ L_2(\Gamma))$-Dirichlet Boundary Terms*, Appl. Math. Optim. **10**, 1983, pp. 275–286.
20. I. Lasiecka, R. Triggiani: *Feedback Semigroups and Cosine Operators for Boundary Feedback Parabolic and Hyperbolic Equations*, J. of Diff. Eq. **47**,

1983, pp. 246–272.

21. I. Lasiecka, J.L. Lions, R. Triggiani: *Non Homogeneous Boundary Value Problems for Second Order Hyperbolic Operators*, J. Math. pures et appl. **65**, 1986, pp. 149–192.
22. J.L. Lions: *Optimal Control of Systems Governed by Partial Differential Equations*, Die Grundlehren der mathematischen Wissenschaften in Einzeldarstellungen **170**, Springer–Verlag, 1971.
23. J.L. Lions, E. Magenes: *Non Homogeneous Boundary Value Problems and Applications*, Vol. I and II, Die Grundlehren der mathematischen Wissenschaften in Einzeldarstellungen **181** and **182**, Springer–Verlag, 1972.
24. A.J. Pritchard, A. Wirth: *Unbounded Control and Observation Systems and their Duality*, SIAM J. Control & Optim. **16**, 1978, pp. 535–545.
25. D.L. Russell: *A Unified Boundary Controllability Theory for Hyperbolic and Parabolic Partial Differential Equations*, Studies in Applied Mathematics **52**, 1973, pp. 189–211.
26. D.L. Russell: *Some Remarks on Transfer Function Methods for Infinite Dimensional Linear Systems*, Research Report, Virginia Polytechnic Institute and State University, October 1988.
27. D. Salamon: *Control and Observation of Neutral Systems*, Pitman, London, 1984.
28. D. Salamon: *Infinite Dimensional Systems with Unbounded Control and Observation: A Functional Analytic Approach*, Transactions of the A.M.S. **300**, 1987, pp. 383–431.
29. D. Salamon: *Realization Theory in Hilbert Space*, Math. Systems Theory **21**, 1989, *p.* 147–164.
30. T. Seidman: *Observation and Prediction for One–Dimensional Diffusion Equations*, J. Math. An. Appl. **51**, 1975, pp. 165–175.
31. D. Washburn: *A Bound on the Boundary Input Map for Parabolic Equations with Applications to Time Optimal Control*, SIAM J. Control & Optim. **17**, 1979, pp. 652–671.
32. G. Weiss: *Admissibility of Input Elements for Diagonal Semigroups on* ℓ^2, Systems and Control Letters **10**, 1988, pp. 79–82. 33. G. Weiss: *Admissibility of Unbounded Control Operators*, SIAM J. Control & Optim. **27**, 1989 (to appear).
34. G. Weiss: *Admissible Observation Operators for Linear Semigroups*, Israel J. Math. **65**, 1989, pp. 17–43.
35. G. Weiss: *The Representation of Regular Linear Systems on Hilbert Spaces*,

Proceedings of the Conference on Distributed Parameter Systems, Vorau, Austria, July 1988, to appear.

36. G. Weiss: *The Representation of Linear Systems on Banach Spaces*, in preparation.

37. Y. Yamamoto: *Realization Theory of Infinite-Dimensional Linear Systems, Part I*, Math. Systems Theory **15**, 1981, pp. 55–77. *Part II*, Math. Systems Theory **15**, 1981, pp. 169–190.

Symmetric Modeling in System Identification

M. Deistler

Technische Universität Wien
Institut für Oekonometrie und Operations Research
Argentinierstrasze 8/119, A-1040 Wien, Austria

The paper is concerned with the realization problem for linear dynamic errors-in-variables models where the component processes of the noise term are mutually uncorrelated. The analysis is based on the second moments of the observations.

1. Introduction

In identification of linear (dynamic) systems the conventional way of noise modeling is to add all noise to the outputs (or which is the same for our purposes to the equations). In econometrics this is called the errors-in-equations (EE) approach. This approach is widely used and is appropriate in particular if the primary aim of modeling is prediction of the outputs from their own past and from the observed inputs. However in a number of applications a more general and more symmetric way of noise modeling is appropriate, where (in principle) all variables may be contaminated by noise. This is called the errors-in-variables (EV) or latent variables approach. This approach also allows for a more general and more symmetric way of system modeling, where the number of equations and the classification of the variables into inputs and outputs is obtained from data rather than from a priori knowledge. The main cases where the EV approach is appropriate are:

(i) If we are interested in the "true" system generating the data (rather than in prediction or in just coding data by system parameters) and we cannot be sure a priori that the observed inputs are not corrupted by noise. For instance the "true" system may relate to a certain "physical" theory [of course, "true" systems are always fictions, but they may be good ones] and the noise may correspond to measurement noise. This is the classical motivation for EV modeling.

(ii) If we want to explain the "essential part" of a high dimensional data vector by a relatively small number of (in general unobserved) variables (or factors). This is the classical motivation for factor analysis or for principal component analysis, the most familiar application of which are mental tests where the data correspond to the test scores and the factors to mental factors.

(iii) If we have no sufficient a priori information about the number of equations in the system, about the classification of the variables into inputs and outputs or about causality directions, we have to perform a more symmetric system modeling, which demands a more symmetric noise modeling in order to avoid "prejudices" (Kalman 1982). This gives a fairly general setting for the identification problem [in the linear case, which is the only case we consider in this paper] where in particular the number of equations may be determined from the data rather than from a priori knowledge.

The statistical analysis for the EV case turns out to be significantly more complicated compared to the EE case, mainly because the structure of the relation between the (population) second moments of the observations and the system parameters is more complicated. One of the central problems in this context is a basic "non-identifiability" in the sense that in general the system parameters are not uniquely determined from the second moments of the observations, since the separation between the system and the noise part is not unique without imposing assumptions which in general are rather restrictive.

The kind and amount of a priori assumptions imposed clearly varies with the problem setting. The two extreme philosophies are either to add as little additional assumptions on the noise as possible in order to avoid "prejudice" (Kalman 1982) and to describe the resulting non-uniqueness in the realization problem or, on the other hand, to impose sufficiently many additional assumptions in order to guarantee identifiability. We here will mainly take the first point of view, which has been emphasized in particular by Kalman (1982, 1983). However, it should be said, that in a number of applications (see e.g. Schneeweiß und Mittag 1986, Fuller 1987), for instance if repeated measurements for fixed true values are possible, to mention one case, identifiability may be obtained without imposing prejudicial assumptions. For the first approach, lack of knowledge concerning the "true" system is caused by two facts; first by sampling errors and second by non-identifiability. In a non-stochastic world, the non-identifiability mentioned above arises from a variation of norms in defining a measure of (mis)fit of a system to data.

Another feature of identification in the EV case is that (in the non Gaussian case) e.g. moments of order unequal to two (and one) may provide additional identifying information (Geary 1942, Reiersøl 1950).

Whereas, for the stable case at least, the theory of identification for the EE case has reached a certain stage of completeness now (see e.g. Hannan and Deistler 1988), due to the problems mentioned above, in identification in the EV case there is still a great number of unsolved problems. Accordingly, in actual applications, despite of their great appeal, in the properly dynamic case at least, such models have been used to a limited extend only.

For the static case, the EV problem has a long history. More than hundred years ago, Adcock (1878) recommended, that in fitting a straight line to a scatter plot when the errors to the variables are uncorrelated and have the same variances, the distance of a point to the line should be taken orthogonal to the line [rather than parallel to the y-axis as in the typical regression case]. In a more general context, where the variances are unknown, this problem has been analysed by Gini [1921] and for more than two variables by Frisch [1934] and Koopmans [1937]. A second root is factor analysis, which, in the beginning, has been developed and used mainly in psychology, specifically in order to provide models for human ability and behavior. The earliest papers here are Pearson (1901) and Spearman (1904). From an abstract point of view the EV model and the factor model are equivalent. However, in "classical" EV analysis emphasis was given to models with one or with a few linear relations, whereas in factor analysis emphasis was given to models with one or few factors [and thus with many relations, in general].

It was recognized by Reiersøl (1941) and Geary (1943) that correlation in the true inputs may help to identify the slope parameter of a line. Systematic investigation of dynamic EV models on the other hand is quite recent [see Geweke (1977), Sargent and Sims (1977), Brillinger (1981), Maravall (1979), Söderström (1980), Anderson and Deistler (1984), Novak (1985), Picci and Pinzoni (1986)].

2. The Statement of the Problem

We consider a deterministic system of the form

$$w(z)\hat{z}_t = 0 \tag{2.1}$$

where $\hat{z}_t$ is the n-dimensional vector of latent (i.e. in general unobserved) variables, where z is used for the backward-shift operator on $\mathbb{Z}$ (i.e. $z(\hat{z}_t \mid t \in \mathbb{Z}) = (\hat{z}_{t-1} \mid t \in \mathbb{Z})$) and finally where

$$w(z) = \sum_{j=-\infty}^{\infty} W_j z^j \quad ; \quad W_j \in \mathbb{R}^{m \times n} \tag{2.2}$$

We will call w the <u>relation function</u> of the exact relation (2.1). Without restriction of generality we will assume that $m \leq n$ holds and that w contains no linear dependent rows. Clearly systems of the form (2.1) allow for a symmetric treatment of all variables in the sense that neither a classification of the variables into inputs and outputs nor causality directions have to be specified a priori. Systems of this form are studied in detail in Willems (1979) (1986) and Blomberg and Ylinen (1983).

The observations are of the form

$$z_t = \hat{z}_t + u_t \tag{2.3}$$

where u_t is noise.

Here, unless the contrary is stated explicitely we will assume

(i) All processes considered are (wide sense) stationary [In addition all limits of random variables are understood in the sense of mean squares convergence]

(ii) $E\hat{z}_t = Eu_t = 0$

(iii) $E\hat{z}_t u_s' = 0$

(iv) (z_t), $(\hat{z}_t)$ and (u_t) are ARMA processes (i.e. their spectral densities f, $\hat{f}$ and $\tilde{f}$ respectively are rational) and $w(z)$ is a rational matrix. Clearly $Eu_t = 0$ and (iii) are innocent assumptions. Assumption (i) clearly is a restriction of generality (for instance unstable systems are ruled out), nevertheless, for the statistical analysis such an assumption is quite common. Most of the results presented in this paper also hold without imposing (iv); again in the statistical analysis such a rationality assumption is frequently imposed and is justified by the approximation properties of rational functions. The assumption $E\hat{z}_t = 0$ can easily be dropped and is imposed for notational convenience only.

In this paper we assume that the information from the observations consists of their second moments only. However it should be noted, that in this context, in the non-Gaussian case, e.g. higher order moments contain additional information about the system (Akaike 1966, Deistler 1986). In addition, unless the contrary is explicitely stated, we impose the idealizing assumption that the population second moments of the observations are available und thus we do not consider sampling aspects.

Let us repeat that f, $\hat{f}$ and $\tilde{f}$ denote the spectral densities of (z_t), $(\hat{z}_t)$ and (u_t) respectively. Since the rational extension of a rational function from the unit circle to the complex plane ¢ is unique we will identify these functions (where spectral densities are considered as functions of $e^{-i\lambda}$ rather than of λ) with their extensions to ¢. Note that the rational functions form a field. Matrices like f, $\hat{f}$ or w are considered as matrices over such fields. Notions like linear independence, rank, ect. are understood in this sense. Clearly in this sense, e.g. $\hat{f}$ has rank r if and only if $\hat{f}(\lambda)$ has rank r λ-a.e.

From (2.3) we have:

$$f = \hat{f} + \tilde{f} \tag{2.4}$$

In this framework our problem can be stated as follows:

Given f,

(a) characterize the set of all $\hat{f}$ (and $\tilde{f}$) compatible with given f

(b) find the maximum corank of $\hat{f}$, m^* say, among all decompositions (2.4) of f. Sometimes we also use the notation mc(f) for m^*.

(c) Under which additional assumptions are $\hat{f}$ and w unique.

In the first step we analyse the system representation (2.1) and in particular the deterministic realization problem under the fiction that $\hat{z}_t$ has been observed. This problem has been analysed in a more general framework in Willems (1986, 1987).

Clearly, from an abstract point of view, (2.1) means that $\hat{z} = (\ldots\hat{z}'_{-1},\ \hat{z}'_o,\ \hat{z}'_1\ \ldots)$ is contained in a linear subspace $(L_2^n)^{\mathbb{Z}}$, where L_2 is the usual (Hilbert) space of square integrable functions defined on the underlying probability space (Ω,A,P). An alternative concrete representation is the factor model (see e.g. Picci and Pinconi 1986)

$$\hat{z}_t = a(z).f_t \tag{2.5}$$

where

$$a(z) = \sum_{j=-\infty}^{\infty} A_j z^j \quad ; \qquad A_j \in \mathbb{R}^{n\times(n-m)}$$

and where (f_t) is the (n-m) dimensional factor process. In this way the subspace for $\hat{z}$ is expressed as the range rather as the kernel of a certain matrix.

Given $\hat{f}$ and no additional information [exceeding (i)-(iv)] e.g. on higher order moments of $(\hat{z}_t)$ or on an a priori classification into inputs and outputs, every rational matrix w satisfying

$$w.\hat{f} = 0 \tag{2.6}$$

gives rise to a representation (2.1) [In the sense of a weak realization, see Picci and Pinzoni (1986), van Putten and van Schuppen (1983), i.e. a realization which is compatible with the probability structure of the observations]. Here we want to explain by the system as much as possible, in the sense, that the number of equations should be maximal. In other words, given $\hat{f}$, we choose w in a way such that its rows form a basis for the (left) kernel of $\hat{f}$. Clearly then, under our assumptions, every such a basis is compatible with given $\hat{f}$ and thus for given $\hat{f}$, w is unique up to left multiplication by a nonsingular rational matrix. Since w can be written as

$$w = p^{-1}.c$$

where p is a least common denominator polynomial of w, we can always find a polynomial basis c for the kernel of $\hat{f}$. Now let us rearrange the entries of $\hat{z}_t$ is such a way that the first m columns of c (after rearrangement) form a nonsingular submatrix, a say. Partitioning $\hat{z}_t$ accordingly as $(\hat{y}_t', \hat{x}_t')'$ we can write

$$a(z)\hat{y}_t = b(z)\hat{x}_t \tag{2.7}$$

where $\hat{y}_t$ are outputs, $\hat{x}_t$ inputs and (2.7) is a usual ("ARMA") representation of a linear system with transfer function $k = a^{-1}.b$. In this way we have defined the inputs as those components of $\hat{z}_t$ which correspond to a maximal set of linearely independent rows of $\hat{f}$ and this definition is not unique in general. However, in general only f and not $\hat{f}$ is known (at best) and the main complication arises in unraveling $\hat{f}$ from f in (2.4).

Let us first consider the static case, i.e. the case where all processes considered are white noise and where w is constant. In this case we write (2.4) as

$$\Sigma = \hat{\Sigma} + \tilde{\Sigma} \tag{2.8}$$

where $\Sigma, \hat{\Sigma}$ and $\tilde{\Sigma}$ are the covariance matrices of $z_t, \hat{z}_t$ and u_t respectively. It is straightforward to show that if Σ is nonsingular, for an arbitrarely chosen singular covariance matrix $\hat{\hat{\Sigma}}$, $c.\hat{\hat{\Sigma}}$, for suitably chosen $c>0$ is a matrix $\hat{\Sigma}$ compatible with Σ; thus any (static) system w is a solution to the problem. For the special case n=2 this corresponds to the problem of fitting a least squares line into a scatter plot, where the distance of a point to the line is defined by a norm with a unit ball corresponding to an arbitrary ellipse in $\mathbb{R}^2$. Here, as is easily seen, unless all points lie exactly on a straight line, every line can be obtained by a suitable choice of a norm.

Thus, in order to give our problem a certain degree of "non arbitrariness", additional structure on the noise has to be imposed. From the point of view applications such assumptions seem to be the most delicate ones in our context, since, at least in the author's opinion, there is no universally justifiable assumption of this kind. In other words such assumptions have to be justified in the actual application, otherwise they may express prejudice rather than real a priori knowledge.

In this paper we will assume

(v) $\tilde{f}$ is diagonal

i.e. all component processes of the noise process are uncorrelated. By this assumption we cover a reasonably wide range of applications such that an analysis of this kind seems to be justified. One interpretation of the assumption (v) is that it gives a definition of noise, by saying that the common effects are attributed to the system $w(z)$ and the individual effects to the errors. One of the nice features of assumption (v) is that it gives a completely "symmetric" noise model, which is compatible with the symmetric system model (2.1). Two common alternatives to assumption (v) are:

(a) To assume that $\tilde{f}$ is block diagonal, where the blocks correspond to inputs and outputs respectively (see e.g. Picci and Pinzoni 1986). This assumption is more general than (v) in a certain sense, however an a priori classification of the entries of z_t into inputs and outputs is required.

(b) $\hat{f}$ is defined as a best approximation to f with a given rank. Clearly, in general also this assumption may impose "prejudice" by the choice of the approximation criterion. It is one of the flavours of our assumption (v) that the effects of a certain variation in the approximation criteria can be analysed.

The case n=2 is well understood now (see Anderson and Deistler 1984, Anderson 1985, Deistler 1986 and Deistler and Anderson 1989); for the case n = 3 where the problem of determining m* arises for the first time, the reader is referred to Anderson and Deistler (1987); for the static case see Kalman (1982) and Anderson (1984).

This paper is organized as follows: In the next section some general properties of the solution set are derived, where special emphasis is given to the case m* = 1. In section 4 a rather complete analysis for the case m* = n-1 is given. Finally in section 5 some identifiability results for special cases are described.

For any matrix denoted by a capital letter, e.g. A, we use the corresponding indexed lower case letters to denote its entries, e.g. a_{ij}. If x is a vector, we use x_j to denote its j-th element.

3. Some Properties of the Solution Set

In this as well as in the following section , the analysis is given for arbitrary but fixed frequency λ. The basic equation (2.4) then is written as

$$\Sigma = \hat{\Sigma} + D \tag{3.1}$$

where $\Sigma, \hat{\Sigma}$ and D respectively are (constant) complex Hermitean matrices rather than spectral densities. The generalization of the results below for variing λ is straightforward.

For given Σ, the matrices $\hat{\Sigma}$ or D are called feasible if

$$0 \leq \Sigma - D \leq \Sigma \tag{3.2}$$

where $\Sigma - D = \hat{\Sigma}$ is singular and D is diagonal. For simplicity, throughout we will assume

(vi) $\quad \Sigma > 0$

and

(vii) $\quad \sigma_{ij} \neq 0 \qquad i,j = 1,\ldots,n$

Clearly, these assumptions are generically fulfilled.

Due to (vii) and since $\hat{\Sigma}$ is singular, the first row of $\hat{\Sigma}$ can always be expressed as a linear combination of the other rows. Accordingly we call a solution (corresponding to Σ) any vector $x \in \mathbb{C}^n$, with first component x_1 equal to one, such that there is a feasible $\hat{\Sigma}$ satisfying

$$x\hat{\Sigma} = 0 \qquad (3.3)$$

Clearly, every solution corresponds to a (row of a) relation-matrix w in (2.1). The set of all solutions corresponding to a given Σ is called the <u>solution set</u> $\mathcal{L}$; analogously we define $\mathcal{D}$ as the set of all feasible matrices D corresponding to Σ.

Our main aim now is to give a description of the solution set $\mathcal{L}$.

Let $\tilde{S} = \Sigma^{-1}$ and let $S = (\tilde{s}_{ij} . \tilde{s}_{1j}^{-1})_{i,j = 1,\dots,n}$. Thereby we in addition will assume throughout

(viii) $\qquad \tilde{s}_{ij} \neq 0 \quad , \quad i,j = 1,\dots,n$

Then, as is easily seen from

$$s_j \Sigma = (0,\dots,0, \tilde{s}_{1j}^{-1} , 0,\dots,0)$$

the rows, s_j say, of S correspond to the elementary regressions, i.e. to the case where all variables in positions unequal to j are assumed to be observed free of noise. Correspondingly we call s_j the j-th <u>elementary solution.</u>

Now, let us investigate the relation between $\mathcal{L}$ and $\mathcal{D}$ which is defined by

$$x\Sigma = xD; \quad x \in \mathcal{L} , \ D \in \mathcal{D} \qquad (3.4)$$

Let

$$\Sigma = \begin{pmatrix} \sigma_{11}, & \Sigma_{12} \\ \Sigma_{12}', & \Sigma_{22} \end{pmatrix} \ ; \ \hat{\Sigma} = \begin{pmatrix} \hat{\sigma}_n, & \Sigma_{12} \\ \Sigma_{12}', & \hat{\Sigma}_{22} \end{pmatrix} \ ; \ D = \begin{pmatrix} d_n & 0 \\ 0 & D_{22} \end{pmatrix}$$

be partitionings of $\Sigma, \hat{\Sigma}$ and D respectively where $\Sigma_{22}, \hat{\Sigma}_{22}, D_{22} \in \mathbb{C}^{(n-1)x(n-1)}$ and let $x = (1,x^2)$, $x^2 \in \mathbb{C}^{n-1}$. Note that under our assumptions

$$\hat{\Sigma} = \begin{pmatrix} - x^2 \\ I_{n-1} \end{pmatrix} \hat{\Sigma}_{22} \ (-x^{2'} \ , \ I_{n-1})$$

holds and thus $\hat{\Sigma}_{22}$ has the same rank as $\hat{\Sigma}$. Thus for given $D\in\mathcal{D}$, the corresponding vectors x are given by

$$x^2 \hat{\Sigma}_{22} = - \Sigma_{12} \qquad (3.5)$$

In particular x is unique if $\hat{\Sigma}$ has corank 1; more generally if $\hat{\Sigma}$ has corank m, then the set of all corresponding solutions is an m-1 dimensional affine space.

Conversely, for given x, the corresponding matrices D are given by

$$x^2\Sigma_{22} = x^2D_{22} \qquad (3.6)$$

and

$$\sigma_{11} + x^2\Sigma_{12}' = d_{11} \qquad (3.7)$$

Sometimes we use the symbol D_x to indicate that D corresponds to x. In particular, D is uniquely determined for given x if every entry of x is unequal to zero. Let σ_j $(\hat{\sigma}_j)$ denote the j-th row of Σ $(\hat{\Sigma})$. Conversely, if one entry of x, x_j say, is equal to zero, then $\hat{\sigma}_1$ can be expressed as a linear combination of the rows $\hat{\sigma}_i$, $i\neq 1$, $i\neq j$. In particular in this case we may put $d_{jj}=0$; then $\sigma_j = \hat{\sigma}_j$ is linearly independent of the other rows of $\hat{\Sigma}$. On the other hand it is quite straightforward to show that $\hat{\sigma}_j$ can be made linearly dependent by substracting a certain positive number, d_{jj}^{max} say, from σ_{jj}, and every d_{jj} with $0\leq d_{jj}\leq d_{jj}^{max}$ gives rise to a feasible D.

For the next theorem, which is concerned with the case m*=1, see Anderson and Deistler (1986b).

Theorem 3.1

(a) $mc(\Sigma) = 1$ if and only if there is no $x\in\mathcal{L}$ with a zero entry.

(b) For $mc(\Sigma) = 1$, $\mathcal{L}$ is bounded and closed and when considered as a subset of $\mathbb{R}^{2n}$ $(\hat{=} \mathbb{C}^n)$ it is of dimension n-1.

(c) For $mc(\Sigma) = 1$, the relation between $\mathcal{L}$ and $\mathcal{D}$ defined by (3.4) is a homeomorphism.

Proof: (a) has been shown above.

(c): That the relation is bijective has been shown above. The continuity in both directions is easy to see from (3.5) and (3.6).
(b) First it is straightforward to show that $\mathfrak{D}$ is bounded closed and of dimension n-1. The rest follows from (c).

It can also be shown (Anderson and Deistler 1989b) that the solution set shrinks to a singleton if the noise spectrum $\tilde{f}$ goes to zero. Note that in the real case (i.e. when Σ has real entries) a much more satisfactory result than Theorem 3.1 for $m^*=1$ can be obtained (see e.g. Kalman 1982). In particular, for the real case a complete characterization of $\mathcal{L}$ is available.

Let us return to the case of general m^*. In order to further investigate the solution set we consider the question which part of the (complex) line

$$\alpha x + (1-\alpha)y \quad ; \; \alpha \in \mathbb{C}$$

connecting two points $x,y \in \mathcal{L}$ is contained in $\mathcal{L}$. (Note that such a line in $\mathbb{R}^{2n} \mathrel{\hat{=}} \mathbb{C}^n$ is a (real) plane). This is the same as asking for which $\alpha \in \mathbb{C}$ the equation.

$$(\alpha x + (1-\alpha)y)\Sigma = \alpha x D_x + (1-\alpha)yDy = (\alpha x + (1-\alpha)y)D \tag{3.8}$$

can be satisfied for a $D \in \mathfrak{D}$ (i.e. $D \geq 0$, $\Sigma - D \geq 0$).

Here the simplest special case is when $x=s_1$ and $y=s_j$, $j>1$. In this case we have

$$D_1 = D_{s_1} = \operatorname{diag}\{d_{11}^{(1)}, 0, \dots, 0\}$$

$$D_j = D_{s_j} = \operatorname{diag}\{0, \dots, 0, \underset{\substack{\uparrow \\ j\text{-th position}}}{d_{jj}^{(j)}}, 0, \dots, 0\}$$

Then the first equation in the system (3.8) is of the form

$$\alpha d_{11}^{(1)} = d_{11} \tag{3.9}$$

Analogously, equation j in (3.8) is of the form

$$(1-\alpha)s_{jj}d_{jj}^{(j)} = (s_{1j} + (1-\alpha)s_{jj})d_{jj}$$

and thus

$$\frac{1}{1+\frac{\alpha}{1-\alpha}\cdot\frac{s_{j1}}{s_{jj}}}\cdot d_{jj}^{(j)} = d_{jj} \tag{3.10}$$

where $s_{j1}\cdot s_{jj}^{-1} > 0$. From the remaining equations we can put $d_{ii}^{(i)}=0$; $i\neq 1$, $i\neq j$ which gives a feasible choice of D. Now note that $d_{11}^{(1)}$ is maximal in the sense that $d_{11}^{(1)} \geq d_{11}$ holds for any d_{11} corresponding to $D\in\mathcal{D}$ and that the same holds for $d_{jj}^{(j)}$. This is straightforward from det $(\Sigma-D)=0$. Thus (3.9) and (3.10) imply $\alpha\in[0,1]$. Conversely, for every $\alpha\in[0,1]$ in this way we obtain a feasible D, as can be checked from the minors.

In a completely analogous way we can investigate the real plane

$$x_\alpha = \alpha s_i + (1-\alpha)s_j, \quad \alpha\in\mathbb{C}, \qquad i,j > 1$$

In this case for the j-th equation of (3.8) we obtain

$$\frac{1}{1+\frac{\alpha}{1-\alpha}\cdot\frac{s_{ji}}{s_{jj}}}\cdot d_{jj}^{(j)} = d_{jj}$$

and thus $x_\alpha\in\mathcal{L}$ if and only if

$$\frac{\alpha}{1-\alpha}\cdot\frac{s_{ji}}{s_{jj}} \geq 0 \tag{3.11}$$

Thus we can distinguish three cases:

(a) If $\frac{s_{ji}}{s_{jj}} > 0$ then $x_\alpha\in\mathcal{L}$ if and only if $\alpha\in[0,1]$

(b) If $\frac{s_{ji}}{s_{jj}} < 0$ then $x_\alpha\in\mathcal{L}$ if and only if α $(-\infty,0]\cup[1,\infty)$

c) If $\frac{s_{j1}}{s_{jj}}$ is not real then α corresponds to a curve (described by (3.11)) in $\mathbb{R}^2$.

4. The Case $mc(\Sigma) = n-1$

For the case $mc(\Sigma) = n-1$ we can give a rather complete description. For the proofs of the results given in this section the reader is referred to Anderson and Deistler (1989a). Clearly $m^* = n-1$ corresponds to a factor model with a one dimensional factor process. The next theorem is a rather straightforward generalization of a result which is already well known for a long time for the real case (see e.g. Bekker and de Leeuw 1987).

Theorem 4.1. Let $n>3$. The following statements are equivalent:

(a) $mc(\Sigma) = n-1$

(b) There exists a diagonal unitary matrix U such that $U^*\Sigma U = (\tau_{ij})_{i,j=1,\dots,n}$ is real with all entries τ_{ij} positive and satisfying

$$\tau_{ik}\cdot\tau_{jl} - \tau_{il}\cdot\tau_{jk} = 0 \quad ; \quad i,j,k,l \text{ all different} \tag{4.1}$$

$$\tau_{ii}\,\tau_{jk} - \tau_{ik}\,\tau_{ji} \geq 0 \quad ; \quad i,j,k \text{ all different} \tag{4.2}$$

(c) There exists a diagonal unitary matrix U such that $U^*\Sigma^{-1}U = (t_{ij})_{i,j=1,\dots,n}$ is real with all off diagonal elements negative and satisfying

$$t_{ik}t_{jl} - t_{il}t_{jk} = 0 \quad ; \quad i,j,k,l \text{ all different} \tag{4.3}$$

For the case $n=3$, $m^*=2$ see Anderson and Deistler (1987). After this characterization of the case $m^* = n-1$ we give a description of the solution set for this case. By $\mathcal{L}_m$ we denote the set of all solutions $x\in\mathcal{L}$ corresponding to a $\hat{\Sigma}$ with corank m (i.e. to m outputs).

Theorem 4.2. Let $n\geq 3$ and $mc(\Sigma)=m^*=n-1$. Then we have
(a) there is a unique $\tilde{\Sigma}, \tilde{\Sigma}_{m^*}$ say, such that

$$\mathcal{L}_{m^*} = \ker\ (\Sigma - \tilde{\Sigma}_{m^*})$$

holds.

(b) Let

$$x_j = \alpha^* s_1 + (1-\alpha^*) s_j \quad ; \quad \alpha^* = (1 - \frac{s_{13}}{s_{23}})^{-1} \tag{4.4}$$

(for all j>1)

Then all entries of x_j not in positions 1 and j are equal to zero and

$$\mathcal{L}_{m^*} = \sum_{j=2}^{n} \beta_j x_j \quad ; \quad \sum_2^n \beta_j = 1 \tag{4.5}$$

(c) $\emptyset \neq \mathcal{L}_m \subset \mathcal{L}_{m^*} \quad , \quad i < j < n-1$

(d) $\mathcal{L}_1$ can be written as $\mathcal{L}_{1,u} \cup \mathcal{L}_{1,1}$ where

$$\mathcal{L}_{1,u} = \{\alpha s_1 + (1-\alpha) \sum_{j=2}^{n} \beta_j x_j \mid \alpha, \beta_j \in [o,1], \sum_2^n \beta_j = 1\} \tag{4.6}$$

and

$$\mathcal{L}_{1,1} = \{\alpha s_1 + (1-\alpha) \sum_{j=2}^{n} \beta_j s_j \mid \sum_2^n \beta_j = 1, \sum_2^n \beta_j s_j \in \mathcal{L}, \alpha \in [0, \alpha^*]\}$$

Note that, once the classification of the entries of $\hat{z}_t$ into an input and the outputs has been made, $\mathcal{L}_{m^*}$ corresponds to a unique system. This, in general is evidently not true for the case n = 2. It is easy to show that for $n \geq 3$ (Anderson and Deistler 1989a) the set of all Σ with $mc(\Sigma) = n-1$ is a manifold of (real) dimension 3n in the set of all Σ which is of dimension n^2. Once $m^* = n-1$ is known, estimation of the solution set $\mathcal{L}$ and in particular of the uniquely determined system corresponding to m*, are rather straightforward. From (4.4) and (4.5) we see that such a system continuously depends on Σ. Thus the main problem is to estimate m*, which can be done for instance by a procedure using an information criterion.

5. Some Results on Identifiability

In this short section we discuss two (rather special) examples where identifiability is obtained using the rational structure of the spectral densities. Thereby the idea is due to Söderström (1980) (see also Anderson and Deistler 1984).

(i) If we know that (u_t) is white noise we can determine the poles of every main diagonal element $\hat{f}_{ii}$ of $\hat{f}$ since they are also the poles of f_{ii}. If we in addition assume that for the rational functions $\hat{f}_{ii}$ the respective numerator degrees are strictly smaller than the corresponding denominator degrees, then $\hat{f}$ is uniquely determined from f.

(ii) If $\hat{z}_t$ is known to be purely autoregressive and u_t is a moving average process, than again from the poles of the f_{ii} we know the poles of $\hat{f}_{ii}$. If every f_{ii} has at least one pole (within the unit circle) then the corresponding innovation variance can be uniquely determined and we again have identifiability.

I want to thank B.D.O. Anderson and J.H. van Schuppen for valuable discussions.

References

Adcock, R.J. (1878). A problem in least squares. Analyst 5, 53-54

Akaike, H. (1966). On the use of non-Gaussian process in the identification of a linear dynamic system. Annals of the Institute of Statistical Mathematics 18, 269-276

Anderson, B.D.O.(1985). Identification of scalar errors-in-variables models with dynamics. Automatica 21, 709-716

Anderson, B.D.O. and M. Deistler (1984). Identifiability in dynamic errors-in-variables models. Journal of Time Series Analysis 5, 1-13

Anderson, B.D.O. and M. Deistler (1987). Dynamic errors-in-variables systems with three variables. Automatica 23, 611-616

Anderson, B.D.O. and M. Deistler (1989a). Identification of dynamic systems from noisy data: The case m*=1. Mimeo

Anderson, B.D.O. and M. Deistler (1989b). Identification of dynamic systems from noisy data: Some further results. Mimeo.

Anderson, T.W. (1984). Estimating linear statistical relationships. Annals of Statistics 12, 1-45

Bekker, P. and J. de Leeuw (1987). The rank of reduced dispersion matrices, Psychometrika 52, 125-135

Blomberg, H. and R. Ylinen (1983). Algebraic theory for multivariate linear systems. Academic Press, London

Brillinger, D.R. (1981). Time Series: Data analysis and theory. Expanded Edition. Holden Day, San Francisco

Deistler, M. (1986). Linear errors-in-variables models. In (S.Bittanti, ed.) Time series and linear systems, Lecture Notes in Control and Information Sciences, Springer-Verlag, Berlin, 37-67

Deistler, M. and B.D.O. Anderson (1989). Linear dynamic errors-in-variables models: Some structure theory. To appear in: Journal of Econometrics 41.

Frisch, R. (1934). Statistical confluence analysis by means of complete regression systems. Publication No. 5, University of Oslo, Economic Institute

Fuller, W.A. (1987). Measurement error models. Wiley, New York

Geary, R.C. (1942). Inherent relations between random variables. Proceedings of the Royal Irish Academy, Sec. A, 47, 63-76

Geary, R.C. (1943). Relations between statistics: The general and the sampling problem when the samples are large. Proceedings of the Royal Irish Academy. Sec. A, 49, 177-196

Geweke, J. (1977). The dynamic factor analysis of econometric time series models. In (Aigner, D.Jand A.S., Goldberger eds.) Latent variables in socioeconomic models. North Holland, Amsterdam

Gini, C. (1921). Sull' interpolazione di una retta quando i valori della variabile indipendente sono affetti da errori accidentali. Metron 1, 63-82

Hannan, E.J. and M. Deistler (1988). The statistical theory of linear systems. Wiley, New York

Kalman, R.E. (1982). System identification from noisy data, in (A. Bednarek and L. Cesari, eds.) Dynamical Systems II, a University of Florida International Symposium, Academic Press, New York

Kalman, R.E. (1983). Identifiability and modeling in econometrics, in (Krishnaiah, P.R., ed.) Developments in Statistics, Vol.4, Academic Press, New York

Koopmans, T.C. (1937). Linear regression analysis of economic time series. Netherlands Economic Institute, Haarlem

Maravall, A. (1979). Identification in dynamic shock-error models, Springer-Verlag, Berlin

Nowak, E. (1985). Global identification of the dynamic shock-error model. Journal of Econometrics 27, 211-219

Pearson, K. (1901). On lines and planes of closest fit to systems of points in space. Philosophical Magazine 2, 559-572

Picci, G. and S. Pinzoni (1986). Dynamic factor-analysis models for stationary processes. IMA J. Math. Control and Information 3, 185-210

Reiersøl, O. (1941). Confluence analysis by means of lag moments and other methods of confluence analysis. Econometrica 9, 1-24

Reiersøl, O. (1950). Identifiability of a linear relation between variables which are subject to error. Econometrica 18, 375-389

Sargent, T.J. and C.A. Sims (1977). Businesscycle modeling without pretending to have too much a priori economic theory. In (Sims, C.A. ed.) New Methods in Business Cycle Research. Fe. Reserve Bank, Minneapolis

Schneeweiß, H. und H.J. Mittag (1986). Lineare Modelle mit fehlerbehafteten Daten. Physica Verlag, Heidelberg

Söderström, T. (1980). Spectral decompositions with application to identification. In (Archetti, F. and M. Dugiani, eds.) Numerical Techniques for Stochastic Systems, North Holland P.C., Amsterdam

Spearman, C. (1904). General intelligence, objectively determined and measured. American Journal of Psychology 15, 201-293

van Putten, C. and J.H. van Schuppen (1983). The weak and strong Gaussian probabilistic realization problem. Journal of Multivariate Analysis 13, 118-137

Willems, J.C. (1979). System theoretic models for the analysis of physical systems. Richerche di Automatica 10, 71-106

Willems, J.C. (1986). From time series to linear systems - Part I, Automatica 22, 561-580

Willems, J.C. (1987). From time series to linear systems - Part III, Automatica 23, 87-115

Elements of Factorization Theory From a Polynomial Point of View

P. A. Fuhrmann

Department of Mathematics, Ben-Gurion Univsersity of the Negev
Beer Sheva, Israel

This paper is dedicated to Jan C. Willems

Abstract

The paper outlines a coherent development of factorization theory in the framework of polynomial model theory. Starting from the most elementary factorizations of polynomial matrices we build up the connections to invariant subspace theory, factorizations of transfer functions, Wiener-Hopf factorizations. We pass on to spectral factorizations of polynomial matrices and rational functions and the connection with the analysis of the algebraic Riccati equation. Finally we study inner/outer factorizations for a class of transfer functions and the derivation of state space formulas.

The aim throughout is to highlight the logical interconnections and the technique rather than the derivation of the most general results.

1 INTRODUCTION

Factorization theory is a typical mathematical subject where the interesting questions are of the inverse problem type. Once a multiplication operation is defined the question of the representation of an object as a product becomes a natural question. This becomes even more interesting when we look for a complete factorization into irreducible factors, or alternatively if certain constraints are put on the factors. Factorization problems are quite often intimately connected to structural properties and to a variety of canonical forms.

In this paper we want to outline the factorization problems most often encountered in the area of systems theory. We will start from the most elementary factorization problem, that of nonsingular polynomial matrices, and study its relation to the analysis of invariant subspaces. We will see what symmetry constraints add to this problem. This will relate Hamiltonian, or parahermitian symmetry to special classes of invariant subspaces of symplectic spaces.

Passing from the polynomial level to that of rational functions we analyze the factorization problem for causally invertible transfer functions. We will give a short description of Wiener-Hopf factorizations.

Our main emphasis will be on the circle of ideas centered around spectral factorizations and the algebraic Riccati quation. The central importance of this set of problems in systems theory needs no elaboration. It is a central tool in optimal control and filtering theory. Our approach is based on the spectral factorization of polynomial matrices. Finally we focus on inner/outer factorizations of rational matrix functions.

Throughout we try to outline the connection between external data, polynomial matrices, transfer functions and factorizations etc., and the internal data given in state space terms. The machinery that makes this connection clearest is the theory of polynomial models developed in Fuhrmann [1976-1988].

The history of the various factorizations is both old and rich. There is no intention of trying at a complete account or a full referencing. The following is the barest of outlines. The connection between polynomial matrices and module theory is quite old and can be found for example in McDuffee [1956] where further references to early contributors can be found. The connection between factorization of polynomial matrices and invariant subspaces derives from the work of Livsic and Brodskii in operator theory. In the context of polynomial models it can be found in Antoulas [1979] and Fuhrmann [1979].

The standard result on factorization of rational transfer functions was derived in Bart, Gohberg, Kaashoek and Van Dooren [1980], and a fuller account is given in Bart, Gohberg, and Kaashoek [1979]. Our development takes a polynomial approach and is based on Shamir and Fuhrmann [1984]. The analysis of spectral and inner-outer factorizations owes a lot to, one might say it is actually based on, Francis [1988] and Chen and Francis [1988]. In fact it arose out of trying to understand the intuition behind some of the state space formulas.

Finally it is a pleasure to acknowledge the influence of Jan Willems' fundamental and pathbreaking paper, Willems [1971]. It is here that we see most clearly the convergence of spectral factorizations and state space methods, namely the Riccati equation, as alternative tools in optimization problems. In view of his long and many outstanding contributions it is hard to believe that Jan has only now reached the ripe old age of fifty. On this occasion this modest contribution is dedicated to him.

2 POLYNOMIAL MODELS

Our starting point is this basic result about free modules.

Theorem 2.1 *Let R be a principal ideal domain and M a free left $R-$module with n basis elements. Then every $R-$submodule N of M is free and has at most n basis elements.*

If V is a finite dimensional vector space over a field F then $V[z]$, the space of vector polynomials is a free finitely generated module over the polynomial ring $F[z]$. Throughout we will assume a basis has been chosen and thus V will be identified with F^n and similarly $V[z]$ with $F^n[z]$. Also we will identify $F^n[z]$ and $(F[z])^n$ and speak of its elements as polynomial vectors. Similarly elements of $F^{m\times n}[z]$ will be referred to as polynomial matrices. Because of the nature of the factorization results we are interested in, and for the consistency of notation, we will identify the field F with the real field $\mathbf{R}$, noting that some of the results hold in greater generality.

The next theorem, whose easy proof follows directly from Theorem 2.1, is the basic representation theorem for submodules.

Theorem 2.2 *A subset M of $\mathbf{R}^n[z]$ is a submodule of $\mathbf{R}^n[z]$ if and only if $M = D\mathbf{R}^n[z]$ for some D in $\mathbf{R}^{n\times n}[z]$.*

The following is the basic theorem that relates submodule inclusion to factorization.

Theorem 2.3 *Let $M = D\mathbf{R}^n[z]$ and $N = E\mathbf{R}^n[z]$ then $M \subset N$ if and only if $D = EG$ for some G in $\mathbf{R}^{n\times n}[z]$.*

Let π_+ and π_- denote the projections of $\mathbf{R}^m((z^{-1}))$ the space of truncated Laurent series on $\mathbf{R}^m[z]$ and $z^{-1}\mathbf{R}^m[[z^{-1}]]$, the space of formal power series vanishing at infinity, respectively. Since

$$\mathbf{R}^m((z^{-1})) = \mathbf{R}^m[z] \oplus z^{-1}\mathbf{R}^m[[z^{-1}]] \tag{1}$$

π_+ and π_- are complementary projections. Given a nonsingular polynomial matrix D in $\mathbf{R}^{m\times m}[z]$ we define two projections π_D in $\mathbf{R}^m[z]$ and π^D in $z^{-1}\mathbf{R}^m[[z^{-1}]]$ by

$$\pi_D f = D\pi_- D^{-1} f \qquad for\ f \in \mathbf{R}^m[z] \tag{2}$$

$$\pi^D f = \pi_- D^{-1}\pi_+ Dh \qquad for\ h \in z^{-1}\mathbf{R}^m[[z^{-1}]] \tag{3}$$

and define two linear subspaces of $\mathbf{R}^m[z]$ and $z^{-1}\mathbf{R}^m[[z^{-1}]]$ by

$$X_D = Im\pi_D. \tag{4}$$

and

$$X^D = Im\pi^D. \tag{5}$$

An element f of $\mathbf{R}^m[z]$ belongs to X_D if and only if $\pi_+ D^{-1} f = 0$, i.e. if and only if $D^{-1}f$ is a strictly proper rational vector function.

We turn X_D into an $\mathbf{R}[z]$–module by defining

$$p \cdot f = \pi_D pf \qquad for\ p \in \mathbf{R}[z],\ f \in X_D. \tag{6}$$

Since $Ker\pi_D = D\mathbf{R}^m[z]$ it follows that X_D is isomorphic to $\mathbf{R}^m[z]/D\mathbf{R}^m[z]$.

Theorem 2.4 *With the previously defined module structure, X_D is isomorphic to the quotient module $\mathbf{R}^n[z]/D\mathbf{R}^n[z]$.*

In X_D we will focus on a special map S_D, an abstraction of the classical companion matrix, which corresponds to the action of the identity polynomial z, i.e.,

$$S_D f = \pi_D z f \qquad for \qquad f \in D.$$

Thus the module structure in X_D is identical to the module structure induced by S_D through $p \cdot f = p(S_D)f$. With this definition the study of S_D is identical to the study of the module structure of X_D. In particular the invariant subspaces of S_D are just the submodules of X_D which we proceed to investigate.

The interpretation of factorization of polynomial matrices on the level of polynomial models is described next.

Theorem 2.5 *A subset M of X_D is a submodule, or equivalently an S_D invariant subspace, if and only if $M = D_1 X_{D_2}$ for some factorization $D = D_1 D_2$ with $D_i \in \mathbf{R}^{n \times n}[z]$.*

We summarize now the connection between the geometry of invariant subspaces and the arithmetic of polynomial matrices.

Theorem 2.6 *Let M_i, $i = 1, \dots, s$ be submodules of X_D, having the representations $M_i = E_i X_{F_i}$, that correspond to the factorizations*

$$D = E_i F_i.$$

Then the following statements are true.

(i) $M_1 \subset M_2$ if and only if $E_1 = E_2 R$, i.e. if and only if E_2 is a left factor of E_1.

(ii) $\cap_{i=1}^{s} M_i$ has the representation $E_\nu X_{F_\nu}$ with E_ν the least common right multiple (l.c.r.m.) of the E_i and F_ν the g.c.r.d. of the F_i.

(iii) $M_1 + \cdots + M_s$ has the representation $E_\mu X_{F_\mu}$ with E_μ the greatest common left divisor (g.c.l.d.) of the E_i and F_μ the l.c.l.m. of all the F_i.

Corollary 2.1 *Let $D = E_i F_i$, for $i = 1, \dots, s$. Then*

(i) We have

$$X_D = E_1 X_{F_1} + \cdots + E_s X_{F_s}$$

if and only if the E_i are left coprime.

(ii) We have $\cap_{i=1}^{s} E_i X_{F_i} = 0$ if and only if the F_i are right coprime.

(iii) The decomposition

$$X_D = E_1 X_{F_1} \oplus \cdots \oplus E_s X_{F_s}$$

is a direct sum if and only if $D = E_i F_i$ for all i, the E_i are left coprime and the F_i are right coprime.

The next result summarizes the relation between factorization and the spectral decomposition of linear maps.

Theorem 2.7 *Let $D(z) \in \mathbf{R}^{n \times n}[z]$ be nonsingular and let $d(z) = det D(z)$ be its characteristic polynomial. Suppose d has a factorization $d = e_1 \cdots e_s$ with the e_i pairwise coprime. Then D admits factorizations*

$$D = D_i E_i$$

with $det D_i = d_i$, $det E_i = e_i$ and such that

$$X_D = D_1 X_{E_1} \oplus \cdots \oplus D_s X_{E_s} \tag{7}$$

Moreover

$$det(S_D | D_i X_{E_i}) = e_i. \tag{8}$$

Denoting by $\tilde{T}$ the transpose of the matrix T we define, for an element $A(z) = \sum_{j=-\infty}^{\infty} A_j z^j$ of $\mathbf{R}^{m\times m}((z^{-1}))$, $\tilde{A}$ by

$$\tilde{A}(z) = \sum_{j=-\infty}^{\infty} \tilde{A}_j z^j. \tag{9}$$

In $\mathbf{R}^m((z^{-1})) \times \mathbf{R}^m((z^{-1}))$ we define a symmetric bilinear form $[f,g]$ by

$$[f,g] = \sum_{j=-\infty}^{\infty} \tilde{f}_j g_{-j-1} \tag{10}$$

where $f(z) = \sum_{j=-\infty}^{\infty} f_j z^j$ and $g(z) = \sum_{j=-\infty}^{\infty} g_j z^j$. As both f and g are truncated Laurent series, it is clear that the sum in (6) is well defined, containing only a finite number of nonzero terms. We denote by T^* the adjoint of a map T relative to the bilinear form of (6), i.e. T^* is the unique map that satisfies

$$[Tf,g] = [f,T^*g] \tag{11}$$

for all $f,g \in \mathbf{R}^m((z^{-1}))$. We use this global bilinear form to obtain a concrete representation of X_D^*, the dual space of X_D.

Theorem 2.8 *The dual space of X_D, to be denoted by X_D^*, can be identified with $X_{\tilde{D}}$ under the pairing*

$$< f,g >= [D^{-1}f,g] \tag{12}$$

for $f \in X_D$ and $g \in X_{\tilde{D}}$. Moreover the module structures of X_D and $X_{\tilde{D}}$ are related through

$$S_D^* = S_{\tilde{D}}. \tag{13}$$

The following result shows how the characterization of submodules of X_D and their relation to factorizations is reflected by duality.

Theorem 2.9 *Let $M \subset X_D$ be a submodule, represented as $M = EX_G$ for some factorization $D = EG$ into nonsingular factors. Then the orthogonal subspace $M^\perp$ is a submodule of $X_{\tilde{D}}$ and is given by $M^\perp = \tilde{G}X_{\tilde{E}}$.*

Recall that a vector space V is *symplectic* if it is equipped with a nondegenerate, alternating bilinear form. A linear map H in a symplectic space V is *Hamiltonian* if $H^* = -H$ relative to this form. A map R in V is a *symplectic* if it is invertible and leaves the form invariant.

The canonical example of a symplectic space is $\mathbf{R}^{2n}$ with the bilinear form induced by

$$J = \begin{pmatrix} 0 & I \\ -I & 0 \end{pmatrix}. \tag{14}$$

A Hamiltonian map H in this case is given by

$$H = \begin{pmatrix} A & P \\ Q & -\tilde{A} \end{pmatrix} \tag{15}$$

with P, Q symmetric. R is symplectic if and only if $\tilde{R}JR = R$.

Contrary to inner product spaces, there are self orthogonal elements and subspaces in symplectic spaces. A subspace $\mathcal{L}$ of a symplectic space V is called *Lagrangian* if it is a maximal self orthogonal subspace.

We modify now the previous approach to duality to accommodate the study of symplectic spaces. To this end we introduce now a global alternating form on $\mathbf{R}^m((z^{-1})) \times \mathbf{R}^m((z^{-1}))$ in the following way. We define now a new bilinear form on $\mathbf{R}^m((z^{-1})) \times \mathbf{R}^m((z^{-1}))$ by

$$\{f,g\} = [\tau f, g] \quad for \quad f,g \in \mathbf{R}^m((z^{-1})) \tag{16}$$

where $\tau : \mathbf{R}^m((z^{-1})) \longrightarrow \mathbf{R}^m((z^{-1}))$ is defined by

$$(\tau f)(z) = f(-z). \tag{17}$$

Lemma 2.1 *The bilinear form defined by (16) on* $\mathbf{R}^m((z^{-1})) \times \mathbf{R}^m((z^{-1}))$ *is alternating, i.e.*

$$\{f,g\} = -\{g,f\}. \tag{18}$$

Given a rational matrix function Φ we define its *Hamiltonian* or *parahermitian* conjugate, Φ_* by

$$\Phi_*(z) = \tilde{\Phi}(-z). \tag{19}$$

This implications of this symmetry were originally studied by Brockett and Rahimi [1972]. An extensive study of Hamiltonian symmetry and its associated realization theory can be found in Fuhrmann [1984].

Given a map $Z : \mathbf{R}^m((z^{-1})) \longrightarrow \mathbf{R}^m((z^{-1}))$ we will denote by Z^∇ the map, assuming it exists, which satisfies

$$\{Zf,g\} = \{f, Z^\nabla g\} \tag{20}$$

for all $f,g \in \mathbf{R}^m((z^{-1}))$. Since the form $[\ ,\ ]$, and hence also $\{\ ,\ \}$, is nondegenerate the map Z^∇, if it exists, is unique. The map Z^∇ will be called the *Hamiltonian adjoint* of Z.

Assume now that we have two nondegenerate bilinear forms on $V \times V^*$ and $V^* \times V$ which satisfy $<< x,y >> = - << y,x >>$ for all $x \in V$ and $y \in V^*$. Note that while the two forms are distinct we do not distinguish between them, as it is always clear from the context which form we use. We say x is *orthogonal* to y if $<< x,y >> = 0$. Given a subset $M \subset V$ we define $M^\perp$ as usual by

$$M^\perp = \{y \in V^* \mid << x,y >> = 0 \quad for\ all\ x \in M\}. \tag{21}$$

Full submodules of $\mathbf{R}^m[z]$ are submodules of the form $D\mathbf{R}^m[z]$ with D a nonsingular polynomial matrix. It is of interest to characterize their orthogonal complements. The result should be compared with Theorem 2.9.

We can use these isomorphisms to define a pairing between X_D and $X_{(D_*)}$ by

Theorem 2.10 *The dual space of* X_D*, to be denoted by* X_D^∇*, can be identified with* $X_{(D_*)}$ *under the pairing*

$$<< f,g >> = \{D^{-1}f, g\} = [JD^{-1}f, g] \tag{22}$$

for all $f \in X_D$ *and* $g \in X_{(D_*)}$*. Moreover the module structures of* X_D *and* $X_{(D_*)}$ *are related through*

$$S_D^* = -S_{(D_*)}. \tag{23}$$

We note at this point that if D is a nonsingular Hamiltonian symmetric polynomial matrix, i.e. if $D_* = D$, then X_D with the metric of (22) is a symplectic space. Much use of this will be made later.

The pairing of elements of X_D and X_{D_*} given by (22) allows us to compute, for a subset V of X_D, the annihilator $V^\perp$, i.e., the set of all $g \in X_{D_*}$ such that $<< f, g >> = 0$ for all $f \in V$. Since $S_D^* = -S_{D_*}$, it follows that if $V \subset X_D$ is a submodule, then so is $V^\perp$ in X_{D_*}. However, we know that submodules are related to factorizations of D into nonsingular factors. The annihilator of V can be concretely identified.

Theorem 2.11 *Let $V \subset X_D$ be a submodule. Then $V^\perp$ is a submodule of $X_{(D_*)}$. Moreover if $V = EX_F$ where $D = EF$ is a factorization of D into nonsingular factors, then*

$$V^\perp = F_* X_{E_*}. \tag{24}$$

If we assume $\tilde{D}(-z) = D(z)$, i.e. $D_* = D$, then X_D is a symplectic space and S_D a Hamiltonian map. For details see Fuhrmann [1984]. One way to obtain Hamiltonian symmetric polynomial matrices is to study those which have a factorization of the form $D(z) = \tilde{E}(-z)E(z)$ or $D = E_* E$.

Theorem 2.12 *Let $D = E_* E$. Then $E_* X_E$ is a Lagrangian S_D-invariant subspace of X_D.*

Definition 2.1 *Let $P_* = P$ be a Hamiltonian symmetric polynomial matrix. A factorization*

$$P = N_* \Sigma N \tag{25}$$

with a signature matrix Σ is called a Hamiltonian symmetric factorization. *It is called an* unmixed factorization *if*

$$\sigma(N) \cap \sigma(N_*) = \emptyset \tag{26}$$

or equivalently if $\det N$ and $\det N_*$ are relatively prime.

The following theorem, a special case of an unmixed factorization of a Hamiltonian symmetric polynomial matrix, is central to the whole theory of spectral factorizations. It goes back to the work of Jacubovich [1970] and Coppel [1972]. A more accessible account is Gohberg, Lancaster and Rodman [1982] where Coppel's proof is reproduced.

Theorem 2.13 *Let $Q \in \mathbf{R}^{n \times n}[z]$ be a polynomial matrix such that $Q(it) = \tilde{Q}(it) > 0$ for all real t. Then there exists a real polynomial matrix P such that*

$$Q(z) = P(z)P_*(z).$$

Moreover, P can be chosen such that all zeroes of $\det P$ lie in the open left (right) half plane.

3 REALIZATION THEORY

As usual, given a proper rational matrix G we will say a system (A, B, C, D) is a *realization* of G if

$$G = D + C(zI - A)^{-1}B.$$

We will use the notation $G = [A, B, C, D]$. We will be interested in realizations associated with rational functions having the following representations

$$G(z) = V(z)T(z)^{-1}U(z) + W(z). \tag{27}$$

Our approach to the analysis of these systems is to associate with each representation of the form (27), a state space realization in the following way. We choose X_T as the state space and define the triple (A, B, C), with $A : X_T \longrightarrow X_T$, $B : \mathbf{R}^m \longrightarrow X_T$, and $C : X_T \longrightarrow \mathbf{R}^p$ by

$$\begin{cases} A = S_T \\ B\xi = \pi_T U\xi, \\ Cf = (VT^{-1}f)_{-1} \\ D = G(\infty). \end{cases} \tag{28}$$

We call this the *associated realization* to the polynomial matrix P given by

$$P = \begin{pmatrix} T & U \\ -V & W \end{pmatrix}. \tag{29}$$

Theorem 3.1 *The system given by (28) is a realization of $G = VT^{-1}U + W$. This realization is reachable if and only if T and U are left coprime and observable if and only if T and V are right coprime.*

The following result as well as its dual, due to Hautus and Heymann [1978], are extremely useful.

Theorem 3.2 *Let (A, C) be an observable pair, $G(z) = C(zI - A)^{-1}$ be the corresponding state to output transfer function and let*

$$G(z) = T(z)^{-1}U(z).$$

be a left coprime matrix fraction representation. Then, given any polynomial matrix N, $T^{-1}N$ is strictly proper if and only if there exists a constant matrix K for which $N(z) = U(z)K$. This is equivalent to the columns of U being a basis for X_T.

Theorem 3.3 *Let $G = ND^{-1}$ have the reachable realization (A, B, C) and let $G' = MD^{-1}$. Then G' has a realization (A, B, C_0) for some C_0.*

Let $G_1 = [A_1, B_1, C_1, D_1]$ and $G_2 = [A_2, B_2, C_2, D_2]$ be two transfer functions realized in the state spaces X_1 and X_2 respectively. If the number of inputs of the second system equals the number of outputs of the first we can feed those outputs to the second system. This gives rise to the *series coupling* and the corresponding transfer function is

$$G_2G_1 = \left[\begin{pmatrix} A_1 & 0 \\ B_2C_1 & A_2 \end{pmatrix}, \begin{pmatrix} B_1 \\ B_2D_1 \end{pmatrix}, \begin{pmatrix} D_2C_1 & C_2 \end{pmatrix}, D_2D_1\right].$$

We will use also the notation $G_2G_1 = [A_2, B_2, C_2, D_2] \times [A_1, B_1, C_1, D_1]$.

Definition 3.1 *We say that* (A,B,C,D) *acting in* $\mathbf{R}^{2n}$ *is a* Hamiltonian system *if, with* J *defined by (14), we have*

$$\begin{aligned} \tilde{A}J &= -JA \\ JB &= \tilde{C} \\ D &= \tilde{D} \end{aligned} \tag{30}$$

The main result concerning Hamiltonian realizations is the following by Brockett and Rahimi [1972].

Theorem 3.4 *(i) A transfer function* G *has a Hamiltonian realization if and only if it is Hamiltonian symmetric.*

(ii) Two minimal Hamiltonian realizations of G *are symplectically equivalent.*

A modification of representation (27) leads to Hamiltonian realizations via polynomial models.

Theorem 3.5 *Let* G*, assumed to be Hamiltonian symmetric, strictly proper and rational, have the representation*

$$G(z) = X_*(z)Q(z)^{-1}X(z) + P(z) \tag{31}$$

where X, P *and* Q *are polynomial matrices, with* Q *nonsingular and Hamiltonian symmetric. Then, in the symplectic space* X_Q *the associated realization to (31) is Hamiltonian. It is minimal if and only if* X *and* Q *are left coprime.*

Theorem 3.6 *Let* G *be a* $p \times m$ *strictly proper rational transfer function having the representation*

$$G(z) = V(z)T(z)^{-1}U(z) + W(z). \tag{32}$$

Let the associated realization (A,B,C)*, in the state space* X_T*, be defined by*

$$\begin{aligned} A &= S_T \\ Bu &= \pi_T U u \quad u \in \mathbf{R}^m \\ Cf &= (VT^{-1}f)_{-1} \quad for\ f \in X_T. \end{aligned} \tag{33}$$

Then the Hamiltonian adjoint of the realization (A,B,C) *associated with this representation is* $(-A^\nabla, C^\nabla, B^\nabla)$ *given by*

$$A^\nabla = S_{T_*} = -S_T \tag{34}$$

$$B^\nabla f = (U_*(T_*)^{-1}f)_{-1} \tag{35}$$

and

$$C^\nabla u = \pi_{(T_*)}V_* u \tag{36}$$

and it is the realization associated with

$$G_*(z) = U_*(z)(T_*(z))^{-1}V_*(z) + W_*(z). \tag{37}$$

∎

We pass now to the polynomial characterization of conditioned invariant subspaces. We will need this in the study of the factorization problem.

Given a pair (C, A), a subspace V of the state space X is called a *conditioned invariant subspace* if there exists a linear transformation L such that

$$(A + LC)V \subset V. \tag{38}$$

Theorem 3.7 *Let (A, B, C) be the observable realization associated with the transfer function $G(z) = T(z)^{-1}U(z)$. Then a subspace $V \subset X_T$ is a conditioned invariant subspace if and only if*

$$V = E_1 X_{F_1} \tag{39}$$

where $T_1 = E_1 F_1$ is such that $T_1^{-1}T$ is a bicausal isomorphism.

The following results will be needed in the study of the factorization problem.

Theorem 3.8 *(i) Let M_1, M_2 be two conditioned invariant subspaces of X_T with the representations $M_i = E_i X_{F_i}$. Then $M_1 \subset M_2$ if and only if $E_1 = E_2 Y$ for some polynomial matrix Y.*

(ii) Let $M_1, M_2 \subset X_T$ be conditioned invariant subspaces and let $M_i = E_i X_{F_i}$, $i = 1, 2$. Let $M = M_1 \cap M_2$. Then

$$M = X_T \cap EF^p[z] \tag{40}$$

where E is the l.c.r.m. of E_1, E_2.

(iii) Let $M_1, M_2 \subset X_T$ be conditioned invariant subspaces, and let $M_i = E_i X_{F_i}$. Let M be the smallest conditioned invariant subspace containing both M_1 and M_2. Then $M = X_T \cap EF^p[z]$ where E is a g.c.l.d. of E_1 and E_2.

(iv) Let T and S be $p \times p$ nonsingular polynomial matrices. Then

$$X_T \cap SF^p[z] = \{0\} \tag{41}$$

if and only if all the right Wiener-Hopf factorization indices at infinity of $S^{-1}T$ are nonpositive, see Fuhrmann and Willems [1979], or equivalently if there exists a unimodular matrix U such that $US^{-1}T$ is proper rational.

4 INVERSE SYSTEMS

The following result is well known. While it has a trivial state space proof we still find it of interest to obtain a polynomial model proof of this result. This is done in order to have the polynomial approach represented in a unified and self contained form.

Theorem 4.1 *Let $G = [A, B, C, I]$ be a normalized bicausal isomorphism. Assume the realization is minimal. Then $G^{-1}(z) = [A^\times, B, -C, I]$ where*

$$A^\times = A - BC.$$

Proof: Let $T^{-1}U$ be a left coprime factorization of π_-G. Then

$$G = I + T^{-1}U = T^{-1}(T+U) = T^{-1}D \tag{42}$$

with D defined by

$$D = T + U.$$

By our assumption $T^{-1}D$ is a bicausal isomorphism, which implies that the polynomial models X_D and X_T contain the same elements. Since (42) implies

$$G^{-1} = D^{-1}T = D^{-1}(D-U) = I - D^{-1}U,$$

the shift realizations of G and G^{-1} have the same state space. Let (A, B, C) and (A_1, B_1, C_1) be the shift realizations of π_-G and π_-G^{-1} respectively, i.e.

$$\begin{aligned} A &= S_T, \\ B\xi &= U\xi = \pi_T U\xi \\ Cf &= (T^{-1}f)_{-1} \end{aligned}$$

and

$$\begin{aligned} A_1 &= S_D, \\ B_1\xi &= -U\xi = \pi_D T\xi \\ C_1 f &= (D^{-1}f)_{-1}. \end{aligned}$$

Clearly $B_1 = -B$ and, since for $f \in X_D$

$$C_1 f = (D^{-1}f)_{-1} = (D^{-1}TT^{-1}f)_{-1} = (T^{-1}f)_{-1} = Cf,$$

by the fact that the constant term of $D^{-1}T$ is the identity, it follows that $C_1 = C$. Finally recall that given $f \in X_T$

$$S_T f = zf - T\xi_f$$

for some constant vector ξ_f. Clearly ξ_f is given by

$$\xi_f = \pi_+ z T^{-1} f = (T^{-1}f)_{-1}.$$

Now

$$\begin{aligned} S_D f &= S_{T+U} f = zf - (T+U)((T+U)^{-1}f)_{-1} \\ &= zf - (T+U)((T+U)^{-1}TT^{-1}f)_{-1} \\ &= zf - (T+U)(T^{-1}f)_{-1} \\ &= (zf - T(T^{-1}f)_{-1}) - U(T^{-1}f)_{-1} \\ &= S_T f - U(T^{-1}f)_{-1} \end{aligned}$$

or

$$A_1 = A - BC = A^{\times}.$$

Thus we have recovered the theorem modulo the trivial isomorphism given by $-I$.

∎

We pass on to the study of one sided inverses. Of special interest will be the characterization of the class of transfer functions for which the singularities of the inverse system are localized in some special subsets of the complex plane.

We recall that H^∞ is the Hardy space of all bounded analytic functions in the right half plane Π_+. By RH^∞ we will denote the real subspace of all real rational functions with all poles in the open left half plane Π_-. For a state space approach to the result below, see for instance Bengtsson [1974].

Lemma 4.1 *Assume $G \in RH^\infty$ has minimal realization $[A,B,C,D]$. Let D^+ be any right inverse of D. Then the following statements are equivalent.*

(i) G has a right inverse in RH^∞.

(ii) D is surjective and for some F we have

$$C + DF = 0 \tag{43}$$

and

$$X_+(A+BF) = \{0\}. \tag{44}$$

(iii) D is surjective and

$$X_+(A - BD^+C) \subset < A - BD^+C, BKerD > . \tag{45}$$

If F is such that (ii) is satisfied then a right inverse G^+ is given by $G = [A + BF, -BD^+, F, D^+]$.

Proof:

$(i) \Rightarrow (iii) \Rightarrow (ii)$. Assume G has a right inverse in RH^∞. Then G has full row rank in Π_+. In particular D is surjective. Let $G(z) = D + N(z)E(z)^{-1}$. Then to G is associated the polynomial system matrix

$$P = \begin{pmatrix} E(z) & I \\ N(z) & D \end{pmatrix}$$

and it has full row rank for $z \in \Pi_+$. This implies the same for the polynomial system matrix

$$P' = \begin{pmatrix} E(z) - D^+N(z) & I - D^+D \\ N(z) & D \end{pmatrix} = \begin{pmatrix} I & -D^+ \\ 0 & I \end{pmatrix} \begin{pmatrix} E(z) & I \\ N(z) & D \end{pmatrix}.$$

Now the fact that the first row of P' is of full rank in Π_+ means, by a generalization of the Hautus controllability test, that the pair $(S_{E-D^+N}, \pi_{E-D^+N} \cdot (I - D^+D))$ is stabilizable. On the other hand this pair is, by the Hautus Heymann [1978] polynomial characterization of state feedback, isomorphic to $(A - BD^+C, B(I - D^+D))$. Now the stabilizability of this pair is equivalent to

$$X_+(A - BD^+C) \subset < A - BD^+C, BIm(I - D^+D) > .$$

and since $Im(I - D^+D) = KerD$ there exists a feedback map K such that $A - BD^+C + B(I - D^+D)K$ is stable. Putting $F = -D^+C + (I - D^+D)K$, then $A + BF$ is clearly stable. Moreover

$$\begin{aligned} C + DF &= C - DD^+C + (D - DD^+D)K \\ &= (I - DD^+)(C + DK) = 0. \end{aligned}$$

$(ii) \Rightarrow (i)$ Assume D is surjective and there exists an F for which (43) and (44) hold. Let D^+ be any right inverse of D. Equality (43) implies

$$D(F + D^+C) = 0$$

i.e. $Im(F + D^+C) \subset KerD = Im(I - D^+D)$. Now define $G^+(z)$ by

$$G^+(z) = (I - F(zI - A - BF)^{-1}B)D^+.$$

Clearly $G^+(z) \in RH^\infty$. We will show that $GG^+ = I$. As

$$GG^+ = D[A, B, C, I] \times [A + BF, B, F, I]D^+$$

it suffices to show that

$$[A, B, C, I] \times [A + BF, B, F, I] = I.$$

Using the series connection we have

$$[A, B, C, I] \times [A + BF, B, F, I]$$

$$= \left[\begin{pmatrix} A + BF & 0 \\ BF & A \end{pmatrix}, \begin{pmatrix} B \\ B \end{pmatrix}, \begin{pmatrix} F & D^+C \end{pmatrix}, I\right]$$

Using the similarity matrix $\begin{pmatrix} I & 0 \\ -I & I \end{pmatrix}$ the previous series connection is isomorphic to

$$\left[\begin{pmatrix} A + BF & 0 \\ 0 & A \end{pmatrix}, \begin{pmatrix} B \\ 0 \end{pmatrix}, \begin{pmatrix} 0 & C \end{pmatrix}, I\right]$$

Hence $GG^+ = I$. ∎

5 TRANSFER FUNCTION FACTORIZATION

Given (A_i, B_i, C_i), $i = 1, 2$, canonical realizations of G_i with $\delta(G_i) = dimX_i$ the McMillan degree of G_i, then we have a realization of G_2G_1 in the state space $X_1 \oplus X_2$. This realization may not be canonical, but it gives an upper bound on the McMillan degree of a product of rational matrices. Specifically we have the inequality

$$\delta(G_2G_1) \leq \delta(G_1) + \delta(G_2).$$

We tackle now the inverse problem. Namely given a transfer function, when can it be factored into the product of two transfer functions. We will assume throughout this section that the transfer function is proper with costant term equal to the identity. We will call such a transfer function a *normalized bicausal isomorphism.* In particular the inverse of a normalized bicausal isomorphism is also a normalized bicausal isomorphism.

In much the same way as factorizations of polynomial matrices were related to invariant subspaces and controlled or conditioned invariant subspaces one expects some such relation in the case of factorizations of transfer functions. As we shall see this indeed turns out to be the case.

We pass now to the basic theorem concerning factorization of transfer functions due to Bart, Gohberg, Kaashoek and Van Dooren [1980].

Theorem 5.1 (Bart, Gohberg, Kaashoek and Van Dooren) *Let $G = [A, B, C, I]$ be a rational normalized bicausal isomorphism with the realization, in the state space X, assumed to be minimal. Then a necessary and sufficient condition for G to admit a factorization*

$$G = G_2 G_1 \tag{46}$$

with G_i normalized bicausal isomorphisms is that

$$X = M_1 \oplus M_2 \tag{47}$$

with M_1 an $A-$invariant subspace, M_2 an $A^\times-$invariant, where $A^\times$ is defined by

$$A^\times = A - BC. \tag{48}$$

Proof: We base our analysis of the factorization problem on polynomial models and this yields a concrete representation of the factors in polynomial terms. A representation of the factors in state space terms was derived in Bart, Gohberg, Kaashoek and Van Dooren [1980].

Assume (46) is a minimal factorization such that both G_i are normalized bicausal isomorphisms. Let G_1 and G_2 have the following left coprime factorizations.

$$G_1 = T_2^{-1}\overline{D}_1 \tag{49}$$

and

$$G_2 = \overline{T}_1^{-1} D_2. \tag{50}$$

Hence $G = T_2^{-1}\overline{D}_1\overline{T}_1^{-1}D_2$. Since the factorization (46) is minimal it follows that $\overline{D}_1, \overline{T}_1$ are right coprime. Let $T_1^{-1}D_1$ be a left coprime factorization of $\overline{D}_1\overline{T}_1^{-1}$ then

$$G = T_2^{-1}T_1^{-1}D_1D_2 = (T_1T_2)^{-1}(D_1D_2) = T^{-1}D \tag{51}$$

and

$$G^{-1} = (D_1D_2)^{-1}(T_1T_2). \tag{52}$$

For the shift realization of G, based on the factorization in (51), the subspace $T_1X_{T_2}$ of X_T is S_T-invariant. Similarly $D_1X_{D_2}$ is S_D-invariant. However since

$$S_D f = S_T f - \pi_T D(T^{-1}f)_{-1}, \tag{53}$$

$D_1X_{D_2}$ is a conditioned invariant subspace. Note that here S_T and S_D correspond to A and $A^\times$ respectively.

We will show now the direct sum representation

$$X_T = T_1X_{T_2} \oplus D_1X_{D_2}, \tag{54}$$

and as a first step we show that

$$T_1X_{T_2} \cap D_1X_{D_2} = \{0\}. \tag{55}$$

We know, by Theorem 3.8, that

$$T_1X_{T_2} \cap D_1X_{D_2} = X_T \cap E\mathbf{R}^p[\cdot] \tag{56}$$

where E is a l.c.r.m. of T_1 and D_1. Thus $E = T_1A = D_1B$ for a pair A, B of right coprime polynomial matrices. Since

$$T_1^{-1}D_1 = AB^{-1} = \overline{D}_1\overline{T}_1^{-1} \tag{57}$$

it follows that for some unimodular matrix U, $A = \overline{D}_1U$ and $B = \overline{T}_1U$. Hence

$$E^{-1}T = E^{-1}T_1T_2 = A^{-1}T_2 = U^{-1}(\overline{D}_1^{-1}T_2). \tag{58}$$

Thus $E^{-1}T$ is the product of a unimodular matrix and a bicausal isomorphism and so all its right Wiener-Hopf factorization indices are zero. By Theorem 3.8 this implies (55).

To show equality (54) we use a dimensionality argument. From the fact that $\overline{T}_1^{-1}D_2$ is a bicausal isomorphism it follows that $\deg\det\overline{T}_1 = \deg\det D_2$. Also the equality $T_1^{-1}D_1 = \overline{D}_1\overline{T}_1^{-1}$ of the two matrix fraction representations implies $\deg\det\overline{T}_1 = \deg\det T_1$. So taking into consideration (55) we have

$$\begin{aligned} &\dim(T_1X_{T_2} + D_1X_{D_2}) = \deg\det T_2 + \deg\det D_2 \\ &= \deg\det T_2 + \deg\det\overline{T}_1 = \deg\det T_2 + \deg\det T_1 \\ &= \deg\det T_1T_2 = \dim X_T. \end{aligned} \tag{59}$$

Hence equality follows.

Conversely assume now that the normalized bicausal isomorphism G is represented by the left coprime matrix fraction $G = T^{-1}D$ such that $T = T_1T_2$ and $D = D_1D_2$. Furthermore assume

$$X_T = T_1X_{T_2} \oplus D_1X_{D_2}. \tag{60}$$

Then we have to show that G has a factorization $G = G_2G_1$ with G_i also normalized bicausal isomorphisms. Applying Theorem 3.8 it follows from the equality

$$X_T = T_1X_{T_2} + D_1X_{D_2} \tag{61}$$

that T_1 and D_1 are left coprime. Let E be the l.c.r.m. of T_1 and D_1. Then, as $T_1X_{T_2} \cap D_1X_{D_2} = \{0\}$, it follows that $X_T \cap E\mathbf{R}^p[z] = \{0\}$ and hence the right Wiener-Hopf factorization indices are nonpositive. Let

$$E = T_1\overline{D}_1 = D_1\overline{T}_1. \tag{62}$$

Since E is the l.c.r.m. then $\overline{D}_1, \overline{T}_1$ are right coprime. Now

$$\begin{aligned} &\deg\det E = \deg\det T_1 + \deg\det\overline{D}_1 = \\ &\deg\det T_1 + \deg\det D_1 = \deg\det T_1 + \deg\det T_2. \end{aligned}$$

Here the equality $\deg\det D_1 = \deg\det T_2$ was a consequence of the direct sum representation (54). Thus necessarily $E^{-1}T_1T_2$ has trivial right factorization indices. Thus

$$E^{-1}T_1T_2 = U\Gamma \tag{63}$$

for some unimodular U and bicausal isomorphism Γ which without loss of generality we can assume normalized.

Since $X_{T_1T_2} = X_{D_1D_2}$ also $E^{-1}D_1D_2 = U_1\Gamma_1$ as before. However

$$E^{-1}T_1T_2 = (T_1\overline{D}_1)^{-1}T_1T_2 = \overline{D}_1^{-1}T_2 = U\Gamma$$

and

$$E^{-1}D_1D_2 = (D_1\overline{T}_1)^{-1}D_1D_2 = \overline{T}_1^{-1}D_2 = U_1\Gamma_1.$$

The equality $T_1\overline{D}_1 = D_1\overline{T}_1$ implies

$$T_1^{-1}D_1 = \overline{D}_1\overline{T}_1^{-1}$$

and hence

$$G = T_2^{-1}T_1^{-1}D_1D_2 = (T_2^{-1}\overline{D}_1)(\overline{T}_1^{-1}D_2) = \Gamma^{-1}U^{-1}U_1\Gamma_1.$$

This in turn implies $\Gamma G\Gamma_1^{-1} = U^{-1}U_1$. Thus the factorization $G = G_2G_1$ has been established with $G_1 = \Gamma_1$ and $G_2 = \Gamma$. ∎

A remark is in order. A l.c.r.m. is only defined up to a right unimodular factor. Thus without loss of generality we can, by (63), assume $E^{-1}T_1T_2$ is a normalized bicausal isomorphism. Thus $X_E = X_{T_1T_2}$ and since

$$E = T_1\overline{D}_1 = D_1\overline{T}_1$$

with T_1, D_1 left coprime and $\overline{T}_1, \overline{D}_1$ right coprime we have

$$X_E = T_1X_{\overline{D}_1} \oplus D_1X_{\overline{T}_1}.$$

This means that on X_T we can redefine the $\mathbf{R}[z]$−module structure so that both $T_1X_{T_2}$ and $D_1X_{D_2}$ both become submodules. Thus the invariant subspace $T_1X_{T_2}$ and the conditioned invariant subspace $D_1X_{D_2}$ both considered as conditioned invariant subspaces are compatible. This should come as no surprise as we know that two linearly independent controlled invariant subspaces are compatible. Hence also two conditioned invariant subspaces whose sum equals the whole state space are compatible in the sense that for some constant H both are $(A + HC)$−invariant, as was the case here.

We focus now on a special case of Wiener-Hopf factorizations.

Definition 5.1 *Let $G \in RH^\infty$ be such that G, G^{-1} have no poles on $i\mathbf{R}$. A factorization*

$$G = G_+G_-$$

such that G_-, G_-^{-1} are stable and G_+, G_+^{-1} are antistable is called a canonical factorization.

The following theorem has been proved in Bart, Gohberg and Kaashoek [1979].

Theorem 5.2 *A normalized bicausal isomorphism G with minimal realization $[A, B, C, I]$ has a canonical factorization if and only if $X_-(A^\times)$ and $X_+(A)$ are complementary subspaces.*

Proof: Assume $G = G_+G_-$ is a canonical factorization. Such a factorization is automatically minimal. Let $G_+ = T_+^{-1}\overline{D}_+$ and $G_- = \overline{T}_-^{-1}D_-$ be a left coprime factorization.

Let $\overline{D}_+\overline{T}_-^{-1} = T_-^{-1}D_+$ with the last one a left coprime factorization. Clearly T_- is stable and D_+ antistable. Thus

$$G_+G_- = T_+^{-1}T_-^{-1}D_+D_- = (T_-T_+)^{-1}(D_+D_-)$$

By Theorem 5.1 we must have

$$X_{T_-T_+} = T_-X_{T_+} \oplus D_+X_{D_-} \tag{64}$$

But $T_-X_{T_+} = X_+(A)$ and $D_+X_{D_-} = X_-(A^\times)$.

Conversely, assume $X = X_+(A) \oplus X_-(A^\times)$. Thus a factorization exists. Let $G = T^{-1}D$. Factor $T = T_-T_+$ and $D = D_+D_-$ such that D_-, T_- are stable and T_+, D_+ are antistable. Such factorizations exist. Also $T_-X_{T_+} = X_+(A)$ and $D_+X_{D_-} = X_-(A^\times)$. By our assumption $X_{T_-T_+} = T_-X_{T_+} \oplus D_+X_{D_-}$ so by Theorem 5.1 this implies a factorization $G = G_+G_-$. ∎

6 SPECTRAL FACTORIZATION

In this section we study the problem of spectral factorization and its relation to the algebraic Riccati equation. Our main object is to obtain state space formulas for the spectral factors.

Definition 6.1 *Given a Hamiltonian symmetric normalized bicausal isomorphism Φ we define a* symmetric factorization *of Φ as a factorization*

$$\Phi(z) = V_*(z)V(z) \tag{65}$$

such that V is also a normalized bicausal isomorphism. The factorization is called minimal *if $\delta(\Phi) = 2\delta(V)$, i.e. if there are no pole-zero cancellations between V and V_*.*

Theorem 6.1 *Let G be a Hamiltonian symmetric transfer function having no poles or zeroes on iR and such that $G(\infty) > 0$. Then $G(z)$ admits a spectral factorization of the form*

$$G(z) = \tilde{G}_-(-z)G_-(z) \tag{66}$$

with G_-, G_-^{-1} stable.
This is called a spectral factorization *and G_- is called a* spectral factor.

Proof: Let d be the characteristic polynomial of G, i.e. the characteristic polynomial of the generator in any minimal realization of G, and let $d = d_+d_-$ be its factorization into a stable and an antistable factors. Let $G = MD^{-1}$ be a right coprime factorization of G. Then D admits a factorization $D = D_+D_-$ with $d_+ = detD_+$ and $d_- = detD_-$. Let $\overline{D}_+^{-1}N$ be a left coprime factorization of MD_+^{-1}. So

$$G(z) = MD_+^{-1}D_-^{-1} = \overline{D}_+^{-1}ND_-^{-1} \tag{67}$$

From the fact that G is Hamiltonian we obtain

$$\overline{D}_+(z)^{-1}N(z)D_-(z)^{-1} = \tilde{D}_-(-z)^{-1}\tilde{N}(-z)\tilde{\overline{D}}_+^{-1}(-z) \tag{68}$$

It follows that, up to a unimodular factor which we incorporate in the polynomial matrix N, we have

$$\overline{D}_+(z) = \tilde{D}_-(-z) \tag{69}$$

and hence

$$G(z) = \tilde{D}_-(-z)^{-1} N D_-(z)^{-1} \tag{70}$$

Again, the fact that G is Hamiltonian implies $\tilde{N}(-z) = N(z)$.

Now the assumption that G is invertible on iR implies that N is invertible there, i.e. there are no imaginary zeroes. Since

$$N(it) = \tilde{D}_-(-it) G(it) D_-(it) \tag{71}$$

is positive, it follows from Theorem 2.13, that N has a spectral factorization

$$N(z) = \tilde{E}_-(-z) E_-(z). \tag{72}$$

With $G_-(z) = E_-(z) D_-(z)^{-1}$, this in turn leads to the spectral factorization (66). Clearly the boundedness of G at infinity implies that of G_-.

We quote next the main results on the Riccati equation needed for our purposes. Assume now that we are given the the algebraic Riccati equation (ARE)

$$\tilde{A}X + XA - XB\tilde{B}X + Q = 0. \tag{73}$$

With the Riccati equation we associate the Hamiltonian matrix

$$H = \begin{pmatrix} A & -B\tilde{B} \\ -Q & -\tilde{A} \end{pmatrix} \tag{74}$$

as well as the rational matrix function $\Phi(z)$ defined by

$$\Phi(z) = I - \tilde{B}(zI + \tilde{A})^{-1} Q (zI - A)^{-1} B. \tag{75}$$

Clearly Φ is an $n \times n$ causally invertible rational matrix function satisfying $\Phi(\infty) = I$, i.e. Φ is a normalized bicausal isomorphism. Moreover it is easily established that Φ is Hamiltonian symmetric.

Lemma 6.1 *(i) For Φ defined by*

$$\Phi(z) = I - \tilde{B}(zI + \tilde{A})^{-1} Q (zI - A)^{-1} B. \tag{76}$$

we have

$$\Phi(z) = I - \begin{pmatrix} 0 & \tilde{B} \end{pmatrix} \begin{pmatrix} zI - A & 0 \\ Q & zI + \tilde{A} \end{pmatrix}^{-1} \begin{pmatrix} B \\ 0 \end{pmatrix} \tag{77}$$

or

$$\Phi = \left[\begin{pmatrix} A & 0 \\ -Q & -\tilde{A} \end{pmatrix}, \begin{pmatrix} B \\ 0 \end{pmatrix}, \begin{pmatrix} 0 & \tilde{B} \end{pmatrix}, I \right]. \tag{78}$$

(ii) For Φ as above we have

$$\Phi(z)^{-1} = I + \begin{pmatrix} 0 & \tilde{B} \end{pmatrix} \begin{pmatrix} zI - A & B\tilde{B} \\ Q & zI + \tilde{A} \end{pmatrix}^{-1} \begin{pmatrix} B \\ 0 \end{pmatrix} \tag{79}$$

i.e.

$$\Phi^{-1} = \left[H, \begin{pmatrix} B \\ 0 \end{pmatrix}, \begin{pmatrix} 0 & \tilde{B} \end{pmatrix}, I \right]. \tag{80}$$

Observe also that Φ defined by Equation (75) has a realization in $\mathbf{R}^{2n}$. In particular its McMillan degree $\delta(\Phi)$ is bounded by $2n$. Our standard assumption throughout this paper is that $\delta(\Phi) = 2n$. In particular this implies the assumption of reachability of the pair (A, B), for otherwise $(zI - A)^{-1}B$ is not a (left) coprime factorization and we could reduce common left factors. Similarly, if Q is nonnegative definite, one can write $Q = C\tilde{C}$. In any such factorization (A, C) is necessarily an observable pair.

Under these assumptions we can assume without loss of generality that the spectra of A and H are disjoint. We will not dwell on this since all the details can be found in Molinari's paper [1973b].

Observe that the realizations of Φ and Φ^{-1} given by (77) and (79) respectively are Hamiltonian realizations in the sense of Brockett and Rahimi [1972]. In this connection see also Fuhrmann [1984].

Definition 6.2 *A real symmetric solution K of ARE is called an* unmixed solution *if*

$$\sigma(A - B\tilde{B}K) \cap \sigma(-\tilde{A} + KB\tilde{B}) = \emptyset. \tag{81}$$

The concept of unmixed solution was introduced by Shayman [1983] as a generalization of extremal solutions of the ARE studied by Willems [1971].

We are ready now to state the following theorem, for a proof see Fuhrmann [1985].

Theorem 6.2 *Let the matrices A, B and Q be given with (A, B) reachable and Q symmetric. Let Φ be defined by*

$$\Phi(z) = I - \tilde{B}(zI + \tilde{A})^{-1}Q(zI - A)^{-1}B. \tag{82}$$

Let H be the Hamiltonian matrix

$$H = \begin{pmatrix} A & -B\tilde{B} \\ -Q & -\tilde{A} \end{pmatrix} \tag{83}$$

and let J be given by

$$J = \begin{pmatrix} 0 & I \\ -I & 0 \end{pmatrix}. \tag{84}$$

Then there exists a bijective correspondence between the following:

(i) Real symmetric solutions of the ARE.

(ii) Minimal symmetric factorizations of $\Phi(z)$.

(iii) Hamiltonian symmetric factorizations of $(zI - H)J$.

(iv) Lagrangian H−invariant subspaces of $\mathbf{R}^{2n}$.

(v) Invariant subspaces of $A - B\tilde{B}K$ for any unmixed real symmetric solution K of the ARE.

(vi) Polynomial matrix solutions N of the Polynomial Riccati Equation *(PRE)*

$$N_*N = D_*D + H_*QH \tag{85}$$

for which ND^{-1} is an normalized bicausal isomorphism and where D, H are right coprime polynomial matrices satisfying

$$(zI - A)^{-1}B = H(z)D(z)^{-1}. \tag{86}$$

Some remarks concerning the history of this theorem are appropriate. The equivalence of (i) and (ii) is essentially contained in Anderson [1969] and Willems [1971] and is fully presented in Molinari [1973b]. That (i) implies (iv) is due to Potter [1966] and others. That (iv) implies (i) as well as the equivalence of (iv) and (v) is due to Shayman [1983]. A result similar to the implication of (iv) by (i) appears in Lancaster and Rodman [1980], The equivalence to (vi) is from Fuhrmann [1985].

Theorem 6.3 *Assume (A, B) is reachable. Then the following statements are equivalent:*

(i) The algebraic Riccati equation has a real symmetric stabilizing solution $\hat{X}$, i.e. $\sigma(A - B\tilde{B}\hat{X}) \subset \Pi_-$.

(ii) The algebraic Riccati equation has a real symmetric antistabilizing solution $\check{X}$, i.e. $\sigma(A - B\tilde{B}\check{X}) \subset \Pi_+$.

(iii) For Φ defined by we have

$$\Phi(i\omega) > \varepsilon \tilde{B}(-i\omega I - \tilde{A})^{-1}(i\omega I - A)^{-1}B \tag{87}$$

for some $\varepsilon > 0$.
(iv) For the Hamiltonian matrix H we have $\sigma(H) \cap iR = \emptyset$.

Clearly the reachability of the pair (A, B) is not necessary for the existence of a real symmetric stabilizing solution. The following result makes this precise.

Theorem 6.4 *The algebraic Riccati equation has a real symmetric stabilizing solution if and only if the following two conditions hold:*

(i) The pair (A, B) is stabilizable.

(ii) For the Hamiltonian matrix H we have $\sigma(H) \cap iR = \emptyset$.

Next we obtain a state space representation of the spectral factors.

Theorem 6.5 *Let $G = [A, B, C, D]$ be a Hamiltonian symmetric transfer function having no poles or zeroes on iR and such that $G(\infty) > 0$. Let*

$$G(z) = D + H_+(z) + H_-(z) \tag{88}$$

be such that H_+ and H_- are strictly proper and stable and antistable respectively. Let $H = [A_-, B_-, C_-]$ with the realization assumed minimal. Then the right spectral factor of G is given by $G_- = [A_-, B_-, D^{-\frac{1}{2}}(C_- - \tilde{B}_- X), D^{\frac{1}{2}}]$ where $D^{\frac{1}{2}}$ is the positive square root of D and X is the unique stabilizing solution of the algebraic Riccati equation

$$X(A_- - B_- D^{-1}C_-) + (\tilde{A}_- - \tilde{C}_- D^{-1}\tilde{B}_-)X + XB_- D^{-1}\tilde{B}_- X + \tilde{C}_- D^{-1}C_- = 0, \tag{89}$$

Proof: Let $G(\infty) = D$. Let d be the characteristic polynomial of G, and let $d = d_+ d_-$ be its factorization into a stable and an antistable factors respectively. Let $G(z) = D + H_+(z) + H_-(z)$ be a partial fraction decomposition of G. So H_- and H_+ are strictly proper and have their poles in the left and right half planes respectively. As G is Hamiltonian clearly $\tilde{D} = D$ and $H_+(z) = \tilde{H}_-(-z)$. Let $H_- = N_- D_-^{-1}$ be a right coprime factorization of H_-. Associate with H_- a minimal realization

$$H_-(z) = C_-(zI - A_-)^{-1} B_-. \tag{90}$$

Clearly

$$H_+(z) = -\tilde{B}_-(zI + \tilde{A}_-)^{-1} \tilde{C}_-. \tag{91}$$

Now

$$\begin{aligned} G(z) &= D + \tilde{D}_-^{-1} \tilde{N}_- + N_- D_-^{-1} \\ &= \tilde{D}_-^{-1} \{\tilde{D}_- D D_- + \tilde{N}_- D_- + \tilde{D}_- N_-\} D_-^{-1} \\ &= \tilde{D}_-^{-1} N D_-^{-1} \end{aligned} \tag{92}$$

Obviously N is Hamiltonian symmetric.

Now the assumption that G is invertible on $i\mathbf{R}$ implies that N is invertible there too, i.e. there are no imaginary zeroes. Since

$$N(it) = \tilde{D}_-(-it) G(it) D_-(it) \tag{93}$$

is positive, it follows from Theorem 2.13, that N has a spectral factorization

$$N(z) = \tilde{E}_-(-z) E_-(z). \tag{94}$$

With $G_-(z) = E_-(z) D_-(z)^{-1}$, this in turn leads to the spectral factorization (66). Clearly the boundedness of G at infinity implies that of G_-.

Now $G(z)$ is the parallel connection of $H_+(z)$ and $H_-(z)$ and has constant term D, so it has a realization

$$\left[\begin{pmatrix} A_- & 0 \\ 0 & -\tilde{A}_- \end{pmatrix}, \begin{pmatrix} B_- \\ \tilde{C}_- \end{pmatrix}, \begin{pmatrix} C_- & -\tilde{B}_- \end{pmatrix}, D \right]. \tag{95}$$

On the other hand we proved G has a factorization

$$G(z) = G_+(z) G_-(z) = \tilde{G}_-(-z) G_-(z) \tag{96}$$

where $G_-(z) = E_-(z) D_-(z)^{-1}$. Comparing this with $H_-(z) = N_-(z) D_-(z)^{-1}$ it follows that G_- has a realization $[A_-, B_-, C_0, D^{\frac{1}{2}}]$ where C_0 has to be determined. This realization implies the realization $[-\tilde{A}_-, \tilde{C}_0, -\tilde{B}_-, D^{\frac{1}{2}}]$ for $\tilde{G}_-(-z)$. Hence G as a series coupling of G_- and $\tilde{G}_-(-z)$ has a realization

$$\left[\begin{pmatrix} A_- & 0 \\ \tilde{C}_0 C_0 & -\tilde{A}_- \end{pmatrix}, \begin{pmatrix} B_- \\ \tilde{C}_0 D^{\frac{1}{2}} \end{pmatrix}, \begin{pmatrix} D^{\frac{1}{2}} C_0 & -\tilde{B}_- \end{pmatrix}, D \right]. \tag{97}$$

As both realizations (95) and (97) are canonical they are, by the state space isomorphism theorem, isomorphic. Based on the study of skew primeness, we will look for a map of the form $\begin{pmatrix} I & 0 \\ X & I \end{pmatrix}$, with X symmetric, intertwining the two realizations. The intertwining condition reduces to the following two matrix equations:

$$\begin{pmatrix} I & 0 \\ X & I \end{pmatrix} \begin{pmatrix} A_- & 0 \\ \tilde{C}_0 C_0 & -\tilde{A}_- \end{pmatrix} = \begin{pmatrix} A_- & 0 \\ 0 & -\tilde{A}_- \end{pmatrix} \begin{pmatrix} I & 0 \\ X & I \end{pmatrix}, \tag{98}$$

as well as

$$\begin{pmatrix} C_- & -\tilde{B}_- \end{pmatrix} \begin{pmatrix} I & 0 \\ X & I \end{pmatrix} = \begin{pmatrix} D^{\frac{1}{2}} C_0 & -\tilde{B}_- \end{pmatrix}. \tag{99}$$

The third relation follows from the previous one by symmetry.

Thus we must have

$$\tilde{A}_- X + X A_- = -\tilde{C}_0 C_0 \tag{100}$$

and

$$C_- - \tilde{B}_- X = D^{\frac{1}{2}} C_0 \tag{101}$$

Thus $C_0 = D^{-\frac{1}{2}}(C_- - \tilde{B}_- X)$ and substituting this back into equation (100) we obtain the algebraic Riccati equation

$$X(A_- - B_- D^{-1} C_-) + (\tilde{A}_- - \tilde{C}_- D^{-1} \tilde{B}_-) X + X B_- D^{-1} \tilde{B}_- X + \tilde{C}_- D^{-1} C_- = 0, \tag{102}$$

i.e. X must be a solution to the ARE that corresponds to the Hamiltonian matrix

$$H = \begin{pmatrix} A_- - B_- D^{-1} C_- & -B_- D^{-1} \tilde{B}_- \\ \tilde{C}_- D^{-1} C_- & -(\tilde{A}_- - \tilde{C}_- D^{-1} \tilde{B}_-) \end{pmatrix} \tag{103}$$

Now, from the realization (95) it follows that G^{-1} is realized by

$$\left[\begin{pmatrix} A_- - B_- D^{-1} C_- & -B_- D^{-1} \tilde{B}_- \\ \tilde{C}_- D^{-1} C_- & -(\tilde{A}_- - \tilde{C}_- D^{-1} \tilde{B}_-) \end{pmatrix}, \begin{pmatrix} B_- D^{-1} \\ -\tilde{C}_- D^{-1} \end{pmatrix}, \begin{pmatrix} D^{-1} C_- & -D^{-1} \tilde{B}_- \end{pmatrix}, D^{-1} \right] \tag{104}$$

However $G^{-1}(z) = D_-(\tilde{E}_- E_-)^{-1} \tilde{D}_-$ has a Hamiltonian realization in $X_{\tilde{E}_- E_-}$ with generator $S_{\tilde{E}_- E_-}$, so the two realizations are symplectically equivalent. Since the last map has a Lagrangian invariant subspace, namely $\tilde{E}_- X_{E_-}$, so has the matrix H. Hence by Theorem 6.2, the corresponding Riccati equation is solvable. Since $G_- = E_- D_-^{-1}$ is a spectral factor both G and G^{-1} are stable.

Now as G_- has the realization $[A_-, B_-, C_0, D^{\frac{1}{2}}]$ it follows that

$$G_-^{-1} = [A_- - B_- D^{-\frac{1}{2}} C_0, B_- D^{-\frac{1}{2}}, D^{-\frac{1}{2}} C_0, D^{-\frac{1}{2}}].$$

The stability of G_-^{-1} means that $A_- - B_- D^{-\frac{1}{2}} C_0$ is also stable. However it is easily computed that

$$\begin{pmatrix} A_- - B_- D^{-1} C_- & -B_- D^{-1} \tilde{B}_- \\ \tilde{C}_- D^{-1} C_- & -(\tilde{A}_- - \tilde{C}_- D^{-1} \tilde{B}_-) \end{pmatrix} \begin{pmatrix} I & 0 \\ X & I \end{pmatrix}$$

$$= \begin{pmatrix} I & 0 \\ X & I \end{pmatrix} \begin{pmatrix} A_- - B_- D^{-1} C_0 & -B_- D^{-1} \tilde{B}_- \\ 0 & -(\tilde{A}_- - \tilde{C}_0 D^{-1} \tilde{B}_-) \end{pmatrix}$$

so that $\begin{pmatrix} I \\ X \end{pmatrix}$ is a basis matrix for $X_-(H)$. In fact if $\begin{pmatrix} U \\ V \end{pmatrix}$ is any basis matrix for $X_-(H)$ then necessarily U is invertible and $X = VU^{-1}$. ∎

An obvious way to construct Hamiltonian symmetric transfer functions is to study transfer functions of the form $\tilde{G}(-z)G(z)$ where G is an arbitrary real transfer function. The next theorem tackles spectral factorizations of functions given in this form.

Theorem 6.6 *Let $G \in RH^\infty$ be an proper transfer function, and assume $G(it)$ is injective for all real t. Further assume G has a realization $[A, B, C, D]$. Let $\underline{D} = \tilde{D}D$. Then G_*G has a spectral factorization $G_{-*}G_-$, with the spectral factor G_- realized by*

$$\left[A, B, \underline{D}^{-\frac{1}{2}}(\tilde{D}C - \tilde{B}X), \underline{D}^{\frac{1}{2}}\right] \tag{105}$$

and X is the stabilizing solution of the algebraic Riccati equation corresponding to the Hamiltonian matrix

$$\begin{pmatrix} A - B\underline{D}^{-1}\tilde{D}C & -B\underline{D}^{-1}\tilde{B} \\ -(\tilde{C}C - \tilde{C}D\underline{D}^{-1}\tilde{D}C) & -(\tilde{A} - \tilde{C}D\underline{D}^{-1}\tilde{B}) \end{pmatrix} \tag{106}$$

Proof: Since we assume G is left invertible it follows that $G(\infty) = D$ is injective. Clearly $\underline{D}$ is invertible. Let

$$G = NE_-^{-1} \tag{107}$$

be a right coprime factorization. As $G \in RH^\infty$ it follows that E_- is a stable polynomial matrix. Thus $\tilde{G}G = \tilde{E}_-^{-1}\tilde{N}NE_-^{-1}$ and a spectral factorization of $\tilde{G}G$ is obtained from a spectral factorization of $\tilde{N}N$, i.e. $\tilde{N}N = \tilde{N}_-N_-$. Such a factorization exists by Theorem 2.13 as we assume $N(it)$ is injective for all real t which implies $\tilde{N}N > 0$ on the imaginary axis. Thus

$$G_-(z) = N_-(z)E_-(z)^{-1} \tag{108}$$

and hence,comparing (107) and (108) and as

$$\tilde{G}_-(\infty)G_-(\infty) = \tilde{G}(\infty)G(\infty) = \tilde{D}D = \underline{D}$$

we must have $G_-(\infty) = \underline{D}^{\frac{1}{2}}$ and, applying Theorem 3.3, G_- is realized by $[A, B, C_0, \underline{D}^{\frac{1}{2}}]$, i.e.

$$G_-(z) = \underline{D}^{\frac{1}{2}} + C_0(zI - A)^{-1}B.$$

This implies

$$\tilde{G}_-(-z)G_-(z) = (\underline{D}^{\frac{1}{2}} - \tilde{B}(zI + \tilde{A})^{-1}\tilde{C}_0)(\underline{D}^{\frac{1}{2}} + C_0(zI - A)^{-1}B) \tag{109}$$

A product of transfer functions is realized by the series coupling and so $\tilde{G}_-(-z)G_-(z)$ is realized by

$$\left[\begin{pmatrix} A & 0 \\ \tilde{C}_0C_0 & -\tilde{A} \end{pmatrix}, \begin{pmatrix} B \\ \tilde{C}_0\underline{D}^{\frac{1}{2}} \end{pmatrix}, \begin{pmatrix} \underline{D}^{\frac{1}{2}}C_0 & -\tilde{B} \end{pmatrix}, \underline{D}\right]. \tag{110}$$

But on the other hand the equality $G_*G = G_{-*}G_-$ shows that there is an equivalent realization given by

$$\left[\begin{pmatrix} A & 0 \\ \tilde{C}C & -\tilde{A} \end{pmatrix}, \begin{pmatrix} B \\ \tilde{C}D \end{pmatrix}, \begin{pmatrix} \tilde{D}C & -\tilde{B} \end{pmatrix}, \tilde{D}D\right]. \tag{111}$$

The two realizations (110) and (111) are isomorphic. The map $\begin{pmatrix} I & 0 \\ X & I \end{pmatrix}$ defines an isomorphism of the two realizations if and only if

$$\begin{pmatrix} I & 0 \\ X & I \end{pmatrix}\begin{pmatrix} A & 0 \\ \tilde{C}_0C_0 & -\tilde{A} \end{pmatrix} = \begin{pmatrix} A & 0 \\ \tilde{C}C & -\tilde{A} \end{pmatrix}\begin{pmatrix} I & 0 \\ X & I \end{pmatrix} \tag{112}$$

and

$$\begin{pmatrix} \tilde{D}C & -\tilde{B} \end{pmatrix} \begin{pmatrix} I & 0 \\ X & I \end{pmatrix} = \begin{pmatrix} \underline{D}^{\frac{1}{2}}C_0 & -\tilde{B} \end{pmatrix}. \tag{113}$$

hold. These two matrix equations reduce to

$$\tilde{A}X + XA + \tilde{C}_0C_0 - \tilde{C}C = 0 \tag{114}$$

and

$$\tilde{D}C - \tilde{B}X = \underline{D}^{\frac{1}{2}}C_0. \tag{115}$$

Solving the last equation for C_0 yields

$$C_0 = \underline{D}^{-\frac{1}{2}}(\tilde{D}C - \tilde{B}X). \tag{116}$$

which upon substituting back into equation (114) leads to

$$\begin{aligned} 0 &= \tilde{A}X + XA - \tilde{C}C + \tilde{C}_0C_0 \\ &= \tilde{A}X + XA - \tilde{C}C + (\tilde{C}D - XB)\underline{D}^{-1}(\tilde{D}C - \tilde{B}X). \end{aligned} \tag{117}$$

This is the algebraic Riccati equation

$$X(A - B\underline{D}^{-1}\tilde{D}C) + (\tilde{A}\_\tilde{C}D\underline{D}^{-1}\tilde{B})X - \tilde{C}C + \tilde{C}D\underline{D}^{-1}\tilde{D}C + XB\underline{D}^{-1}\tilde{B}X = 0. \tag{118}$$

This algebraic Riccati equation is associated with the Hamiltonian matrix

$$\begin{pmatrix} A - B\underline{D}^{-1}\tilde{D}C & -B\underline{D}^{-1}\tilde{B} \\ -(\tilde{C}C - \tilde{C}D\underline{D}^{-1}\tilde{D}C) & -(\tilde{A} - \tilde{C}D\underline{D}^{-1}\tilde{B}) \end{pmatrix} \tag{119}$$

From an inspection of the realization (111) it is clear that H is the generator of a realization of $(\tilde{G}(-z)G(z))^{-1}$. But

$$(\tilde{G}(-z)G(z))^{-1} = E_-(-z)(\tilde{N}(-z)N(z))^{-1}\tilde{E}_-(z).$$

So H is symplectically equivalent to $S_{\tilde{N}N}$. The last map clearly has an invariant Lagrangian subspace and therefore so has H. Thus, by Theorem 6.2, the algebraic Riccati equation (118) has a symmetric solution. Moreover there exists a unique stabilizing solution i.e. one for which

$$A - B\underline{D}^{-1}\tilde{D}C - B\underline{D}^{-1}\tilde{B}X$$

is stable. ∎

The next theorem discusses a special case of Theorem 6. In this case the algebraic Riccati equation reduces to a homogeneous one which is equivalent to the Liapunov equation.

Theorem 6.7 *Let G be a rational proper transfer function with realization $[A, B, C, I]$. Assume A is stable and $A^{\times} = A - BC$ is antistable. Then G_*G has a spectral factorization $G_{-*}G_-$, with the spectral factor G_- realized by*

$$\left[A, B, C - \tilde{B}X, I\right] \tag{120}$$

and X is the unique stabilizing solution of the homogeneous algebraic Riccati equation

$$(\tilde{A} - \tilde{C}\tilde{B})X + X(A - BC) + XB\tilde{B}X = 0. \tag{121}$$

Proof: Let

$$G(z) = N_+(z)E_-^{-1}(z) \tag{122}$$

then $G(z)^{-1} = E_-(z)N_+(z)^{-1}$. By the assumptions of the theorem E_- is stable and N_+ is antistable. A spectral factorization of G_*G is obtained from a spectral factorization $\tilde{N}_-N_-$ of $\tilde{N}_+N_+$. Since $\tilde{N}_+X_{N_+}$ is just the antistable invariant subspace of $S_{\tilde{N}_+N_+}$ then the spectral factorization problem reduces to finding the complementary stable invariant subspace. This reduces to a Liapunov equation, which we proceed to derive. From (122) it follows that G_- is realized by $[A, B, C_0, I]$. Computing $\tilde{G}(-z)G(z) = \tilde{G}_-(-z)_-G(z)$ we have two realizations of G_*G given by

$$\left[\begin{pmatrix} A & 0 \\ \tilde{C}C & -\tilde{A} \end{pmatrix}, \begin{pmatrix} B \\ \tilde{C} \end{pmatrix}, \begin{pmatrix} C & -\tilde{B} \end{pmatrix}, I\right]. \tag{123}$$

and

$$\left[\begin{pmatrix} A & 0 \\ \tilde{C}_0C_0 & -\tilde{A} \end{pmatrix}, \begin{pmatrix} B \\ \tilde{C}_0 \end{pmatrix}, \begin{pmatrix} C_0 & -\tilde{B} \end{pmatrix}, I\right]. \tag{124}$$

respectively. The map $\begin{pmatrix} I & 0 \\ X & I \end{pmatrix}$ provides an isomorphism of the two realizations if and only if the two matrix equations

$$\begin{pmatrix} I & 0 \\ X & I \end{pmatrix}\begin{pmatrix} A & 0 \\ \tilde{C}_0C_0 & -\tilde{A} \end{pmatrix} = \begin{pmatrix} A & 0 \\ \tilde{C}C & -\tilde{A} \end{pmatrix}\begin{pmatrix} I & 0 \\ X & I \end{pmatrix} \tag{125}$$

and

$$\begin{pmatrix} C & -\tilde{B} \end{pmatrix}\begin{pmatrix} I & 0 \\ X & I \end{pmatrix} = \begin{pmatrix} C_0 & -\tilde{B} \end{pmatrix}. \tag{126}$$

hold. These two matrix equations reduce to

$$C_0 = C - \tilde{B}X \tag{127}$$

and

$$\tilde{A}X + XA + \tilde{C}_0C_0 - \tilde{C}C = 0 \tag{128}$$

Substituting (127) into (128) we obtain the homogeneous Riccati equation

$$(\tilde{A} - \tilde{C}\tilde{B})X + X(A - BC) + XB\tilde{B}X = 0. \tag{129}$$

But, as $A - BC$ is antistable, there exists, by Liapunov's theorem, a positive definite solution Y to the equation

$$Y(\tilde{A} - \tilde{C}\tilde{B}) + (A - BC)Y = B\tilde{B} \tag{130}$$

Letting $X = Y^{-1}$ we get a solution to the homogeneous algebraic Riccati equation. ∎

Assume $G \in RH^\infty$. Certainly if $\gamma > \|G\|_\infty$ then $\gamma^2 I - G_*G > 0$ on iR. Thus a spectral factorization exists. The following result gives a state space formula for the spectral factor of G.

Theorem 6.8 *Let $G \in RH^\infty$ have the realization $[A,B,C,D]$ and let $\gamma > \|G\|_\infty$. Let $\underline{D} = (\gamma^2 I - \tilde{D}D)^{\frac{1}{2}}$. Then G_*G has a spectral factorization $G_{-*}G_-$, with the spectral factor G_- realized by*

$$\left[A, B, \underline{D}^{-1}(\tilde{D}C - \tilde{B}X), \underline{D}\right] \tag{131}$$

where X is the unique stabilizing solution of the algebraic Riccati equation associated with the Hamiltonian matrix

$$\begin{pmatrix} A - B\underline{D}^{-2}\tilde{D}C & -B\underline{D}^{-2}\tilde{B} \\ -(\tilde{C}C - \tilde{C}D\underline{D}^{-2}\tilde{D}C) & -(\tilde{A} - \tilde{C}D\underline{D}^{-2}\tilde{B}) \end{pmatrix} \tag{132}$$

Proof: Assume $G(z) = N(z)E_-(z)^{-1}$ and by our assumption E_- is stable. As

$$\gamma^2 I - \tilde{G}(-z)G(z) = \gamma^2 I - (D - \tilde{B}(zI + \tilde{A})^{-1}\tilde{C})(D + C(zI - A)^{-1}B) \tag{133}$$

it follows that $\gamma^2 I - \tilde{G}(-z)G(z)$ is realized by

$$\left[\begin{pmatrix} A & 0 \\ \tilde{C}C & -\tilde{A} \end{pmatrix}, \begin{pmatrix} B \\ \tilde{C}D \end{pmatrix}, \begin{pmatrix} -\tilde{D}C & \tilde{B}_- \end{pmatrix}, \gamma^2 I - \tilde{D}D\right]. \tag{134}$$

On the other hand we have also

$$\begin{aligned} \gamma^2 I - G_*G &= \gamma^2 I - (E_{-*})^{-1}N_*NE_-^{-1} \\ &= (E_{-*})^{-1}[\gamma^2 E_{-*}E_- - N_*N]E_-^{-1} \end{aligned} \tag{135}$$

and as the middle factor is positive on iR we have a spectral factorization

$$\gamma^2 E_{-*}E_- - N_*N = N_{-*}N_-. \tag{136}$$

Hence

$$\gamma^2 I - G_*G = [(E_{-*})^{-1}N_{-*}][N_-E_-^{-1}] = G_{-*}G_-. \tag{137}$$

From the representation $G_- = N_-E_-^{-1}$ it follows that G_- has a realization $G_-(z) = \underline{D} + C_0(zI - A)^{-1}B$ with $\underline{D} = (\gamma^2 I - \tilde{D}D)^{\frac{1}{2}}$. So

$$\gamma^2 I - \tilde{G}(-z)G(z) = G_{-*}(z)G_-(z) = (\underline{D} - \tilde{B}(zI + \tilde{A})^{-1}\tilde{C}_0)(\underline{D} + C_0(zI - A)^{-1}B). \tag{138}$$

This implies the realization

$$\left[\begin{pmatrix} A & 0 \\ \tilde{C}_0C_0 & -\tilde{A} \end{pmatrix}, \begin{pmatrix} B \\ \tilde{C}_0\underline{D} \end{pmatrix}, \begin{pmatrix} \underline{D}C_0 & -\tilde{B} \end{pmatrix}, \underline{D}\right] \tag{139}$$

for $\gamma^2 I - G_*G$. The two realizations are isomorphic and we try an isomorphism of the form $\begin{pmatrix} I & 0 \\ X & I \end{pmatrix}$. This map will indeed be an isomorphism provided the following equalities

$$\begin{pmatrix} I & 0 \\ X & I \end{pmatrix}\begin{pmatrix} A & 0 \\ \tilde{C}_0C_0 & -\tilde{A} \end{pmatrix} = \begin{pmatrix} A & 0 \\ \tilde{C}C & -\tilde{A} \end{pmatrix}\begin{pmatrix} I & 0 \\ X & I \end{pmatrix} \tag{140}$$

and

$$\begin{pmatrix} \tilde{D}C_0 & -\tilde{B} \end{pmatrix}\begin{pmatrix} I & 0 \\ X & I \end{pmatrix} = \begin{pmatrix} \underline{D}C_0 & -\tilde{B} \end{pmatrix} \tag{141}$$

hold. These two matrix equations imply

$$\tilde{A}X + XA + \tilde{C}_0C_0 - \tilde{C}C = 0 \tag{142}$$

and

$$\tilde{D}C - \tilde{B}X = \underline{D}C_0. \tag{143}$$

Solving the last equation for C_0 yields

$$C_0 = \underline{D}^{-1}(\tilde{D}C - \tilde{B}X). \tag{144}$$

which upon substituting back into equation (142) leads to

$$X(A - B\underline{D}^{-2}\tilde{D}C) + (\tilde{A}_- - \tilde{C}D\underline{D}^{-2}\tilde{B})X - \tilde{C}C + \tilde{C}D\underline{D}^{-2}\tilde{D}C + XB\underline{D}^{-2}\tilde{B}X = 0, \tag{145}$$

i.e. this is the algebraic Riccati equation associated with the Hamiltonian matrix

$$H = \begin{pmatrix} A - B\underline{D}^{-2}\tilde{D}C & -B\underline{D}^{-2}\tilde{B} \\ -(\tilde{C}C - \tilde{C}D\underline{D}^{-2}\tilde{D}C) & -(\tilde{A} - \tilde{C}D\underline{D}^{-2}\tilde{B}) \end{pmatrix} \tag{146}$$

Now H is symplectically equivalent to the map $S_{\gamma^2 E_{-*}E_- - N_*N} = S_{N_{-*}N_-}$. Clearly $N_{-*}X_{N_-}$ is a Lagrangian invariant subspace and $S_{N_{-*}N_-} \mid N_{-*}X_{N_-}$, which is isomorphic to S_{N_-}, is stable.. On the other hand in the isomorphism of H and $S_{N_{-*}N_-}$ the Lagrangian subspace $N_{-*}X_{N_-}$ corresponds to the image space $\mathcal{M}$ of $\begin{pmatrix} I \\ X \end{pmatrix}$. Now a standard calculation shows that $H \mid \mathcal{M}$ is isomorphic to $A - B\underline{D}^{-2}\tilde{D}C - B\underline{D}^{-2}\tilde{B}X$. So the last map is stable and X is the stabilizing solution. ∎

7 INNER-OUTER FACTORIZATIONS

Inner-outer factorizations of analytic functions belonging to vectorial Hardy spaces play an important role both in operator theory as well as its applications in the analysis of infinite dimensional systems and in modern H^∞ control theory, more specifically in the various forms of the model matching problem.

In this section we study a special case of inner-outer factorizations in the case that the function to be factored is in RH^∞. We will deal here only with the square case and we modify the definitions accordingly.

Thus a function $G \in RH^\infty$ will be called *inner* if $G_*G = I$ on the imaginary axis, i.e. it is unitary there. A function $G \in RH^\infty$ will be called *outer* if $GH^2(\mathbf{C}^n)$ is dense in $H^2(\mathbf{C}^n)$. However we will restrict our usage of the term outer to functions G that have an inverse in $H^\infty(\mathbf{C}^n)$. This restriction excludes the possibility of zeroes on the imaginary axis.

Under these terms the product of an inner function and an outer function is invertible in the right half plane with the exception of at most a finite number of points.

The next theorem focuses on factorization resuls and state space formulas for the factors.

Let $G \in RH^\infty$ be a rational matrix, which we assume to satisfy $G(\infty) = I$ and let G have a realization $[A, B, C, I]$. Let $G = NE_-^{-1}$ be a right coprime factorization. As

$G \in RH^\infty$, $detE_-$ has all its zeroes in the open left half plane, i.e. $detE_-$ is a stable polynomial. Let $N = D_+N_-$ be a factorization of N such that $detD_+$ has its zeroes in the right half plane whereas $detN_-$ has them in the left half plane. Let $G_i = D_+\overline{D}_-^{-1}$ be the inner factor of G. Then we can write

$$G_i = (D_+\overline{D}_-^{-1})(\overline{D}_-N_-E_-^{-1}).$$

Now

$$\begin{aligned} G_i &= D_+\overline{D}_-^{-1} \\ &= (D_+N_-)(N_-^{-1}\overline{D}_-^{-1}) \\ &= (D_+N_-)(\overline{D}_-N_-)^{-1}. \end{aligned} \tag{147}$$

Note that the associated realization corresponding to the last matrix fraction representation of G_i is reachable but not necessarily observable. Since G_i is a bicausal isomorphism we have

$$G_i^{-1} = (\overline{D}_-N_-)(D_+N_-)^{-1}. \tag{148}$$

Comparing (148) with

$$G^{-1} = (E_-)(D_+N_-)^{-1}. \tag{149}$$

and noting that $G^{-1} = [A - BC, B, -C, I]$, it follows that $G_i^{-1} = [A - BC, B, C_0, I]$. Using again the realization of inverse system it follows that $[A - B(C + C_0), B, -C_0, I]$ realizes G_i. Putting $F = -(C + C_0)$ the inner function G_i is realized by $(A + BF, B, C + F, I)$. Next we proceed to characterize F.

Since G_i is an inner function, it satisfies $G_i^{-1} = G_{i*}$. Now from the realization $[A + BF, B, C + F, I]$ of G_i it follows that G_{i*} is realized by $[-(\widetilde{A + BF}), (\widetilde{C + F}), -\tilde{B}, I]$. On the other hand G_i^{-1} is, by the characterization of inverse systems, realized by

$$[A + BF - B(C + F), B, -(C + F), I] = [A - BC, B, -(C + F), I].$$

We look for a possible isomorphism X between the two systems, i.e. an invertible map for which the diagram

$$\begin{array}{ccccccc} & & & R^m & & & \\ & B & \swarrow & & \searrow & \tilde{C} + \tilde{F} & \\ & & & X & & & \\ & R^n & & \longrightarrow & & R^n & \\ A - BC & \downarrow & & & & \downarrow & -(\tilde{A} + \tilde{F}\tilde{B}) \\ & & & X & & & \\ & R^n & & \longrightarrow & & R^n & \\ & & \searrow & & \swarrow & & \\ & -(C + F) & & R^m & & -\tilde{B} & \end{array} \tag{150}$$

is commutative.

The commutativity of this diagram is equivalent to the equations

$$X(A - BC) + (\tilde{A} + \tilde{F}\tilde{B})X = 0 \tag{151}$$

and

$$-\tilde{B}X = -(C + F). \tag{152}$$

Thus $F = -C + \tilde{B}X$ which substituted back into equation (151) leads to

$$\begin{aligned} &X(A - BC) + (\tilde{A} + (XB - \tilde{C})\tilde{B})X \\ &= X(A - BC) + (\tilde{A} - \tilde{C}\tilde{B})X + XB\tilde{B}X = 0. \end{aligned} \tag{153}$$

The algebraic Riccati equation (153) is associated with the Hamiltonian matrix

$$H = \begin{pmatrix} A - BC & \tilde{B}B \\ 0 & -(\tilde{A} - \tilde{C}\tilde{B}) \end{pmatrix}. \tag{154}$$

However

$$\begin{pmatrix} A - BC & \tilde{B}B \\ 0 & -(\tilde{A} - \tilde{C}\tilde{B}) \end{pmatrix} = \begin{pmatrix} A & 0 \\ -\tilde{C}C & -\tilde{A} \end{pmatrix} - \begin{pmatrix} B \\ -\tilde{C} \end{pmatrix} \begin{pmatrix} C & \tilde{B} \end{pmatrix}. \tag{155}$$

But if we compute the series realization of G_*G we obtain

$$\left[\begin{pmatrix} A & 0 \\ \tilde{C}C & -\tilde{A} \end{pmatrix}, \begin{pmatrix} B \\ -\tilde{C} \end{pmatrix}, \begin{pmatrix} C & \tilde{B} \end{pmatrix}, I \right]. \tag{156}$$

This implies that $(G_*G)^{-1}$ has a realization

$$\left[\begin{pmatrix} A - BC & \tilde{B}B \\ 0 & -(\tilde{A} - \tilde{C}\tilde{B}) \end{pmatrix}, \begin{pmatrix} B \\ \tilde{C} \end{pmatrix}, -\begin{pmatrix} C & \tilde{B} \end{pmatrix}, I \right]. \tag{157}$$

But on the other hand $(G_*G)^{-1} = E_-(N_*N)^{-1}E_{-*}$. Thus the algebraic Riccati equation (153) is solvable if and only if N_*N has a spectral factorization. Clearly such a factorization exists if $N_*N > 0$ on iR. This is guaranteed if we assume the injectivity of $N(z)$ on the imaginary axis.

Theorem 7.1 *Let the square rational proper transfer function $G \in RH^\infty$ be injective on iR have a realization $[A, B, C, I]$. Then G_*G has a spectral factorization $G_{-*}G_-$, with the spectral factor G_- realized by*

$$\left[A, B, C - \tilde{B}X, I\right] \tag{158}$$

and X is the unique stabilizing solution of the homogeneous algebraic Riccati equation

$$(\tilde{A} - \tilde{C}\tilde{B})X + X(A - BC) + XB\tilde{B}X = 0 \tag{159}$$

associated with the Hamiltonian matrix

$$\begin{pmatrix} A - BC & B\tilde{B} \\ 0 & -(\tilde{A} - \tilde{C}\tilde{B}) \end{pmatrix}. \tag{160}$$

An inner-outer factorization exists. The outer and inner factors are given by

$$G_o(z) = G_-(z) \tag{161}$$

and

$$G_i(z) := [A + BF, B, C + F, I] \tag{162}$$

respectively, where $F = -C + \tilde{B}X$.

∎

References

[1979] A. C. Antoulas, "A polynomial matrix approach to F mod G-invariant subspaces", Doctoral Dissertation, Dept. of Mathematics, ETH Zurich.

[1979] H. Bart, I. Gohberg and M. A. Kaashoek, *Minimal Factorization of Matrix and Operator Functions*, Birkhauser, Basel.

[1980] H. Bart, I. Gohberg, M. A. Kaashoek and P. Van Dooren, "Factorizations of transfer functions", *SIAM J. Contr. Optim.*, 18, 675-696.

[1974] G. Bengtsson, "Minimal system inverses for linear multivariable systems", *J. Math. Anal. Appl.* 46, 261-274.

[1972] R.W.Brockett and A.Rahimi, "Lie Algebras and Linear Differential Equations", in *Ordinary Differential Equations*, (L. Weiss, Ed.) Academic Press, New York, 1972.

[1988] T. Chen and B. Francis, "Spectral and inner-outer factorizations of rational matrices", to appear.

[1974a] W. A. Coppel, "Matrix quadratic equations", *Bull. Austr. Math. Soc.*, 10, 377-401.

[1974b] W. A. Coppel, "Matrices of rational functions", *Bull. Austral. Math. Soc.* 11, 89-113.

[1980] E. Emre, "Nonsingular factors of polynomial matrices and (A,B)-invariant subspaces", *SIAM J. Contr. Optimiz.* 18,288-296.

[1980] E. Emre and M. L. J. Hautus, "A polynomial characterization of (A,B)-invariant and reachability subspaces", *SIAM J. Contr. Optimiz.*, 18, 420-436.

[1988] B. Francis, "H^∞ Control Theory", Springer.

[1976] P. A. Fuhrmann, "Algebraic system theory: An analyst's point of view", *J. Franklin Inst.*, 301,521-540.

[1977] P. A. Fuhrmann, "On strict system equivalence and similarity", *Int. J. Contr.* 25,5-10.

[1978] P. A. Fuhrmann, "Simulation of linear systems and factorization of matrix polynomials", *Int. J. Contr.*, 28,689-705.

[1979] P. A. Fuhrmann, "Linear feedback via polynomial models", *Int. J. Contr.* 30,363-377.

[1981] P. A. Fuhrmann, "Duality in polynomial models with some applications to geometric control theory," *Trans. Aut. Control*, AC-26,284-295.

[1983] P. A. Fuhrmann, "On symmetric rational transfer functions", *Linear Algebra and Appl.*, 50,167-250.

[1983] P. A. Fuhrmann, "A matrix Euclidean algorithm and matrix continued fractions", *Systems and Control Letters*, 3, 263-271.

[1984] P. A. Fuhrmann, "On Hamiltonian transfer functions", *Lin. Alg. Appl.*, 84, 1-93.

[1985] P. A. Fuhrmann, "The algebraic Riccati equation - a polynomial approach", *Systems and Control Letters*, , 369-376.

[1979] P. A. Fuhrmann and J. C. Willems, "Factorization indices at infinity for rational matrix functions", *Integral Equat. and Oper. Theory*, 2,287-301.

[1980] P. A. Fuhrmann and J. C. Willems, "A study of (A,B)-invariant subspaces via polynomial models", *Int. J. Contr.* 31,467-494.

[1959] F. R. Gantmacher, *Matrix Theory*, Chelsea, New York.

[1982] I. Gohberg, P. Lancaster and L. Rodman, "Factorization of selfadjoint matrix polynomials with constant signature", *Lin. and Multilin. Alg.*, 11, 209-224.

[1978] M. L. J. Hautus and M. Heymann, "Linear feedback-an algebraic approach", *SIAM J. Control* 16,83-105.

[1989] U. Helmke and P. A. Fuhrmann, "Bezoutians", to appear, *Lin. Alg. Appl.*.

[1970] V. A. Jacubovich, "Factorization of symmetric matrix polynomials", *Soviet Math. Dokl.*, 11, 1261-1264.

[1982] P. P. Khargonekar and E. Emre, "Further results on polynomial characterization of (F,G)-invariant and reachability subspaces", *IEEE Trans. Aut. Control*, 27, 352-366.

[1980] P. Lancaster and L. Rodman, "Existence and uniqueness theorems for the algebraic Riccati equation", *SIAM J. Cont.*.

[1956] C. C. MacDuffee, *The Theory of Matrices* , Chelsea, New York.

[1971] K. Martensson, "On the matrix Riccati equation", *Inform. Sci.* 3,17-49.

[1973a] B. P. Molinari, "The stabilizing solution of the algebraic Riccati equation", *SIAM J. Cont.*, 11,262-271.

[1973b] B. P. Molinari, "Equivalence relations for the algebraic Riccati equation", *SIAM J. Cont.*, 11,272-285.

[1966] J. E. Potter, "Matrix quadratic solutions", *SIAM J. Appl. Math.*, 14,496-501.

[1985] T. Shamir and P. A. Fuhrmann, "Minimal factorizations of rational matrix functions in terms of polynomial models", , *Lin. Alg. Appl.*, 68, 67-91.

[1983] M. A. Shayman, "Geometry of the algebraic Riccati equation, Part I", *SIAM J. Cont.*, 21,375-394.

[1971] J. C. Willems, "Least squares stationary optimal control and the algebraic Riccati equation", *Trans. Automat. Contr.*, 16,621-634.

[1966] D. C. Youla and P. Tissi, "N-port synthesis via reactance extraction- part I", *IEEE Inter. Convention Record*, 183-205.

A State Space Approach to H_∞ Optimal Control

K. Glover

Department of Engineering, University of Cambridge
Trumpington Street, Cambridge CB2 1PZ, United Kingdom

J. C. Doyle

Department of Electrical Engineering
California Institute of Technology
Pasadena, CA 91125, USA

Abstract

Simple state-space formulae are derived for all controllers solving a standard $\mathcal{H}_\infty$ problem: for a given number $\gamma \geq 0$, find all controllers such that the $\mathcal{H}_\infty$ norm of the closed-loop transfer function is $< \gamma$. Under these conditions, a parametrization of all controllers solving the problem is given as a linear fractional transformation (LFT) on a contractive, stable free parameter. The state dimension of the coefficient matrix for the LFT equals that of the plant, and has a separation structure reminiscent of classical LQG (i.e., $\mathcal{H}_2$) theory. Indeed, the whole development is very reminiscent of earlier $\mathcal{H}_2$ results, especially those of Willems (1971). This paper directly generalizes the results in Doyle, Glover, Khargonekar, and Francis, 1989, and Glover and Doyle, 1988. Some aspects of the optimal case ($\leq \gamma$) are considered.

1 Introduction

1.1 Overview

The $\mathcal{H}_\infty$ norm defined in the frequency-domain for a stable transfer matrix $G(s)$ is

$$\|G\|_\infty := \sup_\omega \overline{\sigma}[G(j\omega)] \quad (\overline{\sigma} := \text{maximum singular value})$$

The problem of analysis and synthesis of control systems using this norm arises in a number of ways. We assume the reader either is familiar with the engineering motivation for these problems, or is interested in the results of this paper for some other reason. This paper considers particular $\mathcal{H}_\infty$ optimal control problems that are direct generalizations of those considered in Doyle, Glover, Khargonekar, and Francis (1989), and Glover and Doyle (1988), hereafter referred to as DGKF and GD, respectively.

The basic block diagram used in this paper is

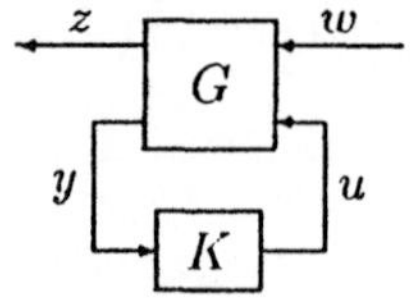

where G is the generalized plant and K is the controller. Only finite dimensional linear time-invariant (LTI) systems and controllers will be considered in this paper. The generalized plant G contains what is usually called the plant in a control problem plus all weighting functions. The signal w contains all external inputs, including disturbances, sensor noise, and commands, the output z is an error signal, y is the measured variables, and u is the control input. The diagram is also referred to as a linear fractional transformation (LFT) on K and G is called the coefficient matrix for the LFT. The resulting closed loop transfer function from w to z is denoted by $T_{zw} = \mathcal{F}_\ell(G, K)$.

The main $\mathcal{H}_\infty$ output feedback results of this paper as described in the abstract are presented in Section 4. The proofs of these results exploit the "separation" structure of the controller. If full information (x and w) is available, then the central controller is simply a gain matrix F_∞, obtained through finding a certain stable invariant subspace of a Hamiltonian matrix. Also, the optimal output estimator is an observer whose gain is obtained in a similar way from a dual Hamiltonian matrix. These special cases are described in Section 3. In the general output feedback case the controller can be interpreted as an optimal estimator for $F_\infty x$. Furthermore, the two Hamiltonians involved in this solution can be associated with full information and output estimation problems.

The proofs of these results are constructed out of a series of lemmas, several of which have some independent interest, particularly those involving state-space characterizations of mixed Hankel-Toeplitz operators. A possible contribution of this paper, beyond the new formulae and theorems, may be some of this technical machinery, most of which is developed in Section 2. The result is that the proofs of both the theorems and the lemmas leading to them are quite short. Furthermore, the development is reasonably self-contained, and the primary background required is a knowledge of elementary aspects of state-space theory, $\mathcal{L}_2$ spaces, and operators on $\mathcal{L}_2$, including projections and adjoints. More specialized knowledge about the connections between Riccati equations, spectral factorization, and Hamiltonian matrices would also be useful.

As mentioned, this paper is a direct generalization of DGKF, and contains a substantial repetition of material. Roughly speaking, we prove those results in GD which were stated without proof, using DGKF machinery, which considered a less general problem. An alternative approach in relaxing some of the assumptions in DGKF is to use loop-shifting techniques as in Zhou and Khargonekar (1988), GD, and more completely in Safonov *et al.* (1989). We also organize this paper much differently than DGKF. The results are presented in a conventional bottom-up linear order, with lemmas and theorems followed by their proofs, which in turn only use lemmas and theorems already proven. Readers interested in pursuing all the details of the proofs may find it more convenient than DGKF. This paper lacks the tutorial flavor of DGKF and the explicit connections with the more familiar $\mathcal{H}_2$ problem, although the $\mathcal{H}_2$ theory will be found lurking at every corner.

We also consider some aspects of generalizations to the $\leq$ case, primarily to indicate the problems encountered in the optimal case. A detailed derivation of the necessity the generalized conditions for the Full Information problem is given. In keeping with the style of GD and DGKF, we don't present a complete treatment of the $\leq$ case, but leave it for yet another day. Complete derivations of the optimal output feedback case can be found in Glover *et al.* (1989) using different techniques.

1.2 Historical perspective

This section is not intended as a review of the literature in $\mathcal{H}_\infty$ theory, nor even an attempt to outline the work that most closely touches on this paper. For a bibliography and review of the early $\mathcal{H}_\infty$ literature, the interested reader might see [Francis, 1987] and [Francis and Doyle, 1987], and an historical account of the results leading up to those in this paper may be found in DGKF. Instead, we will offer a slightly revisionist history, which lacks some factual accuracy, but has the advantage of more clearly emphasizing state-space methods and, more specifically, Willems' central role in $\mathcal{H}_\infty$ theory. This mildly fictionalized reconstruction tells things as they could have been, if only we'd been more clever, and thus contains a certain truth as valuable as that of a more factually accurate accounting. Besides, "historical perspectives" are often revisionist anyway, we're just admitting to it.

Zames' (1981) original formulation of $\mathcal{H}_\infty$ optimal control theory was in an input-output setting. Most solution techniques available at that time involved analytic functions (Nevanlinna-Pick interpolation) or operator-theoretic methods [Sarason, 1967; Adamjan *et al.*, 1978; Ball and Helton, 1983]. Indeed, $\mathcal{H}_\infty$ theory seemed to many to signal the beginning of the end for the state-space methods which had dominated control for the previous 20 years. Unfortunately, the standard frequency-domain approaches to $\mathcal{H}_\infty$ started running into significant obstacles in dealing with multi-input-output (MIMO) systems, both mathematically and computationally, much as the $\mathcal{H}_2$ theory of the 1950's had.

Not surprisingly, the first solution to a general rational MIMO $\mathcal{H}_\infty$ optimal control problem, presented in [Doyle, 1984], relied heavily on state-space methods, although more as a computational tool than in any essential way. The steps in this solution were as follows: parametrize all internally-stabilizing controllers via [Youla *et al.*, 1976]; obtain realizations of the closed-loop transfer matrix; convert the resulting model-matching problem into the equivalent 2×2-block general distance or best approximation problem involving mixed Hankel-Toeplitz operators; reduce to the Nehari problem (Hankel only); solve the Nehari problem by the procedure of Glover (1984). Both [Francis, 1987] and [Francis and Doyle, 1987] give expositions of this approach, which will be referred to as the "1984" approach.

In a mathematical sense, the 1984 procedure "solved" the $\mathcal{H}_\infty$ optimal control problem. Unfortunately, it involved a peculiar patchwork of techniques and the associated complexity of computation was substantial, involving several Riccati equations of increasing dimension, and formulae for the resulting controllers tended to be very complicated and have high state dimension. Nevertheless, much of the subsequent work in $\mathcal{H}_\infty$ control theory focused on the 2×2-block problems, either in the model-matching or general distance forms. This continued to provide a context for a stimulating interchange with operator theory, the benefits of which will hopefully continue to accrue. But from a control perspective, the $\mathcal{H}_\infty$ theory seemed once again to be headed into a cul-de-sac, but now with a Q in the corner. The solution has turned out to involve an even more radical emphasis on state-space theory.

In addition to providing controller formulae that are simple and expressed in terms of plant data, the methods in DGKF and this paper are a fundamental departure from the earlier work described above. In particular, the Youla parametrization and the resulting 2×2-block model-matching problem of the 1984 solution are avoided entirely; replaced by a more purely state-space approach involving observer-based compensators, a pair of 2×1 block problems, and a separation argument. The operator theory still plays a central role (as does Redheffer's work [Redheffer, 1960] on linear fractional transformations), but its use is more straightforward. The key to this was a return to simple and familiar state-space tools, in the style of Willems (1971), such as completing the square, and the connection between frequency domain inequalities (e.g. $\|G\|_\infty < 1$), Riccati equations, and spectral factorization. In essence, one only needed to think about how Willems would do it, and the rest is simply technical detail.

The state-space theory of $\mathcal{H}_\infty$ can be carried much further, by generalizing time-invariant to time-varying, infinite horizon to finite horizon, and finite dimensional to infinite dimensional. A flourish of activity has begun on these problems and the already numerous results indicate, not surprisingly, that many of the results of this paper generalize *mutatis mutandis*, to these cases. In fact, a cynic might express a sense of *déjà vu*, that despite all the rhetoric, $\mathcal{H}_\infty$ theory has come to look much like LQG, circa 1970 (or even more specifically, LQ differential games). A more charitable view might be that current $\mathcal{H}_\infty$ theory, rather than ending the reign of state-space, reaffirms the power of its computational machinery and the wisdom of its visionaries, exemplified by Jan Willems.

1.3 Notation

The notation is fairly standard. The Hardy spaces $\mathcal{H}_2$ and $\mathcal{H}_2^\perp$ consist of square-integrable functions on the imaginary axis with analytic continuation into, respectively, the right and left half-plane. The Hardy space $\mathcal{H}_\infty$ consists of bounded functions with analytic continuation into the right half-plane. The Lebesgue spaces $\mathcal{L}_2 = \mathcal{L}_2(-\infty,\infty)$, $\mathcal{L}_{2+} = \mathcal{L}_2[0,\infty)$, and $\mathcal{L}_{2-} = \mathcal{L}_2(-\infty,0]$ consist, respectively of square-integrable functions on $(-\infty,\infty)$, $[0,\infty)$, and $(-\infty,0]$, and $\mathcal{L}_\infty$ consists of bounded functions on $(-\infty,\infty)$. As interpreted in this paper, $\mathcal{L}_\infty$ will consist of functions of frequency, $\mathcal{L}_{2+}$ and $\mathcal{L}_{2-}$ functions of time, and $\mathcal{L}_2$ will be used for both.

We will make liberal use of the Hilbert space isomorphism, via the Laplace transform and the Paley-Wiener theorem, of $\mathcal{L}_2 = \mathcal{L}_{2+} \oplus \mathcal{L}_{2-}$ in the time-domain with $\mathcal{L}_2 = \mathcal{H}_2 \oplus \mathcal{H}_2^\perp$ in the frequency-domain and of $\mathcal{L}_{2+}$ with $\mathcal{H}_2$ and $\mathcal{L}_{2-}$ with $\mathcal{H}_2^\perp$. In fact, we will normally not make any distinction between a time-domain signal and its transform. Thus we may write $w \in \mathcal{L}_{2+}$ and then treat w as if $w \in \mathcal{H}_2$. This style streamlines the development, as well as the notation, but when any possibility of confusion could arise, we will make it clear whether we are working in the time- or frequency- domain.

All matrices and vectors will be assumed to be complex. A transfer matrix in terms of state-space data is denoted

$$\left[\begin{array}{c|c} A & B \\ \hline C & D \end{array}\right] := C(sI - A)^{-1}B + D$$

For a matrix $M \in \mathcal{C}^{p\times r}$, M' denotes its conjugate transpose, $\bar{\sigma}(M) = \rho(M'M)^{1/2}$ denotes its maximum singular value, $\rho(M)$ denotes its spectral radius (if $p = r$), and $M^{\dagger}$ denotes the Moore-Penrose psuedoinverse of M. Im denotes image, ker denotes kernel, and $G^{\sim}(s) := G(-\bar{s})'$. For operators, Γ^* denotes the adjoint of Γ. The prefix $\mathcal{B}$ denotes the open unit ball and the prefix $\mathcal{R}_c$ denotes complex-rational.

The orthogonal projections P_+ and P_- map $\mathcal{L}_2$ to, respectively, $\mathcal{H}_2$ and $\mathcal{H}_2^{\perp}$ (or $\mathcal{L}_{2+}$ and $\mathcal{L}_{2-}$). For $G \in \mathcal{L}_\infty$, the Laurent or multiplication operator $M_G : \mathcal{L}_2 \rightarrow \mathcal{L}_2$ for frequncy-domain $w \in \mathcal{L}_2$ is defined by $M_G w = Gw$. The norms on $\mathcal{L}_\infty$ and $\mathcal{L}_2$ in the frequency-domain were defined in Section 1.1. Note that both norms apply to matrix or vector-valued functions. The unsubscripted norm $\|\bullet\|$ will denote the standard Euclidean norm on vectors. We will omit all vector and matrix dimensions throughout, and assume that all quantities have compatible dimensions.

1.4 Problem statement

Consider the system described by the block diagram

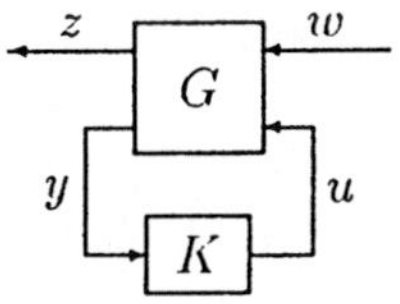

Both G and K are complex-rational and proper, K is constrained to provide internal stability. We will denote the transfer functions from w to z as T_{zw} in general and for a feedback connection (LFT) as above we also write $T_{zw} = \mathcal{F}_\ell(G, K)$. This section discusses the assumptions on G that will be used. In our application we shall have state models of G and K. Then *internal stability* will mean that the states of G and K go to zero from all initial values when $w = 0$.

Since we will restrict our attention exclusively to proper, complex-rational controllers which are stabilizable and detectable, these properties will be assumed throughout. Thus the term controller will be taken to mean a controller which satisfies these properties. Controllers that have the additional property of being internally-stabilizing will be said to be *admissible*. Although we are taking everything to be complex, in the special case where the original data is real (e.g. G is real-rational) then all the of the results (such as K) will also be real.

The problem to be considered is to find all admissible $K(s)$ such that $\|T_{zw}\|_\infty < \gamma$ ($\leq \gamma$). The realization of the transfer matrix G is taken to be of the form

$$G(s) = \left[\begin{array}{c|cc} A & B_1 & B_2 \\ \hline C_1 & D_{11} & D_{12} \\ C_2 & D_{21} & 0 \end{array}\right] = \left[\begin{array}{c|c} A & B \\ \hline C & D \end{array}\right]$$

compatible with the dimensions $z(t) \in \mathcal{C}^{p_1}$, $y(t) \in \mathcal{C}^{p_2}$, $w(t) \in \mathcal{C}^{m_1}$, $u(t) \in \mathcal{C}^{m_2}$, and the state $x(t) \in \mathcal{C}^n$. The following assumptions are made:

(A1) (A, B_2) is stabilizable and (C_2, A) is detectable

(A2) D_{12} is full column rank with $\begin{bmatrix} D_{12} & D_\perp \end{bmatrix}$ unitary and D_{21} is full row rank with $\begin{bmatrix} D_{21} \\ \tilde{D}_\perp \end{bmatrix}$ unitary.

(A3) $\begin{bmatrix} A - j\omega I & B_2 \\ C_1 & D_{12} \end{bmatrix}$ has full column rank for all ω.

(A4) $\begin{bmatrix} A - j\omega I & B_1 \\ C_2 & D_{21} \end{bmatrix}$ has full row rank for all ω.

Assumption (A1) is necessary for the existence of stabilizing controllers. The assumptions in (A2) mean that the penalty on $z = C_1x + D_{12}u$ includes a nonsingular, normalized penalty on the control u, and that the exogenous signal w includes both plant disturbance and sensor noise, and the sensor noise weighting is normalized and nonsingular. Relaxation of (A2) leads to singular control problems.

Assumption (A3) relaxes the DGKF assumptions that (C_1, A) is detectable and $D_{12}'C_1 = 0$, and (A4) relaxes (A, B_1) stabilizable and $B_1 D_{21}' = 0$. Assumptions (A3) and (A4) are made for a technical reason: together with (A1) it guarantees that the two Hamiltonian matrices in the corresponding $\mathcal{H}_2$ problem belong to *dom*(*Ric*). It is tempting to suggest that (A3) and (A4) can be dropped, but they are, in some sense, necessary for the methods in this paper to be applicable. A further discussion of the assumptions and their possible relaxation will be discussed in Section 5.2.

It can be assumed, without loss of generality, that $\gamma = 1$ since this is achieved by the scalings $\gamma^{-1}D_{11}$, $\gamma^{-1/2}B_1$, $\gamma^{-1/2}C_1$, $\gamma^{1/2}B_2$, $\gamma^{1/2}C_2$, and $\gamma^{-1}K$. This will be done implicitly for many of the proofs and statements of this paper.

2 Preliminaries

This section reviews some mathematical preliminaries, in particular the computation of the various norms of a transfer matrix G. Consider the transfer matrix

$$G(s) = \left[\begin{array}{c|c} A & B \\ \hline C & D \end{array}\right] \tag{2.1}$$

with A stable (i.e., all eigenvalues in the left half-plane).

The norm $\|G\|_\infty$ arises in a number of ways. Suppose that we apply an input $w \in \mathcal{L}_2$ and consider the output $z \in \mathcal{L}_2$. Then a standard result is that $\|G\|_\infty$ is the induced norm of the multiplication operator M_G, as well as the Toeplitz operator $P_+M_G : \mathcal{H}_2 \to \mathcal{H}_2$.

$$\|G\|_\infty = \sup_{w \in \mathcal{BL}_2} \|z\|_2 = \sup_{w \in \mathcal{BL}_{2+}} \|P_+ z\|_2 = \sup_{w \in \mathcal{BH}_2} \|P_+ M_G w\|_2$$

The rest of this section involves additional characterizations of the norms in terms of state-space descriptions. Section 2.1 collects some basic material on the Riccati equation and the Riccati operator which play an essential role in the development of both theories. Section 2.3 reviews some results on Hankel operators and introduces the 2×1-block mixed Hankel-Toeplitz operator result that will play a key role in the $\mathcal{H}_\infty$ FI problem.

Section 2.4 includes two lemmas on characterizing inner transfer functions and their role in certain LFT's and Section 2.5 considers the stabilizability and detectability of feedback systems.

2.1 The Riccati operator

Let A, Q, R be complex $n \times n$ matrices with Q and R Hermitian. Define the $2n \times 2n$ Hamiltonian matrix

$$H := \begin{bmatrix} A & R \\ Q & -A' \end{bmatrix}$$

If we begin by assuming H has no eigenvalues on the imaginary axis, then it must have n eigenvalues in Re $s < 0$ and n in Re $s > 0$. Consider the two n-dimensional spectral subspaces $\mathcal{X}_-(H)$ and $\mathcal{X}_+(H)$: the former is the invariant subspace corresponding to eigenvalues in Re $s < 0$; the latter, to eigenvalues in Re $s > 0$. Finding a basis for $\mathcal{X}_-(H)$, stacking the basis vectors up to form a matrix, and partitioning the matrix, we get

$$\mathcal{X}_-(H) = \mathrm{Im} \begin{bmatrix} X_1 \\ X_2 \end{bmatrix} \tag{2.2}$$

where $X_1, X_2 \in \mathcal{C}^{n \times n}$, and

$$H \begin{bmatrix} X_1 \\ X_2 \end{bmatrix} = \begin{bmatrix} X_1 \\ X_2 \end{bmatrix} T_X, \quad \mathrm{Re}\ \lambda_i(T_X) < 0\ \forall\, i \tag{2.3}$$

If X_1 is nonsingular, or equivalently, if the two subspaces

$$\mathcal{X}_-(H), \quad \mathrm{Im} \begin{bmatrix} 0 \\ I \end{bmatrix} \tag{2.4}$$

are complementary, we can set $X := X_2 X_1^{-1}$. Then X is uniquely determined by H, i.e., $H \mapsto X$ is a function, which will be denoted Ric; thus, $X = Ric(H)$. We will take the domain of Ric, denoted $dom(Ric)$, to consist of Hamiltonian matrices H with two properties, namely, H has no eigenvalues on the imaginary axis and the two subspaces in (2.4) are complementary. For ease of reference, these will be called the stability property and the complementarity property, respectively. The following well-known results give some properties of X as well as verifiable conditions under which H belongs to $dom(Ric)$. See, for example, Section 7.2 in [Francis, 1987], Theorem 12.2 in [Wonham, 1985], and [Kucera, 1972].

Lemma 2.1 *Suppose $H \in dom(Ric)$ and $X = Ric(H)$. Then*

(a) X is Hermitian

(b) X satisfies the algebraic Riccati equation

$$A'X + XA + XRX - Q = 0$$

(c) $A + RX$ is stable

Lemma 2.2 *Suppose H has no imaginary eigenvalues, R is either positive semi-definite or negative semi-definite, and (A, R) is stabilizable. Then $H \in dom(Ric)$.*

Lemma 2.3 *Suppose H has the form*

$$H = \begin{bmatrix} A & -BB' \\ -C'C & -A' \end{bmatrix}$$

with (A, B) stabilizable . Then $H \in dom(Ric)$, $X = Ric(H) \geq 0$, and $\ker(X) \subset \mathcal{X} :=$ *stable unobservable subspace.*

By stable unobservable subspace we mean the intersection of the stable invariant subspace of A with the unobservable subspace of (A, C). Note that if $(C, -A)$ is detectable, then $Ric(H) > 0$. Also, note that $\ker(X) \subset \mathcal{X} \subset \ker(C)$, so that the equation $XM = C'$ always has a solution for M, for example the least-squares solution given by $X^{\dagger}C'$.

We may extend the domain of Ric by relaxing the stability requirement. Even if H has eigenvalues on the imaginary axis, it must have at least n eigenvalues in $\mathrm{Re}\, s \leq 0$. Suppose that we now choose some n-dimensional invariant subspace, again denoted by $\mathcal{X}_-(H)$, corresponding to n eigenvalues in $\mathrm{Re}\, s \leq 0$ and a corresponding basis as in (2.2), but now satisfying

$$H \begin{bmatrix} X_1 \\ X_2 \end{bmatrix} = \begin{bmatrix} X_1 \\ X_2 \end{bmatrix} T_X, \quad \mathrm{Re}\, \lambda_i(T_X) \leq 0 \; \forall \, i \tag{2.5}$$

This subspace is not uniquely determined by H, but if it still satisfies the complementarity property, then we can set $X := X_2 X_1^{-1}$ as before, if this X is also Hermitian. We may thus define a new map $\overline{Ric}$, whose domain $dom(\overline{Ric})$ will be taken to consist of Hamiltonian matrices H with the property that an $\mathcal{X}_-(H)$ exists satisfying the complementarity condition and with the resulting $X := X_2 X_1^{-1}$ Hermitian. To show that this is actually a map, we have to verify that X is uniquely determined, which is not always the case. In fact, the conditions under which $\overline{Ric}$ is actually a map are intimately connected with the conditions on existence of $\mathcal{H}_\infty$ optimal controllers. It turns out that for the cases of interest in the present paper, whenever H is in $dom(\overline{Ric})$, the subspace will be uniquely determined. Thus whenever $\overline{Ric}$ is needed, it will be a well-defined map, but this must be proven. Fortunately, these cases can essentially be reduced to spectral factorization problems and standard theory can be applied (e.g. Gohberg, Lancaster and Rodman (1986)).

We may further extend the domain of $\overline{Ric}$ by relaxing the complementarity condition. The minimal requirement we will place on $\mathcal{X}_-(H)$ is that (2.5) hold and that

$$X_1' X_2 = X_2' X_1 \tag{2.6}$$

is Hermitian. Note that this condition also does not depend on the particular choice of basis taken in (2.2). It is convenient to define $dom(\overline{\overline{Ric}})$ to be the set of those H for which a subspace $\mathcal{X}_-(H)$ exists and satisfies (2.5) and (2.6). Once again, the map $\overline{\overline{Ric}}$, from $dom(\overline{\overline{Ric}})$ to n dimensional subspaces of $\mathcal{C}^{2n}$ (this is a Grassman manifold) does not always exist as the subspace is not uniquely determined by H.

The same remarks about $\overline{Ric}$ as a map apply here to $\overline{\overline{Ric}}$. These have been introduced in order to treat the optimal case, but their use will be limited as this case is not analysed in detail. Note that $dom(Ric) \subset dom(\overline{Ric}) \subset dom(\overline{\overline{Ric}})$. Also, if $H \in dom(Ric)$ then $\overline{Ric}$ and $\overline{\overline{Ric}}$ are obviously well-defined maps and $Ric(H) = \overline{Ric}(H)$.

2.2 Computing $\mathcal{H}_\infty$ norm

For the transfer matrix $G(s)$ in (2.1), with A stable, define the Hamiltonian matrix

$$H := \begin{bmatrix} A + BR^{-1}D'C & BR^{-1}B' \\ -C'(I - DD')^{-1}C & -A + BR^{-1}D'C' \end{bmatrix} \tag{2.7}$$

$$= \begin{bmatrix} A & 0 \\ -C'C & -A' \end{bmatrix} + \begin{bmatrix} B \\ -C'D \end{bmatrix} R^{-1} \begin{bmatrix} D'C & B' \end{bmatrix} \tag{2.8}$$

where $R = I - D'D$. The following lemma is essentially from [Anderson, 1967], [Willems, 1971], and [Boyd *et al.*, 1989].

Lemma 2.4 *I. The following conditions are equivalent:*

(a) $\|G\|_\infty < 1$

(b) H *has no eigenvalues on the imaginary axis*

(c) $H \in dom(Ric)$

(d) $H \in dom(Ric)$ *and* $Ric(H) \geq 0$ *(*$Ric(H) > 0$ *if* (C, A) *is observable)*

II. The following conditions are equivalent:

(a) $\|G\|_\infty \leq 1$ *and* $\bar{\sigma}(D) < 1$

(b) $H \in dom(\overline{Ric})$

(c) $H \in dom(\overline{Ric})$ *and* $\overline{Ric}(H)$ *is unique with* $\overline{Ric}(H) \geq 0$ *(*$\overline{Ric}(H) > 0$ *if* (C, A) *is observable)*

Proof From

$$(I - G^{\sim}G)(s) = \left[\begin{array}{cc|c} A & 0 & -B \\ -C'C & -A' & C'D \\ \hline D'C & B' & R \end{array}\right]$$

it is immediate that H is the A-matrix of $(I - G^{\sim}G)^{-1}$. It is easy to check using the PBH test that this realization has no uncontrollable or unobservable modes on the imaginary axis. Thus H has no eigenvalues on the imaginary axis iff $(I - G^{\sim}G)^{-1}$ has no poles there, i.e., $(I - G^{\sim}G)^{-1} \in \mathcal{RL}_\infty$. So to prove the equivalence of (Ia) and (Ib) it suffices to prove that

$$\|G\|_\infty < 1 \Leftrightarrow (I - G^{\sim}G)^{-1} \in \mathcal{RL}_\infty$$

If $\|G\|_\infty < 1$, then $I - G(j\omega)^*G(j\omega) > 0$, $\forall\omega$, and hence $(I - G^{\sim}G)^{-1} \in \mathcal{RL}_\infty$. Conversely, if $\|G\|_\infty \geq 1$, then $\bar{\sigma}[G(j\omega)] = 1$ for some ω, i.e., 1 is an eigenvalue of $G(j\omega)^*G(j\omega)$, so $I - G(j\omega)^*G(j\omega)$ is singular. Thus (Ia) and (Ib) are equivalent.

The equivalence of (Ib) and (Ic) follows from Lemma 2.2, and the equivalence of (Ic) and (Id) follows from Lemma 2.1 and standard results for solutions of Lyapunov equations.

The proof of part II is more involved and is given by the established results on spectral factorization as in Gohberg *et al.*(1986), since $I - G^{\sim}G \geq 0$ for all $s = j\omega$. ∎

In part II it was assumed that $\bar{\sigma}(D) < 1$ so that the Hamiltonian matrix could be defined. Alternatives that avoid this are to consider Linear Matrix Inequalities or the deflating subspaces of matrix pencils. This is discussed more in Section 5.2.5.

Lemma 2.4 suggests the following way to compute an $\mathcal{H}_\infty$ norm: select a positive number γ; test if $\|G\|_\infty < \gamma$ by calculating the eigenvalues of H; increase or decrease γ accordingly; repeat. Thus $\mathcal{H}_\infty$ norm computation requires a search, over either γ or ω. We should not be surprised by similar characteristics of the $\mathcal{H}_\infty$-optimal control problem. A somewhat analogous situation occurs for matrices with the norms $\|M\|_2^2 = \text{trace}(M^*M)$ and $\|M\|_\infty = \bar{\sigma}[M]$. In principle, $\|M\|_2^2$ can be computed exactly with a finite number of operations, as can the test for whether $\bar{\sigma}(M) < \gamma$ (e.g. $\gamma^2 I - M^*M > 0$), but the value of $\bar{\sigma}(M)$ cannot. To compute $\bar{\sigma}(M)$ we must use some type of iterative algorithm.

2.3 Mixed Hankel-Toeplitz Operators

It will be useful to characterize some additional induced norms of $G(s)$ in (2.1) and its associated differential equation

$$\begin{aligned} \dot{x} &= Ax + Bw \\ z &= Cx + Dw \end{aligned} \tag{2.9}$$

with A stable. We will prove several lemmas that will be useful in the rest of the paper. It is convenient to describe all the results in the frequency-domain and give all the proofs in time-domain.

Consider first the problem of using an input $w \in \mathcal{L}_{2-}$ to maximize $\|P_+ z\|_2^2$. This is exactly the standard problem of computing the Hankel norm of G (i.e., the induced norm of the Hankel operator $P_+ M_G : \mathcal{H}_2^\perp \to \mathcal{H}_2$), and can be expressed in terms of the Gramians L_c and L_o

$$AL_c + L_cA' + BB' = 0 \qquad A'L_o + L_oA + C'C = 0 \tag{2.10}$$

Although this result is well-known, we will include a time-domain proof similar in technique to the proofs of the new results in this paper.

Lemma 2.5 $\displaystyle \sup_{w \in \mathcal{BL}_{2-}} \|P_+ z\|_2^2 = \sup_{w \in \mathcal{BH}_2^\perp} \|P_+ M_G w\|_2^2 = \rho(L_o L_c)$

Proof Assume (A, B) is controllable; otherwise, restrict attention to the controllable subspace. Then L_c is invertible and $w \in \mathcal{L}_{2-}$ can be used to produce any $x(0) = x_0$ given $x(-\infty) = 0$. The proof is in two steps. First,

$$\inf_{w \in \mathcal{L}_{2-}} \left\{ \|w\|_2^2 \;\middle|\; x(0) = x_0 \right\} = x_0' L_c^{-1} x_0 \tag{2.11}$$

To show this, we can differentiate $x(t)' L_c^{-1} x(t)$ along the solution of (2.9) for any given input w as follows:

$$\frac{d}{dt}(x' L_c^{-1} x) = \dot{x}' L_c^{-1} x + x' L_c^{-1} \dot{x} = x'(A' L_c^{-1} + L_c^{-1} A)x + w' B' L_c^{-1} x + x' L_c^{-1} Bw$$

Use of (2.10) to substitute for $A'L_c^{-1} + L_c^{-1}A$ and completion of the squares give

$$\frac{d}{dt}(x'L_c^{-1}x) = \|w\|^2 - \|w - B'L_c^{-1}x\|^2$$

Integration from $t = -\infty$ to $t = 0$ with $x(-\infty) = 0$ and $x(0) = x_0$ gives

$$x_0'L_c^{-1}x_0 = \|w\|_2^2 - \|w - B'L_c^{-1}x\|_2^2 \leq \|w\|_2^2$$

If $w(t) = B'e^{-A't}L_c^{-1}x_0 = B'L_c^{-1}e^{(A+BB'L_c^{-1})t}x_0$ on $(-\infty, 0]$, then $w = B'L_c^{-1}x$ and equality is achieved, thus proving (2.11).

Second, given $x(0) = x_0$ and $w = 0$, the norm of $z(t) = Ce^{At}x_0$ can be found from

$$\|P_+z\|_2^2 = \int_0^\infty x_0'e^{A't}C'Ce^{At}x_0 dt = x_0'L_o x_0$$

These two results can be combined as in Section 2 of [Glover, 1984]:

$$\sup_{w\in\mathcal{BL}_{2-}} \|P_+z\|_2^2 = \sup_{0\neq w\in\mathcal{L}_{2-}} \frac{\|P_+z\|_2^2}{\|w\|_2^2} = \max_{x_0\neq 0}\frac{x_0'L_o x_0}{x_0'L_c^{-1}x_0} = \rho(L_oL_c) \qquad \blacksquare$$

If $\|G\|_\infty < 1$ then by Lemmas 2.1 and 2.4, the Hamiltonian matrix H in (2.8) is in $dom(Ric)$, $X = Ric(H) \geq 0$, $A + BB'X$ is stable and

$$A'X + XA + C'C + (XB + C'D)R^{-1}(B'X + D'C) = 0 \tag{2.12}$$

Similarly, if $\bar\sigma(D) < 1$ and $\|G\|_\infty \leq 1$ then by Lemma 2.4, the Hamiltonian matrix H in (2.8) is in $dom(\overline{Ric})$, $X = \overline{Ric}(H) \geq 0$, $A + BB'X$ has eigen values in the closed left half plane and (2.12) holds.

The following lemma offers additional consequence of bounds on $\|G\|_\infty$. In fact, this simple time-domain characterization and its proof form the basis for the entire development to follow.

Lemma 2.6 *I. Suppose $\|G\|_\infty < 1$ and $x(0) = x_0$. Then*

$$\sup_{w\in\mathcal{L}_{2+}} (\|z\|_2^2 - \|w\|_2^2) = x_0'Xx_0$$

and the sup is achieved.

II. Suppose that $\|G\|_\infty \leq 1$, $\bar\sigma(D) < 1$, and $x(0) = x_0$. Then

$$\sup_{w\in\mathcal{L}_{2+}} (\|z\|_2^2 - \|w\|_2^2) = x_0'Xx_0$$

Proof: We can differentiate $x(t)'Xx(t)$ as above, use the Riccati equation (2.12) to substitute for $A'X + XA$, and complete the squares to get

$$\frac{d}{dt}(x'Xx) = -\|z\|^2 + \|w\|^2 - \|R^{-1/2}[Rw - (B'X + D'C)x]\|^2$$

If $w \in \mathcal{L}_{2+}$, then $x \in \mathcal{L}_{2+}$, so integrating from $t = 0$ to $t = \infty$ gives

$$\|z\|_2^2 - \|w\|_2^2 = x_0'Xx_0 - \|R^{-1/2}[Rw - (B'X + D'C)x]\|_2^2 \leq x_0'Xx_0 \tag{2.13}$$

For Part I, if we let $w = -R^{-1}(B'X + D'C)x = B'Xe^{[A+BR^{-1}(B'X+D'C)]t}x_0$, then $w \in \mathcal{L}_{2+}$ because $A + BR^{-1}(B'X + D'C)$ is stable. Thus the inequality in (2.13) can be made an equality and the proof is complete. Note that the sup is achieved for a w which is a linear function of the state.

For Part II, $A + BR^{-1}(B'X + D'C)$ may have imaginary axis eigenvalues, hence the inequality in (2.13) is still valid, but may not give the supremum. A sequence of functions w_ϵ can however be constructed to approach the supremum by considering $X_\epsilon = Ric(H_\epsilon)$ where

$$H_\epsilon = \begin{bmatrix} A & 0 \\ -C'C & -A' \end{bmatrix} + \begin{bmatrix} B \\ -C'D \end{bmatrix} (R + \epsilon^2 I)^{-1} \begin{bmatrix} D'C & B' \end{bmatrix}$$

Then for $w_\epsilon = (R + \epsilon^2 I)^{-1}(B'X_\epsilon + D'C)x$

$$\|z\|_2^2 - \|w_\epsilon\|_2^2 = x_0' X_\epsilon x_0 + \epsilon^2 \|w_\epsilon\|_2^2 \leq x_0' X x_0$$

Finally taking the limit as $\epsilon \to 0$ gives the result by uniqueness of $X = \lim_{\epsilon \to 0} X_\epsilon$. ∎

Now suppose that the input is partitioned so that $B = \begin{bmatrix} B_1 & B_2 \end{bmatrix}$, $D = \begin{bmatrix} D_1 & D_2 \end{bmatrix}$, $G(s) = \begin{bmatrix} G_1(s) & G_2(s) \end{bmatrix}$, and w is partitioned conformally. Then $\|G_2\|_\infty < 1$ iff

$$H_W := \begin{bmatrix} A & 0 \\ -C'C & -A' \end{bmatrix} + \begin{bmatrix} B_2 \\ -C'D_2 \end{bmatrix} R_2^{-1} \begin{bmatrix} D_2'C & B_2' \end{bmatrix}$$

is in $dom(Ric)$, where $R_2 := I - D_2'D_2$. Similarly, $\bar{\sigma}(D_2) < 1$ and $\|G\|_\infty \leq 1$ iff $H_W \in dom(\overline{Ric})$. In either case, define $W = \overline{Ric}(H_W)$, which will be unique, and let

$$w \in \mathcal{W} := \left\{ \begin{bmatrix} w_1 \\ w_2 \end{bmatrix} \middle| w_1 \in \mathcal{H}_2^\perp, w_2 \in \mathcal{L}_2 \right\} \tag{2.14}$$

We are interested in a test for $\sup_{w \in B\mathcal{W}} \|P_+ z\|_2 < 1 \quad (\leq 1)$, or equivalently

$$\sup_{w \in B\mathcal{W}} \|\Gamma w\|_2 < 1 \quad (\leq 1) \tag{2.15}$$

where $\Gamma = P_+[M_{G_1}\ M_{G_2}] : \mathcal{W} \to \mathcal{H}_2$ is a mixed Hankel-Toeplitz operator:

$$\Gamma \begin{bmatrix} w_1 \\ w_2 \end{bmatrix} = P_+ \begin{bmatrix} G_1 & G_2 \end{bmatrix} \begin{bmatrix} w_1 \\ w_2 \end{bmatrix} \qquad w_1 \in \mathcal{H}_2^\perp, \quad w_2 \in \mathcal{L}_2$$

Note that Γ is the sum of the Hankel operator $P_+ M_G P_-$ with the Toeplitz operator $P_+ M_{G_2} P_+$. The following lemma generalizes Lemma 2.4 ($B_1 = 0, D_1 = 0$) and Lemma 2.5 ($B_2 = 0, D_2 = 0$).

Lemma 2.7 *I. (2.15) holds with $<$ iff the following two conditions hold:*

(i) $H_W \in dom(Ric)$

(ii) $\rho(W L_c) < 1$

II. (2.15) holds with $\leq$ iff the following two conditions hold:

(i) $H_W \in dom(\overline{Ric})$

(ii) $\rho(WL_c) \leq 1$

Proof As in Lemma 2.5, assume (A, B) is controllable; otherwise, restrict attention to the controllable subspace. By Lemma 2.4, condition (i) is necessary for (2.15) for both cases, so we will prove that given condition (i), (2.15) holds iff condition (ii) holds. By definition of $\mathcal{W}$, if $w \in \mathcal{W}$ then

$$\|P_+z\|_2^2 - \|w\|_2^2 = \|P_+z\|_2^2 - \|P_+w_2\|_2^2 - \|P_-w\|_2^2$$

Note that the last term only contributes to $\|P_+z\|_2^2$ through $x(0)$. Thus if L_c is invertible, then Lemma 2.6 and (2.11) yield

$$\sup_{w\in\mathcal{W}} \left\{ \|P_+z\|_2^2 - \|w\|_2^2 \;\middle|\; x(0) = x_0 \right\} \;=\; x_0'Wx_0 - x_0'L_c^{-1}x_0 \tag{2.16}$$

For part I we will prove the equivalent statement that $\rho(WL_c) \geq 1$ iff $\sup_{w\in\mathcal{BW}} \|\Gamma w\|_2 \geq 1$. The supremum is achieved in (2.16) for some $w \in \mathcal{W}$ that can be constructed from the previous lemmas. Since $\rho(WL_c) \geq 1$ iff $\exists\, x_0 \neq 0$ such that the right-hand side of (2.16) is ≥ 0, we have, by (2.16), that $\rho(WL_c) \geq 1$ iff $\exists\, w \in \mathcal{W}$, $w \neq 0$ such that $\|P_+z\|_2^2 \geq \|w\|_2^2$. But this is true iff $\sup_{w\in\mathcal{BW}} \|\Gamma w\|_2 \geq 1$.

For part II, note that (2.15) holds with $\leq$ iff

$$\sup_{w\in\mathcal{BW}} \|\Gamma w\|_2^2 - \|w\|_2^2 \leq 0$$

which by (2.16) is iff $\rho(WL_c) \leq 1$. ∎

The FI proof of Section 3.3 will make use of the adjoint $\Gamma^* : \mathcal{H}_2 \to \mathcal{W}$, which is given by

$$\Gamma^* z = \begin{bmatrix} P_-(G_1^\sim z) \\ G_2^\sim z \end{bmatrix} = \begin{bmatrix} P_-G_1^\sim \\ G_2^\sim \end{bmatrix} z \tag{2.17}$$

where $P_-Gz := P_-(Gz) = (P_-M_G)z$. That the expression in (2.17) is actually the adjoint of Γ is easily verified from the definition of the inner product on vector-valued $\mathcal{L}_2$, expressed in the frequency-domain as

$$< x_1, x_2 > := \frac{1}{2\pi}\int_{-\infty}^{\infty} x_1(j\omega)^* x_2(j\omega)\,d\omega \tag{2.18}$$

The adjoint of $\Gamma : \mathcal{W} \to \mathcal{H}_2$ is the operator $\Gamma^* : \mathcal{H}_2 \to \mathcal{W}$ such that $< z, \Gamma w > = < \Gamma^* z, w >$ for all $w \in \mathcal{W}$, $z \in \mathcal{H}_2$. Directly using the definition in (2.18), we get

$$\begin{aligned} < z, \Gamma w > &= < z, P_+(G_1w_1 + G_2w_2) > = < z, G_1w_1 > + < z, G_2w_2 > \\ &= < P_-(G_1^\sim z), w_1 > + < G_2^\sim z, w_2 > \\ &= < \Gamma^* z, w > \end{aligned}$$

2.4 LFT's and inner matrices

A transfer function G in $\mathcal{RH}_\infty$, is called *inner* if $G^\sim G = I$, and hence $G(j\omega)^* G(j\omega) = I$ for all ω. Note that G inner implies that G has at least as many rows as columns. For G inner, and any $q \in \mathcal{C}^m$, $w \in \mathcal{L}_2$, then $\|G(j\omega)q\| = \|q\|$, $\forall\omega$, and $\|Gw\|_2 = \|w\|_2$. Because of these norm preserving properties inner matrices will be central to several of the proofs. In this section we give a characterization of inner functions and some properties of linear fractional transformations. First, we present a state-space characterization of inner transfer functions analogous to Lemma 2.4 that is well-known and simple to verify (see [Anderson, 1967], [Wonham, 1985], [Glover, 1984]).

Lemma 2.8 *Suppose* $G = \left[\begin{array}{c|c} A & B \\ \hline C & D \end{array}\right]$ *with* (C, A) *detectable and* $L_o = L_o'$ *satisfies*

$$A'L_o + L_o A + C'C = 0.$$

Then

(a) $L_o \geq 0$ *iff* A *is stable*

(b) $D'C + B'L_o = 0$ *implies* $G^\sim G = D'D$

(c) $L_o \geq 0$, (A, B) *controllable, and* $G^\sim G = D'D$ *implies* $D'C + B'L_o = 0$.

The next lemma considers linear fractional transformations with inner matrices and is based on the work of Redheffer (1960).

Lemma 2.9 *Consider the following feedback system,*

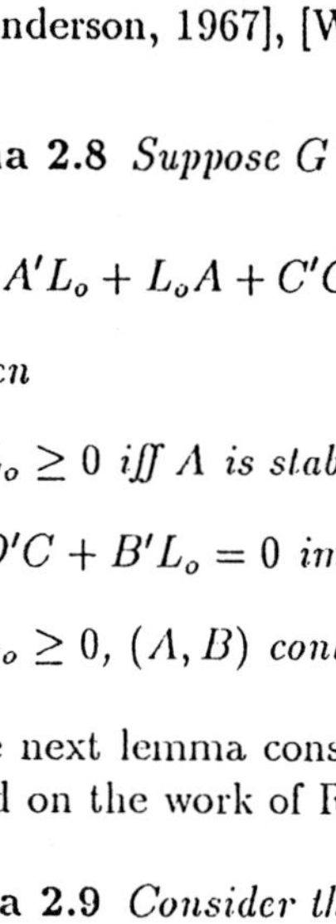

$$P = \begin{bmatrix} P_{11} & P_{12} \\ P_{21} & P_{22} \end{bmatrix} \in \mathcal{RH}_\infty$$

Suppose that $P^\sim P = I$, $P_{21}^{-1} \in \mathcal{RH}_\infty$, *and* Q *is a proper rational matrix. Then the following are equivalent:*

(a) The system is internally stable and well-posed, and $\|T_{zw}\|_\infty < 1$.

(b) $Q \in \mathcal{RH}_\infty$ *and* $\|Q\|_\infty < 1$.

Proof (b) $\Rightarrow$ (a). Internal stability and well-posedness follow from $P, Q \in \mathcal{RH}_\infty$, $\|P_{22}\|_\infty \leq 1$, $\|Q\|_\infty < 1$, and a small gain argument. To show that $\|T_{zw}\|_\infty < 1$ consider the closed-loop system at any frequency $s = j\omega$ with the signals fixed as complex constant vectors. Let $\|Q\|_\infty =: \epsilon < 1$ and note that $T_{wr} = P_{21}^{-1}(I - P_{22}Q) \in \mathcal{RH}_\infty$. Also let $\kappa := \|T_{wr}\|_\infty$. Then $\|w\| \leq \kappa\|r\|$, and P inner implies that $\|z\|^2 + \|r\|^2 = \|w\|^2 + \|v\|^2$. Therefore,

$$\|z\|^2 \leq \|w\|^2 + (\epsilon^2 - 1)\|r\|^2 \leq [1 - (1 - \epsilon^2)\kappa^{-2}]\|w\|^2$$

which implies $\|T_{zw}\|_\infty < 1$.

(a) $\Rightarrow$ (b). To show that $\|Q\|_\infty < 1$ suppose there exist a (real or infinite) frequency ω and a constant nonzero vector r such that at $s = j\omega$, $\|Qr\| \geq \|r\|$. Then setting $w = P_{21}^{-1}(I - P_{22}Q)r$, $v = Qr$ gives $v = T_{vw}w$. But as above, P inner implies that $\|z\|^2 + \|r\|^2 = \|w\|^2 + \|v\|^2$ and hence $\|z\|^2 \geq \|w\|^2$, which is impossible since $\|T_{zw}\|_\infty < 1$. It follows that $\bar{\sigma}(Q(j\omega)) < 1$ for all ω, i.e., $\|Q\|_\infty < 1$, since Q is rational.

Finally, Q has a right-coprime factorization $Q = NM^{-1}$ with $N, M \in \mathcal{RH}_\infty$. We shall show that $M^{-1} \in \mathcal{RH}_\infty$. Since $T_{vw}P_{21}^{-1} = Q(I - P_{22}Q)^{-1}$ it has the right-coprime factorization $T_{vw}P_{21}^{-1} = N(M - P_{22}N)^{-1}$ But since $T_{vw}P_{21}^{-1} \in \mathcal{RH}_\infty$, so does $(M - P_{22}N)^{-1}$. This implies that the winding number of $\det(M - P_{22}N)$, as s traverses the Nyquist contour, equals zero. Furthermore, since $\det(M - \alpha P_{22}N) \neq 0$ for all α in [0,1] and all $s = j\omega$ (this uses the fact that $\|P_{22}\|_\infty \leq 1$ and $\|Q\|_\infty < 1$), we have that the winding number of $\det M$ equals zero too. Therefore, $Q \in \mathcal{RH}_\infty$ and the proof is complete. ■

2.5 LFT's and stability

In this section, we consider the stabilizability and detectability of feedback systems. The proofs in this section are very routine and use standard techniques, principally the PBH test for controllability or observability, so they will only be sketched.

Recall the realization of G from Section 1.4 and suppose that $A \in \mathcal{C}^{n \times n}$, and that z, y, w and u have dimension p_1, p_2, m_1, and m_2, respectively. Thus $C_1 \in \mathcal{C}^{p_1 \times n}$, $B_2 \in \mathcal{C}^{n \times m_2}$, and so on. Now suppose we apply a controller K with stabilizable and detectable realization to G to obtain T_{zw}. For the following lemma, we do not need the assumptions from Section 1.4 on G for the output feedback problem.

Lemma 2.10 *The feedback connection of the realizations for G and K is,*

(a) detectable if rank $\begin{bmatrix} A - \lambda I & B_2 \\ C_1 & D_{12} \end{bmatrix} = n + m_2$ *for all* $Re\lambda \geq 0$.

(b) stabilizable if rank $\begin{bmatrix} A - \lambda I & B_1 \\ C_2 & D_{21} \end{bmatrix} = n + p_2$ *for all* $Re\lambda \geq 0$.

Proof Form the closed-loop state-space matrices and perform a PBH test for controllability and observability. It is easily checked that any unobservable or uncontrollable modes must occur at λ violating the above rank conditions (see Limebeer and Halikias (1988) or Glover(1989) for more details), hence giving the results. ■

3 Full Information and Full Control Problems

In this section we discuss four problems from which the output feedback solutions will be constructed via a separation argument. These special problems are central to the whole approach taken in this paper, and as we shall see, they are also important in their own right. All pertain to the standard block diagram,

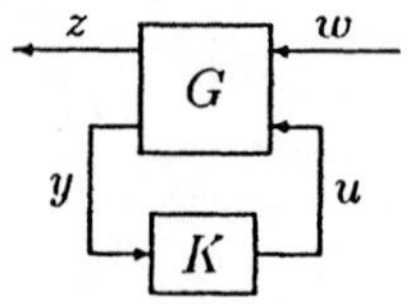

but with different structures for G. The problems are labeled

FI. Full information

FC. Full control

DF. Disturbance feedforward (to be considered in section 4.1)

OE. Output estimation (to be considered in section 4.1)

FC and OE are natural duals of FI and DF, respectively. The DF solution can be easily obtained from the FI solution, as shown in Section 4.1. The output feedback solutions will be constructed out of the FI and OE results. A dual derivation could use the FC and DF results.

The FI and FC problems are not, strictly speaking, special cases of the output feedback problem, as they do not satisfy all of the assumptions. Each of the four problems inherits certain of the assumptions A1–A4 from Section 1.4 as appropriate. The terminology and assumptions will be discussed in the subsections for each problem. In each of the four cases, the results are necessary and sufficient conditions for the existence of a controller such that $\|T_{zw}\|_\infty < \gamma$ and the family of all controllers such that $\|T_{zw}\|_\infty < \gamma$. In all cases, K must be admissible.

The $\mathcal{H}_\infty$ solution involves two Hamiltonian matrices, H_∞ and J_∞ which are defined as follows:

$$R := D_{1\bullet}'D_{1\bullet} - \begin{bmatrix} \gamma^2 I_{m_1} & 0 \\ 0 & 0 \end{bmatrix}, \quad \text{where} \quad D_{1\bullet} := [D_{11} \;\; D_{12}]$$

$$\tilde{R} := D_{\bullet 1}D_{\bullet 1}' - \begin{bmatrix} \gamma^2 I_{p_1} & 0 \\ 0 & 0 \end{bmatrix}, \quad \text{where} \quad D_{\bullet 1} := \begin{bmatrix} D_{11} \\ D_{21} \end{bmatrix}$$

$$H_\infty := \begin{bmatrix} A & 0 \\ -C_1'C_1 & -A' \end{bmatrix} - \begin{bmatrix} B \\ -C_1'D_{1\bullet} \end{bmatrix} R^{-1} \begin{bmatrix} D_{1\bullet}'C_1 & B' \end{bmatrix} \tag{3.1}$$

$$J_\infty := \begin{bmatrix} A' & 0 \\ -B_1B_1' & -A \end{bmatrix} - \begin{bmatrix} C' \\ -B_1D_{\bullet 1}' \end{bmatrix} \tilde{R}^{-1} \begin{bmatrix} D_{\bullet 1}B_1' & C \end{bmatrix} \tag{3.2}$$

If $H_\infty \in dom(\overline{Ric})$ then let X_1, X_2 be any matrices such that

$$H_\infty \begin{bmatrix} X_1 \\ X_2 \end{bmatrix} = \begin{bmatrix} X_1 \\ X_2 \end{bmatrix} T_X, \quad X_1'X_2 = X_2'X_1, \quad \text{Re}\,\lambda_i(T_X) \le 0 \;\forall\, i \tag{3.3}$$

Similarly if $J_\infty \in dom(\overline{Ric})$ then let Y_1, Y_2 be any matrices such that

$$H_\infty \begin{bmatrix} Y_1 \\ Y_2 \end{bmatrix} = \begin{bmatrix} Y_1 \\ Y_2 \end{bmatrix} T_Y, \quad Y_1'Y_2 = Y_2'Y_1 \quad \text{Re}\,\lambda_i(T_Y) \le 0 \;\forall\, i \tag{3.4}$$

Further if in addition $H_\infty \in dom(\overline{Ric})$ and/or $J_\infty \in dom(\overline{Ric})$ then define,

$$X_\infty := X_2 X_1^{-1}, \qquad Y_\infty := Y_2 Y_1^{-1} \tag{3.5}$$

Finally define the 'state feedback' and 'output injection' matrices as

$$F := \begin{bmatrix} F_1 \\ F_2 \end{bmatrix} := -R^{-1}\left[D_{1\bullet}'C_1 + B'X_\infty\right] \tag{3.6}$$

$$L := \begin{bmatrix} L_1 & L_2 \end{bmatrix} := -[B_1 D_{\bullet 1}' + Y_\infty C']\tilde{R}^{-1} \tag{3.7}$$

3.1 Problem FI: Full Information

In the FI special problem G has the following form.

$$G(s) = \left[\begin{array}{c|cc} A & B_1 & B_2 \\ \hline C_1 & D_{11} & D_{12} \\ \begin{bmatrix} I \\ 0 \end{bmatrix} & \begin{bmatrix} 0 \\ I \end{bmatrix} & \begin{bmatrix} 0 \\ 0 \end{bmatrix} \end{array}\right] \tag{3.8}$$

It is seen that the controller is provided with *Full Information* since $y = \begin{pmatrix} x \\ w \end{pmatrix}$. In some cases, a suboptimal controller may exist which uses just the state feedback x, but this will not always be possible. While the state feedback problem is more traditional, we believe that the full information problem is more fundamental and more natural than the state feedback problem, once one gets outside the pure $\mathcal{H}_2$ setting.

The assumptions relevant to the FI problem which are inherited from the output feedback problem are

(A1) (A, B_2) is stabilizable.

(A2) D_{12} is full column rank with $\begin{bmatrix} D_{12} & D_\perp \end{bmatrix}$ unitary.

(A3) $\begin{bmatrix} A - j\omega I & B_2 \\ C_1 & D_{12} \end{bmatrix}$ has full column rank for all ω.

The results for the Full Information case are as follows:

Theorem 3.1 *Suppose G is given by (3.8) and satisfies A1-A3. Then*

(a) $\exists K$ such that $\|T_{zw}\|_\infty < 1 \Leftrightarrow H_\infty \in dom(Ric)$, $Ric(H_\infty) \geq 0$

(b) If $\bar{\sigma}(D_{\perp}'D_{11}) < 1$ then $\exists K$ such that $\|T_{zw}\|_\infty \leq 1 \Leftrightarrow H_\infty \in dom(\overline{Ric})$, $X_1'X_2 = X_2'X_1 \geq 0$. X_1 and X_2 are defined in (3.3).

(c) All admissible $K(s)$ such that $\|T_{zw}\|_\infty < 1$ are given by

$$K(s) = \begin{bmatrix} -Q(s) & I \end{bmatrix} \begin{bmatrix} T_1 & 0 \\ T_2 & I \end{bmatrix} \begin{bmatrix} F_1 & -I \\ F_2 & 0 \end{bmatrix}$$

for $Q \in \mathcal{R}_c\mathcal{H}_\infty$, $\|Q\|_\infty < 1$.

Note that the sufficiency proof for part (b) is omitted. We will prove the FI results and the FC results follow by duality.

3.2 Motivation of the proofs for Problem FI

We will first motivate the proof by considering a completion of the squares assuming that $X_\infty \geq 0$ and exists. Let us factor

$$R = \begin{bmatrix} T_1' & T_2' \\ 0 & I \end{bmatrix} \begin{bmatrix} -I & 0 \\ 0 & I \end{bmatrix} \begin{bmatrix} T_1 & 0 \\ T_2 & I \end{bmatrix} \tag{3.9}$$

$$=: T'JT \tag{3.10}$$

$$= \begin{bmatrix} -T_1'T_1 + T_2'T_2 & T_2' \\ T_2 & I \end{bmatrix} = \begin{bmatrix} D_{11}'D_{11} - I & D_{11}'D_{12} \\ D_{12}'D_{11} & I \end{bmatrix} \tag{3.11}$$

$$\Rightarrow T_2 = D_{12}'D_{11}, \quad T_1'T_1 = I - D_{11}'D_\perp D_\perp' D_{11} \tag{3.12}$$

Now $I - D_{11}'D_\perp D_\perp' D_{11} > 0$ since at $s = \infty$

$$\mathcal{F}_l(G, K)(\infty) = D_{11} + D_{12}K(\infty)\begin{bmatrix} 0 \\ I \end{bmatrix}$$

and $1 > \overline{\sigma}(\mathcal{F}_l(G,K)(\infty)) \geq \overline{\sigma}(D_\perp' D_{11})$.

Now consider the Riccati equation for X_∞,

$$\begin{bmatrix} X_\infty & -I \end{bmatrix} H_\infty \begin{bmatrix} I \\ X_\infty \end{bmatrix} = 0$$

$$\Rightarrow X_\infty A + A'X_\infty + C_1'C_1 - F'RF = 0 \tag{3.13}$$

and observe that

$$\left(\begin{bmatrix} w \\ u \end{bmatrix} - Fx\right)' R \left(\begin{bmatrix} w \\ u \end{bmatrix} - Fx\right)$$

$$= \begin{bmatrix} w' & u' \end{bmatrix} D_{1\bullet}'D_{1\bullet} \begin{bmatrix} w \\ u \end{bmatrix} - w'w + x'(C_1'D_{1\bullet} + X_\infty B)\begin{bmatrix} w \\ u \end{bmatrix}$$

$$+ \begin{bmatrix} w' & u' \end{bmatrix}(D_{1\bullet}'C_1 + B'X_\infty)x + x'F'RFx$$

$$= z'z - x'C_1'C_1x - w'w + x'X_\infty B\begin{bmatrix} w \\ u \end{bmatrix} + \begin{bmatrix} w' & u' \end{bmatrix} B'X_\infty x + x'F'RFx$$

$$= z'z - w'w + x'X_\infty Ax + x'A'X_\infty x + z'X_\infty B\begin{bmatrix} w \\ u \end{bmatrix} + \begin{bmatrix} w' & u' \end{bmatrix} B'X_\infty x$$

$$= z'z - w'w + \tfrac{d}{dt}(x'X_\infty x)$$

Integrating from $t = 0$ to ∞ with $x(0) = x(\infty) = 0$ gives

$$\|z\|_2^2 - \|w\|_2^2 = \|T_2w + u - \begin{bmatrix} T_2 & I \end{bmatrix} Fx\|_2^2 - \|T_1(w - F_1x)\|_2^2. \tag{3.14}$$

Hence to obtain $\|z\|_2 < \|w\|_2$ we require $\|T_2w + u - \begin{bmatrix} T_2 & I \end{bmatrix} Fx\|_2 < \|T_1(w - F_1x)\|_2$ and we see that in some sense the "worst w" is F_1x, whereas the "best u" is $-T_2w + \begin{bmatrix} T_2 & I \end{bmatrix} Fx$. Notice that in the case $T_2 \neq 0$ ($\Rightarrow D_{11} \neq 0$) the natural full information controller uses both w and x.

3.3 Proofs for Problem FI: Necessity

(a) If there exists an admissible controller such that $\|T_{zw}\|_\infty < 1$, then

$$H_\infty \in dom(Ric), \quad Ric(H_\infty) \geq 0 \tag{3.15}$$

(b) If there exists an admissible controller such that $\|T_{zw}\|_\infty \leq 1$, then

$$H_\infty \in dom(\overline{Ric}), \quad X_1'X_2 = X_2'X_1 \geq 0. \tag{3.16}$$

We will prove a slightly stronger result, but before that, we need some preliminary results. Let us first consider

$$H_\infty = \begin{bmatrix} A & 0 \\ -C_1'C_1 & A' \end{bmatrix} - \begin{bmatrix} B \\ -C_1'D_{1\bullet} \end{bmatrix} T^{-1}J^{-1}T'^{-1} \begin{bmatrix} D_{1\bullet}'C_1, & B' \end{bmatrix}$$

where T and J are given in (3.10). Note that

$$\begin{aligned} D_{1\bullet}T^{-1} &= \begin{bmatrix} D_\perp D_\perp' D_{11}T_1^{-1}, & D_{12} \end{bmatrix} \\ BT^{-1} &= \begin{bmatrix} (B_1 - B_2T_2)T_1^{-1}, & B_2 \end{bmatrix} =: \begin{bmatrix} \tilde{B}_1 & B_2 \end{bmatrix} \qquad (3.17) \\ I - D_{1\bullet}R^{-1}D_{1\bullet}' &= I - D_{12}D_{12}' + D_\perp D_\perp' D_{11}T_1^{-1}T_1'^{-1}D_{11}'D_\perp D_\perp' \\ &= D_\perp(I + D_\perp' D_{11}(T_1'T_1)^{-1}D_{11}'D_\perp)D_\perp' \\ &= D_\perp S^{-1}D_\perp' \qquad (3.18) \end{aligned}$$

where

$$S := I - D_\perp' D_{11}D_{11}'D_\perp > 0.$$

Hence

$$H_\infty = \begin{bmatrix} N & -B_2B_2' + \tilde{B}_1\tilde{B}_1' \\ -C_1'D_\perp S^{-1}D_\perp' C_1 & -N' \end{bmatrix}$$

where

$$N := A - B_2D_{12}'C_1 + \tilde{B}_1T_1'^{-1}D_{11}'D_\perp D_\perp' C_1$$

Next we will show that we can assume without loss of generality that the pair $(D_\perp' C_1, -N)$ is detectable. This simplifies the technical details of the proof. Thus suppose that the pair $(D_\perp' C_1, -N)$ is not detectable or equivalently that $(D_\perp' C_1, -A + B_2D_{12}'C_1)$ is not detectable. That is, $(A - B_2D_{12}'C_1)$ has stable modes that are not observable from $D_\perp' C_1$ (note that modes of $(A - B_2D_{12}'C_1)$ on the imaginary axis are observable from $D_\perp' C_1$ by $A3$). If we now change state coordinates so that

$$\left[\begin{array}{c|c} A & B \\ \hline C_1 & D_{1\bullet} \end{array}\right] = \left[\begin{array}{cc|cc} A_{11} & A_{12} & B_{11} & B_{21} \\ A_{21} & A_{22} & B_{12} & B_{22} \\ \hline C_{11} & C_{12} & D_{11} & D_{12} \end{array}\right]$$

with $A_{12} - B_{21}D'_{12}C_{12} = 0$, $D'_\perp C_{12} = 0$, $(D'_\perp C_{11}, -A_{11} + B_{21}D'_{12}C_{11})$ detectable and $(A_{22} - B_{22}D'_{12}C_{12})$ stable, then the state equations for the system with controller $K = \left[\begin{array}{c|c} \hat{A} & \hat{B} \\ \hline \hat{C} & \hat{D} \end{array}\right]$ can be written as

$$\begin{aligned}
\dot{x}_1 &= A_{11}x_1 + B_{11}w + B_{21}(D'_{12}C_{12}x_2 + u) \\
z &= C_{11}x_1 + D_{11}w + D_{12}(D'_{12}C_{12}x_2 + u) \\
\dot{x}_2 &= A_{22}x_2 + A_{21}x_1 + B_{12}w + B_{22}u \\
\dot{\hat{x}} &= \hat{A}\hat{x} + \hat{B}_1x_1 + \hat{B}_2x_2 + \hat{B}_3w \\
u + D'_{12}C_{12}x_2 &= \hat{C}\hat{x} + \hat{D}_1x_1 + \hat{D}_2x_2 + \hat{D}_3w + D'_{12}C_{12}x_2
\end{aligned}$$

If the controller, K, is admissible with $\|\mathcal{F}_\ell(G, K)\|_\infty < 1$ (≤ 1), then the above state equations show that the subsystem $G_1 = \left[\begin{array}{c|cc} A_{11} & B_{11} & B_{21} \\ \hline C_{11} & D_{11} & D_{12} \end{array}\right]$ also has an admissible controller, K_1 (given by the final three equations above), which satisfies $\|\mathcal{F}_\ell(G_1, K_1)\|_\infty < 1$ (≤ 1). Furthermore, suppose we can find a suitable stable invariant subspace

$$\begin{bmatrix} X_{11} \\ X_{21} \end{bmatrix}$$

for the Hamiltonian for G_1 then

$$\begin{bmatrix} X_{11} & 0 \\ 0 & I \\ X_{21} & 0 \\ 0 & 0 \end{bmatrix}$$

will be suitable for G since $(A_{22} - B_{22}D'_{12}C_{12})$ is stable. We will therefore assume that $(D'_\perp C_1, -A + B_2D'_{12}C_1)$ is detectable for the remainder of the necessity proof.

The proof also requires a preliminary change of variables to

$$\nu := u - F_0 x$$

This change of variables will neither change internal stability nor the achievable norm since the states can be measured. The matrix F_0 is the optimal state feedback matrix for a corresponding $\mathcal{H}_2$ problem as given below. By Lemma 2.3 the Hamiltonian matrix

$$H_0 := \begin{bmatrix} A - B_2D'_{12}C_1 & -B_2B'_2 \\ -C'_1D_\perp D'_\perp C_1 & -(A - B_2D'_{12}C_1)' \end{bmatrix}$$

belongs to $dom(Ric)$ since (A, B_2) is stabilizable, and $X_0 := Ric(H_0) > 0$ since $(D'_\perp C_1, -A + B_2D'_{12}C_1)$ is detectable. Define

$$F_0 := -(D'_{12}C_1 + B'_2X_0), \quad A_{F_0} := A + B_2F_0, \quad C_{1F_0} := C_1 + D_{12}F_0$$

$$G_c(s) := \left[\begin{array}{c|c} A_{F_0} & B_1 \\ \hline C_{1F_0} & D_{11} \end{array}\right]$$

Suppose $D_\perp$ is any matrix making $[D_{12}\ D_\perp]$ an orthogonal matrix, and define

$$\begin{bmatrix} U & U_\perp \end{bmatrix} = \left[\begin{array}{c|cc} A_{F_0} & B_2 & -X_0^{-1}C_1'D_\perp \\ \hline C_{1F_0} & D_{12} & D_\perp \end{array}\right] \tag{3.19}$$

Then the transfer function from w and ν to z becomes

$$z = \left[\begin{array}{c|cc} A_{F_0} & B_1 & B_2 \\ \hline C_{1F_0} & D_{11} & D_{12} \end{array}\right] \begin{bmatrix} w \\ \nu \end{bmatrix} = G_c w + U\nu \tag{3.20}$$

The last result needed for the proof is the following lemma which is easily proven using Lemma 2.8 by obtaining a state-space realization, and then eliminating uncontrollable states using a little algebra involving the Riccati equation for X_0.

Lemma 3.2 *$[U\ U_\perp]$ is square and inner and a realization for $G_c^\sim \begin{bmatrix} U & U_\perp \end{bmatrix}$ is*

$$G_c^\sim \begin{bmatrix} U & U_\perp \end{bmatrix} = \left[\begin{array}{c|cc} A_{F_0} & B_2 & -X_0^{-1}C_1'D_\perp \\ \hline B_1'X_0 + D_{11}'C_{1F_0} & D_{11}'D_{12} & D_{11}'D_\perp \end{array}\right] \in \mathcal{RH}_2 \tag{3.21}$$

This implies that U and $U_\perp$ are each inner, and both $U_\perp^\sim G_c$ and $U^\sim G_c$ are in $\mathcal{RH}_2^\perp$.

We are now ready to state and prove the main result.

Proposition 3.3 *I. If* $\sup\limits_{w\in\mathcal{BL}_{2+}} \min\limits_{\nu\in\mathcal{L}_{2+}} \|z\|_2 < 1$ *then $H_\infty \in dom(Ric)$ and $Ric(H_\infty) > 0$.*

II. If $\sup\limits_{w\in\mathcal{BL}_{2+}} \min\limits_{\nu\in\mathcal{L}_{2+}} \|z\|_2 \le 1$ *then $H_\infty \in dom(\overline{Ric})$ and $X_1'X_2 = X_2'X_1 \ge 0$. X_1 and X_2 are defined in (3.3).*

Proof of Proposition Since $[U\ U_\perp]$ is square and inner by Lemma 3.2, $\|z\|_2 = \|[U\ U_\perp]^\sim z\|_2$, and

$$\begin{bmatrix} U & U_\perp \end{bmatrix}^\sim z = \begin{bmatrix} U^\sim G_c w + \nu \\ U_\perp^\sim G_c w \end{bmatrix}$$

Since $\nu \in \mathcal{H}_2$, its optimal value is $\nu = -P_+ U^\sim G_c w$ and the hypotheses of the proposition imply that

$$\sup_{w\in\mathcal{BH}_2} \left\| \begin{bmatrix} P_-(U^\sim G_c w) \\ U_\perp^\sim G_c B_1 w \end{bmatrix} \right\|_2 < 1 \quad (\le 1)$$

Mixed Hankel-Toeplitz operators of this type were considered in Section 2.3. We can define the adjoint operator $\Gamma^* : \mathcal{L}_{2+} \to \mathcal{W}$ ($\mathcal{W}$ from (2.14)) by

$$\Gamma^* w = \begin{bmatrix} P_-(U^\sim G_c w) \\ U_\perp^\sim G_c w \end{bmatrix} = \begin{bmatrix} P_- U^\sim \\ U_\perp^\sim \end{bmatrix} G_c w$$

of the operator $\Gamma : \mathcal{W} \to \mathcal{H}_2$ given by

$$\Gamma \begin{bmatrix} q_1 \\ q_2 \end{bmatrix} = P_+(G_c^\sim(Uq_1 + U_\perp q_2)) = P_+ G_c^\sim \begin{bmatrix} U & U_\perp \end{bmatrix} \begin{bmatrix} q_1 \\ q_2 \end{bmatrix}$$

Hence

$$\sup_{q \in \mathcal{BW}} \|\Gamma q\|_2 < 1 \qquad (\leq 1)$$

This is just the condition (2.15), so from Lemmas 2.3 and 2.7 and (3.21) we have that

$$\|G_c^{\sim} U_{\perp}\|_{\infty} < 1 \qquad (\leq 1)$$

and hence $H_W \in dom(Ric)$ (or $H_W \in dom(\overline{Ric})$) where (substituting for the Riccati equation for X_0 and noting that $B_1' X_0 + D_{11}' C_{1F_0} = T_1' \tilde{B}_1' X_0 + D_{11}' D_{\perp} D_{\perp}' C_1$, see (3.17),

$$\begin{aligned}
(H_W)_{11} &= A_{F_0} + (-X_0^{-1} C_1' D_{\perp}) D_{\perp}' D_{11} (T_1' T_1)^{-1} (T_1' \tilde{B}_1' X_0 + D_{11}' D_{\perp} D_{\perp}' C_1) \\
&= -X_0^{-1} (A - B_2 D_{12}' C_1)' X_0 - X_0^{-1} C_1' D_{\perp} D_{\perp}' C_1 \\
&\qquad - X_0^{-1} C_1' D_{\perp} D_{\perp}' D_{11} T_1^{-1} \tilde{B}_1' X_0 \\
&\qquad - X_0^{-1} C_1' D_{\perp} S^{-1} D_{\perp}' D_{11} D_{11}' D_{\perp} D_{\perp}' C_1 \\
&= -X_0^{-1} N' X_0 - X_0^{-1} C_1' D_{\perp} S^{-1} D_{\perp}' C_1
\end{aligned}$$

$$(H_W)_{12} = X_0^{-1} C_1' D_{\perp} S^{-1} D_{\perp}' C_1 X_0^{-1}$$

$$\begin{aligned}
(H_W)_{21} &= -(X_0 \tilde{B}_1 T_1 + C_1' D_{\perp} D_{\perp}' D_{11})(T_1' T_1)^{-1} (T_1' \tilde{B}_1' X_0 + D_{11}' D_{\perp} D_{\perp}' C_1) \\
&= -X_0 \tilde{B}_1 \tilde{B}_1' X_0 - N' X_0 - X_0 N + X_0 B_2 B_2' X_0 - C_1' D_{\perp} S^{-1} D_{\perp}' C_1
\end{aligned}$$

It is now immediate that

$$H_W = T^{-1} H_{\infty} T \text{ where } T = \begin{bmatrix} -I & X_0^{-1} \\ -X_0 & 0 \end{bmatrix}, \quad T^{-1} = \begin{bmatrix} 0 & -X_0^{-1} \\ X_0 & -I \end{bmatrix}$$

The appropriate stable invariant subspace for H_W will be $\mathrm{Im} \begin{bmatrix} I \\ W \end{bmatrix}$ and hence that for H_{∞} will be

$$\mathrm{Im}\, T \begin{bmatrix} I \\ W \end{bmatrix} = \mathrm{Im} \begin{bmatrix} I - X_0^{-1} W \\ X_0 \end{bmatrix}$$

Moreover Lemma 2.7 will give that $\rho(W X_0^{-1}) < 1$ (≤ 1) and hence $X_0 > W$ $(X_0 \geq W)$ giving that

$$(I - X_0^{-1} W)' X_0 = X_0 - W > 0 \quad (\geq 0)$$

or

$$X_{\infty} = X_0 (X_0 - W)^{-1} X_0 > 0$$

in case (a). This completes the necessity proof for both parts (a) and (b).

■

3.4 Proofs for Problem FI: Sufficiency

All admissible $K(s)$ such that $\|T_{zw}\|_\infty < 1$ are given by

$$K(s) = \begin{bmatrix} -Q(s) & I \end{bmatrix} \begin{bmatrix} T_1 & 0 \\ T_2 & I \end{bmatrix} \begin{bmatrix} F_1 & -I \\ F_2 & 0 \end{bmatrix}$$

for $Q \in \mathcal{R}_c\mathcal{H}_\infty$, $\|Q\|_\infty < 1$.

Note that this contains the *if* part of *(a)*.

Before beginning the proof, we will perform a change of variables suggested by Section 3.2. Firstly change the input variable to

$$v = u + T_2 w - \begin{bmatrix} T_2 & I \end{bmatrix} Fx$$

with the corresponding controller

$$K_{\text{tmp}}(s) = K(s) + \begin{bmatrix} -\begin{bmatrix} T_2 & I \end{bmatrix} F & T_2 \end{bmatrix}$$

and state equations

$$\begin{aligned} \dot{x} &= A_F x + (B_1 - B_2 T_2) w + B_2 v \\ z &= C_{1F} x + D_\perp D_\perp' D_{11} w + D_{12} v \end{aligned}$$

where

$$A_F := (A + B_2 \begin{bmatrix} T_2 & I \end{bmatrix} F); \quad C_{1F} = C_1 + D_{12} \begin{bmatrix} T_2 & I \end{bmatrix} F$$

Also define the new feedback variable

$$r := T_1(w - F_1 x)$$

Now suppose

$$K_{\text{tmp}}(s) = Q(s) T_1 \begin{bmatrix} -F_1 & I \end{bmatrix};$$

that is

$$v = Qr$$

This gives the following feedback configuration in which one would expect from (3.14) that $P^\sim P = I$ since $\|z\|_2^2 - \|w\|_2^2 = \|v\|_2^2 - \|r\|_2^2$ and this is now proven together with the stability of A_F.

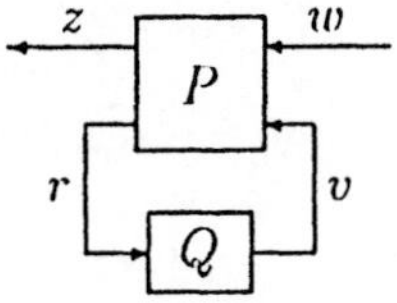

$$P = \left[\begin{array}{c|cc} A_F & B_1 - B_2 T_2 & B_2 \\ \hline C_{1F} & D_\perp D_\perp' D_{11} & D_{12} \\ -T_1 F_1 & T_1 & 0 \end{array}\right] \qquad (3.22)$$

Lemma 3.4 $P \in \mathcal{R}_c\mathcal{H}_\infty$, $P^\sim P = I$ *and* $P_{21}^{-1} \in \mathcal{R}_c\mathcal{H}_\infty$.

Proof The observability Gramian of P is X_∞ since

$$\begin{aligned}
&A'_F X_\infty + X_\infty A_F + C'_{1F}C_{1F} + F'\begin{bmatrix} T'_1 \\ 0 \end{bmatrix}\begin{bmatrix} T_1 & 0 \end{bmatrix} F \\
&= A'X_\infty + X_\infty A + C'_1 C_1 + F'\begin{bmatrix} T'_2 \\ I \end{bmatrix}(B'_2 X_\infty + D'_{12}C_1) \\
&\quad +(X_\infty B_2 + C'_1 D_{12})\begin{bmatrix} T_2 & I \end{bmatrix} F \\
&\quad +F'\left\{\begin{bmatrix} T'_2 \\ I \end{bmatrix}\begin{bmatrix} T_2 & I \end{bmatrix} + \begin{bmatrix} T'_1 T_1 & 0 \\ 0 & 0 \end{bmatrix}\right\} \\
&= F'\left\{-\begin{bmatrix} T'_2 T_2 & T'_2 \\ T_2 & I \end{bmatrix} + \begin{bmatrix} T'_1 T_1 & 0 \\ 0 & 0 \end{bmatrix} + R\right\} F \\
&= 0
\end{aligned}$$

where we have used the identity $-B'_2 X_\infty - D'_{12}C_1 = \begin{bmatrix} T_2 & I \end{bmatrix} F$. Furthermore, since $X_\infty \geq 0$ and (F_1, A_F) is detectable (note $A_F + (B_1 - B_2 T_2)F_1 = A + BF$ is stable since $X_\infty = \mathrm{Ric}(H_\infty)$) we have that A_F is stable by Lemma 2.8(a). Also

$$\begin{aligned}
&\begin{bmatrix} D'_{11}D_\perp D'_\perp & T'_1 \\ D'_{12} & 0 \end{bmatrix}\begin{bmatrix} C_1 + D_{12}\begin{bmatrix} T_2 & I \end{bmatrix} F \\ -\begin{bmatrix} T_1, & 0 \end{bmatrix} F \end{bmatrix} + \begin{bmatrix} B'_1 - T'_2 B'_2 \\ B'_2 \end{bmatrix} X_\infty \\
&= \begin{bmatrix} I & -T'_2 \\ 0 & I \end{bmatrix}(D'_{1\bullet}C_1 + RF + B'X_\infty) \\
&= 0
\end{aligned}$$

Hence by Lemma 2.8(b),

$$\begin{aligned}
P^\sim P &= \begin{bmatrix} D'_{11}D_\perp D'_\perp & T'_1 \\ D'_{12} & 0 \end{bmatrix}\begin{bmatrix} D_\perp D'_\perp D_{11} & D_{12} \\ T_1 & 0 \end{bmatrix} \\
&= \begin{bmatrix} I & 0 \\ 0 & I \end{bmatrix}
\end{aligned}$$

as claimed. It is also easily shown that $P_{21}^{-1} \in \mathcal{RH}_\infty$ since its poles are $\lambda_i(A + BF)$. ∎

The proof of sufficiency for Theorem 3.1(a) and the class of all controllers given in Theorem 3.1(c) can now be completed. Let K be any admissible controller such that $\|T_{zw}\|_\infty < 1$. Then $T_{vw} \in \mathcal{RH}_\infty$ and $T_{zw} = P_{11} + P_{12}T_{vw}$. Now define $Q = (I + T_{vw}P_{21}^{-1}P_{22})^{-1}T_{vw}P_{21}^{-1}$ so that $Q(I - P_{22}Q)^{-1}P_{21} = T_{vw}$ and $T_{zw} = P_{11} + P_{12}Q(I - P_{22}Q)^{-1}P_{21}$. Since P_{22} is strictly proper all the above are well-posed and Q is real-rational and proper. Hence Lemma 2.9 implies that $Q \in \mathcal{RH}_\infty$ with $\|Q\|_\infty < 1$. This verifies that all transfer functions T_{vw} and hence T_{uw}, can be represented in this way. ∎

Remark

In the optimal case of part (*b*) the proof of sufficiency is more delicate and to illustrate the difficulty the following example is given. Let

$$
G = \left[\begin{array}{c|c c}
1 & 1 & 1 \\
\hline
\begin{bmatrix} 1 \\ 0 \\ 1 \\ 0 \end{bmatrix} & \begin{bmatrix} 0 \\ 0 \\ 0 \\ 1 \end{bmatrix} & \begin{bmatrix} 0 \\ 1 \\ 0 \\ 0 \end{bmatrix}
\end{array}\right]
$$

then,

$$
H_\infty = \begin{bmatrix} 1 & 0 \\ -1 & -1 \end{bmatrix}, \quad \begin{bmatrix} X_1 \\ X_2 \end{bmatrix} = \begin{bmatrix} 0 \\ 1 \end{bmatrix}, \quad T_X = -1, \quad X_1' X_2 = 0 \geq 0.
$$

An optimal controller is given by

$$
u = Fx - w, \Rightarrow \; \dot{x} = (F+1)x, \; z_1 = x, \; z_2 = u = Fx - w,
$$

where $F + 1 < 0$ but F is otherwise arbitrary. Clearly for this controller $x = 0$ and hence $z_1 = 0, \; z_2 = -w$.

If the controller for the suboptimal case with $\gamma^{-2} = 1 - \epsilon^2$ is applied (see DGKF item FI.5), then,

$$
\begin{aligned}
X_\infty &= \frac{1 + \sqrt{1 + \epsilon^2}}{\epsilon^2} \\
K(s) &= \begin{bmatrix} -X_\infty - Q(s)(1 - \epsilon^2)X_\infty & Q(s) \end{bmatrix}
\end{aligned}
$$

An admissible optimal controller is obtained as $\epsilon \to 0$ iff $Q(s) = -1$, in which case $K(s) \to \begin{bmatrix} -(1 + \sqrt{1 + \epsilon^2}) & -1 \end{bmatrix}$.

3.5 Problem FC: Full Control

The FC problem has G given by,

$$
G(s) = \left[\begin{array}{c|c c}
A & B_1 & \begin{bmatrix} I & 0 \end{bmatrix} \\
C_1 & D_{11} & \begin{bmatrix} 0 & I \end{bmatrix} \\
C_2 & D_{21} & \begin{bmatrix} 0 & 0 \end{bmatrix}
\end{array}\right] \tag{3.23}
$$

and is the dual of the Full Information case: the G for the FC problem has the same form as the transpose of G for the FI problem. The term *Full Control* is used because the controller has full access to both the state through output injection and to the output z. The only restriction on the controller is that it must work with the measurement y. The assumptions that the FC problem inherits from the output feedback problem are just the dual of those in the FI problem:

(A1) (C_2, A) is detectable

(A2) D_{21} is full row rank with $\begin{bmatrix} D_{21} \\ \tilde{D}_\perp \end{bmatrix}$ unitary.

(A4) $\begin{bmatrix} A - j\omega I & B_1 \\ C_2 & D_{21} \end{bmatrix}$ has full row rank for all ω.

Necessary and sufficient conditions for the FC case are given in the following corollary. The family of all controllers can be obtained from the dual of Theorem 3.1 but these will not be required in the sequel and are hence omitted.

Corollary 3.5 *Suppose G is given by (3.23) and satisfies A1, A2 and A4. Then*

(a) $\exists K$ *such that* $\|T_{zw}\|_\infty < 1 \;\Leftrightarrow\; J_\infty \in dom(Ric),\ Ric(J_\infty) \geq 0$

(b) $\exists K$ *such that* $\|T_{zw}\|_\infty \leq 1 \;\Leftrightarrow\; J_\infty \in dom(\overline{Ric}),\ X_1'X_2 = X_2'X_1 \geq 0$. *$X_1$ and X_2 are defined in (3.4).*

4 Main Results: Output feedback

The solution to the Full Information problem of section 3 is used in this section to solve the output feedback problem. Firstly in Theorem 4.2 a so-called disturbance feedforward problem is solved. In this problem one component of the disturbance, w_2, can be estimated exactly from y using an observer, and the other component of the disturbance, w_1, does not affect the state or the output. The conditions for the existence of a controller satisfying a closed-loop $\mathcal{H}_\infty$-norm constraint is then identical to the FI case.

The solution to the general output feedback problem can then be derived from the transpose of Theorem 4.1 (Corollary 4.3) by a suitable change of variables which is based on X_∞ and the completion of the squares argument given in Section 3.2 and the characterization of all solutions given in Section 3.4.

The main result is now stated in terms of the matrices defined in section 3 involving the solutions of the X_∞ and Y_∞ Riccati equations together with the "state feedback" and "output injection" matrices F and L. It will further be convenient to additionally assume unitary changes of coordinates on w and z have been carried out to give the following partitions of D, F_1 and L_1.

$$\left[\begin{array}{c|c} & F' \\ \hline L' & D \end{array}\right] = \left[\begin{array}{c|ccc} & F_{11}' & F_{12}' & F_2' \\ \hline L_{11}' & D_{1111} & D_{1112} & 0 \\ L_{12}' & D_{1121} & D_{1122} & I \\ L_2' & 0 & I & 0 \end{array}\right] \tag{4.1}$$

Theorem 4.1 *Suppose G satisfies the assumptions A1–A4 of section 1.4.*

(a) There exists an admissible controller $K(s)$ such that $\|\mathcal{F}_\ell(G,K)\|_\infty < \gamma$ (i.e. $\|T_{zw}\|_\infty < \gamma$) if and only if

   (i) $\gamma > max(\bar{\sigma}[D_{1111}, D_{1112},], \bar{\sigma}[D_{1111}', D_{1121}'])$

   (ii) $H_\infty \in dom(Ric)$ *with* $X_\infty = Ric(H_\infty) \geq 0$

   (iii) $J_\infty \in dom(Ric)$ *with* $Y_\infty = Ric(J_\infty) \geq 0$

   (iv) $\rho(X_\infty Y_\infty) < \gamma^2$.

(b) Given that the conditions of part (a) are satisfied, then all rational internally stabilizing controllers $K(s)$ satisfying $\|\mathcal{F}_\ell(G,K)\|_\infty < \gamma$ are given by

$$K = \mathcal{F}_\ell(K_a, \Phi) \qquad \textit{for arbitrary } \Phi \in \mathcal{R}_c\mathcal{H}_\infty \qquad \textit{such that} \quad \|\Phi\|_\infty < \gamma$$

where

$$K_a = \left[\begin{array}{c|cc} \hat{A} & \hat{B}_1 & \hat{B}_2 \\ \hline \hat{C}_1 & \hat{D}_{11} & \hat{D}_{12} \\ \hat{C}_2 & \hat{D}_{21} & 0 \end{array}\right]$$

$$\hat{D}_{11} = -D_{1121}D'_{1111}(\gamma^2 I - D_{1111}D'_{1111})^{-1}D_{1112} - D_{1122},$$

$\hat{D}_{12} \in \mathcal{C}^{m_2 \times m_2}$ *and* $\hat{D}_{21} \in \mathcal{C}^{p_2 \times p_2}$ *are any matrices (e.g. Cholesky factors) satisfying*

$$\hat{D}_{12}\hat{D}'_{12} = I - D_{1121}(\gamma^2 I - D'_{1111}D_{1111})^{-1}D'_{1121},$$

$$\hat{D}'_{21}\hat{D}_{21} = I - D'_{1112}(\gamma^2 I - D_{1111}D'_{1111})^{-1}D_{1112},$$

and

$$\hat{B}_2 = Z_\infty^{-1}(B_2 + L_{12})\hat{D}_{12},$$

$$\hat{C}_2 = -\hat{D}_{21}(C_2 + F_{12}),$$

$$\hat{B}_1 = -Z_\infty^{-1}L_2 + \hat{B}_2\hat{D}_{12}^{-1}\hat{D}_{11},$$

$$\hat{C}_1 = F_2 + \hat{D}_{11}\hat{D}_{21}^{-1}\hat{C}_2,$$

$$\hat{A} = A + BF + \hat{B}_1\hat{D}_{21}^{-1}\hat{C}_2,$$

where

$$Z_\infty = (I - \gamma^{-2}Y_\infty X_\infty).$$

■

(Note that if $D_{11} = 0$ then the formulae are considerably simplified.)

The proof of this main result is via some special problems that are simpler special cases of the general problem and can be derived from the FI and FC problems. A separation type argument can then give the solution to the general problem from these special problems. It can be assumed, without loss of generality, that $\gamma = 1$ since this is achieved by the scalings $\gamma^{-1}D_{11}$, $\gamma^{-1/2}B_1$, $\gamma^{-1/2}C_1$, $\gamma^{1/2}B_2$, $\gamma^{1/2}C_2$, $\gamma^{-1}X_\infty$, $\gamma^{-1}Y_\infty$ and $\gamma^{-1}K$. All the proofs will be given for the case $\gamma = 1$.

4.1 Disturbance Feedforward

In the Disturbance Feedforward problem one component of the disturbance, w_1, does not affect the state or the output. The other component of the disturbance, w_2 (and hence the state x), can be estimated exactly from y using an observer. The conditions for the existence of a controller satisfying a closed-loop $\mathcal{H}_\infty$-norm constraint is then identical to the Full Information case.

Theorem 4.2 (Disturbance Feedforward)
Theorem 4.1 is true under the additional assumptions that

$$B_1\tilde{D}'_\perp = 0; \quad A - B_1D'_{21}C_2 \text{ is stable.} \tag{4.2}$$

In this case,

$$Y_\infty = 0,\ Z = I,\ L = -\left[\ 0,\ \ B_1D'_{21}\ \right]$$

Proof

(a) The necessity of the conditions is immediate from Theorem 3.1 since the existence of an output feedback controller implies the existence of a state feedback controller. Further, the additional condition $\overline{\sigma}(D_{11}\tilde{D}'_{\perp}) < 1$ is clearly necessary by considering $s = \infty$. Theorem 3.1 also shows that all controllers satisfying $\|\mathcal{F}_\ell(G,K)\|_\infty < 1$ are given by

$$\begin{aligned} u &= Q(s)T_1(w - F_1x) + T_2(F_1x - w) + F_2x \\ r &= T_1(w - F_1x) \\ v &= Qr \end{aligned}$$

For any $Q \in \mathcal{RH}_\infty$, $\|Q\|_\infty < 1$. Also the transfer function T_{uw} is obtained from the block diagram

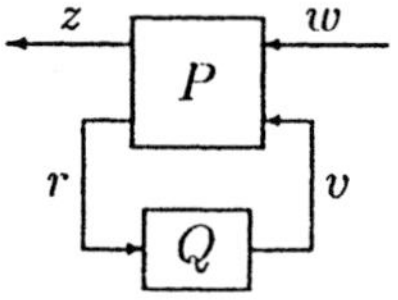

as $T_{vw} = (I - QP_{22})^{-1}QP_{21}$ (4.3)

and hence

$$u = ((I - QP_{22})^{-1}QP_{21} - T_2)w + T_2F_1x + F_2x \tag{4.4}$$

We need to find a $Q(s)$ that can be written as an output feedback. The assumption of (4.1) will give the following realization for G,

$$G = \left[\begin{array}{c|ccc} A & 0 & B_{12} & B_2 \\ \hline C_{11} & D_{1111} & D_{1112} & 0 \\ C_{12} & D_{1121} & D_{1122} & I \\ C_2 & 0 & I & 0 \end{array}\right]$$

$$w = \begin{bmatrix} w_1 \\ w_2 \end{bmatrix}$$

Hence w_1 affects z but neither x nor y and we must firstly find a $Q(s)$ in (4.4) such that T_{uw_1} is zero. Since T_{xw_1} is zero we need

$$T_{uw_1} = (I - QP_{22})^{-1}(Q(P_{21} + P_{22}T_2) - T_2)\tilde{D}'_{\perp} = 0 \tag{4.5}$$

Using the state space realization that for $\begin{bmatrix} P_{21} & P_{22} \end{bmatrix}$ in (3.22) gives

$$[P_{21} + P_{22}T_2]\tilde{D}'_{\perp} = T_1\tilde{D}'_{\perp}$$

$$\Rightarrow QT_1\tilde{D}'_{\perp} = T_2\tilde{D}'_{\perp} \tag{4.6}$$

Again without loss of generality we can assume that

$$T_1 = \begin{bmatrix} T_{11} & T_{12} \\ 0 & T_{13} \end{bmatrix}$$

where

$$T_1'T_1 = I - D_{11}'D_\perp D_\perp' D_{11}$$

and hence

$$\begin{aligned}
T_{11}'T_{11} &= I - D_{11\,11}'D_{11\,11} \\
T_2\tilde{D}_\perp' &= D_{12}'D_{11}\tilde{D}_\perp' = D_{11\,21} \\
T_{11}'T_{12} &= -D_{11\,11}'D_{11\,12} \\
T_{13}'T_{13} &= I - D_{11\,12}'D_{11\,11}T_{11}^{-1}T_{11}'^{-1}D_{11\,11}'D_{11\,12} - D_{11\,12}'D_{11\,12} =: \hat{D}_{21}'\hat{D}_{21}
\end{aligned}$$

Hence (4.6) implies that for $Q = \begin{bmatrix} Q_1 & Q_2 \end{bmatrix}$,

$$Q_1 = D_{11\,21}T_{11}^{-1}$$

and $QQ^\sim < I$ implies that

$$\begin{aligned}
Q_2Q_2^\sim &< I - D_{11\,21}(T_{11}'T_{11})^{-1}D_{11\,21}' \\
&= (I + D_{11\,21}(I - D_{11\,11}'D_{11\,11} - D_{11\,21}'D_{11\,21})^{-1}D_{11\,21}')^{-1} \\
&:= \hat{D}_{12}\hat{D}_{12}'
\end{aligned}$$

where the indicated inverses exist by *(a)(i)*. Hence

$$Q_2 = \hat{D}_{12}Q_3 \text{ for } Q_3 \in R\mathcal{H}_\infty,\ \|Q_3\|_\infty < 1.$$

We have hence shown that all controllers can be written as feedback from w_2 and x by substituting for Q into (4.4) as

$$\begin{aligned}
u &= \begin{bmatrix} D_{11\,21}T_{11}^{-1}, & \hat{D}_{12}Q_3 \end{bmatrix} \begin{bmatrix} T_{11} & T_{12} \\ 0 & T_{13} \end{bmatrix} \left(\begin{bmatrix} w_1 \\ w_2 \end{bmatrix} - F_1x \right) \\
&\quad + \begin{bmatrix} D_{11\,21}, & D_{11\,22} \end{bmatrix} (F_1x - w) + F_2x \\
&= \hat{D}_{12}Q_3\hat{D}_{21}(w_2 - D_{21}F_1x) + D_{1121}T_{11}^{-1}[T_{12}w_2 - [T_{11}T_{12}]F_1x] \\
&\quad + \begin{bmatrix} D_{11\,21}, & D_{11\,22} \end{bmatrix} F_1x - D_{11\,22}w_2 + F_2x \\
&= \hat{D}_{12}Q_3\hat{D}_{21}(w_2 - D_{21}F_1x) \\
&\quad + (-D_{1121}(I - D_{11\,11}'D_{11\,11})^{-1}D_{11\,11}'D_{11\,12} - D_{11\,22})(w_2 - D_{21}F_1x) \\
&\quad + F_2x \\
&= (\hat{D}_{11} + \hat{D}_{12}Q_3\hat{D}_{21})(w_2 - F_{12}x) + F_2x.
\end{aligned}$$

This gives the complete family of controllers in terms of x and w_2. The disturbance, w_2, and state x can be exactly estimated from the measurement, y, by means of an observer as follows,

$$\begin{aligned}
\dot{\hat{x}} &= A\hat{x} + B_{12}\hat{w}_2 + B_2u \\
\hat{w}_2 &= -C_2\hat{x} + y \\
u &= \hat{D}_{11}(\hat{w}_2 - F_{12}\hat{x}) + \hat{D}_{12}p + F_2\hat{x} \\
p &= Q_3q \\
q &= \hat{D}_{21}\hat{w}_2 - \hat{D}_{21}F_{12}\hat{x}
\end{aligned}$$

It follows that

$$(\dot{x} - \dot{\hat{x}}) = (A - B_{12}C_2)(x - \hat{x})$$

and hence for $x(0) = \hat{x}(0) = 0$, $\hat{x}(t) = x(t)$ and $\hat{w}(t) = w(t)$ for all $t \geq 0$. Furthermore, internal stability will follow from the stability of $A - B_{12}C_2$.

Finally it is straightforward to verify that this family of controllers corresponds exactly to those of Theorem 4.1 with $Y_\infty = 0$, $Z = I$, and since

$$L = -B_1 D'_{\bullet 1} \tilde{R}^{-1} = -\begin{bmatrix} 0 & B_{12} \end{bmatrix}$$

and $0 = \mathrm{Ric}(J_\infty)$.

■

The transpose of Theorem 4.2 can now be stated to obtain another special case of Theorem 4.1.

Corollary 4.3 *(Output Estimation)*
Theorem 4.1 is true under the additional assumptions that

$$D'_\perp C_1 = 0, \; A - B_2 D'_{12} C_1 \text{ is stable.}$$

In this case

$$X_\infty = 0, \; Z = I, \; F = -\begin{bmatrix} 0 \\ D'_{12}C_1 \end{bmatrix}$$

■

4.2 Converting Output Feedback to Output Estimation

The output feedback case when the disturbance, w, cannot be estimated from the output is reduced to the case of Corollary 4.3 by a suitable change of variables. Since we showed in (3.14) that

$$\|z\|_2^2 - \|w\|_2^2 = \|v\|_2^2 - \|r\|_2^2$$

where

$$\begin{aligned} v &= u + T_2 w - \begin{bmatrix} T_2, & I \end{bmatrix} F x \\ r &= T_1(w - F_1 x) \end{aligned}$$

We will perform the change of variables with v replacing z and r replacing w. Hence

$$\begin{aligned} \dot{x} &= (A + B_1 F_1)x + B_1 T_1^{-1} r + B_2 u \\ v &= u + T_2 T_1^{-1} r - F_2 x \\ y &= C_2 x + D_{21} T_1^{-1} r + D_{21} F_1 x \end{aligned}$$

and the transfer matrix from $\begin{pmatrix} r \\ u \end{pmatrix}$ to $\begin{pmatrix} v \\ y \end{pmatrix}$ is

$$G_{vyru}(s) := \left[\begin{array}{c|cc} A+B_1F_1 & B_1T_1^{-1} & B_2 \\ \hline -F_2 & T_2T_1^{-1} & I \\ C_2+D_{21}F_1 & D_{21}T_1^{-1} & 0 \end{array}\right] \tag{4.7}$$

Similarly substituting v for u in the equation for G gives that the transfer function from $\begin{pmatrix} w \\ v \end{pmatrix}$ to $\begin{pmatrix} z \\ r \end{pmatrix}$ is P as defined in (3.22). We can show with a little algebra the equivalence of the first two of the following block diagrams, with T_{vr} given by the third one.

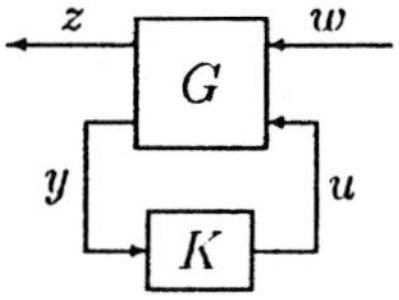

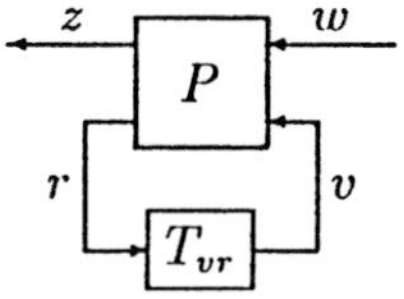

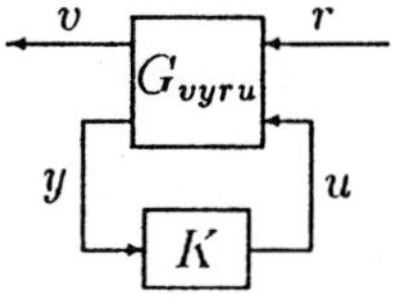

Lemma 4.4 *Let G satisfy A1–A4, and assume that X_∞ exists and $X_\infty \geq 0$. Then the following are equivalent:*

(a) K internally stabilizes G and $\|\mathcal{F}_\ell(G,K)\|_\infty < 1$,

(b) K internally stabilizes G_{vyru} and $\|\mathcal{F}_\ell(G_{vyru},K)\|_\infty < 1$,

(c) K internally stabilizes G_{tmp} and $\|\mathcal{F}_\ell(G_{tmp},K)\|_\infty < 1$,

where G_{vyru} is given by (4.7) and

$$G_{tmp} := \left[\begin{array}{c|cc} A+B_1F_1 & B_1 & B_2 \\ \hline -D_{12}F_2 & D_{11} & D_{12} \\ C_2+D_{21}F_1 & D_{21} & 0 \end{array}\right].$$

Proof

(a) ⇔ (b) Referring to the above block diagram for P and T_{vr}, it is seen by Lemma 2.9 that $T_{zw} \in \mathcal{RH}_\infty$ with $\|T_{zw}\|_\infty < 1$ iff $T_{vr} \in \mathcal{RH}_\infty$ with $\|T_{vr}\|_\infty < 1$. (Recall that $P^\sim P = I$, $P \in \mathcal{RH}_\infty$, and $P_{21}^{-1} \in \mathcal{RH}_\infty$). In order to prove internal stability of both systems we note that this is equivalent to the realizations being stabilizable and detectable. The realization of T_{vr} is detectable since the system zeros of $(G_{vyru})_{12}$ are the eigenvalues of $A+BF$ (see Lemma 2.10). Further the realisation of T_{vr} is stabilizable from r iff the realisation of T_{zw} is stabilizable from w since they are related by state feedback. Finally if the realisation of T_{vr} is internally stable with $\|T_{vr}\|_\infty < 1$ then the above block diagram for $T_{zw} = \mathcal{F}_\ell(P, \mathcal{F}_\ell(G_{vyru}, K))$ is internally stable by a small gain argument and hence so is that for $\mathcal{F}_\ell(G,K)$.

(b) ⇔ (c) Internal stability of both systems is equivalent since the closed-loop A-matrices are identical. Further note that

$$G_{tmp} = \begin{bmatrix} D_\perp & D_{12} & 0 \\ 0 & 0 & I \end{bmatrix} \begin{bmatrix} D_\perp' D_{11} T_1^{-1} & 0 \\ \multicolumn{2}{c}{G_{vyru}} \end{bmatrix} \begin{bmatrix} T_1 & 0 \\ 0 & I \end{bmatrix}$$

and recall that

$$T_1'T_1 = I - D_{11}'D_\perp D_\perp' D_{11}.$$

Hence

$$\mathcal{F}_\ell(G_{tmp}, K) = \begin{bmatrix} D_\perp & D_{12} \end{bmatrix} \begin{bmatrix} D_\perp' D_{11} \\ \mathcal{F}_\ell(G_{vyru}, K)T_1 \end{bmatrix}$$

and

$$I - \mathcal{F}_\ell(G_{tmp}, K)^\sim \mathcal{F}_\ell(G_{tmp}, K) = T_1'(I - \mathcal{F}_\ell(G_{vyru}, K)^\sim \mathcal{F}_\ell(G_{vyru}, K))T_1$$

hence giving the equivalence of (b) and (c).

■

The importance of the above constructions for G_{vyru} and G_{tmp} is that they satisfy the assumptions for the output estimation problem (Corollary 4.3) since $A + BF$ is stable. Hence we are now able to prove Theorem 4.1.

Proof of Theorem 4.1 (Output Feedback)

(a). The necessity of the conditions will be first proved. Let K be a proper controller satisfying $\|T_{zw}\|_\infty < 1$, then the controller $K \begin{bmatrix} C_2 & D_{21} \end{bmatrix}$ solves the full information problem and hence (ii) holds. Similarly $\begin{bmatrix} B_2 \\ D_{12} \end{bmatrix} K$ solves the full control problem and hence (iii) holds. From Lemma 4.4 K stabilizes G_{tmp} with $\|\mathcal{F}_\ell(G_{tmp}, K)\|_\infty < 1$, which satisfies the assumptions for the output estimation problem of Corollary 4.3, since $A4$ implies that

$$\text{rank} \begin{bmatrix} A + B_1F_1 - j\omega I & B_1 \\ C_2 + D_{21}F_1 & D_{21} \end{bmatrix} = n + m_1$$

Hence we require

$$J_{tmp} \in dom(Ric)$$

and

$$Y_{tmp} := Ric(J_{tmp}) \geq 0$$

where

$$\begin{aligned} J_{tmp} &= \begin{bmatrix} A' + F_1'B_1' & 0 \\ -B_1B_1' & -A - B_1F_1 \end{bmatrix} \\ &\quad - \begin{bmatrix} -F_2'D_{12}' & C_2' + F_1'D_{21}' \\ -B_1D_{11}' & -B_1D_{21}' \end{bmatrix} \tilde{R}^{-1} \begin{bmatrix} D_{11}B_1' & -D_{12}F_2 \\ D_{21}B_1' & C_2 + D_{21}F_1 \end{bmatrix} \end{aligned}$$

We claim that

$$J_\alpha := \begin{bmatrix} I & -X_\infty \\ 0 & I \end{bmatrix} J_\infty \begin{bmatrix} I & X_\infty \\ 0 & I \end{bmatrix} = J_{tmp}$$

where J_∞ was defined in (3.2) as

$$J_\infty := \begin{bmatrix} A' & 0 \\ -B_1B_1' & -A \end{bmatrix} - \begin{bmatrix} C' \\ -B_1D_{\bullet 1}' \end{bmatrix} \tilde{R}^{-1} \begin{bmatrix} D_{\bullet 1}B_1' & C \end{bmatrix}$$

To verify this claim let

$$\begin{aligned} M &:= \begin{bmatrix} -D_{12}F_2 \\ C_2 + D_{21}F_1 \end{bmatrix} & (4.8) \\ E &:= D_{\bullet 1}B_1'X_\infty + C - M \\ \hat{N} &:= D_{1\bullet}F + C_1 \\ & & (4.9) \end{aligned}$$

Substituting for $B_1'X_\infty$ from (3.7) gives

$$\begin{aligned} B_1'X_\infty &= F_1 - D_{11}'\hat{N} & (4.10) \\ E &= D_{\bullet 1}(F_1 - D_{11}'\hat{N}) + \begin{bmatrix} C_1 + D_{12}F_2 \\ -D_{21}F_1 \end{bmatrix} \\ &= \begin{bmatrix} I - D_{11}D_{11}' \\ -D_{21}D_{11}' \end{bmatrix} \hat{N} \end{aligned}$$

and hence

$$-\tilde{R}^{-1}E = \begin{bmatrix} I \\ 0 \end{bmatrix} \hat{N} \qquad (4.11)$$

Now consider the claim component by component. Clearly

$$(J_\alpha)_{21} = (J_\infty)_{21} = (J_{\text{tmp}})_{21}$$

Secondly

$$\begin{aligned} (J_\alpha)_{22} &= -A - B_1B_1'X_\infty + B_1D_{\bullet 1}'\tilde{R}^{-1}(M + E) \\ &= -A + B_1D_{\bullet 1}'\tilde{R}^{-1}M - B_1(B_1'X_\infty + D_{11}'\hat{N}) \\ &= (J_{\text{tmp}})_{22} \end{aligned}$$

by (4.10).

Finally

$$\begin{aligned} (J_\alpha)_{12} - (J_{\text{tmp}})_{12} &= X_\infty A + A'X_\infty + X_\infty B_1B_1'X_\infty \\ &\quad - (M' + E')\tilde{R}^{-1}(M + E) + M'\tilde{R}^{-1}M \end{aligned}$$

Substitute from (3.13):

$$\begin{aligned} X_\infty A + A'X_\infty &= -C_1'C_1 + F'RF \\ &= -(\hat{N}' - F'D_{1\bullet}')(\hat{N} - D_{1\bullet}F) + F'D_{1\bullet}'D_{1\bullet}F - F_1'F_1 \\ &= -\hat{N}'\hat{N} + F'D_{1\bullet}'\hat{N} + \hat{N}'D_{1\bullet}F - F_1'F_1 \end{aligned}$$

Equation (4.10) gives

$$X_\infty B_1 B_1' X_\infty = (F_1' - \hat{N}' D_{11})(F_1 - D_{11}' \hat{N})$$

and (4.11) and (4.8) give

$$\begin{aligned} -M'\tilde{R}^{-1}E - E'\tilde{R}^{-1}M - E'\tilde{R}^{-1}E &= -F_2' D_{12}' \hat{N} - \hat{N}' D_{12} F_2 \\ &\quad - \hat{N}'(D_{11}D_{11}' - I)\hat{N} \end{aligned}$$

Adding these three expressions gives $(J_\alpha)_{12} = (J_{tmp})_{12}$ and the claim that $J_\alpha = J_{tmp}$ is verified.

Since

$$J_\infty \begin{bmatrix} I \\ Y_\infty \end{bmatrix} = \begin{bmatrix} I \\ Y_\infty \end{bmatrix} (A' + C'L'),$$

we have

$$J_{tmp} \begin{bmatrix} I - X_\infty Y_\infty \\ Y_\infty \end{bmatrix} = \begin{bmatrix} I - X_\infty Y_\infty \\ Y_\infty \end{bmatrix} (A' + C'L')$$

and

$$Y_{tmp} := \mathrm{Ric}(J_{tmp}) = Y_\infty (I - X_\infty Y_\infty)^{-1} \geq 0.$$

It is readily verified that this implies and is implied by *(iv)*, that $\rho(X_\infty Y_\infty) < 1$. To see this, consider $Y_\infty = \begin{bmatrix} Y_{\infty 1} & 0 \\ 0 & 0 \end{bmatrix}$, $Y_{\infty 1} > 0$, and note that $Y_{\infty 1}^{-1} - X_{\infty 1} > 0$; conversely note that $X_\infty Y_\infty = (I + Y_{tmp} X_\infty)^{-1} Y_{tmp} X_\infty$ and hence $Y_{tmp} \geq 0$ implies $\rho(X_\infty Y_\infty) < 1$.

Therefore the necessity of the condition is proven. Sufficiency also follows immediately because of the equivalence of the G and G_{tmp} problems.

(b) Characterization of all solutions

To characterize all controllers for G we just need to characterize all controllers for G_{tmp} using Corollary 4.3, with $Y_{tmp} = Y_\infty Z_\infty'^{-1}$ where

$$\begin{aligned} Z_\infty &:= (I - Y_\infty X_\infty) \\ L_{tmp} &= -(B_1 D_{\bullet 1}' + Y_{tmp} M')\tilde{R}^{-1} \\ &= -Z_\infty^{-1}(B_1 D_{\bullet 1}' + Y_\infty(-X_\infty B_1 D_{\bullet 1}' + M'))\tilde{R}^{-1} \\ &= -Z_\infty^{-1}(B_1 D_{\bullet 1}' + Y_\infty(C' - E'))\tilde{R}^{-1} \\ &= Z_\infty^{-1} L - Z_\infty^{-1} Y_\infty \begin{bmatrix} \hat{N}' & 0 \end{bmatrix} \\ F_{tmp} &= \begin{bmatrix} 0 \\ F_2 \end{bmatrix}; \quad X_{tmp} = 0; \quad Z_{tmp} = I \end{aligned}$$

We can now substitute in the formulae of Theorem 4.1 with G_{tmp} and the above $(\bullet)_{tmp}$ values to obtain the class of controllers.

$$\begin{aligned} \hat{B}_2 &= (B_2 + Z_\infty^{-1} L_1 D_{12} - Z_\infty^{-1} Y_\infty \hat{N}' D_{12})\hat{D}_{12} \\ &= Z_\infty^{-1}(B_2 - Y_\infty X_\infty B_2 + L_1 D_{12} - Y_\infty(F' D_{1\bullet}' + C_1')D_{12})\hat{D}_{12} \\ &= Z_\infty^{-1}(B_2 + L_1 D_{12})\hat{D}_{12} \end{aligned}$$

by (3.7). The expressions for $\hat{C}_1$, $\hat{C}_2$, $\hat{B}_1$ and $\hat{A}$ are then obtained by a direct transcription of the above expressions and are hence omitted. This completes the proof. ∎

5 Generalizations

In this section we indicate how the results of section 4 can be extended to more general cases. Firstly the optimal case is considered when a variety of new phenomena are encountered. Secondly the removal of assumptions A1–A4 is discussed. Finally some comments are included for the case when the optimal $\mathcal{H}_\infty$-norm is necessarily achieved at $s = \infty$.

5.1 The Optimal Case

In the optimal case any combination of the conditions of Theorem 4.1(a) may be violated. In order that the Hamiltonian matrices H_∞ and J_∞ can be defined we will assume that condition $(a)(i)$ in Theorem 4.1 is satisfied and will state the result proven in Glover *et al.* (1989).

Firstly if H_∞, J_∞ $\in dom(\overline{\overline{Ric}})$ then there exist $\begin{bmatrix} X_1 \\ X_2 \end{bmatrix}$ satisfying equation (3.3) and $\begin{bmatrix} Y_1 \\ Y_2 \end{bmatrix}$ satisfying equation (3.4). In the optimal case X_1 and/or Y_1 may be singular so that $X_\infty := X_2X_1^{-1}$ and $Y_\infty := Y_2Y_1^{-1}$ may not exist, and if these inverses exist $Z_\infty := I - \gamma^{-2}Y_\infty X_\infty$ may be singular. In order to avoid taking these inverses we will modify the definitions of the 'state-feedback' matrix, F in (3.7), and the 'output injection' matrix, L in (3.7), as follows.

$$
\begin{aligned}
F^o &:= -R^{-1}[D_{1\bullet}'C_1X_1 + B'X_2] \\
L^o &:= -[Y_1'B_1D_{\bullet 1}' + Y_2'C']\tilde{R}^{-1}
\end{aligned}
$$

Furthermore as in (4.1) we assume that D, F^o, and L^o have been transformed and partitioned as follows.

$$
\left[\begin{array}{c|c} & F^{o\prime} \\ \hline L^{o\prime} & D \end{array}\right] = \left[\begin{array}{c|ccc} & F_{11}^{o\ \prime} & F_{12}^{o\ \prime} & F_2^{o\prime} \\ \hline L_{11}^{o\ \prime} & D_{1111} & D_{1112} & 0 \\ L_{12}^{o\ \prime} & D_{1121} & D_{1122} & I \\ L_2^{o\prime} & 0 & I & 0 \end{array}\right] \tag{5.1}
$$

The solution to the output feedback problem in the optimal case can now be stated (Glover *et al.* (1989)).

Theorem 5.1 *Suppose G satisfies the assumptions A1–A4 of section 1.4 and*

$$\gamma > \max(\bar{\sigma}[D_{1111}, D_{1112},], \bar{\sigma}[D_{1111}', D_{1121}']).$$

(a) There exists an admissible controller $K(s)$ such that $\|\mathcal{F}_\ell(G,K)\|_\infty \le \gamma$ (i.e. $\|T_{zw}\|_\infty \le \gamma$) if and only if

(i) $H_\infty \in dom(\overline{\overline{Ric}})$ with X_1, X_2 satisfying (3.3) such that $X_1'X_2 \ge 0$.

(ii) $J_\infty \in dom(\overline{\overline{Ric}})$ with Y_1, Y_2 satisfying (3.4) such that $Y_1'Y_2 \ge 0$.

(iii) $\begin{bmatrix} X_2'X_1 & \gamma^{-1}X_2'Y_2 \\ \gamma^{-1}Y_2'X_2 & Y_2'Y_1 \end{bmatrix} \ge 0$

(b) Given that the conditions of part (a) are satisfied, then all rational internally stabilizing controllers $K(s)$ satisfying $\|\mathcal{F}_\ell(G,K)\|_\infty \le \gamma$ are given by

$$K = \mathcal{F}_\ell(K_a, \Phi) \quad \text{for arbitrary } \Phi \in \mathcal{R}_c\mathcal{H}_\infty$$

$$\text{such that } \quad \|\Phi\|_\infty \le \gamma, \quad \det(I - (K_a)_{22}(\infty)\Phi(\infty)) \ne 0.$$

where

$$K_a := \begin{bmatrix} \hat{D}_{11} & \hat{D}_{12} \\ \hat{D}_{21} & 0 \end{bmatrix} + \begin{bmatrix} \hat{C}_1^o \\ \hat{C}_2^o \end{bmatrix} \left(s\hat{E} - \hat{A}\right)^{\#} \begin{bmatrix} \hat{B}_1^o & \hat{B}_2^o \end{bmatrix}$$

 # denotes a suitable pseudo inverse, $\hat{D}_{ij}$ are defined in Theorem 4.1, and

$$\begin{aligned}
\hat{B}_2^o &:= (Y_1'B_2 + L_{12}^o)\hat{D}_{12} \\
\hat{C}_2^o &:= -\hat{D}_{21}(C_2X_1 + F_{12}^o) \\
\hat{B}_1^o &:= -L_2^o + \hat{B}_2^o\hat{D}_{12}^{-1}\hat{D}_{11} \\
\hat{C}_1^o &:= F_2^o + \hat{D}_{11}\hat{D}_{21}^{-1}\hat{C}_2^o \\
\hat{A}^o &:= \hat{E}T_X + \hat{B}_1^o\hat{D}_{21}^{-1}\hat{C}_2^o = T_Y'\hat{E} + \hat{B}_2^o\hat{D}_{12}^{-1}\hat{C}_1^o \\
\hat{E} &:= Y_1'X_1 - \gamma^{-2}Y_2'X_2
\end{aligned}$$

■

The descriptor form of the equations for the controllers has been used as proposed for optimal Hankel-norm approximation by Safonov *et al.* (1987). At optimality $\hat{E}$ will typically be singular and the state-space equations of Theorem 4.1 are not possible. Moreover the matrix $(s\hat{E} - \hat{A}^o)$ may be singular for all s, but the transfer function $K_a(s)$ nevertheless remains uniquely defined. The condition that $\det(I - (K_a)_{22}(\infty)\Phi(\infty)) \ne 0$ is required so that this LFT is well-posed. It is often the case that all the controllers can be characterized by $\Phi = M_1\Phi_1M_2$ for non-square constant matrices, $M_1'M_1 = I$ and $M_2M_2' = I$, with $\Phi_1 \in \mathcal{R}_c\mathcal{H}_\infty$ such that $\|\Phi_1\|_\infty \le \gamma$.

The optimal case may also occur when H_∞ or J_∞ have eigen-values on the imaginary axis but H_∞, $J_\infty \in dom(Ric)$. In this case Theorem 5.1 can give regular state-space equations with $\hat{E}$, X_1, and Y_1 all invertible. The stable invariant subspace of H_∞ or J_∞ will only be unique when the additional constraint that $X_1'X_2$ and $Y_1'Y_2$ are Hermitian is included, and this requires some special purpose algorithms (see Section 5.2.5).

5.2 Relaxing Assumptions A1-A4

5.2.1 Relaxing A3 and A4

Suppose that,

$$G = \left[\begin{array}{c|cc} 0 & 0 & 1 \\ \hline 0 & 0 & 1 \\ 1 & 1 & 0 \end{array}\right]$$

which violates both A3 and A4 and corresponds to the robust stabilization of an integrator. If the controller $u = -\epsilon x$, for $\epsilon > 0$ is used then

$$T_{zw} = \frac{-\epsilon s}{s+\epsilon}, \quad \text{with } \|T_{zw}\|_\infty = \epsilon$$

Hence the norm can be made arbitrarily small as $\epsilon \to 0$, but $\epsilon = 0$ is not admissible since it is not stabilizing. This may be thought of as a case where the $\mathcal{H}_\infty$-optimum is not achieved on the set of admissible controllers. Of course, for this system, $\mathcal{H}_\infty$ *optimal* control is a silly problem, although the suboptimal case is not obviously so.

If one simply drops the requirement that controllers be admissible and removes assumptions A3 and A4, then the formulae in this paper will yield $u = 0$ for both the optimal controller and the suboptimal controller with $\Phi = 0$. This illustrates that assumptions A3 and A4 are necessary for the techniques in this paper to be directly applicable. An alternative is to develop a theory which maintains the same notion of admissibility, but relaxes A3 and A4. The easiest way to do this would be to pursue the suboptimal case introducing ϵ perturbations so that A3 and A4 are satisfied.

5.2.2 Relaxing A1

If assumption A1 is violated, then it is obvious that no admissible controllers exist. Suppose A1 is relaxed to allow unstabilizable and/or undetectable modes on the $j\omega$ axis, and internal stability is also relaxed to also allow closed-loop $j\omega$ axis poles, but A2-A4 is still satisfied. It can be easily shown that under these conditions the closed-loop $\mathcal{H}_\infty$ norm cannot be made finite, and in particular, that the unstabilizable and/or undetectable modes on the $j\omega$ axis must show up as poles in the closed-loop system.

5.2.3 Violating A1 *and* either or both of A3 and A4

Sensible control problems can be posed which violate A1 *and* either or both of A3 and A4. For example, cases when A has modes at $s = 0$ which are unstabilizable through B_2 and/or undetectable through C_2 arise when an integrator is included in a weight on a disturbance input or an error term. In these cases, either A3 or A4 are also violated, or the closed-loop $\mathcal{H}_\infty$ norm cannot be made finite. In many applications such problems can be reformulated so that the integrator occurs inside the loop (essentially using the internal model principle), and is hence detectable and stabilizable.

An alternative approach to such problems which could potentially avoid the problem reformulation would be pursue the techniques in this paper, but relax internal stability to the requirement that all closed-loop modes be in the closed left half plane. Clearly, to have finite $\mathcal{H}_\infty$ norm these closed-loop modes could not appear as poles in T_{zw}. The formulae given in this paper will often yield controllers compatible with these assumptions. The user would then have to decide whether closed-loop poles on the imaginary axis were due to weights and hence acceptable or due to the problem being poorly posed as in the above example.

A third alternative is to again introduce ϵ perturbations so that A1, A3 and A4 are satisfied. Roughly speaking, this would produce sensible answers for sensible problems, but the behaviour as $\epsilon \to 0$ could be problematic.

5.2.4 Relaxing A2

In the cases that either D_{12} is not full column rank or D_{21} is not full row rank then improper controllers can give bounded $\mathcal{H}_\infty$-norm for T_{zw}, although will not be admissible as defined in section 1.4. Such singular filtering and control problems have been well-studied in $\mathcal{H}_2$ theory and many of the same techniques go over to the $\mathcal{H}_\infty$-case (e.g. Willems(1981), Willems *et al.*(1986) and Hautus and Silverman(1983)). In particular the structure algorithm of Silverman (1969) could be used to make the terms D_{12} and D_{21} full rank by the introduction of suitable differentiators in the controller.

5.2.5 Behaviour at $s = \infty$

It has been assumed in Theorem 5.1 that

$$\gamma > \max(\bar{\sigma}[D_{1111}, D_{1112},], \bar{\sigma}[D'_{1111}, D'_{1121}])$$

and a necessary condition for a solution is that this holds with $\geq$. If equality holds then one or both of the Hamiltonian matrices cannot be defined. This corresponds to the case

$$\inf_{K(\infty)} \bar{\sigma}(\mathcal{F}_\ell(G(\infty), K(\infty))) = 1$$

where $K(\infty)$ is just considered to be an arbitrary matrix. If

$$\inf_{K(j\omega)} \bar{\sigma}(\mathcal{F}_\ell(G(j\omega), K(j\omega))) < 1, \quad \text{for some } \omega = \omega_o$$

then a bilinear transformation from the right half plane to the right half plane that moves the point $j\omega_o$ to ∞ will enable the Hamiltonians to be defined. One of them will however have an eigen value at the point on the imaginary axis to which the point at ∞ has been transformed.

A more intricate situation arises when

$$\inf_{K(j\omega)} \bar{\sigma}(\mathcal{F}_\ell(G(j\omega), K(j\omega))) = 1, \quad \forall\, \omega.$$

Here the corresponding J-factorization problem (see Green *et al.*(1988)) or spectral factorization problem is rank deficient for all ω. The theory of spectral factorization for such cases can be derived via the solutions to a Linear Matrix Inequality (Willems(1971)), or via the stable deflating subspace of the zero pencil (see Van Dooren (1981) and Clements and Glover(1989)).

Acknowledgement The second author would like to thank the first author for his meticulous attention to the technical minutiae and the first author would like to thank the second author for his careful typing of parts of the manuscript. Both authors gratefully acknowledge financial support from AFOSR, NASA, NSF, ONR (USA), and SERC(UK).

References

Adamjan, V.M., D.Z. Arov, and M.G. Krein (1978). "Infinite block Hankel matrices and related extension problems," *AMS Transl.*, vol. 111, pp. 133-156.

Anderson, B.D.O. (1967). "An algebraic solution to the spectral factorization problem," *IEEE Trans. Auto. Control*, vol. AC-12, pp. 410-414.

Ball, J.A. and J.W. Helton (1983). "A Beurling-Lax theorem for the Lie group U(m,n) which contains most classical interpolation theory," *J. Op. Theory*, vol. 9, pp. 107-142.

Boyd, S., V. Balakrishnan, and P. Kabamba (1989). "A bisection method for computing the $\mathcal{H}_\infty$ norm of a transfer matrix and related problems," *Math. Control, Signals, and Systems*, vol. 2, no. 3, pp. 207-220.

Clements, D.J. and K. Glover (1989) "Spectral Factorization via Hermitian Pencils", to appear *Linear Algebra and its Applications*, Linear Systems Special Issue.

Doyle, J.C. (1984). "Lecture notes in advances in multivariable control," *ONR/Honeywell Workshop*, Minneapolis.

Doyle, J.C., K. Glover, P.P. Khargonekar, B.A. Francis (1988). " State-space solutions to standard $\mathcal{H}_2$ and $\mathcal{H}_\infty$ control problems," *IEEE Trans. Auto. Control*, vol. AC-34, no. 8. A preliminary version appeared in *Proc. 1988 American Control Conference*, Atlanta, June, 1988.

Francis, B.A. (1987). *A course in $\mathcal{H}_\infty$ control theory*, Lecture Notes in Control and Information Sciences, vol. 88, Springer-Verlag, Berlin.

Francis, B.A. and J.C. Doyle (1987). "Linear control theory with an $\mathcal{H}_\infty$ optimality criterion," *SIAM J. Control Opt.*, vol. 25, pp. 815-844.

Glover, K. (1984). "All optimal Hankel-norm approximations of linear multivariable systems and their $\mathcal{L}_\infty$-error bounds," *Int. J. Control*, vol. 39, pp. 1115-1193.

Glover, K. (1989). "Tutorial on Hankel-norm approximation," to appear in *From Data to Model* (J.C. Willems ed.), Springer-Verlag, 1989.

Glover, K. and J. Doyle (1988). "State-space formulae for all stabilizing controllers that satisfy an $\mathcal{H}_\infty$ norm bound and relations to risk sensitivity," *Systems and Control Letters*, vol. 11, pp. 167-172.

Glover, K. and D. Mustafa (1989). "Derivation of the Maximum Entropy $\mathcal{H}_\infty$-controller and a State-space formula for its Entropy," *Int. J. Control*, to appear.

Glover, K., D.J.N. Limebeer, J.C. Doyle, E.M. Kasenally, and M.G. Safonov (1988). "A characterization of all solutions to the four block general distance problem," submitted to *SIAM J. Control Opt.*.

Gohberg, I., P. Lancaster and L. Rodman (1986), "On the Hermitian solutions of the symmetric algebraic Riccati equation," SIAM J. Control and Optim., vol. 24, no. 6, pp. 1323-1334.

Green, M., K. Glover, D.J.N. Limebeer and J.C. Doyle (1988), "A J-spectral factorization approach to H_∞ control," submitted to *SIAM J. Control Opt.*.

Hautus, M.L.J. and L.M. Silverman (1983). "System structure and singular control." *Linear Algebra Applic.*, vol. 50, pp 369-402.

Kucera V. (1972), "A contribution to matrix quadratic equations," *IEEE Trans. Auto. Control* , AC-17, No. 3, 344-347.

Limebeer, D.J.N. and G.D. Halikias (1988). "A controller degree bound for $\mathcal{H}_\infty$-optimal control problems of the second kind," *SIAM J. Control Opt.*, vol. 26, no. 3, pp. 646-677.

Mustafa, D. and K. Glover (1988). "Controllers which satisfy a closed-loop $\mathcal{H}_\infty$ norm bound and maximize an entropy integral," *Proc. 27th IEEE Conf. on Decision and Control*, Austin, Texas.

Redheffer, R.M. (1960). "On a certain linear fractional transformation," *J. Math. and Physics*, vol. 39, pp. 269-286.

Safonov, M.G., R.Y. Chiang and D.J.N. Limebeer (1987), Hankel model reduction without balancing: a descriptor approach, *Proc. 26th IEEE Conf. Dec. and Cont.*, Los Angeles.

Safonov, M.G., D.J.N. Limebeer, and R.Y. Chiang (1989). "Simplifying the $\mathcal{H}_\infty$ theory via loop shifting, matrix pencil and descriptor concepts," submitted to *Int. J. Control.* A preliminary version appeared in *Proc. 27th IEEE Conf. Decision and Control*, Austin, Texas.

Sarason, D. (1967). "Generalized interpolation in $\mathcal{H}_\infty$," *Trans. AMS.*, vol. 127, pp. 179-203.

Van Dooren, P. (1981). "A generalized eigenvalue approach for solving Riccati equations", *SIAM J. Sci. Comput.*, 2, pp. 121-135.

Silverman, L.M. (1969). "Inversion of multivariable linear systems," *IEEE Trans. Auto. Control*, vol. AC-14, pp 270-276.

Willems, J.C. (1971). "Least-squares stationary optimal control and the algebraic Riccati equation," *IEEE Trans. Auto. Control*, vol. AC-16, pp. 621-634.

Willems, J.C. (1981). "Almost invariant subspaces: an approach to high gain feedback design – Part I: almost controlled invariant subspaces." *IEEE Trans. Auto. Control*, vol. AC-26, pp235-252.

Willems, J.C., A. Kitapci and L.M. Silverman(1986). "Singular optimal control: a geometric approach." *SIAM J. Control Optim.*, vol. 24, pp 323-337.

Youla, D.C., H.A. Jabr, and J.J. Bongiorno (1976). "Modern Wiener-Hopf design of optimal controllers: part II," *IEEE Trans. Auto. Control*, vol. AC-21, pp. 319-338.

Wonham, W.M. (1985). *Linear Multivariable Control: A Geometric Approach*, third edition, Springer-Verlag, New York.

Zames, G. (1981). "Feedback and optimal sensitivity: model reference transformations, multiplicative seminorms, and approximate inverses," *IEEE Trans. Auto. Control*, vol. AC-26, pp. 301-320.

Decentralized Control of Large Scale Systems

M. Ikeda

Department of Systems Engineering
Kobe University, Kobe 657, Japan

Abstract: This paper is concerned with the decentralized stabilization problem for interconnected systems. Following the history of research, we first survey time-domain methods for decentralized state feedback and observers. We then outline an s-domain method of designing decentralized output feedback, which is an application of the recent factorization approach.

1 INTRODUCTION

In this paper, we refer to a system as a large scale system if it is more appropriate to consider the system as an interconnection of small subsystems than dealing with it as a whole. "Large scale" does not necessarily mean high dimensionality of the system here. There are many actual systems, for example, energy systems, socioeconomic systems, transportation systems, water systems, etc. (e.g., Siljak, 1978; Singh and Title, 1978; Jamshidi, 1983; Geering and Mansour, 1987; Tamura and Yoshikawa, 1989), to which interconnected modeling is suitable.

When we consider such large scale systems, we usually acknowledge the autonomy of subsystems as desirable explicitly or implicitly. Rather often in practical systems described by interconnected models, autonomy of individual subsystems is a crucial requirement. If we would like to preserve the autonomy in control systems, the control strategy we can apply is restricted to be a decentralized one with an information structure constraint conformable to the subsystems. Each subsystem is controlled locally using its own available information.

One of the most fundamental problems in control is stabilization. In the case of decentralized control for interconnected systems, it is commonly required that not only the overall closed-loop system is stable, but also the closed-loop subsystems are autonomously stable. Furthermore, stability is desired even when perturbations occur in interconnections among subsystems. This kind of robust stability is called connective stability (Siljak, 1978).

A large number of papers have been published concerning connective stabilization. The objective of this paper is to make an overview of those results for future research. When we require stability of subsystems as well as stability of the overall system, which is the case of this paper, the well-known fixed mode condition (Wang and Davison, 1973) for decentralized stabilization is necessary, but not sufficient. The readers who are less interested in stability of subsystems should consult other surveys written in the context of fixed modes (e.g., Davison, 1987; Yoshikawa, 1989).

The control law employed first for connective stabilization was state feedback in individual subsystems (Davison, 1974), which is reasonable if we look back into the history of modern control theory. The state feedback has been considered extensively until the middle of the eighties (Siljak and Vukucevic, 1977; Ikeda, Umefuji and Kodama, 1978; Sezer and Siljak, 1981; Ikeda, Siljak and Yasuda, 1983; Shi and Gao, 1986). Much effort has been made to broaden the class of stabilizable systems. All the obtained stabilizability conditions are sufficient ones.

The decentralized state feedback has been considered under the assumption that the states of subsystems are available. If we cannot measure the state of a subsystem directly, we need to estimate it from the input and output data using an observer. For autonomy of the subsystem the state of which is to be estimated, the associated observer has to compute the estimate using local informations only. Since the subsystem is connected with other subsystems, its behavior is influenced by the behaviors of adjacent subsystems which are unknown to the local observer. Therefore, the state estimation is not an easy task in the decentralized case. Disturbance suppressing observers are adopted to construct output feedback controllers utilizing the results obtained by state feedback (Viswanadham and Ramakrishna, 1982; Willems and Ikeda, 1984; Ikeda and Willems, 1987). Conditions for the existence of such observers restrict the class of stabilizable systems.

Recently, a new method of designing decentralized controllers of output feedback type has been proposed (Tan and Ikeda, 1987; Ikeda and Tan, 1989) using the proper stable factorization approach (Vidyasagar, 1985). The most significant result of the factorization approach is the parametrization of all centralized stabilizing controllers for a given system. We apply this result to each subsystem to define a local stabilizing controller with an unspecified parameter. Then, we tune the parameter to make the overall closed-loop system connectively stable. This can be done under a certain condition.

2 SYSTEM DESCRIPTION

The large scale system we deal with is a so-called input-output decentralized system (Siljak, 1978) described by

$$\mathsf{S}: \quad \begin{aligned} \dot{x}_i &= A_i x_i + B_i u_i + \sum_{j=1}^{N} A_{ij} x_j \\ y_i &= C_i x_i, \qquad i = 1, 2, ..., N, \end{aligned} \tag{2.1}$$

which is composed of N subsystems

$$\mathsf{S}_i: \quad \begin{aligned} \dot{x}_i &= A_i x_i + B_i u_i \\ y_i &= C_i x_i, \qquad i = 1, 2, ..., N \end{aligned} \tag{2.2}$$

interacting with each other through the static interconnection $\sum_{j=1}^{N} A_{ij} x_j$. In (2.1) and (2.2), x_i is the state, u_i is the input, and y_i is the output of the subsystem S_i, which are of appropriate dimensions. The matrices A_i, B_i, C_i and A_{ij} are constant and also of appropriate dimensions. We assume that the pair (A_i, B_i) is stabilizable and (C_i, A_i) is detectable.

In some cases, we describe the interconnection matrices A_{ij} as

$$A_{ij} = G_i E_{ij} H_j, \tag{2.3}$$

where the matrices G_i and H_j are defined so that

$$\begin{aligned} \mathrm{im} G_i &= \mathrm{im} A_{i1} + \mathrm{im} A_{i2} + ... + \mathrm{im} A_{iN} \\ \ker H_j &= \ker A_{1j} \cap \ker A_{2j} \cap \ldots \cap \ker A_{Nj}, \end{aligned} \tag{2.4}$$

and E_{ij} is selected for (2.3) to hold. Then, we rewrite the overall system S as

$$\mathsf{S}: \quad \begin{aligned} \dot{x}_i &= A_i x_i + B_i u_i + \sum_{j=1}^{N} G_i E_{ij} H_j x_j \\ y_i &= C_i x_i, \qquad i = 1, 2, \ldots, N \end{aligned} \tag{2.5}$$

and the subsystems S_i as

$$\begin{aligned} \mathsf{S}_i: \quad \dot{x}_i &= A_i x_i + B_i u_i + G_i v_i \\ y_i &= C_i x_i \\ w_i &= H_i x_i, \qquad i = 1, 2, ..., N \end{aligned} \tag{2.6}$$

which are connected via

$$v_i = \sum_{j=1}^{N} E_{ij} w_j, \qquad i = 1, 2, ..., N, \tag{2.7}$$

where v_i is the interconnection input of the subsystem S_i and w_j is the interconnection output of S_j. For convenience, we also use the compact form

$$\begin{aligned} \mathsf{S}: \quad \dot{x} &= A_D x + B_D u + G_D E_C H_D x \\ y &= C_D x \end{aligned} \tag{2.8}$$

where

$$\begin{aligned} A_D &= \operatorname{diag}\{A_1,\ A_2,\ \ldots,\ A_N\}, & B_D &= \operatorname{diag}\{B_1,\ B_2,\ \ldots,\ B_N\} \\ C_D &= \operatorname{diag}\{C_1,\ C_2,\ \ldots,\ C_N\}, & G_D &= \operatorname{diag}\{G_1,\ G_2,\ \ldots,\ G_N\} \\ H_D &= \operatorname{diag}\{H_1,\ H_2,\ \ldots,\ H_N\}, & E_C &= [E_{ij}] \\ x &= [x_1^t\ x_2^t\ \ldots\ x_N^t]^t, & u &= [u_1^t\ u_2^t\ \ldots\ u_N^t]^t \\ y &= [y_1^t\ y_2^t\ \ldots\ y_N^t]^t. \end{aligned} \tag{2.9}$$

Throughout this paper, we assume that the subsystems are not overlapping with each other, that is, no part of the given system is shared by a number of subsystems in common. There are, however, actual large scale systems for which it is reasonable to consider decompositions into overlapping subsystems. For manipulating such unorthodox decompositions of systems, a suitable mathematical framework is the inclusion principle (Ikeda and Siljak, 1980a; Ikeda, Siljak and White, 1981, 1984). Using the principle we can treat overlapping subsystems as non-overlapping ones.

3 STABILITY CONDITIONS

A typical scheme of stability analysis for large scale systems is as follows (e.g., Siljak, 1978; Araki, 1978; Michel and Miller, 1977). We first disconnect all subsystems, and ensure their stability by applying common techniques developed for centralized systems. Then, using aggregated informations about the subsystems and interconnections, we define and test a condition for stability of the overall system. The conditions obtained in this way usually require a certain matrix to be an M-matrix (Araki, 1975), and are satisfied when the magnitudes of interconnections are sufficiently small. In this sense, the conditions are of the small gain type (Zames, 1966). For the discussions of following sections, we present two stability conditions here, one in the time domain and the other in the s domain.

Let us consider the large scale system

$$\mathsf{S}: \quad \dot{x}_i = A_i x_i + \sum_{j=1}^{N} A_{ij} x_j, \quad i = 1, 2, ..., N, \tag{3.1}$$

which is the system of (2.1) with no subsystem inputs nor outputs. We assume that each subsystem

$$\mathsf{S}_i: \quad \dot{x}_i = A_i x_i \tag{3.2}$$

is stable. Then, there exists a positive definite matrix P_i which satisfies the Liapunov equation

$$A_i^t P_i + P_i A_i = -I, \tag{3.3}$$

where I is an identity matrix of a proper dimension. (We may consider any positive definite matrix Q_i instead of I in the right hand side of (3.3). However, I produces the least restrictive stability condition (Patel and Toda, 1980; Ikeda and Siljak, 1987).) The matrix P_i defines a Liapunov function

$$\mathbf{v}_i(x_i) = x_i^t P_i x_i \tag{3.4}$$

for the subsystem S_i of (3.2), and the time derivative is

$$\dot{\mathbf{v}}_i(x_i) = -x_i^t x_i. \tag{3.5}$$

We consider the linear combination of $\mathbf{v}_i(x_i)$,

$$\mathbf{v}(x) = \sum_{i=1}^{N} d_i \mathbf{v}_i(x_i), \tag{3.6}$$

as a candidate Liapunov function for the overall system S of (3.1), where d_i are positive numbers, which are not yet specified. The time derivative of $\mathbf{v}(x)$ with respect to (3.1) is computed and can be majorized as

$$\begin{aligned} \dot{\mathbf{v}}(x) &= -\sum_{j=1}^{N} d_i (x_i^t x_i - \sum_{i=1}^{N} 2 x_i^t P_i A_{ij} x_j) \\ &\leq -\bar{x}^t (W^t \bar{P} D + D \bar{P} W) \bar{x}, \end{aligned} \tag{3.7}$$

where $W = [w_{ij}]$ is an aggregated matrix defined by

$$w_{ij} = \begin{cases} 1/2\lambda_{\mathrm{M}}(P_i) - \|A_{ii}\|, & i = j \\ -\|A_{ij}\|, & i \neq j, \end{cases} \tag{3.8}$$

and

$$\begin{aligned} \bar{x} &= [\ \|x_1\|\ \|x_2\|\ \ldots\ \|x_N\|\ ]^t \\ \overline{P} &= \mathrm{diag}\{\lambda_{\mathrm{M}}(P_1),\ \lambda_{\mathrm{M}}(P_2),\ \ldots,\ \lambda_{\mathrm{M}}(P_N)\} \\ D &= \mathrm{diag}\{d_1,\ d_2,\ \ldots,\ d_N\}. \end{aligned} \tag{3.9}$$

Here, we use the Euclidean vector norm and the corresponding induced matrix norm. The notation λ_{M} means the maximum eigenvalue of the indicated matrix. The Liapunov's stability theory (e.g., Siljak, 1969) implies that we can conclude stability of the overall system if the matrix $W^t\overline{P}D + D\overline{P}W$ is positive definite.

To state a stability theorem obtained in this way, we introduce the notion of M-matrix, which is a kind of quasi diagonal dominant matrices.

DEFINITION 3.1 *A matrix W with nonpositive off-diagonal elements is said to be an M-matrix if its leading principal minors are all positive.*

The most important property of an M-matrix for stability analysis of large scale systems is given by (Tartar, 1971; Araki and Kondo, 1972; Araki, 1975):

LEMMA 3.1 *If W is an M-matrix, then there exists a positive diagonal matrix D such that $W^tD + DW$ is positive definite.*

When W defined in (3.8) is an M-matrix, $\overline{P}W$ is also an M-matrix. Then, this lemma immediately implies the following stability condition.

THEOREM 3.1 *If the matrix W of* (3.8) *is an M-matrix, then the system* S *of* (3.1) *is stable.*

We note that the M-matrix property of W is preserved even when the interconnection matrices A_{ij} change, provided their norms do not increase. This means that stability guaranteed by Theorem 3.1 is robust to the perturbations in couplings among subsystems if they do not become stronger.

We also note that Theorem 3.1 holds even if the testing matrix W of (3.8) is defined using any upper bounds of the norms of interconnection matrices instead of the norms themselves. However, the resultant stability condition is more conservative.

Now, we present another stability condition in the s domain. Let us consider the system description

$$\mathsf{S}: \quad \dot{x} = A_D x + G_D E_C H_D x, \tag{3.10}$$

which is (2.8) without input and output. We again assume stability of the subsystems, that is, A_D is a stable matrix. The characteristic polynomial of this system is computed as

$$\begin{aligned} \det(sI - A_D - G_D E_C H_D) &= \det(sI - A_D)\det(I - (sI - A_D)^{-1} G_D E_C H_D) \\ &= \det(sI - A_D)\det(I - H_D(sI - A_D)^{-1} G_D E_C), \end{aligned} \tag{3.11}$$

which implies that if $(I - H_D(sI - A_D)^{-1}G_D E_C)$ is nonsingular for all s in Re $s \geq 0$, then this polynomial does not have unstable roots and we can conclude stability of S.

To derive a condition for $(I - H_D(sI - A_D)^{-1}G_D E_C)$ to be nonsingular in Re $s \geq 0$, we consider the contrary situation. If it is singular for some s' in Re $s \geq 0$, then there exists a nonzero vector $y = [y_1^t \ y_2^t \ \dots \ y_N^t]^t$ such that

$$\begin{bmatrix} y_1 \\ y_2 \\ \vdots \\ y_N \end{bmatrix} = \begin{bmatrix} T_1(s')E_{11} & T_1(s')E_{12} & \dots & T_1(s')E_{1N} \\ T_2(s')E_{21} & T_2(s')E_{22} & \dots & T_2(s')E_{2N} \\ \vdots & \vdots & \vdots & \vdots \\ T_N(s')E_{N1} & T_N(s')E_{N2} & \dots & T_N(s')E_{NN} \end{bmatrix} \begin{bmatrix} y_1 \\ y_2 \\ \vdots \\ y_N \end{bmatrix} \tag{3.12}$$

where $T_i(s) = H_i(sI - A_i)^{-1}G_i$ is the transfer matrix of the subsystem S_i of (2.6) from the interconnection input v_i to the interconnection output w_i. Taking the norm of each component and using the maximum modulus principle for analytic functions, we obtain from this equation, the vector inequality

$$V\bar{y} \leq 0, \tag{3.13}$$

where $V = [v_{ij}]$ is defined by

$$v_{ij} = \begin{cases} 1 - \|T_i\| \ \|E_{ii}\|, & i = j \\ -\|T_i\| \ \|E_{ij}\|, & i \neq j \end{cases} \tag{3.14}$$

and

$$\begin{aligned} \|T_i\| &= \sup_{\omega} \|T_i(j\omega)\| \\ \bar{y} &= [\ \|y_1\| \ \ \|y_2\| \ \ \dots \ \ \|y_N\| \ ]^t. \end{aligned} \tag{3.15}$$

Here, we use the L_1, L_2, or L_∞ vector norm for y_i and the corresponding induced matrix norm for $T_i(j\omega)$ and E_{ij}.

Now, we use the following property of M-matrices (Araki and Kondo, 1972).

LEMMA 3.2 *If V is an M-matrix, then it has an inverse matrix whose elements are all nonnegative.*

This implies that when V is an M-matrix, we have $\bar{y} \leq 0$, which is seen by multiplying the inequality (3.13) by V^{-1} from the left. This contradicts nonnegativity of the nonzero vector $\bar{y}$, and hence the equation (3.12) does not hold. Thus, we obtain the following stability condition.

THEOREM 3.2 *If the matrix V of (3.14) is an M-matrix, then the system S of (3.10) is stable.*

Stability guaranteed by this theorem also is robust to the perturbations of interconnection matrices in the same sense as the robustness implied by Theorem 3.1. This theorem is still valid even if in the definition (3.14) of the testing matrix V, the norms of interconnection matrices are replaced by their upper bounds.

Stability analysis for interconnected systems has been a major subject of large-scale system theory in the seventies. The decomposition-aggregation scheme adopted here has been a common technique. Great efforts have been made to obtain less restrictive stability conditions. Readers who are interested in those results should refer to, for example, Michel and Miller (1977), Siljak(1978), and Araki(1978).

4 DECENTRALIZED STATE FEEDBACK

We start our discussion of decentralized stabilization by state feedback control under the assumption that the subsystems are controllable and their states are available for control synthesis. In this case, we can arbitrarily assign the eigenvalues with large negative real parts in each closed-loop subsystem. It might seem that such stabilization of subsystems may always result in stability of the overall system. This is not true, however (Ikeda and Siljak, 1979). The validity depends on the interconnection patterns among subsystems. Many efforts have been made to define and broaden the class of interconnection structures for which appropriate eigenvalue assignment in the subsystems stabilizes the overall system (e.g., Sezer and Siljak, 1981; Shi and Gao, 1986).

For simplicity, we consider the case in which the subsystems S_i of (2.2) are single-input ones and each one is described in the companion form

$$A_i = \begin{bmatrix} 0 & 1 & 0 & \cdots & 0 & 0 \\ 0 & 0 & 1 & \cdots & 0 & 0 \\ \vdots & \vdots & \vdots & \vdots & \vdots & \vdots \\ 0 & 0 & 0 & \cdots & 0 & 1 \\ -a_1^i & -a_2^i & -a_3^i & \cdots & -a_{n_i-1}^i & -a_{n_i}^i \end{bmatrix}, \quad B_i = \begin{bmatrix} 0 \\ 0 \\ \vdots \\ 0 \\ 1 \end{bmatrix}. \tag{4.1}$$

where n_i is the dimension of S_i. We apply the local state feedback

$$u_i = K_i x_i \tag{4.2}$$

to S_i and obtain the closed-loop subsystem

$$\mathsf{S}_i^c: \quad \dot{x}_i = (A_i + B_i K_i)x_i. \tag{4.3}$$

The feedback gain K_i is determined so that the matrix $(A_i + B_i K_i)$ has a set of distinct real eigenvalues defined by

$$\Phi_i = \{-\alpha^{\nu_i}\sigma_k^i, \; k = 1, 2, \cdots, n_i\} \tag{4.4}$$

where $\alpha \geq 1, \nu_i > 0$ and $0 < \sigma_1^i < \sigma_2^i < \cdots < \sigma_{n_i}^i$. The positive number α is common to all subsystems, and is used to change the subsystem eigenvalues so that the overall closed-loop system

$$\mathsf{S}^c: \quad \dot{x}_i = (A_i + B_i K_i)x_i + \sum_{j=1}^{N} A_{ij}x_j, \quad i = 1, 2, ..., N, \tag{4.5}$$

becomes stable. ν_i are introduced to adjust the rates of eigenvalues among subsystems when α is changed, which are selected appropriately later. σ_k^i can be arbitrarily fixed to determine the nominal patterns of the subsystem-eigenvalues. Although the choice of subsystem-eigenvalues defined by Φ_i of (4.4) may restrict the class of stabilizable large scale systems, it is appropriate in order to define the structures of interconnection matrices A_{ij} for which there always exist local feedback gains K_i which stabilize the overall closed-loop system S^c of (4.5).

To investigate stability of the closed-loop system S^c of (4.5), we transform the state space as

$$\tilde{x}_i = (\Gamma_i)^{-1}x_i, \quad i = 1, 2, ..., N \tag{4.6}$$

to obtain

$$\tilde{\mathsf{S}}^c: \quad \dot{\tilde{x}}_i = \Lambda_i\tilde{x}_i + \sum_{j=1}^{N} \Xi_{ij}\tilde{x}_j, \quad i = 1, 2, ..., N, \tag{4.7}$$

where Γ_i is the Vandermonde matrix (e.g., Kailath, 1980) defined by the set Φ_i of (4.4), and

$$\begin{aligned} \Lambda_i &= \Gamma_i^{-1}(A_i + B_i K_i)\Gamma_i \\ &= -\mathrm{diag}\{\alpha^{\nu_i}\sigma_1^i, \; \alpha^{\nu_i}\sigma_2^i, \; \ldots, \; \alpha^{\nu_i}\sigma_{n_i}^i\} \\ \Xi_{ij} &= \Gamma_i^{-1}A_{ij}\Gamma_j. \end{aligned} \tag{4.8}$$

The matrix Γ_i can be factorized (Siljak and Vukcevic, 1977) as

$$\begin{aligned}\Gamma_i &= \Pi_i \Psi_i \\ \Pi_i &= \text{diag}\{1,\ \alpha^{\nu_i},\ \ldots,\ \alpha^{\nu_i(n_i-1)}\}, \\ \Psi_i &= \begin{bmatrix} 1 & 1 & \cdots & 1 \\ -\sigma_1^i & -\sigma_2^i & \cdots & -\sigma_{n_i}^i \\ \vdots & \vdots & \vdots & \vdots \\ (-\sigma_1^i)^{n_i-1} & (-\sigma_2^i)^{n_i-1} & \cdots & (-\sigma_{n_i}^i)^{n_i-1} \end{bmatrix}.\end{aligned} \tag{4.9}$$

We now apply Theorem 3.1 to the closed-loop system $\tilde{S}^c$ of (4.7). In this case, the solution P_i to the Liapunov equation of (3.3) with $A_i = \Lambda_i$ is

$$P_i = (1/2)\text{diag}\{(1/\sigma_1^i)\alpha^{-\nu_i},\ (1/\sigma_2^i)\alpha^{-\nu_i},\ \ldots,\ (1/\sigma_{n_i}^i)\alpha^{-\nu_i}\} \tag{4.10}$$

and $\lambda_M(P_i) = (1/2\sigma_1^i)\alpha^{-\nu_i}$. To calculate the testing matrix W of (3.8), we need also the norms of the interconnection matrices Ξ_{ij}, which are majorized as

$$\|\Xi_{ij}\| \leq \|\Psi_i^{-1}\|\|\Psi_j\|\|\Pi_i^{-1}A_{ij}\Pi_j\|. \tag{4.11}$$

We note that in the right hand side, only the last norm depends on α.

For computing an upper bound of the norm $\|\Pi_i^{-1}A_{ij}\Pi_j\|$ in (4.11) when α is changed, we characterize the structure of the original interconnection matrix $A_{ij} = [a_{pq}^{ij}]$ by employing the integers $1 \leq p_1^{ij} < p_2^{ij} < \ldots < p_{\kappa_{ij}}^{ij} \leq n_i$ and $1 \leq q_1^{ij} < q_2^{ij} < \ldots < q_{\kappa_{ij}}^{ij} \leq n_j$ which are defined as follows.

(i) If $a_{1q}^{ij} \neq 0$ for some q, then $p_1^{ij} = 1$. If $a_{1q}^{ij} = 0$ for all q, then p_1^{ij} is the largest integer such that $a_{pq}^{ij} = 0$ for all $p < p_1^{ij}$ and q.

(ii) q_k^{ij} is the smallest integer such that $a_{pq}^{ij} = 0$ for all $p \leq p_k^{ij}$ and $q > q_k^{ij}, k = 1, 2, \ldots, \kappa_{ij}$.

(iii) p_k^{ij} is the largest integer such that $a_{pq}^{ij} = 0$ for all $p < p_k^{ij}$ and $q > q_{k-1}^{ij}, k = 2, 3, \ldots, \kappa_{ij}$.

The pairs (p_k^{ij}, q_k^{ij}), $k = 1, 2, \ldots, \kappa_{ij}$, define a boundary for nonzero elements of A_{ij} as

$A_{ij} =$

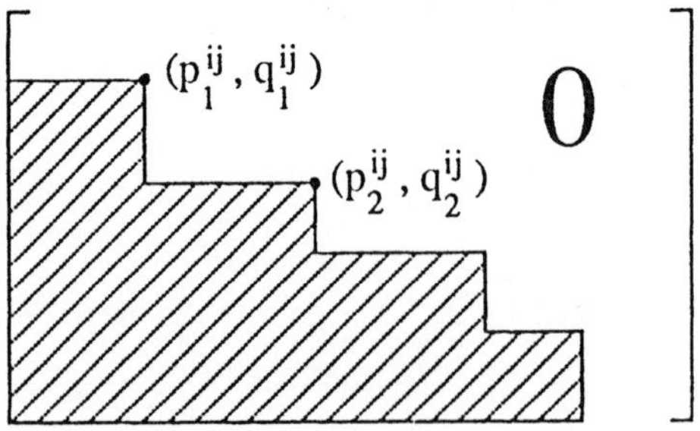

(4.12)

where the shadowed portion indicates the area of nonzero elements allowed. By an easy calculation, we obtain

$$\|\Pi_i^{-1}A_{ij}\Pi_j\| \leq \sum_{p=1}^{n_i}\sum_{q=1}^{n_j} | a_{pq}^{ij} | \alpha^{\max[-\nu_i(p_k^{ij}-1)+\nu_j(q_k^{ij}-1)]}. \tag{4.13}$$

In the right hand side of (4.13), the maximum is taken with respect to k.

Using this majorization and $\lambda_M(P_i) = (1/2\sigma_1^i)\alpha^{-\nu_i}$, we define the matrix $W = [w_{ij}]$ of (3.8) for testing stability of the closed-loop system S^c of (4.7) as

$$w_{ij} = \begin{cases} \sigma_1^i\alpha^{\nu_i} - \xi_{ii}\alpha^{\max \nu_i(-p_k^{ii}+q_k^{ii})}, & i = j \\ -\xi_{ij}\alpha^{\max[-\nu_i(p_k^{ij}-1)+\nu_j(q_k^{ij}-1)]}, & i \neq j \end{cases} \tag{4.14}$$

where

$$\xi_{ij} = \|\Psi_i^{-1}\|\|\Psi_j\|\sum_{p=1}^{n_i}\sum_{q=1}^{n_j} | a_{pq}^{ij} | . \tag{4.15}$$

Theorem 3.1 implies that if this W is an M-matrix, then the closed-loop system $\tilde{\mathsf{S}}^c$ of (4.7) is stable, and hence the given open-loop system S of (2.1) is decentrally stabilizable. Under a certain condition on the pairs (p_k^{ij}, q_k^{ij}) defined for the interconnection matrices A_{ij} of S, we can make W an M-matrix by increasing the positive α. To present the condition, we use a directed graph which describes the interconnection pattern among the subsystems in S. In the graph, which we denote by G, node i represents the subsystem S_i and the directed (oriented) branch from node j to node i means that there is a connection from S_j to S_i.

We note that the leading principal minors of W are composed of multiplications of w_{ij} along the directed loops in G. Therefore, the minors are all positive for a sufficiently large α so that W is an M-matrix if

$$\sum_{i,j} \max_k[-\nu_i p_k^{ij} + \nu_j(q_k^{ij} - 1)] < 0 \tag{4.16}$$

holds with respect to all the directed loops in G. Thus, we state the following (Sezer and Siljak, 1981):

THEOREM 4.1 *The system* S *of* (2.1) *is decentrally stabilizable if there exist positive numbers* $\nu_i, i = 1, 2, ..., N$, *which satisfy the inequalities* (4.16) *along all the directed loops in* G.

It seems that testing the condition of this theorem is not an easy task. Actually, we need to employ linear programming techniques to treat general cases (Sezer and Siljak,

1981). However, there are nontrivial classes of interconnection matrices for which testing is not needed. For example, if all A_{ij} are of the forms

$A_{ij} =$ 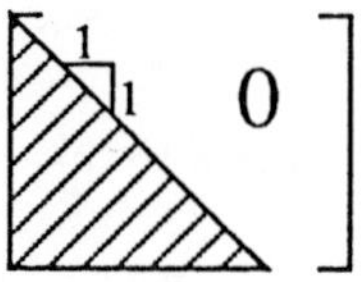

$A_{ij} =$ 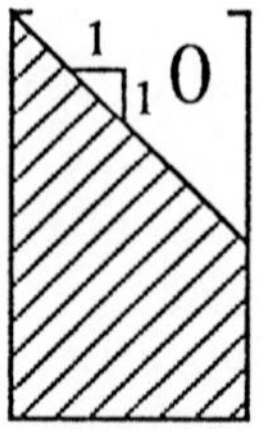

, (4.17)

where the shadowed portions indicate the area of nonzero elements allowed and $a_{pq}^{ij} = 0,\ p < q$, then these conditions always hold for $\nu_i = 1,\ i = 1, 2, \ldots, N$. Another example is described by lower triangular forms as

$A_{ij} =$ 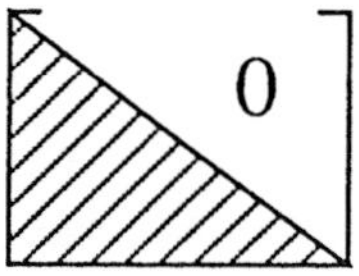

$A_{ij} =$ 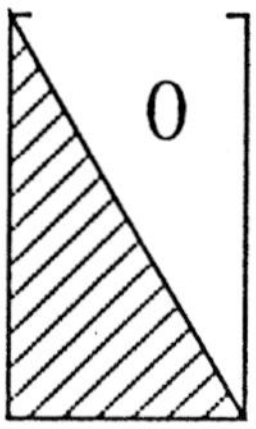

, (4.18)

for which these conditions hold for $\nu_i = n_i, \quad i = 1, 2, ..., N$. More general examples are seen in Sezer and Siljak (1981).

We have considered only single-input subsystems to obtain Theorem 4.1. The theorem can be generalized for the case of multi-input subsystems as follows. We describe each multi-input subsystem in the controllability canonical form (Luenberger, 1967), and reduce it to a set of single-input components by employing preliminary local state feedback (Ikeda and Siljak, 1980b). Then, we consider the single-input components as subsystems, and decompose interconnection matrices correspondingly. Thus, we can apply the theorem.

The research of decentralized stabilization by state feedback was initiated by Davison (1974). The structure of interconnection matrices he considered explicitly was such that

$$\text{Range}A_{ij} \subset \text{Range}B_i, \quad \text{for all} \ \ i,\ j \tag{4.19}$$

which means that the interconnection effects on each subsystem can be cancelled out by the local control input. This is a special structure included in the class of (4.18).

The class of interconnection matrices with which large-scale systems are decentrally stabilizable, has been broadened by a number of researchers. The structure of (4.17) was presented by Siljak and Vukcevic (1977). Ikeda and Siljak (1980b) indicated that the boundaries of nonzero elements may be upper and more right in some A_{ij} if those of others are lower and more left. Then, Sezer and Siljak (1981) showed that the slopes of the boundaries need not be -1 if negative. The structure of (4.18) is a special case of their result.

These results are obtained by applying state feedback of a high gain type as we did here. The areas where nonzero elements are allowed in interconnection matrices are essentially of triangular forms including the lower left corner. Shi and Gao (1986) used low gain local feedback as well by removing the restriction of positivity on ν_i in (4.4) when we assign eigenvalues in the closed-loop subsystem. In this case, the permissible areas of nonzero elements can be represented as triangles including the lower right, upper right, or upper left corners.

All these results have been obtained using the eigenvalue-assignment method presented in this section. The class of decentrally stabilizable large-scale systems can be broadened more (Willems and Ikeda, 1984) using the system description S of (2.5) and applying the almost disturbance decoupling technique (Willems, 1981) to the subsystems S_i of (2.6). The approach is the dual of that taken in the next section for the state estimation.

Like stabilization, optimization of large-scale systems has been similarly dealt with employing the decentralized state feedback conformable to subsystems (Ozguner, 1975; Ikeda and Siljak, 1982). It has been shown (Ikeda, Siljak, and Yasuda, 1983) that under the conditions of Theorem 4.1, there exists a decentralized control law which is optimal to a quadratic performance index. Although this result is obtained in the context of the inverse regulator problem and we generally cannot choose the performance index arbitrarily, the resultant optimal control system has the same robustness properties as the optimal centralized control system has (Safonov and Athans, 1977).

5 DECENTRALIZED OBSERVER

To implement the decentralized state feedback, we need to know the states of subsystems. When the state of a subsystem is not measurable directly, we have to estimate it from the control input and the available measured output. In the case of centralized control, it is common to use an observer for such a purpose (Luenberger, 1971). The necessary and sufficient condition under which we can obtain the state of a system via an observer is detectability of the system.

In the decentralized control case, to satisfy the information structure constraints, we consider a set of disjoint local observers associated with the individual subsystems.

Then, the application of the observer theory developed for centralized control systems is not as straightforward. Since the essence of the observer is an on-line real-time simulating model, each local observer needs all the information about inputs affecting the behavior of the subsystem the state of which is to be estimated. However, the subsystem has an interconnection input, which is generally not measurable, as well as a control input. The local observer does not have enough information for estimating the state. This severely complicates the decentralized estimation problem.

Somehow, however, we have to estimate the subsystem state from the local control input and measured output. In this situation, what we can do is to consider the interconnection input as an unknown disturbance and try to suppress its influence on the state estimate. We employ an almost disturbance decoupling observer (Willems, 1982) for this purpose.

Let us consider the subsystem description of (2.6),

$$\mathsf{S_i}: \quad \begin{aligned} \dot{x}_i &= A_i x_i + B_i u_i + G_i v_i \\ y_i &= C_i x_i \end{aligned} \tag{5.1}$$

where we omit the interconnection output w_i because it is not necessary for designing an observer. For this subsystem, we consider

$$\mathsf{Ob_i}: \quad \dot{\hat{x}}_i = (A_i - L_i C_i)\hat{x}_i + B_i u_i + L_i y_i, \tag{5.2}$$

which works as an observer when the subsystem $\mathsf{S_i}$ is disconnected from other subsystems and $v_i = 0$, where $\hat{x}_i$ is the estimate of x_i and L_i is a gain matrix such that $(A_i - L_i C_i)$ is stable. Then, the estimation error $e_i = \hat{x}_i - x_i$ is governed by

$$\dot{e}_i = (A_i - L_i C_i)e_i - G_i v_i. \tag{5.3}$$

This equation implies that the interconnection input v_i affects e_i, and in general e_i may not decay as t goes on. Through v_i the state estimation problem is coupled with the control problem, and the separation property between estimation and control does not hold.

Although the estimation error e_i does not decay, we expect that if the error e_i is small enough, then the estimate $\hat{x}_i$ can serve for the real state x_i in the decentralized state feedback. For this purpose, to make the transfer matrix $(sI - A_i + L_i C_i)^{-1} G_i$ from v_i to e_i in (5.3) sufficiently small, we use the following lemma which is implied by the results of Willems (1982), and Hautus and Silverman (1983). The lemma is dual to the result of perfect regulation (Kimura, 1981). We can employ any induced matrix norm to state the following:

LEMMA 5.1 *For any $\epsilon > 0$ there exists a gain matrix L_i such that the matrix $(A_i - L_iC_i)$ is stable and*

$$\|(sI - A_i + L_iC_i)^{-1}G_i\| < \epsilon, \quad \textit{for all } s \textit{ in } \mathrm{Re}\, s \geq 0 \tag{5.4}$$

if

$$\mathrm{rank}\begin{bmatrix} sI - A_i & G_i \\ C_i & 0 \end{bmatrix} = \textit{full column rank}, \quad \textit{for all } s \textit{ in } \mathrm{Re}\, s \geq 0. \tag{5.5}$$

Roughly speaking, this lemma means that unstable modes (if present) in the interconnection input v_i can be suppressed arbitrarily in the estimate $\hat{x}_i$ if the subsystem S_i satisfies the condition (5.5) which is of a minimum phase type. Then, Ob_i of (5.2) works as an observer for S_i, though approximately, even when S_i is connected with other subsystems. We collect such local observers to construct a decentralized observer for the overall system, which we describe using the notations in (2.9) as

$$\mathsf{Ob}: \quad \dot{\hat{x}} = (A_D - L_DC_D)\hat{x} + B_Du + L_Dy \tag{5.6}$$

where

$$\begin{aligned} \hat{x} &= [\hat{x}_1^t \;\; \hat{x}_2^t \;\; \dots \;\; \hat{x}_N^t]^t \\ L_D &= \mathrm{diag}\{L_1,\; L_2,\; \dots,\; L_N\}. \end{aligned} \tag{5.7}$$

Now, we employ this decentralized observer to implement the decentralized state feedback

$$u = K_Dx \tag{5.8}$$

which, we assume, stabilizes the overall system S of (2.8), where

$$K_D = \mathrm{diag}\{K_1,\; K_2,\; \dots,\; K_N\} \tag{5.9}$$

and K_i, $i = 1, 2, \dots, N$, are local state feedback gains. The state estimate $\hat{x}$ generated by Ob is used for the state x in (5.8). Then we have the observer-based output feedback controller

$$\begin{aligned} \dot{\hat{x}} &= (A_D - L_DC_D + B_DK_D)\hat{x} + L_Dy \\ u &= K_D\hat{x}, \end{aligned} \tag{5.10}$$

and the resultant overall closed-loop system is written as

$$\hat{\mathsf{S}}^{\mathsf{C}}: \quad \begin{bmatrix} \dot{x} \\ \dot{\hat{x}} \end{bmatrix} = \begin{bmatrix} A_D + G_DE_CH_D & B_DK_D \\ L_DC_D & A_D - L_DC_D + B_DK_D \end{bmatrix}\begin{bmatrix} x \\ \hat{x} \end{bmatrix}. \tag{5.11}$$

In this case, stability of $\hat{\mathsf{S}}^{\mathsf{C}}$ is not automatically implied. This is a significant difference from the centralized case.

To investigate stability of the overall closed-loop system $\hat{S}^c$, we calculate the characteristic polynomial of the system matrix as

$$\det\begin{bmatrix} sI - A_D - G_D E_C H_D & -B_D K_D \\ -L_D C_D & sI - A_D + L_D C_D - B_D K_D \end{bmatrix}$$

$$\begin{aligned} = \; & \det(sI - A_D - B_D K_D - G_D E_C H_D)\det(sI - A_D + L_D C_D) \\ & \cdot\det[I + E_C H_D(sI - A_D - B_D K_D - G_D E_C H_D)^{-1} B_D K_D \\ & \cdot(sI - A_D + L_D C_D)^{-1} G_D]. \end{aligned} \tag{5.12}$$

Since K_D is assumed to be a stabilizing state feedback gain for the overall system, the first determinant of the right hand side has no root in Re $s \geq 0$. We can say the same thing for the second determinant because $(A_D - L_D C_D)$ is stable. It is also seen that when the local observer gain L_i in Ob_i is chosen to satisfy (5.4) with a positive ϵ such that

$$\epsilon < [\sup_{\mathrm{Re}\, s \geq 0} \|E_C H_D(sI - A_D - B_D K_D - G_D E_C H_D)^{-1} B_D K_D\|]^{-1}, \tag{5.13}$$

the third determinant does not become 0 in Re $s \geq 0$. Thus, stability of the closed-loop system $\hat{S}^c$ of (5.11) is concluded. We note here that the choice of the observer gain L_D is not independent of the state feedback gain K_D. We can now state:

THEOREM 5.1 *If the subsystems* S_i *of* (5.1) *satisfy the condition* (5.5) *in Lemma* 5.1, *then there exists a decentralized observer* Ob *of* (5.6) *which provides local state estimates that can be used in stabilizing decentralized state feedback.*

We note that the condition (5.5) implies detectability of the pair (C_i, A_i), but the converse is not valid. Hence, the condition for a decentralized observer to exist is more restrictive than in the centralized case. A way of relaxing the condition is to estimate the function $K_i x_i$ instead of the full state x_i. Once a stabilizing gain $K_D = \mathrm{diag}\{K_1, K_2, \ldots, K_N\}$ of decentralized state feedback is determined, it suffices to estimate $K_i x_i$ suppressing the influence of the interconnection input v_i on this linear function. The existence condition for such an observer has been given (Ikeda and Willems, 1987) in terms of the infimal complementary detectability subspace (Willems and Commault, 1981) and the infimal almost complementary observability subspace (Willems, 1982).

In closing this section, we mention that if in addition to (5.5), the condition

$$\mathrm{rank} C_i G_i = \mathrm{rank} G_i \tag{5.14}$$

holds in the subsystem S_i, there exists a local (minimal order) observer which completely rejects the influence of the interconnection input on the state estimate (Kudva,

Viswanadham, and Ramakrishna, 1980). Then, we have no problem in implementing the local state feedback (Viswanadham and Ramakrishna, 1982).

6 DECENTRALIZED OUTPUT FEEDBACK CONTROLLER

It is well known that in the centralized control case, the design of a stabilizing output feedback controller can be decomposed into the stabilizing state feedback design and the stable observer design, which can be carried out independently. This is not true in the case of decentralized control, and the two design problems are coupled together as mentioned in the previous section. Therefore, it is not so easy to determine the state feedback gains and observer gains in local observer-based state-feedback controllers in order to stabilize the overall system. From this point of view, an appropriate approach to the design of decentralized output feedback controllers would be that which is based on the factorization of transfer matrices of given systems (Vidyasagar, 1985). The most fundamental and significant result of this approach is the parametrization of all (centralized) stabilizing controllers for a given system. In this section, we outline how the factorization approach can be utilized in the design of stabilizing decentralized controllers. The underlining idea is as follows. We first define local stabilizing controllers for individual subsystems, which have unspecified parameters. Then, we tune the local parameters to stabilize the overall closed-loop system (Tan and Ikeda, 1987; Ikeda and Tan, 1989).

In this section, we say that a rational matrix in s with real coefficients is stable if it is analytic in the closed right half complex plane $\mathbf{C}_+$ (excluding $s = \infty$). By $\mathbf{R}_s$ and $\mathbf{R}_{ps}$ we denote the sets of stable and proper stable rational matrices, respectively.

To apply the factorization approach, we represent the subsystem S_i of (2.6) by the transfer matrix

$$\mathsf{S}_i: \quad \begin{bmatrix} w_i \\ y_i \end{bmatrix} = \begin{bmatrix} Z_{11}^i & Z_{12}^i \\ Z_{21}^i & Z_{22}^i \end{bmatrix} \begin{bmatrix} v_i \\ u_i \end{bmatrix} \tag{6.1}$$

where $Z_{pq}^i(p, q = 1, 2)$ are defined as

$$\begin{aligned} Z_{11}^i &= H_i(sI - A_i)^{-1}G_i, & Z_{12}^i &= H_i(sI - A_i)^{-1}B_i \\ Z_{21}^i &= C_i(sI - A_i)^{-1}G_i, & Z_{22}^i &= C_i(sI - A_i)^{-1}B_i \end{aligned} \tag{6.2}$$

and we use the same notations u_i, y_i, v_i, w_i in the s domain as in the time domain. Since we have assumed stabilizability of (A_i, B_i) and detectability of (C_i, A_i), stabilization of the submatrix Z_{22}^i implies stabilization of the whole S_i. For this purpose, we factorize the strictly proper Z_{22}^i as

$$Z_{22}^i = N_i D_i^{-1} = \widetilde{D}_i^{-1} \widetilde{N}_i, \tag{6.3}$$

where $N_i, D_i \in \mathbf{R}_{ps}$ and $\widetilde{N}_i, \widetilde{D}_i \in \mathbf{R}_{ps}$ satisfy

$$\begin{bmatrix} Q_i & P_i \\ -\widetilde{N}_i & \widetilde{D}_i \end{bmatrix} \begin{bmatrix} D_i & -\tilde{P}_i \\ N_i & \tilde{Q}_i \end{bmatrix} = \begin{bmatrix} I & 0 \\ 0 & I \end{bmatrix} \tag{6.4}$$

for some $P_i, Q_i, \tilde{P}_i, \tilde{Q}_i \in \mathbf{R}_{ps}$. This is called a doubly coprime factorization. Then, the set of all stabilizing output feedback controllers for $\mathsf{S_i}$ is given as

$$\mathsf{LC_i}: \quad u_i = K_i(R_i)y_i, \tag{6.5}$$

where $K_i(R_i)$ is the gain transfer matrix defined by

$$K_i(R_i) = -(\tilde{P}_i + D_iR_i)(\tilde{Q}_i - N_iR_i)^{-1} \tag{6.6}$$

and $R_i \in \mathbf{R}_{ps}$ is arbitrary (Vidyasagar, 1985). We select later the parameter matrix R_i appropriately to stabilize the overall system.

When we apply the local stabilizing controller $\mathsf{LC_i}$ of (6.5) to the subsystem $\mathsf{S_i}$ of (6.1), the transfer matrix T_i from the interconnection input v_i to the interconnection output w_i of the resultant closed-loop system is calculated using (6.3) and (6.4) as

$$T_i(R_i) = T_1^i - T_2^iR_iT_3^i \tag{6.7}$$

which is an affine function of the parameter R_i, where

$$\begin{aligned} T_1^i &= Z_{11}^i - Z_{12}^i\tilde{P}_i\widetilde{D}_iZ_{21}^i \\ T_2^i &= Z_{12}^iD_i \\ T_3^i &= \widetilde{D}_iZ_{21}^i. \end{aligned} \tag{6.8}$$

The matrices T_1^i, T_2^i, and T_3^i belong to $\mathbf{R}_{ps}$ because $T_i(R_i)$ is stable for any $R_i \in \mathbf{R}_{ps}$. Then, we define the matrix $V = [v_{ij}]$ as

$$v_{ij} = \begin{cases} 1 - \|T_i(R_i)\| \, \|E_{ii}\|, & i = j \\ -\|T_i(R_i)\| \, \|E_{ij}\|, & i \neq j \end{cases} \tag{6.9}$$

where the norm is the same one as in (3.14). We employ Theorem 3.2 to state the following:

LEMMA 6.1 *When the local controllers* $\mathsf{LC_i}$ *of* (6.5) *with the gain transfer matrices of* (6.6) *are applied to the overall system* S *of* (2.5), *the resultant closed-loop system is stable if the matrix* V *of* (6.9) *is an M-matrix.*

For the matrix V defined by (6.9) to be an M-matrix, we need to choose the parameters $R_i \in \mathbf{R}_{\mathbf{ps}}$ so that $\|T_i(R_i)\|$ are sufficiently small. If V is not an M-matrix even for the infimum of $\|T_i(R_i)\|$ with respect to R_i, it can never be made so by changing R_i. Therefore, for testing purpose, it might seem better to define V using the infimums instead of $\|T_i(R_i)\|$ in (6.9). In the case where $\|T_i(R_i)\|$ is the function norm induced from the L_2 vector norm, the calculation of the infimum is reduced to the H_∞ optimization problem (Vidyasagar, 1985; Francis, 1987). However, this requires some computation efforts. In the cases of norms induced from the L_1 and L_∞ vector norms, we know little about the infimums at present.

To present a practical, though more conservative, stabilizability condition, we note that if there exists an $X^i \in \mathbf{R_s}$ such that the equation

$$T_1^i = T_2^i X^i T_3^i \tag{6.10}$$

holds, then

$$\inf_{R_i \in \mathbf{R}_{\mathbf{ps}}} \|T_i(R_i)\| = 0. \tag{6.11}$$

This is obvious in case X^i is proper and we can set $R_i = X^i$. When X^i is not proper, we use a proper approximation (Francis, 1987) to define R_i and conclude (6.11). Although the matrices T_1^i, T_2^i and T_3^i in (6.10) are defined by (6.8) using the factorization of Z_{22}^i, which is not unique (Vidyasagar, 1985), it can be shown that the existence of X^i is independent of the choice of the factorization (Ikeda and Tan, 1989). Employing a particular factorization (Nett, Jacobson, and Balas, 1984)

$$\begin{aligned}
N_i &= C_i(sI - A_K^i)^{-1}B_i, & D_i &= K_i(sI - A_K^i)^{-1}B_i + I \\
\widetilde{N}_i &= C_i(sI - A_L^i)^{-1}B_i, & \widetilde{D}_i &= C_i(sI - A_L^i)^{-1}L_i + I \\
P_i &= K_i(sI - A_L^i)^{-1}L_i, & Q_i &= -K_i(sI - A_L^i)^{-1}B_i + I \\
\widetilde{P}_i &= K_i(sI - A_K^i)^{-1}L_i, & \widetilde{Q}_i &= -C_i(sI - A_K^i)^{-1}L_i + I,
\end{aligned} \tag{6.12}$$

where $A_K^i = A_i + B_iK_i$, $A_L^i = A_i + L_iC_i$ and K_i, L_i are matrices such that A_K^i, A_L^i are stable, we can show the following lemma (Ikeda and Tan, 1989). The lemma is written in terms of the matrices A_i, B_i, C_i, G_i, H_i of the given open-loop subsystem S_i, and implies that we do not need any factorization of Z_{22}^i to see whether we can make the norm $\|T_i(R_i)\|$ arbitrarily small.

LEMMA 6.2 *The equation* (6.10) *has a solution* X^i *in* $\mathbf{R_s}$ *if and only if the equation*

$$\begin{bmatrix} sI - A_i & B_i \\ H_i & 0 \end{bmatrix} \begin{bmatrix} Y_{11}^i & Y_{12}^i \\ Y_{21}^i & Y_{22}^i \end{bmatrix} \begin{bmatrix} sI - A_i & G_i \\ C_i & 0 \end{bmatrix} = \begin{bmatrix} sI - A_i & G_i \\ H_i & 0 \end{bmatrix} \tag{6.13}$$

has a solution $(Y_{11}^i, Y_{12}^i, Y_{21}^i, Y_{22}^i)$ *in* $\mathbf{R_s}$.

The solution X^i of the equation (6.10) can be represented using the solution $(Y^i_{11}, Y^i_{12}, Y^i_{21}, Y^i_{22})$ of (6.13) in a simple form, if it exists (Ikeda and Tan, 1989). If

$$\begin{aligned} \operatorname{rank}\begin{bmatrix} sI - A_i & G_i \\ C_i & 0 \end{bmatrix} &= \mathit{full\ column\ rank}, \qquad \mathit{for\ all\ s\ in}\ \operatorname{Re} s \geq 0 \\ \operatorname{rank}\begin{bmatrix} sI - A_i & B_i \\ H_i & 0 \end{bmatrix} &= \mathit{full\ row\ rank}, \qquad \mathit{for\ all\ s\ in}\ \operatorname{Re} s \geq 0, \end{aligned} \tag{6.14}$$

then (6.13) can obviously be solved using the pseudoinverses of these matrices. We note that the first condition of (6.14) is the same as (5.5) which guaranteed the existence of an almost disturbance decoupling observer. The second condition of (6.14) has been known as the condition for perfect regulation (Kimura, 1981).

A way of investigating solvability of the general two-sided matrix equation (6.13) is transformation of the coefficient polynomial matrices in the left hand side into the Smith form. This transformation reduces the matrix equation to a set of scalar equations, which is equivalent to the original equation. The scalar equations are much more tractable, and the solutions can be computed readily.

Now, we recall the directed graph G defined in Section 4, which describes the interconnection pattern of the given large scale system S of (2.5). If the equation (6.13) is solvable in $\mathbf{R}_s$, then we remove all the branches which go into or go out of node i. We refer to this graph as $\tilde{\mathsf{G}}$, and present a graph theoretic condition for decentralized stabilizability (Ikeda and Tan, 1989).

THEOREM 6.1 *If there is no directed loop in the graph* $\tilde{\mathsf{G}}$*, then the system* S *of* (2.5) *is decentrally stabilizable.*

This theorem can be shown as follows. The k-th leading principal minor of the matrix V defined by (6.9) can be expressed as $1 - *$, where $*$ is composed of products of $\|T_i(R_i)\|$ and $\|E_{ij}\|$ along the directed loops in the subgraph of G containing the nodes $1, 2, \ldots, k$ with branches among them. The condition of Theorem 6.1 means that there is at least one $\|T_i(R_i)\|$ in each product, which can be made arbitrarily small by choosing R_i appropriately. Thus, $*$ can be made small as well in order to make V to be an M-matrix. Then, Lemma 6.1 concludes this theorem.

In this section, we discussed whether we can stabilize a subsystem and at the same time, can make the norm of the transfer matrix $T_i(R_i)$ from the interconnection input to the interconnection output arbitrarily small as described by (6.11). If we consider the interconnection input as an external disturbance and the interconnection output as a controlled output in each decoupled subsystem, this stabilization problem is identical to the almost disturbance decoupling problem with stability by measurement feedback (Weiland and Willems, 1989). Actually, the condition (6.13) of Lemma 6.2 has been obtained for such disturbance decoupling (Shimizu and Ikeda, 1986). We can give an equivalent condition in terms of almost invariant subspaces, which has been derived in the same context (Weiland and Willems, 1989).

7 CONCLUDING REMARKS

It does not seem that in the second half of the eighties, a lot of researchers are interested in decentralized control problems for large scale systems. In the author's opinion, one of the reasons is that there are not many actual control problems which motivate the research of this particular area at present. In the seventies, this area was excited by a large number of actual problems in diverse fields such as energy systems, transportation systems, socioeconomic systems, water systems, etc.. The control object which may stimulate this area in the near future would be flexible large space structures.

Another reason is that although some problems formulated in the seventies have been solved to some extent, they were tractable ones and the contributions of the solutions to actual problems have not been much appreciated. To make significant contributions, we need to reformulate the decentralized control problems including the stage of modeling. The modeling of large scale systems should be different from the centralized cases, and the synthesis of decentralized controllers is not separable from the modeling. In the case of large space structures, we can obtain precise subsystem models, but the collection of them may not necessarily form a precise overall model. Decentralized control strategies are more suitable to such systems than centralized ones.

Without doubt, the necessity of decentralized control will increase in the future as the systems we deal with become larger and more complex. Development of mathematical system theory for large scale systems is really expected.

ACKNOWLEDGEMENT

The author is grateful for many useful discussions on the topic of this paper with Prof. D. D. Siljak of Santa Clara University, Santa Clara, California, U.S.A., Prof. M. E. Sezer of Bilkent University, Ankara, Turkey, and Prof. K. Yasuda and Mr. H. -L. Tan of Kobe University, Kobe, Japan.

REFERENCES

[1] Araki, M. (1975). Applications of M-matrices to the stability problems of composite dynamical systems, *J. Mathematical Analysis and Applications*, 52, 309-321.

[2] Araki, M. (1978). Stability of large-scale nonlinear systems: Quadratic order theory of composite system method using M-matrices, *IEEE Trans. Automatic Control*, AC-23, 129-142.

[3] Araki, M., and B. Kondo (1972). Stability and transient behavior of composite nonlinear systems, *IEEE Trans. Automatic Control*, AC-17, 537-541.

[4] Davison, E. J. (1974). The decentralized stabilization and control of a class of unknown nonlinear time-varying systems, *Automatica*, 10, 309-316.

[5] Davison, E. J. (1987). Control and stabilization of interconnected dynamical systems, *Large-Scale Systems: Theory and Applications 1986*, ed. by H. P. Geering and M. Mansour, IFAC Proceedings Series, 1987, No.11, Pergamon Press, 41-52.

[6] Francis, B. A. (1987). *A Course in H_∞ Control Theory*, Springer Verlag.

[7] Geering, H. P., and M. Mansour, Eds. (1987). *Large-Scale Systems: Theory and Applications 1986*, IFAC Proceedings Series, 1987, No.11, Pergamon Press.

[8] Hautus, M. L. J., and L. M. Silverman (1983). System structure and singular control, *Linear Algebra and Its Applications*, 50, 369-402.

[9] Ikeda, M., and D. D. Siljak (1979). Counterexamples to Fessas' conjecture, *IEEE Trans. Automatic Control*, AC-24, 670.

[10] Ikeda, M., and D. D. Siljak (1980a). Overlapping decompositions, expansions and contractions of dynamic systems, *Large Scale Systems*, 1, 29-38.

[11] Ikeda, M., and D. D. Siljak (1980b). On decentrally stabilizable large-scale systems, *Automatica*, 16, 331-334.

[12] Ikeda, M., and D. D. Siljak (1982). When is a linear decentralized control optimal?, *Proc. 5th International Conference on Analysis and Optimization of Systems*, 419-431.

[13] Ikeda, M., and D. D. Siljak (1987). Stability of reduced-order models via vector Liapunov functions, *Proc. 1987 American Control Conference*, 482-489.

[14] Ikeda, M., D. D. Siljak, and D. E. White (1981). Decentralized control with overlapping information sets, *J. Optimization Theory and Applications*, 34, 279-310.

[15] Ikeda, M., D. D. Siljak, and D. E. White (1984). An inclusion principle for dynamic systems, *IEEE Trans. Automatic Control*, AC-29, 244-249.

[16] Ikeda, M., D. D. Siljak, and K. Yasuda (1983). Optimality of decentralized control for large-scale systems, *Automatica*, 19, 309-316.

[17] Ikeda, M. and H.-L. Tan (1989). Decentralized control for connective stability: A factorization approach, *Proc. China-Japan Joint Symposium on System Control Theory and Its Applications*, 15-20.

[18] Ikeda, M., and O. Umefuji, and S. Kodama (1978). Stabilization of large-scale linear systems, *Systems, Computers, and Control*, 7, 34-41.

[19] Ikeda, M., and J. C. Willems (1987). An observer theory for decentralized control of large-scale interconnected systems, *Large-Scale Systems: Theory and Applications 1986*, ed. by H. P. Geering and M. Mansour, IFAC Proceedings Series, 1987, No.11, Pergamon Press, 329-333.

[20] Jamshidi, M. (1983). *Large-Scale Systems: Modeling and Control*, North-Holland.

[21] Kailath, T. (1980). *Linear Systems*, Prentice-Hall.

[22] Kimura, H. (1981). A new approach to the perfect regulation and the bounded peaking in linear multivariable control systems, *IEEE Trans. Automatic Control*, AC-26, 253-270.

[23] Kudva, P., N. Viswanadham, and A. Ramakrishna (1980). Observers for linear systems with unknown inputs, *IEEE Trans. Automatic Control*, AC-25, 113-115.

[24] Luenberger, D. G. (1967). Canonical forms for linear multivariable systems, *IEEE Trans. Automatic Control*, AC-12, 290-293.

[25] Luenberger, D. G. (1971). An introduction to observers, *IEEE Trans. Automatic Control*, AC-16, 596-602.

[26] Michel, A. N., and R. K. Miller (1977). *Qualitative Analysis of Large Scale Dynamical Systems*, Academic Press.

[27] Nett, C. N., C. A. Jacobson, and M. J. Balas (1984). A connection between state-space and doubly coprime fractional representations, *IEEE Trans. Automatic Control*, AC-29, 831-832.

[28] Ozguner, U. (1975). Local optimization in large scale composite dynamic systems, *Proc. 9th Asilomar Conference on Circuits, Systems, and Computers*, 87-91.

[29] Patel, R. V., and M. Toda (1980). Quantitative measures of robustness for multivariable systems, *Proc. 1980 Joint Automatic Control Conference*, TP8-A.

[30] Safonov, M. G., and M. Athans (1977). Gain and phase margin for multiloop LQG regulators, *IEEE Trans. Automatic Control*, AC-22, 173-179.

[31] Sezer, M. E., and D. D. Siljak (1981). On decentralized stabilization and structure of linear large-scale systems, *Automatica*, 17, 641-644.

[32] Shi, Z.-C., and W.-B. Gao (1986). Stabilization by decentralized control for large-scale interconnected syst ems, *Large Scale Systems*, 10, 147-155.

[33] Shimizu, H., and M. Ikeda (1986). On the almost disturbance decoupling problem: A solvability condition in case internal stability is required, *Proc. 15th SICE Symposium on Control Theory*, 283-286 (In Japanese).

[34] Siljak, D. D. (1969). *Nonlinear Systems*, John Wiley.

[35] Siljak, D. D. (1978). *Large Scale Systems: Stability and Structure*, North-Holland.

[36] Siljak, D. D., and M. B. Vukcevic (1977). Decentrally stabilizable linear and bilinear large-scale systems, *Int. J. Control*, 26, 289-305

[37] Singh, M. G., and A. Titli (1978). *Systems: Decomposition, Optimization, and Control*, Pergamon Press.

[38] Tamura, H., and T. Yoshikawa, Eds. (1989). *Large Scale Systems Control and Decision Making*, Marcel Dekker.

[39] Tan, H.-L., and M. Ikeda (1987). Decentralized stabilization of large-scale interconnected systems: A stable factorization approach, *Proc. 26th IEEE Conference on Decision and Control*, 2295-2300.

[40] Tartar, L. (1971). Une nouvelle caracterisation des M matrices, *RAIRO*, 5, 127-128.

[41] Vidyasagar, M. (1985). *Control System Synthesis: A Factorization Approach*, MIT Press.

[42] Viswanadham, N., and A. Ramakrishna (1982). Decentralized estimation and control for interconnected systems, *Large Scale Systems*, 3, 255-266.

[43] Wang, S. H., and E. J. Davison (1973). On the stabilization of decentralized control systems, *IEEE Trans. Automatic Control*, AC-18, 473-478.

[44] Weiland, S., and J. C. Willems (1989). Almost disturbance decoupling with internal stability, *IEEE Trans. Automatic Control*, AC-34, 277-286.

[45] Willems, J. C. (1981). Almost invariant subspaces: An approach to high gain feedback design - Part 1: Almost controlled invariant subspaces, *IEEE Trans. Automatic Control*, AC-26, 235-252.

[46] Willems, J. C. (1982). Almost invariant subspaces: An approach to high gain feedback design - Part 2: Almost conditionally invariant subspaces, *IEEE Trans. Automatic Control*, AC-27, 1071-1085.

[47] Willems, J. C., and C. Commault (1981). Disturbance decoupling by measurement feedback with stability or pole placement, *SIAM J. Control*, 19, 490-504.

[48] Willems, J. C., and M. Ikeda (1984). Decentralized stabilization of large-scale interconnected systems, *Proc. 6th International Conference on Analysis and Optimization of Systems*, 236-244.

[49] Yoshikawa, T. (1989). Decentralized control systems and fixed modes, *Large Scale Systems Control and Decision Making*, ed. by H. Tamura and T. Yoshikawa, Marcel Dekker.

[50] Zames, G. (1966). On input-output stability of time-varying nonlinear systems, Parts I and II, *IEEE Trans. Automatic Control*, AC-11, 228-238, 465-476.

State Space Approach to the Classical Interpolation Problem and Its Applications

H. Kimura

Department of Mechanical Engineering for Computer-Controlled Machinery
Osaka University
2-1, Yamada-oka, Suita 565, Japan

ABSTRACT

The Pick-Nevanlinna interpolation theory in classical analysis plays an important role in the recent progress of linear system theory in the frequency domain. In this paper, we shall show how the classical interpolation theory which relies heavily on function theoretic properties is described in the algebraic framework of the state space. The notion of conjugation, or more specifically, of J-lossless conjugation is shown to be the state-space representation of the classical interpolation, or its modern versions. Thus, the notion of J-lossless conjugation provides a unified treatment of the H^∞ control problem, as well as of robust stabilization and the model reduction.

1. INTRODUCTION

The transfer function G(s) of a linear time-invariant lumped-parameter system has two distinct aspects. If you regard it as a complex function, you are concerned with the *analytic* aspect of the system. If you regard it as a rational function, you are concerned with the *algebraic* aspect of the systems. These two aspects of linear systems are, of course, closely related and the interplay between them is one of the main sources of the rich structure in system theory. Let us take the stability criteria of linear systems as an example. The Nyquist stability test is essentially based on a function-theoretic property of loop transfer functions, while the Routh stability test is essentially based on the rationality of transfer functions. The analytic or topological procedure of the Nyquist test looks totally different from the algebraic procedure of the Routh test. Sometimes, it is hardly believable for the beginner of control theory that the stability can be checked in such different ways. If he is smart enough, he might sense the logical depth of system theory there.

In classical system theory, the analytic theory of transfer functions was fertile. We can list some fundamental results concerning the analytic aspect of transfer functions, such as the Nyquist stability test, the realizability condition of Wiener-Paley, the Bode formula concerning the relation between the real and the imaginary parts of minimal phase transfer functions, the Fujisawa criterion for ladder networks, to name a few. These results, some of which are extended in the recent book by Freudenberg and Looze[15], did not receive much attention in the far-reaching progress of modern system theory, in which the state-space paradigm dominates.

The revival of analytic system theory gradually took shape in the mid-70's. Perhaps, the multivariable synthesis methods in the frequency domain developed mainly by the British school [32] [33] [38] paved the avenue for the more systematic and sophisticated theories in the frequency domain, such as Hankel-norm model reduction, H^∞ control and robust stabilization.

In the early stage of the developement of these fields, the function-theoretic aspect of transfer functions played a dominant role. It is remarkable that, in these new frequency domain theories, function theoretic properties of transfer functions manifest themselves as *interpolation constraints*. Thus, the classical interpolation theory in classical analysis, which dates back to the beginning of this century, gives the analytical basis of the new system theory in the frequency domain. Especially, the Pick-Nevanlinna interpolation theory almost directly solves the robust stabilization problem. When writing [24], the author was amazed to notice that the result of 70 years ago can be used in a very straightforward way to solve a problem of contemporary technology.

It was in the late 60's when the Pick-Nevanlinna interpolation theory was first brought into system theory. Youla and Saito gave a circuit theoretical proof of the Pick criterion [44]. It is worth mentioning that the celebrated paper on lattice filters by Itakura and Saito [20] appeared four years later which revived the interest in the moment theory developed from the Caratheodory-Fejer interpolation problem, another famous classical interpolation theory related closely to time series analysis.

The paper by Youla and Saito did not seem to receive much attention in circuit theory at that time. However, approximately ten years later, Youla's paper on broad-band matching [43] which gave a motivation for [44] was given a new light by Helton [18] who used a more sophisticated operator-theoretic version of the Pick-Nevanlinna interpolation theory. In the field of control, we had a pioneering work by Tannenbaum who applied initially the Pick-Nevanlinna theory to the problem of robust stabilization [35]. This work was extended later in [23]. The problem of robust stabilizability for SISO system turned out to be almost directly connected to the Pick-Nevanlinna problem [24].

In the field of H^∞ control, the paper by Chang and Pearson [4] solved the so-called one-block problem based on the matrix version of Pick-Nevanlinna theory [5]. Also, a multivariable extension of the robust stabilizability was given based on the same technique in [42]. However, the matrix version of Pick-Nevanlinna theory developed in [5] is not suited for computation and the resultant controller is not minimal. A version of matrix interpolation problem was proposed in [26] to cover these drawbacks.

If we confine our scope to linear time-invariant lumped-parameter systems, we can have another important property of systems in our hands. This is the rationality. In state-space theory, we can neglect cumbersome arguments on the existence of Laplace transform and many function-theoretic properties like spectral factorization can be translated into the much more transparent algebraic properties of systems. The same route can be found for the classical interpolation theory.

In this paper, we shall show how the classical interpolation theory, which relies heavily on function theoretic properties, is described in the algebraic framework of the state space. The notion of conjugation, or more specifically, of J-lossless conjugation is shown to be the state-space representation of the classical interpolation, or its modern versions. Thus, the notion of J-lossless conjugation provides a unified treatment of the H^∞ control problem, as well as of robust stabilization and model reduction.

It should be noted that the operator-theoretic approaches to the interpolation problems have contributed substantially to the development of system theory in the frequency domain. Since Zames' paper, the first breakthrough in H^∞ control theory was done by Francis and Zames [12] which was based on the work of Sarason [39]. The geometric approach by Ball and Helton [2] [3] was extensively used in [13]. Dym [9] proposed another framework of interpolation theory, in which the matrix version was treated in a natural way.

In Sections 2 and 3, the classical treatments are reviewed briefly as faithfully as possible. The lattice structure of the Nevanlinna algorithm is also described. In Section 4, some applications of classical interpolation are briefly discussed. In Section 5, the notion of conjugation is introduced. It is shown in Section 6 that the inner-outer factorization is regarded as a special class of conjugation, the lossless conjugation. The J-lossless conjugation is introduced in Section 7. The relation between the J-lossless conjugation and the interpolation problem is fully discussed. Section 8 is devoted to solving the model-matching problem based on J-lossless conjugation. The complete solution to the one-block problem of H^∞ control is given.

Notations

$$\mathbf{P}_1 := \{f(z)\ ;\ \text{analytic and Re } f(z) \geq 0 \text{ in } |z| \leq 1\}$$

$\mathbf{P}$: = { f(z) ; analytic and Re f(z) ≥ 0 in Re s ≥ 0}
$\mathbf{B}_1$: = {f(z) ; analytic and |f(z)| ≤ 1 in |z| ≤ 1}
$\mathbf{B}$: = {f(s) ; analytic and |f(s)| ≤ 1 in Re s ≥ 0}
$\mathbf{R}_{m\times r}$: The set of real matrices of size m×r.

$\mathbf{RH}^{\infty}_{m\times r}$; The set of rational proper stable matrices of size m×r.

$\mathbf{BH}^{\infty}_{m\times r}$; The subset of $\mathbf{RH}^{\infty}_{m\times r}$ consisting of contractions

$\{A, B, C, D\}: = D + C(sI - A)^{-1}B$

$G^{\sim}(s) := G^T(-s)$, $G^*(s) = \overline{G}^T(\overline{s})$.

2. PICK-NEVANLINNA INTERPOLATION PROBLEM

In 1916, Pick posed the following question [35] : *Let $f(z) \in P_1$ and $w_i = f(z_i)$, $|z_i| < 1$, $i = 1, \cdots, n$. What conditions are imposed on the n pairs of complex numbers (z_i, w_i), $i = 1, \cdots, n$?*

Obviously, the condition Re $w_i > 0$, $i = 1, \cdots, n$, is necessary due to the definition of $\mathbf{P}_1$. The crucial fact is that it is *not* sufficient. An additional condition is imposed which comes from the analyticity of f(z).

In order to answer the above question, Pick derived an integral representation of f(z) in the disk $|z| < r < 1$ which is now known as Schwarz's formula [1, p.168]:

$$f(z) = \frac{1}{2\pi}\int_0^{2\pi} \frac{re^{j\theta} + z}{re^{j\theta} - z} \, \mathrm{Re}\left[f(re^{-j\theta})\right] d\theta + \frac{1}{2\pi}\int_0^{2\pi} \mathrm{Im}\left[f(re^{j\theta})\right] d\theta \tag{2.1}$$

From this representation, it follows that

$$\frac{f(z_i) + \overline{f(z_j)}}{r^2 - z_i\overline{z_j}} = \frac{1}{2\pi}\int_0^{2\pi} \frac{\mathrm{Re}[f(re^{j\theta})]}{(re^{j\theta} - z_i)(re^{-j\theta} - \overline{z_j})} \, d\theta,$$

where r is taken sufficiently close to 1 enough to guarantee that $|z_i| < r$ for each i. Therefore, for all x_i, $i = 1, \cdots, n$, we have

$$\sum_{i,j=1}^{n} \frac{f(z_i) + \overline{f(z_j)}}{r^2 - z_i\overline{z_j}} x_i\overline{x_j} = \frac{1}{\pi}\int_0^{2\pi} \mathrm{Re}[f(re^{j\theta})] \left| \sum_{i=1}^{n} \frac{x_i}{re^{j\theta} - z_j} \right|^2 d\theta > 0.$$

Since $f(z) \in \mathbf{P}_1$, the integrand at the right-hand side is non-negative. Letting $r \to 1$ verifies the inequality

$$P: = \begin{bmatrix} \dfrac{w_1 + \overline{w_1}}{1 - z_1\overline{z_1}} & \cdots & \dfrac{w_1 + \overline{w_n}}{1 - z_1\overline{z_n}} \\ & \cdots & \\ \dfrac{w_n + \overline{w_1}}{1 - z_n\overline{z_1}} & \cdots & \dfrac{w_n + \overline{w_n}}{1 - z_n\overline{z_n}} \end{bmatrix} \geq 0 \tag{2.2}$$

The matrix P defind in (2.2) is usually referred to as the *Pick matrix* . Now, it has been proved that $P \geq 0$ is a necessary condition for the existence of $f(z) \in \mathbf{P}_1$ satisfying the interpolation conditions

$$w_i = f(z_i), \qquad i = 1,2,\cdots,n. \tag{2.3}$$

At first glance, it seems obvious that $P \geq 0$ is also sufficient for the existence of $f(z) \in \mathbf{P}_1$ satisfying (2.3). Pick himself did not consider this problem. Actually, it is far from trivial and an answer was given later by Nevanlinna [34] who derived an algorithm to construct such an f.

Before proceeding to the Nevanlinna construction, we note an important property of the Pick matrix (2.2) which was fully discussed by Pick himself in [35]. Consider a linear fractional transformation

$$z = \frac{a + b\lambda}{c + d\lambda}, \qquad \lambda = \frac{a - cz}{-b + dz}. \tag{2.4}$$

This maps the unit disk in the z-plane to a region in the λ-plane. For instance,

$$z = \frac{1 - \lambda}{1 + \lambda}, \qquad \lambda = \frac{1 - z}{1 + z}. \tag{2.5}$$

maps the unit disk to the right half plane $\lambda + \overline{\lambda} \geq 0$.

Assume that λ_i is mapped to z_i in (2.4). From the identity

$$1 - z_i\overline{z_j} = \frac{K(\lambda_i, \overline{\lambda_j})}{(c + d\lambda_i)(\overline{c} + \overline{d\lambda_j})} \tag{2.6}$$

$$K(\lambda_i, \overline{\lambda_j}) = (|c|^2 - |a|^2) + (\overline{c}d - \overline{a}b)\lambda_i + (c\overline{d} - a\overline{b})\overline{\lambda_j} + (|d|^2 - |b|^2)\lambda_i\overline{\lambda_j},$$

the unit disk in the z-plane is mapped to the region

$$K(\lambda, \overline{\lambda}) \geq 0 \tag{2.7}$$

in the λ-plane.

Also, a linear fractional transformation

$$w = \frac{a_1 + b_1\beta}{c_1 + d_1\beta}, \qquad \beta = \frac{a_1 - c_1 w}{-b_1 + d_1 w}$$

maps the right half plane Re $z \geq 0$ to a region

$$L(\beta, \bar{\beta}) \geq 0 \tag{2.8}$$

in the β-plane, where $L(\beta_i, \bar{\beta}_j)$ is defined as

$$w_i + \bar{w}_j = \frac{L(\beta_i, \bar{\beta}_j)}{(c_1 + d_1\beta_i)(\bar{c}_1 + \bar{d}_1\bar{\beta}_j)}, \tag{2.9}$$

$$L(\beta_i, \bar{\beta}_j) = (a_1\bar{c}_1 + \bar{a}_1 c_1) + (\bar{c}_1 b_1 + \bar{a}_1 d_1)\beta_i$$
$$+(a_1\bar{d}_1 + c_1\bar{b}_1)\bar{\beta}_j + (b_1\bar{d}_1 + \bar{b}_1 d_1)\beta_i\bar{\beta}_j.$$

From (2.6) and (2.9), it follows that

$$\frac{w_i + \bar{w}_j}{1 - z_i\bar{z}_j} = \frac{(c + d\lambda_i)(\bar{c} + \overline{d\lambda_j})}{(c_1 + d_1\beta_i)(\bar{c}_1 + \bar{d}_1\bar{\beta}_j)} \cdot \frac{L(\beta_i, \bar{\beta}_j)}{K(\lambda_i, \bar{\lambda}_j)}$$

Therefore, the inequality (2.2) is equivalent to the inequality

$$\begin{bmatrix} \frac{L(\beta_1, \bar{\beta}_1)}{K(\lambda_1, \bar{\lambda}_1)} & \cdots & \frac{L(\beta_1, \bar{\beta}_n)}{K(\lambda_1, \bar{\lambda}_n)} \\ & \cdots\cdots & \\ \frac{L(\beta_n, \bar{\beta}_1)}{K(\lambda_n, \bar{\lambda}_1)} & \cdots & \frac{L(\beta_n, \bar{\beta}_n)}{K(\lambda_n, \bar{\lambda}_n)} \end{bmatrix} \geq 0. \tag{2.10}$$

Thus, if $f(\lambda)$ is analytic and satisfies $L(f, \bar{f}) \geq 0$ in the region (2.7), the interpolation data $\beta_i = f(\lambda_i)$, $i = 1, \cdots, n$, must satisfy the inequality (2.10).

As an example, take the transformation (2.5) and

$$w = \frac{1 - \beta}{1 + \beta}, \tag{2.11}$$

which maps the half plane Re $w \geq 0$ to the unit disk $|\beta| \leq 1$. Since $K(\lambda_i, \bar{\lambda}_j) = 2(\lambda_i + \bar{\lambda}_j)$ and $L(\beta_i, \bar{\beta}_j) = 2(1 - \beta_i\bar{\beta}_j)$, the inequality (2.10) becomes

$$P = \begin{bmatrix} \frac{1-\beta_1\bar{\beta}_1}{\lambda_1+\bar{\lambda}_1} & \cdots & \frac{1-\beta_1\bar{\beta}_n}{\lambda_1+\bar{\lambda}_n} \\ & \cdots\cdots & \\ \frac{1-\beta_n\bar{\beta}_1}{\lambda_n+\bar{\lambda}_1} & \cdots & \frac{1-\beta_n\bar{\beta}_n}{\lambda_n+\bar{\lambda}_n} \end{bmatrix} \geq 0. \qquad (2.12)$$

The inequality (2.12) is actually a necessary and sufficient condition for the existence of $f(\lambda) \in \mathbf{B}$ satisfying the interpolation conditions $\beta_i = f(\lambda_i)$, Re $\lambda_i \geq 0$, $i=1, \cdots, n$.

3. NEVANLINNA ALGORITHM

Three years after Pick's work [35], a paper by Nevanlinna [34] appeared, in which the converse of Pick's result was extensively discussed. He worked with $\mathbf{B}_1$ instead of $\mathbf{P}_1$ and formulated the problem as follows : *Find a necessary and sufficient condition on the n pairs (z_i, β_i), $|z_i| \leq 1$, $i=1,\cdots,n$, which guarantees the existence of a function $f \in \mathbf{B}_1$ satisfying*

$$\beta_i = f(z_i), \quad i = 1,\cdots,n, \qquad (3.1)$$

Probably, Nevanlinna did not know the result of Pick at that time, because the Pick's paper [35] was not quoted in [34], and his approach was totally different from Pick's approach. Instead, Nevanlinna's paper seemed to be strongly influenced by the work of Schur [40] who gave an alternative proof of the Caratheodory-Toeplitz theorem for the Caratheodory-Fejer interpolation problem.

The Nevanlinna's construction algorithm is essentially sequential, and is based on the Schwarz lemma at an arbitrary point.

Schwarz Lemma Assume that $f(z) \in \mathbf{B}$ and $\beta_1 = f(z_1)$ with $|z_1| < 1$. Then, for any z satisfying

$$|z - z_1| = r|1 - \bar{z}_1 z|, \qquad 0 < r < 1,$$

f(z) satisfies the inequality

$$|f(z) - \beta_1| \leq r|1 - \bar{\beta}_1 f(z)|.$$

The special case $z_1 = \beta_1 = 0$ in the above lemma is usually referred to as the Schwarz lemma in textbooks on complex function theory.

A direct application of this lemma proves that, if $\beta_1 = f(z_1)$ and $f(z) \in \mathbf{B}$, then

$$f_1(z):=\frac{1-\bar{z}_1 z}{z-z_1}\cdot\frac{f(z)-\beta_1}{1-\bar{\beta}_1 f(z)}\in \mathbf{B}_1 \tag{3.2}$$

Solving the above identity with respect to f(z) yields

$$f(z)=\frac{B_1(z)f_1(z)+\beta_1}{\bar{\beta}_1 B_1(z)f_1(z)+1},\quad B_1(z)=\frac{z-z_1}{1-\bar{z}_1 z} \tag{3.3}$$

Since $B_1(z)\in \mathbf{B}_1$, $f(z)\in \mathbf{B}_1$ if and only if $f_1(z)\in \mathbf{B}_1$. Moreover, $f(z_1)=\beta_1$ for any choice of $f_1(z)$. Hence, the first interpolation condition of (3.1) is always satisfied irrespective of the selection of $f_1(z)$. In order to satisfy the remaining n-1 interpolation conditions, it is sufficient to choose $f_1(z)$ in (3.3) such that

$$f_1(z_i)=\beta_i^{(2)},\quad i=2,\cdots,n \tag{3.4}$$

$$\beta_i^{(2)}=\frac{1-\bar{z}_1 z_i}{z_i-z_1}\cdot\frac{\beta_i-\beta_1}{1-\bar{\beta}_1\beta_i}$$

Thus, the problem is reduced to a simpler one, in which the number of interpolation conditions is less than the original problem.

Since $f_1(z)\in \mathbf{B}_1$, the inequalities

$$\left|\beta_i^{(2)}\right|\le 1,\quad i=2,\cdots,n \tag{3.5}$$

must be satisfied. If $|\beta_m^{(2)}|=1$ for some m, $f_1(z)\equiv\beta_m^{(2)}$ is the only function in $\mathbf{B}_1$ satisfying (3.5) due to the maximum modulus theorem. Therefore, $\beta_2^{(2)}=\beta_3^{(2)}=\cdots=\beta_n^{(2)}$ must hold in this case for the solvability of the original interpolation problem. If the strict inequalities hold for each i in (3.5), we can apply the same procedure to the simpler interpolation problem (3.4) based on the representation

$$f_2(z):=\frac{1-\bar{z}_2 z}{z-z_2}\cdot\frac{f_1(z)-\beta_2^{(2)}}{1-\bar{\beta}_2^{(2)}f_1(z)}$$

and its inverse

$$f_1(z):=\frac{B_2(z)f_2(z)-\beta_2^{(2)}}{\bar{\beta}_2^{(2)}B_2(z)f_2(z)+1},\quad B_2(z)=\frac{z-z_2}{1-\bar{z}_2 z}.$$

The problem is reduced further to the simpler one

$$f_2(z_i)=\beta_i^{(3)},\quad i=3,\cdots,n \tag{3.6}$$

$$\beta_i^{(3)} := \frac{1 - \bar{z}_3 z_i}{z_i - z_3} \cdot \frac{\beta_i^{(2)} - \beta_2^{(2)}}{1 - \bar{\beta}_2^{(2)} \beta_2^{(i)}} \cdot ,$$

In general, we have the recursion

$$f_j(z) = \frac{B_j(z) f_{j+1}(z) + \rho_j}{\bar{\rho}_j B_j(z) f_{j+1}(z) + 1}, \quad B_j(z) = \frac{1 - \bar{z}_j z}{z - z_j}, \quad j = 1, \cdots, n, \quad (3.7)$$

where we write $\rho_j = \beta_j^{(j)}$ which is computed by another recursion

$$\beta_i^{(j+1)} = \frac{1 - \bar{z}_j z_i}{z_i - z_j} \cdot \frac{\beta_i^{(j)} - \rho_j}{1 - \beta_i^{(j)} \bar{\rho}_j} . \quad (3.8)$$

The recursion can be continued as long as $|\rho_j| < 1$ holds. In that case, we satisfy the interpolation constraints at $j = n$, and therefore, we can choose any $f_{n+1} \in \mathbf{B}_1$.

Nevanlinna showed that the original interpolation problem is solvable if and only if either

$$\left|\beta_j^{(j)}\right| < 1, \qquad j = 1, 2, \cdots, n \quad (3.9a)$$

or

$$\left|\beta_j^{(j)}\right| < 1, \qquad j = 1, 2, \cdots, k - 1 \quad (3.9b)$$

$$\left|\beta_j^{(j)}\right| = 1, \qquad \beta_k^{(k)} = \beta_{k+1}^{(k)} = \cdots = \beta_n^{(k)} .$$

The solvability criterion (3.9) looks totally different from the Pick's criterion which is represented as

$$P = \begin{bmatrix} \frac{1 - \beta_1 \bar{\beta}_1}{1 - z_1 \bar{z}_1} & \cdots & \frac{1 - \beta_1 \bar{\beta}_n}{1 - z_1 \bar{z}_n} \\ & \cdots\cdots & \\ \frac{1 - \beta_n \bar{\beta}_1}{1 - z_n \bar{z}_1} & \cdots & \frac{1 - \beta_n \bar{\beta}_n}{1 - z_n \bar{z}_n} \end{bmatrix} \geq 0. \quad (3.10)$$

in this case. Later, it turned out that the Nevanlinna algorithm is essentially identical to the Cholesky factorization of P in (3.10).

The structure of the Nevanlinna algorithm can be more clearly viewed in the form of the fractional representation

$$f_j(z) = \frac{n_j(z)}{d_j(z)}. \tag{3.11}$$

The recursion (3.7) is represented as

$$\begin{bmatrix} n_j(z) \\ d_j(z) \end{bmatrix} = \Theta_j(z) \begin{bmatrix} n_{j+1}(z) \\ d_{j+1}(z) \end{bmatrix} \tag{3.12}$$

$$\Theta_j(z) = \frac{1}{\sqrt{1-|\rho_j|^2}} \begin{bmatrix} 1 & \rho_j \\ \bar{\rho}_j & 1 \end{bmatrix} \begin{bmatrix} B_j(z) & 0 \\ 0 & 1 \end{bmatrix}. \tag{3.13}$$

Here, we assumed condition (3.9a) holds. The common term $(1-|\rho_j|^2)^{-1/2}$ is a normalizing factor. Since $\overline{B}_j(z^{-1})B_j(z) = 1$, we have

$$\overline{\Theta}_j^T(\overline{z}^{-1}) \begin{bmatrix} 1 & 0 \\ 0 & -1 \end{bmatrix} \Theta_j(z) = \begin{bmatrix} 1 & 0 \\ 0 & -1 \end{bmatrix}.$$

This implies that $\Theta_j(z)$ is a *J-unitary matrix*. Such matrices will be discussed extensively in Section 7. The relation between $f_j(z)$ and $f_{j+1}(z)$ is represented in the form of a generalized lattice section, as shown in Fig. 1.

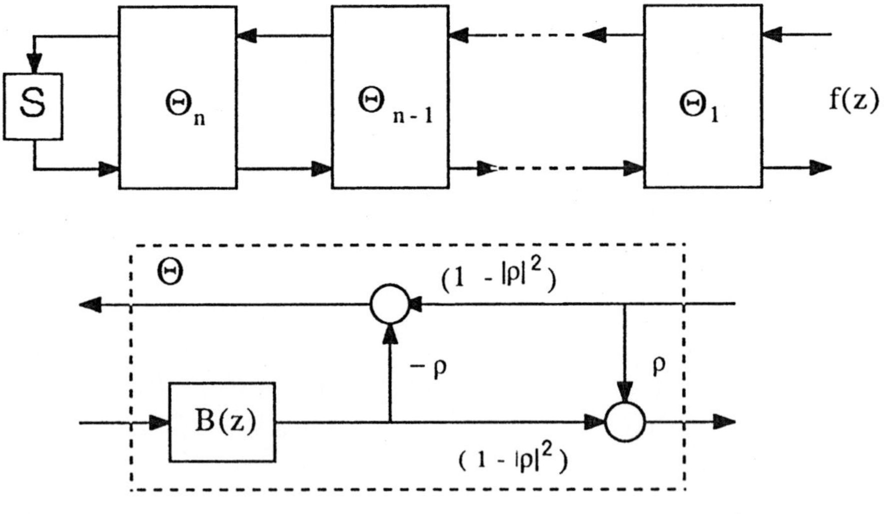

Fig.1 Lattice Structure of Nevanlinna Algorithm

Based on the recursion (3.12), the solution of the Pick-Nevanlinna interpolation problem $f(z) = n(z)/d(z)$ is represented as

$$\begin{pmatrix} n(z) \\ d(z) \end{pmatrix} = \Theta_1(z)\Theta_2(z) \cdots \Theta_n(z) \begin{pmatrix} n_{n+1}(z) \\ d_{n+1}(z) \end{pmatrix},$$

where $f_{n+1}(z) = n_{n+1}(z)/d_{n+1}(z)$ can be freely chosen from $\mathbf{B}_1$. Thus, the formula

$$f(z) = \frac{\theta_{11}(z)S(z) + \theta_{12}(z)}{\theta_{21}(z)S(z) + \theta_{22}(z)} \tag{3.15}$$

$$\Theta(z) = \begin{bmatrix} \theta_{11}(z) & \theta_{12}(z) \\ \theta_{21}(z) & \theta_{22}(z) \end{bmatrix} : = \Theta_1(z)\Theta_2(z) \cdots \Theta_n(z) \tag{3.16}$$

gives a parameterization of all solutions to the Pick-Nevanlinna problem under the condition (3.9a), where $S(z)$ represents an arbitrary element of $\mathbf{B}_1$. The factorization (3.16) was extensively studied in the classical work of Potapov [36].

4. APPLICATIONS

4.1 Broadband Matching

The first serious application of the Pick-Nevanlinna interpolation theory was probably the broadband matching. The problem itself is old, and Youla formulated it as a Pick-Nevanlinna problem as shown below [43]. Later, more general cases were treated by Helton [18][19].

Consider the simple circuit of Fig. 2. The power absorbed by the load is calculated to be

$$P = \frac{R_l}{\left(R_g + R_l\right)^2} E^2 = \left(1 - \left(\frac{R_g - R_l}{R_g + R_l}\right)^2\right) \frac{E^2}{4R_g} \tag{4.1}$$

This implies that the maximum power is absorbed by the load when the load is "matched" to the source, i.e., $R_l = R_g$, and the maximum absorbed power is given by $P_{max} = E^2 / 4R_g$.

In the case where the load impedance is frequency dependent, it is necessary to construct a network N which transfers the maximum power from the source to the load over a wide range of frequencies. This is the problem of broadband matching [10][18][43], which is depicted schematically in Fig. 3. The network N is usually assumed to be lossless and is specified by the output impedance $Z(s)$.

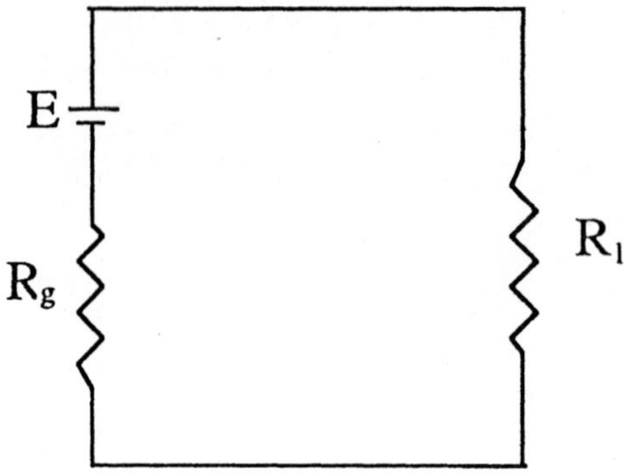

Fig. 2 A Simple Electrical Circuit

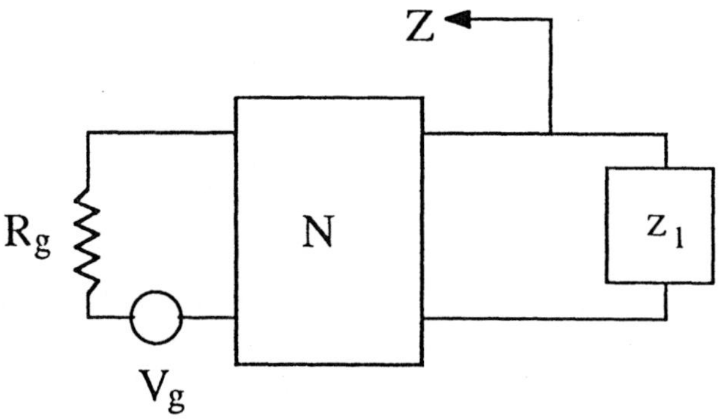

Fig. 3 Broadband Matching

The ratio of the power P absorbed by the load impedance $z_l(s)$ to the maximum power P_{max} absorbed by the ideal but not always realizable impedance at a frequency ω is given by

$$\frac{P}{P_{max}} = 1 - \left| \frac{Z(j\omega) - \bar{z}_l(j\omega)}{Z(j\omega) + z_l(j\omega)} \right|^2, \tag{4.2}$$

which is a generalization of (4.1). See [43] for the derivation. Let

$$u(s) = \frac{Z(s) - z_l(-s)}{Z(s) + z_l(s)} . \tag{4.3}$$

Our task is to choose a passive (hopefully lossless) impedance $Z(s)$ such that $|u(j\omega)|$ is as small as possible for a given frequency range. It is well-known that $Z(s)$ represents an impedance of a passive network, if and only if $Z \in \mathbf{P}$. Since $z_l(-s)$ is not in $\mathbf{P}$, we cannot choose $Z(s)=z_l(-s)$.

Let $\{-\mu_1, -\mu_2, \cdots, -\mu_m\}$ be the set of poles of $z_l(s)$, and define

$$b(s) = \frac{(s-\mu_1)(s-\mu_2)\cdots(s-\mu_m)}{(s+\bar{\mu}_1)(s+\bar{\mu}_2)\cdots(s+\bar{\mu}_m)}. \tag{4.4}$$

From (4.3), it is obvious that

$$v(s) = b(s)u(s) \tag{4.5}$$

is stable. In other words, the unstable poles μ_i, $i=1,\cdots, m$, of $u(s)$ which come from $z_l(-s)$ are cancelled out by the zeros of $b(s)$. Since $b(s)b(-s)=1$, we have

$$1 - v(s)v(-s) = \frac{(Z(s)+Z(-s))(z_l(s)+z_l(-s))}{(Z(s)+z_l(s))(Z(-s)+z_l(-s))}.$$

Since $Z(s) \in \mathbf{P}$ and $z_l(s) \in \mathbf{P}$, we have $|v(j\omega)| \le 1$. This implies that $v(s) \in \mathbf{B}$.

Let λ_i, $i = 1,\cdots, n$, be the numbers satisfying

$$z_l(\lambda_i) + z_l(-\lambda_i) = 0, \qquad \mathrm{Re}\,\lambda_i \ge 0.$$

Due to (4.3), $u(\lambda_i)=1$. Hence,

$$v(\lambda_i) = b(\lambda_i), \qquad i = 1, \cdots, n \tag{4.6}$$

must be satisfied for any choice of $Z(s)$. Thus, the problem is reduced to finding $U \in \mathbf{B}$ which satisfies (4.6). This is exactly the Pick-Nevanlinna problem.

4.2 Robust Stabilization

Consider the closed-loop system of Fig. 4, where $p(s)$ denotes the transfer function of a single-input single-output plant and $c(s)$ the transfer function of the controller. We say that $p(s)$ belongs to a class $\mathbf{A}(p_0, r)$ corresponding to the nominal model $p_0(s)$ and the uncertainty band $r(s)$, if

(i) $|p(j\omega) - p_0(j\omega)| \le |r(j\omega)|$, $\forall\omega$, $r(s)$: stable.

(ii) $p(s)$ and $p_0(s)$ have the same number of unstable poles.

Our purpose is to find a *fixed* controller $c(s)$ such that the closed-loop system of Fig.4 is stable for any plant $p(s)$ in the class $\mathbf{A}(p_0, r)$. If such controller exists, we call it a *robust stabilizer* of $\mathbf{A}(p_0, r)$. The existence of a robust stabilizer will be shown to be reduced to the solvability of a Pick-Nevanlinna interpolation problem.

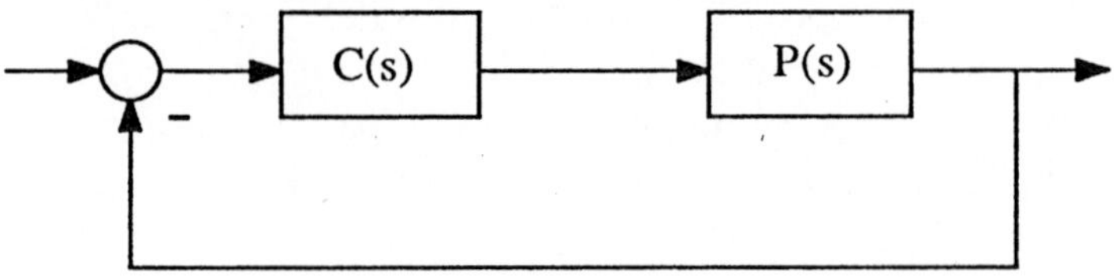

Fig. 4 The unity feedback system

For a given controller, let

$$q(s) := \frac{c(s)}{1 + p_0(s)c(s)} \tag{4.7}$$

Using the Nyquist stability criterion, one can easily show that c(s) is a robust stabilizer of $\mathbf{A}(p_0,r)$, if and only if it is a stabilizer for the nominal plant $p_0(s)$ and satisfies

$$|r(j\omega)q(j\omega)| < 1, \quad \forall\omega. \tag{4.8}$$

The inequality (4.8) was apparenty first derived by Doyle [8]. Due to Zames and Francis [46] , c(s) is a stabilizer for $p_0(s)$ if and only if (i) q(s) is stable, (ii) $1- p_0(s)q(s)$ has zeros at the unstable poles of $p_0(s)$, multiplicities included.

Let $\lambda_1, \lambda_2, \cdots, \lambda_n$ be unstable poles of $p_0(s)$ satisfying $\mathrm{Re}\,\lambda_i > 0$, $i = 1,\cdots,n$. We assume that the λ_i s' are all distinct. Define

$$b(s) = \frac{(s-\lambda_1)(s-\lambda_2)\cdots(s-\lambda_n)}{(s+\bar{\lambda}_1)(s+\bar{\lambda}_2)\cdots(s+\bar{\lambda}_n)} \tag{4.9}$$

From the assumption,

$$u(s) := b(s)p_0(s) \tag{4.10}$$

is also stable. Since q(s) in (4.7) has unstable zeros at λ_i , we see that

$$w(s) := q(s)/b(s)$$

is also stable. Therefore, since r(s) is stable,

$$\varphi(s) := r(s)w(s) \tag{4.11}$$

is stable. Moreover, since $|b(j\omega)|=1$ for each ω, it follows, from (4.8), that $|\varphi(j\omega)| < 1$. Hence, $\varphi \in \mathbf{B}$.

The interpolation constraints come from the condition that $1- p_0(s)q(s) = 1- v(s)w(s)$ vanishes at $s=\lambda_i$, $i=1,\cdots,n$. This implies that

$$\varphi(\lambda_i) = r(\lambda_i)w(\lambda_i) = \beta_i, \qquad i = 1, \cdots, n \tag{4.12}$$

$$\beta_i := r(\lambda_i)/v(\lambda_i).$$

Thus, the problem is reduced to finding a function $\varphi \in \mathbf{B}$ satisfying the interpolation conditions (4.12). This is again exactly the Pick-Nevanlinna problem, and was discussed in [24].

4.3 H^∞-control and Directional Interpolation Problem

Consider the closed-loop system of Fig.5, where

$$\begin{pmatrix} z \\ y \end{pmatrix} = P\begin{pmatrix} v \\ u \end{pmatrix} = \begin{pmatrix} P_{11} & P_{12} \\ P_{21} & P_{22} \end{pmatrix}\begin{pmatrix} v \\ u \end{pmatrix} \tag{4.13}$$

denotes the plant and $u = C(s)y$ the controller. The closed-loop transfer function $\Phi(s)$ from the exogenous signal v to the controlled variable z is given by

$$\Phi = P_{11} + P_{12}C(I - P_{22}C)^{-1}P_{21}. \tag{4.14}$$

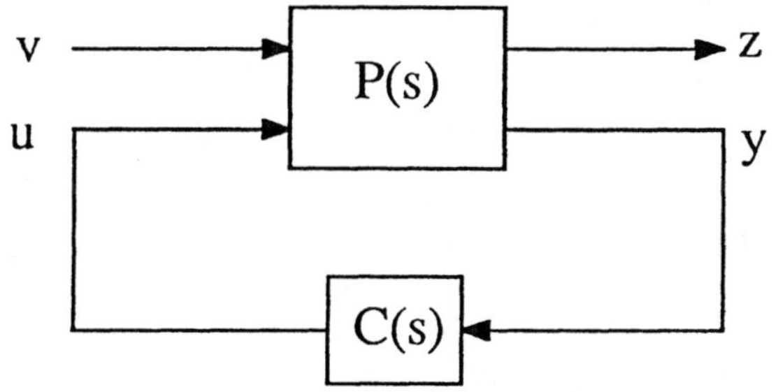

Fig. 5. H^∞ control scheme

The purpose of H^∞control is to find a stabilizing controller C(s) satisfying the norm bound of Φ represented as

$$\|\Phi\|_\infty < \sigma \tag{4.15}$$

for a given $\sigma > 0$.

It is now standard that Φ in (4.14) can be transformed to a simple representation through the parameterization of all stabilizing controllers. Let

$$P_{22} = NM^{-1} = M_1^{-1}N_1$$

be right and left coprime factorizations of P_{22}, respectively, such that

$$\begin{bmatrix} V_1 & -U_1 \\ -N_1 & M_1 \end{bmatrix} \begin{bmatrix} M & U \\ N & V \end{bmatrix} = \begin{bmatrix} I & 0 \\ 0 & I \end{bmatrix}, \tag{4.16}$$

for some U, V, U_1 and V_1. Here, all the matrices in (4.16) are stable and proper. The definition of the coprimeness in this context is found in [14]. Then, it is well-known that any controller that stabilizes the closed-loop system of Fig. 5 is represented as

$$C = (U + MQ)(V + NQ)^{-1} \tag{4.17}$$

for some stable Q. Substituting (4.17) in (4.14) and using (4.16) yield

$$\begin{aligned} \Phi &= T_1 - T_2QT_3 \\ T_1 &:= P_{11} + P_{12}UM_1P_{21}, \\ T_2 &:= P_{12}M, \quad T_3 := -M_1P_{21} \end{aligned} \tag{4.18}$$

Thus, the problem is reduced to finding a stable Q such that

$$\|T_1 - T_2QT_3\|_\infty < 1, \tag{4.19}$$

where we normalized the norm bound σ to unity. This problem is usually referred to as *the model-matching problem.* [14].

The model-matching problem is reduced to a version of the matrix Pick-Nevanlinna interpolation problem under the assumption that

(A_1) both T_2^{-1} and T_3^{-1} exist.

If both T_2^{-1} and T_3^{-1} are stable, the problem is trivially solved by taking $Q = T_2^{-1}T_1T_3^{-1}$.

Let λ_i , $i = 1, 2, \cdots, n_2$ be the unstable poles of T_2^{-1}. Since they are the unstable zeros of T_2, we can find a left annihilator ξ_i^T of $T_2(\lambda_i)$ so that $\xi_i^T T_2(\lambda_i) = 0$. Therefore, each Φ of the form (4.18) must satisfy

$$\xi_i^T \Phi(\lambda_i) = \eta_i^T, \qquad i = 1, 2, \cdots, n_2, \tag{4.20}$$

irrespective of the selection of Q, where $\eta_i^T := \xi_i^T T_1(\lambda_i)$. The condition (4.20) specifies $\Phi(\lambda_i)$ with respect to a given direction. Due to (4.19), Φ must be a contraction. The problem of finding a contraction Φ satisfying

(4.20) was called a *directional interpolation problem* and was extensively discussed in [26]. It was shown in [11][25] that a contraction Φ satisfying (4.20) exists, if and only if

$$P := \begin{bmatrix} \frac{\xi_1^T\bar{\xi}_1-\eta_1^T\bar{\eta}_1}{\lambda_1+\bar{\lambda}_1}, & \cdots & \frac{\xi_1^T\bar{\xi}_n-\eta_1^T\bar{\eta}_n}{\lambda_1+\bar{\lambda}_n} \\ & \cdots\cdots & \\ \frac{\xi_n^T\bar{\xi}_1-\eta_n^T\bar{\eta}_1}{\lambda_n+\bar{\lambda}_1} & \cdots & \frac{\xi_n^T\bar{\xi}_n-\eta_n^T\bar{\eta}_n}{\lambda_n+\bar{\lambda}_n} \end{bmatrix} > 0 \qquad (4.21)$$

Here we write $n_2 = n$ for the sake of notational simplicity. This condition is obviously a generalization of (2.12).

In the same way, at each unstable zero μ_j, $j = 1, \cdots, n_3$ of $T_3(s)$, Φ in (4.18) must satisfy

$$\Phi(\mu_j)\zeta_j = \psi_j, \qquad j = 1, \cdots, n_3 \qquad (4.22)$$

where ζ_j is the left annihilator of $T_3(\mu_j)$ and $\psi_j := T_1(\mu_j)\zeta_j$. The existence condition of a contraction Φ satisfying both (4.20) and (4.22) was first obtained by Limebeer and Anderson [30] as

$$P := \begin{bmatrix} \left\{\frac{\xi_i^T\bar{\xi}_j-\eta_i^T\bar{\eta}_j}{\lambda_i+\bar{\lambda}_j}\right\}, & \left\{\frac{\xi_i^T\psi_j-\eta_i^T\eta_j}{\lambda_i-\mu_j}\right\} \\ \left\{\frac{\bar{\zeta}_i^T\eta_j-\bar{\psi}_i^T\xi_j}{\mu_i-\bar{\lambda}_j}\right\} & \left\{\frac{\bar{\zeta}_i^T\zeta_j-\bar{\psi}_i^T\psi_j}{\mu_i+\bar{\mu}_j}\right\} \end{bmatrix} > 0$$

5. CONJUGATION

Multiplication by an all-pass function can replace the poles of the system by their conjugates. For example, let $g(s) = (\beta s + \gamma)/(s - \alpha)$. The multiplication by an all-pass function $(s - \alpha)/(s + \alpha)$ yields

$$g(s)\cdot\frac{s - \alpha}{s + \alpha} = \frac{\beta s + \gamma}{s + \alpha} \qquad (5.1)$$

Here the pole at $s = \alpha$ of $g(s)$ is replaced by its conjugate $s = -\alpha$. It is important to notice that the multiplication by an all-pass function does not change the $j\omega$-axis gain. This operation has been extensively used in various problems of system theory. We have actually observed that this operation is a crucial step to formulate the problem as an interpolation problem in Section 4. In the broadband matching problem, the unstable poles of u(s) in (4.3) are "conjugated" by b(s) in (4.4) in the definition of v(s) in (4.5). In the robust stabilization problem, the unstable poles of p_o (s) are also "conjugated" by b(s) in (4.9) to yield a stable system u(s) whose gain on the $j\omega$-axis is same as that of the original $p_o(s)$.

Now we formulate this operation in the state space. Let

$$G(s) = \{A, B, C, D\} \tag{5.2}$$

be a state-space realization of a transfer function matrix G(s). We seek a transfer function V(s) such that

$$G(s)V(s) = \{-A^T, *, *, *\}, \tag{5.3}$$

where $*$ denotes a matrix whose exact form is not relevant. The identity (5.3) is a generalization of (5.1), in which the A-matrix of G(s) is replaced by its *conjugate* $-A^T$ in in (5.3). A system V(s) which carries out this job is called *a conjugator* of G(s).

To make the subsequent argument simpler, we make the following assumption :

(A2) The matrix A in (5.2) has no eigenvalue on the $j\omega$-axis.

This assumption is a technical one and can be removed easily by extracting the $j\omega$-mode from A.

In order to characterize the conjugator, we first note that the A-matrix of the conjugator of G(s) must be similar to $-A^T$ because of (A2). Write

$$V(s) = \{-A^T, B_c, C_c, D_c\} \tag{5.4}$$

The well-known product rule is applied to yield

$$G(s)V(s) = \{\begin{bmatrix} A & BC_c \\ 0 & -A^T \end{bmatrix}, \begin{bmatrix} D_c \\ B_c \end{bmatrix}, \begin{bmatrix} C & DC_c \end{bmatrix}, DD_c\}. \tag{5.5}$$

Let X be a matrix satisfying

$$AX + XA^T = -BC_c \tag{5.6}$$

Note that this equation is always solvable uniquely due to the assumption (A2). Now, we choose B_c and D_c as satisfying

$$BD_c = XB_c \tag{5.7}$$

The similarity transformation of the realization (5.4) with the matrix

$$T = \begin{bmatrix} I & -X \\ 0 & I \end{bmatrix}, \quad T^{-1} = \begin{bmatrix} I & X \\ 0 & I \end{bmatrix}$$

verifies that the modes associated with A are uncontrollable. After cancelling out this uncontrollable portion, we get

$$G(s)V(s) = \{-A^T, B_c, CX + DC_c, DD_c\} \tag{5.8}$$

Now we have seen that V(s) in (5.4) with the conditions (5.6) and (5.7) satisfies (5.3). We regard (5.6) and (5.7) as the defining equations of the conjugator of G(s) in (5.2). Note that (5.6) and (5.7) depend only on A and B in (5.2).

DEFINITION 5.1 A system V(s) given in (5.4) is said to be a *conjugator* of G(s) in (5.2) (or of the pair (A, B)), if it satisfies (5.6) and (5.7). The operation performed by a conjugator is called a *conjugation.*

Now, we shall discuss some properties of the conjugator. Assume that there exists $D_c^\dagger$ such that $D_c D_c^\dagger = I$. From (5.7), it follows that $B = XB_cD_c^\dagger$. The equation (5.6) then becomes

$$AX + X(A^T + B_cD_c^\dagger C_c) = 0. \tag{5.9}$$

From (5.7) and (5.9), $A^k B \in Im(X)$ for each k. Hence, X is invertible if (A, B) is controllable. In this case, (5.9) implies that the eigenvalues of $-A^T - B_cD_c^\dagger C_c$, or the zeros of V(s), are identical to the eigenvalues of A. This implies that all the poles of G(s) are cancelled out by the zeros of V(s). This cancellation is obviously necessary for (5.3) to be satisfied.

Next, we assume that both D^{-1} and D_c^{-1} exist. Then, from (5.6) and (5.7), it follows that

$$-A^T - B_c(DD_c)^{-1}(CX + DC_c) = X^{-1}(A - BD^{-1}C)X.$$

In view of (5.8), this identity establishes the following important property of conjugators.

LEMMA 5.2 If both G(s) and V(s) are invertible, the zeros of G(s)V(s) are identical to those of G(s).

6. LOSSLESS CONJUGATION

From the structure of the defining equations (5.6) (5.7) of conjugators, it is clear that various classes of conjugators can be generated corresponding to the various selections of C_c in (5.6). This section is a short digression, in which a class of conjugations called a *lossless conjugation* (the conjugation by a lossless V(s)) is shown to be actually equivalent to the inner-outer factorization of invertible systems.

Let $F(s) \in \mathbf{R}\mathbf{H}^{\infty}_{m \times m}$ which is invertible. The inverse $F(s)^{-1}$ can be represented in the state-space form

$$F(s)^{-1} = \{\begin{bmatrix} A_0 & 0 \\ A_{10} & A_1 \end{bmatrix}, \begin{bmatrix} B_0 \\ B_1 \end{bmatrix}, \begin{bmatrix} C_0 & C_1 \end{bmatrix}, D\}, \tag{6.1}$$

where A_0 is anti-stable (all the eigenvalues have positive real parts) and A_1 is stable. Now, we carry out a conjugation of the pair (A_0, B_0) with the special selection of

$$C_c = -B_0^T. \tag{6.2}$$

The equations (5.6) (5.7) become in this case

$$A_0X + XA_0^T = B_0B_0^T, \quad B_0D_c = XB_c \tag{6.3}$$

If we take an orthogonal matrix as D_c, the conjugator

$$V(s) = \{-A_0^T, X^{-1}B_0D_c, -B_0^T, D_c\} \tag{6.4}$$

is an inner matrix, i.e., V(s) is stable and satisfies

$$V^{\sim}(s)V(s) = I. \tag{6.5}$$

Some manipulations using (6.3) and the product rule yield.

$$K(s) := F(s)^{-1}V(s) =$$

$$\left\{\begin{bmatrix} -A_0^T & 0 \\ A_{10}X - B_1B_0^T & A_1 \end{bmatrix}, \begin{bmatrix} X^{-1}B_0 \\ B_1 \end{bmatrix}, \begin{bmatrix} C_0X - DB_0^T & C_1 \end{bmatrix}, D\right\}$$

and

$$K(s)^{-1} = \{\begin{bmatrix} A_0 & 0 \\ A_{10} & A_1 \end{bmatrix}, -\begin{bmatrix} B_0 \\ B_1 \end{bmatrix}D^{-1}\begin{bmatrix} C_0 & C_1 \end{bmatrix}, \begin{bmatrix} C_0 - DB_0^TX & C_1 \end{bmatrix}, D^{-1}\}$$

Since both $-A_0^T$ and A_1 are stable, so is K(s)·. Also, since the A-matrix of $K(s)^{-1}$ is identical to the A-matrix of $(F(s)^{-1})^{-1} = F(s)$, $K(s)^{-1}$ is also stable. Therefore, the identity

$$F(s) = V(s)K(s)^{-1}$$

represents an inner-outer factorization of F(s), which is obtained without solving a Riccati equation.

7. J-LOSSLESS CONJUGATION

It is well-known that the interpolation problem is suitably treated in the framework of J- lossless systems [2] [3]. In this section, we shall show that the interpolation problem is reduced to the J-lossless conjugation, the conjugation by a J-lossless matrix, in the state space.

A transfer function matrix $\Theta(s)$ of size (m + r)x(m + r) is said to be *J-unitary,* if

$$\Theta^{\sim}(s)J\Theta(s) = J, \tag{7.1}$$

where J is a signature matrix defined as

$$J: = \begin{bmatrix} I_m & 0 \\ 0 & -I_r \end{bmatrix}. \tag{7.2}$$

A J-unitary matrix $\Theta(s)$ is said to be *J-lossless*, if it is J-contractive in the right half plane, i.e.,

$$\Theta^{*}(s)J\,\Theta(s) \leq J, \qquad \mathrm{Re}\,[s] \geq 0. \tag{7.3}$$

The lossless matrix is a special case of J-lossless matrices for which either m = 0 or r = 0, or equivalently, J = I.

Let a controllable pair (A_0, B_0) be given with $A_0 \in R_{n\times n}$ and $B_0 \in R_{n\times(m+r)}$. In the preceding section it was shown that a lossless conjugator of (A_0, B_0) always exists under the condition that A_0 is anti-stable. This is no longer true for J-lossless conjugators. Though A_0 is allowed to be unstable, the pair (A_0, B_0) must satisfy a strong condition in order that there exists a *J-lossless conjugator* (a conjugator which is J-lossless) of (A_0, B_0). The following result gives an existence condition of the J-lossless conjugator, as well as its state-space representation. The proof is found in [27].

THEOREM 7.1 There exists a J-lossless conjugator of a given controllable pair (A_0, B_0), if and only if the equation

$$A_0P_0 + P_0A_0^T = B_0JB_0^T \tag{7.4}$$

has a positive definite solution P_0. In that case, a J-lossless conjugator is given by

$$\Theta(s) = \{-A_0^T, P_0^{-1}B_0D_c, -JB_0^T, D_c\}, \tag{7.5}$$

where D_c is any constant J-unitary matrix.

In order to show the relevance of this theorem to the interpolation problem, assume that A_0 is anti-stable and write

$$B_0 = [L \;\; M], \qquad L \in R_{n\times m}, \qquad M \in R_{n\times r} \tag{7.6}$$

We seek $\Phi(s) \in \mathbf{R}\mathbf{H}^{\infty}_{m\times r}$ such that $L\Phi(s) - M$ is "divisable" by $sI - A_0$, i.e.,

$$L\Phi(s) - M = (sI - A_0)\Psi(s) \tag{7.7}$$

for some $\Psi(s) \in \mathbf{R}\,\mathbf{H}^{\infty}_{m\times r}$. This problem is essentially equivalent to the directional interpolation problem discussed in 4.3. To see this, let λ_i be an eigenvalue of A_0 with x_i^T as its left eigenvector. Writing $\xi_i^T = x_i^T L$, $\eta_i^T = x_i^T M$, we have from (7.7)

$$\xi_i^T \Phi(\lambda_i) = \eta_i^T, \tag{7.8}$$

which is identical to (4.20).

By the pre-multiplication of x_i^T and the post-multiplication of $\bar{x}_j$, the equation (7.4) yields

$$x_i^T P_0 \bar{x}_j = \frac{\xi_i^T \bar{\xi}_j - \eta_i^T \bar{\eta}_j}{\lambda_i + \bar{\lambda}_j}. \tag{7.9}$$

This implies that P_0 is congruent to the Pick matrix (4.21) under the condition that the eigenvalues of A_0 are distinct. Thus, Theorem 7.1 gives a condition for the existence of a contraction $\Phi(s)$ satisfying (7.7). The significance of (7.7) becomes clear in the next section. A similar argument is found in [21] for a scalar interpolation problem.

Now, we shall consider a factorization of a J-lossless system into the product of two J-lossless systems. Assume that a transfer function matrix $G_0(s) = \{A_0, B_0, C_0, D_0\}$ is given in a form of spectral decomposition

$$A_0 = \begin{bmatrix} A_1 & 0 \\ 0 & A_2 \end{bmatrix}, \quad B_0 = \begin{bmatrix} B_1 \\ B_2 \end{bmatrix}, \quad C_0 = \begin{bmatrix} C_1 & C_2 \end{bmatrix}. \tag{7.10}$$

Also, assume that $G_0(s)$ has a J-lossless conjugator $\Theta(s)$. Due to Theorem 7.1, $\Theta(s)$ is represented as

$$\Theta(s) = \left\{ \begin{bmatrix} -A_1^T & 0 \\ 0 & -A_2^T \end{bmatrix}, \; P_0^{-1}\begin{bmatrix} B_1 \\ B_2 \end{bmatrix}, \; -J\begin{bmatrix} B_1^T & B_2^T \end{bmatrix}, \; I_{m+r} \right\} \tag{7.11}$$

where we take $D_c = I_{m+r}$ for the sake of simplicity and P_0 is the positive definite solution of

$$\begin{bmatrix} A_1 & 0 \\ 0 & A_2 \end{bmatrix} P_0 + P_0 \begin{bmatrix} A_1 & 0 \\ 0 & A_2 \end{bmatrix}^T = \begin{bmatrix} B_1 \\ B_2 \end{bmatrix} J \begin{bmatrix} B_1^T & B_2^T \end{bmatrix}. \tag{7.12}$$

Write

$$P_0 = \begin{bmatrix} P_{11} & P_{12} \\ P_{12}^T & P_{22} \end{bmatrix}. \tag{7.13}$$

First, we conjugate the pair (A_1, B_1). Since the equation corresponding to (7.4) is satisfied by P_{11}, where we write

$$\Theta_1 = \{ -A_1^T, P_{11}^{-1}B_1, -J B_1^T, I_{m+r} \} \tag{7.14}$$

according to (7.5). Here, we again take $D_c = I_{m+r}$. The product rule yields

$$G(s)\Theta_1(s) = \{ \begin{bmatrix} A_1 & 0 & -B_1JB_1^T \\ 0 & A_2 & -B_2JB_1^T \\ 0 & 0 & -A_1^T \end{bmatrix}, \begin{bmatrix} B_1 \\ B_2 \\ P_1^{-1}B_1 \end{bmatrix}, \begin{bmatrix} C_1 & C_2 & -DJB_1^T \end{bmatrix}, D \}.$$

Applying the similarity transformation given by

$$T = \begin{bmatrix} I & 0 & -P_{11} \\ 0 & 0 & I \\ 0 & I & -P_{12} \end{bmatrix}, \quad T^{-1} = \begin{bmatrix} I & P_{11} & 0 \\ 0 & P_{12} & I \\ 0 & I & 0 \end{bmatrix}$$

and cancelling out the uncontrollable portion yield

$$G(s)\Theta_1(s) = \{ \begin{bmatrix} -A_1^T & 0 \\ 0 & A_2 \end{bmatrix}, \begin{bmatrix} P_{11}^{-1}B_1 \\ B_2 - P_{12}P_{11}^{-1}B_1 \end{bmatrix}, \begin{bmatrix} C_1' & C_2 \end{bmatrix}, D \}, \tag{7.15}$$

where $C_1' = C_1P_{11} + C_2P_{12} - DJB_1^T$.

Now we conjugate the remaining portion $(A_2, B_2 - P_{12}P_{11}^{-1}B_1)$. Based on (7.12), it is not difficult to see that the Schur complement

$$P_2 = P_{22} - P_{12}^T P_{11}^{-1} P_{12}$$

satisfies the equation

$$A_2P_2 + P_2A_2^T = (B_2 - P_{12}P_{11}^{-1}B_1)J(B_2 - P_{12}P_{11}^{-1}B_1)^T,$$

which corresponds to (7.4) for the pair $(A_2, B_2 - P_{12}P_{11}^{-1}B_1)$. Thus, a J-lossless conjugator of this pair is given by

$$\Theta_2(s) = \{A_2^T, P_2^{-1}(B_2 - P_{12}P_{11}^{-1}B_1), -J(B_2 - P_{12}P_{11}^{-1}B_1)^T, I_{m+r}\}. \quad (7.16)$$

From the product rule, it follows that

$$\Theta_1(s)\Theta_2(s) = \{\begin{bmatrix} -A_1^T & -P_{11}^{-1}B_1JU^T \\ 0 & -A_2^T \end{bmatrix}, \begin{bmatrix} P_{11}^{-1}B_1 \\ P_2^{-1}U \end{bmatrix}, \begin{bmatrix} -JB_1^T & -JU^T \end{bmatrix}, I\},$$

where $U = B_2 - P_{12}P_{11}^{-1}B_1$. The similarity transformation given by

$$T = \begin{bmatrix} I & -P_{11}^{-1}P_{12} \\ 0 & I \end{bmatrix}, \quad T^{-1} = \begin{bmatrix} I & P_{11}^{-1}P_{12} \\ 0 & I \end{bmatrix}$$

applied to $\Theta_1(s)$ and $\Theta_2(s)$ shows that

$$\Theta(s) = \Theta_1(s)\Theta_2(s).$$

Thus, we have established that the original J-lossless conjugator (7.19) is decomposed into the two consecutive J-lossless conjugators $\Theta_1(s)$ and $\Theta_2(s)$ given respectively by (7.14) and (7.16). This factorization clearly represents the factorization (3.16) associated with the Nevanlinna algorithm in the state space. A similar result was found in [16].

8. MODEL MATCHING PROBLEM

In Section 4, we showed that the H^∞ control problem boils down to the model matching problem. It was further shown that the model matching problem was equivalent to a version of the matrix Pick-Nevanlinna problem. In this final section, we solve the model matching problem in the state space by reducing it to the problem of finding a J-lossless conjugator.

Let $T_1 \in RH^\infty_{m \times r}$, $T_2 \in RH^\infty_{m \times m}$ and $T_3 \in RH^\infty_{r \times r}$ be given which are represented in the state space as

$$T_i(s) = \{A_i, B_i, C_i, D_i\}, \qquad i = 1, 2, 3. \qquad (8.1)$$

We assume that

(A_3) $T_2^{-1}(s)$ and $T_3^{-1}(s)$ exist and are anti-stable.

This assumption implies that D_2^{-1} and D_3^{-1} exist and

$$\widehat{A}_2 := A_2 - B_2 D_2^{-1} C_2, \qquad \widehat{A}_3 := A_3 - B_3 D_3^{-1} C_3 \tag{8.2}$$

are anti-stable.

The model matching problem is formulated as follows :

(1) *Determine whether there exists a stable Q such that*

$$\Phi := T_1 - T_2 Q T_3, \qquad \|\Phi\|_\infty < 1. \tag{8.3}$$

(2) *If a solution exists, characterize all such Φ and Q.*

Let

$$L_2 := B_2 D_2^{-1}, \qquad L_3^T := D_3^{-1} C_3. \tag{8.4}$$

Due to the assumption (A_3), both A_2 and A_3 are anti-stable. Since A_1 is stable, the equations

$$\widehat{A}_2 R_2 - R_2 A_1 = L_2 C_1, \quad R_3 \widehat{A}_3 - A_1 R_3 = B_1 L_3^T \tag{8.5}$$

have the unique solutions R_2 and R_3, respectively. Write

$$M_2 := L_2 D_1 + R_2 B_1, \qquad M_3^T := C_1 R_3 + D_1 L_3^T. \tag{8.6}$$

From (8.1) and (8.3), it follows that $L_2 T_2(s) = L_2(D_2 + C_2(sI - A_2)^{-1}B_2) = (sI - \hat{A}_2)(sI - A_2)^{-1}B_2$. Also, due to (8.5), it follows that $L_2 T_1(s) = L_2(D_1 + C_1(sI - A_1)^{-1}B_1) = M_2 - (R_2(sI - A_1) - \hat{A}_2 R_2 + R_2 A_1)(sI - A_1)^{-1}B_1 = M_2 - (sI - \hat{A}_2)R_2(sI - A_1)^{-1}B_1$. Therefore, if $\Phi(s)$ is of the form (8.3), it satisfies

$$L_2 \Phi(s) - M_2 = (sI - \widehat{A}_2)\Psi_2(s), \tag{8.7}$$

where $\Psi_2(s) = -(sI - A_2)^{-1}B_2 Q(s) T_3(s) - R_2(sI - A_1)^{-1}B_1$. This corresponds to the relation (7.7). Analogously, we obtain

$$\Phi(s) L_3^T - M_3^T = \Psi_3(s)(sI - \widehat{A}_3), \tag{8.8}$$

where $\Psi_3(s) = -T_2(s) Q C_3 (sI - A_3)^{-1} - C_1(sI - A_1)^{-1}R_3$. This relation gives the right interpolation constraints (4.22).

The rest of this section is devoted to showing that there exists a stable Q satisfying (8.3), if and only if the pair

$$\left\{ \begin{bmatrix} \hat{A}_2 & 0 \\ 0 & -\hat{A}_3^T \end{bmatrix}, \begin{bmatrix} L_2 & -M_2 \\ M_3 & -L_3 \end{bmatrix} \right\} \tag{8.9}$$

has a J-lossless conjugator. In view of Theorem 7.1, this is equivalent to the condition that the solution P_0 of the equation

$$\begin{pmatrix} \hat{A}_2 & 0 \\ 0 & -\hat{A}_3^T \end{pmatrix} P_0 + P_0 \begin{pmatrix} \hat{A}_2^T & 0 \\ 0 & -\hat{A}_3 \end{pmatrix} = \begin{pmatrix} L_2 & -M_2 \\ M_2 & -L_3 \end{pmatrix} \begin{pmatrix} L_2^T & M_3^T \\ M_2^T & L_3^T \end{pmatrix} \tag{8.10}$$

is positive definite. In that case, a J-lossless conjugator of the pair (8.9) is given by

$$\Theta(s) = \left\{ \begin{pmatrix} -\hat{A}_2^T & 0 \\ 0 & \hat{A}_3 \end{pmatrix}, P_0^{-1} \begin{pmatrix} L_2 & -M_2 \\ M_3 & -L_3 \end{pmatrix}, \begin{pmatrix} L_2 & M_2 \\ M_3 & L_3 \end{pmatrix}^T, I_{m+r} \right\} \tag{8.11}$$

due to (7.5), where we took $D_c = I_{m+r}$ for simplicity.

Let

$$G_2(s) := \begin{bmatrix} T_2(s)^{-1} & 0 \\ 0 & I \end{bmatrix} \begin{bmatrix} -I & T_1(s) \\ 0 & T_3(s) \end{bmatrix}, \tag{8.12}$$

and write

$$\Pi_2(s) := G_2(s)\Theta(s). \tag{8.13}$$

A lengthy but straightforward calculation yields

$$\Pi_2(s) = \{A, B, C, D\} \tag{8.14}$$

$$A = \begin{bmatrix} -\hat{A}_2^T & 0 & 0 \\ B_1 M_2^T & A_1 & 0 \\ -B_3 M_2^T & 0 & A_3 \end{bmatrix}, \quad D = \begin{bmatrix} -D_2^{-1} & D_2^{-1} D_1 \\ 0 & D_3 \end{bmatrix}$$

$$B = \begin{bmatrix} 0 & 0 \\ 0 & -B_1 \\ 0 & B_3 \end{bmatrix} + \begin{bmatrix} I & 0 \\ 0 & -R_3 \\ 0 & -I \end{bmatrix} P_0^{-1} \begin{bmatrix} L_2 & -M_2 \\ M_3 & -L_3 \end{bmatrix}$$

$$C = \begin{bmatrix} D_2^{-1}(C_2P_{11} + L_2^T - D_1M_2^T) & -D_2^{-1}(C_1 + C_2R_2) & 0 \\ -D_3M_2^T & 0 & C_3 \end{bmatrix}$$

If we represent $\Theta(s)$ and $\Pi_2(s)$ in the partitioned forms

$$\Theta(s) = \begin{bmatrix} \Theta_{11}(s) & \Theta_{12}(s) \\ \Theta_{21}(s) & \Theta_{22}(s) \end{bmatrix}, \quad \Pi_2(s) = \begin{bmatrix} \Pi_{11}(s) & \Pi_{12}(s) \\ \Pi_{21}(s) & \Pi_{22}(s) \end{bmatrix}$$

which are consistent with (8.12), we have from (8.13), that $(\Pi_{21}S + \Pi_{22})^{-1}T_3 = (\Theta_{21}S + \Theta_{21})^{-1}$ for each S. It follows that

$$T_1 - T_2(\Pi_{11}S + \Pi_{12})(\Pi_{21}S + \Pi_{22})^{-1}T_3 = (\Theta_{11}S + \Theta_{12})(\Theta_{21}S + \Theta_{22})^{-1}, \quad (8.15)$$

for each S. It is well-known that

$$\Phi = (\Theta_{11}S + \Theta_{12})(\Theta_{21}S + \Theta_{22})^{-1} \qquad (8.16)$$

is a contraction if Θ is J-lossless and S is contractive. From (8.15), Φ given by (8.16) is of the form (8.3) with

$$Q = (\Pi_{11}S + \Pi_{12})(\Pi_{21}S + \Pi_{22})^{-1}. \qquad (8.17)$$

In [27], it was shown that Q is stable. Thus, Φ is a solution to the model matching problem (8.3) for each $S \in \mathbf{B}\,\mathbf{H}^{\infty}_{m \times r}$. We have now established that the positive definiteness of P_0 in (8.10) is sufficient for the solvability of the model matching problem.

Conversely, assume that there exists a contraction Φ of the form (8.3). It was shown in [27] that the solution P_0 of (8.10) can be represented as

$$P_0 = \frac{1}{2\pi}\int_{-\infty}^{\infty} K(j\omega)\begin{bmatrix} L_2 & -M_2 \\ M_3 & -L_3 \end{bmatrix}\begin{bmatrix} I & \Phi \\ \Phi^{\sim} & I \end{bmatrix}\begin{bmatrix} L_2 & -M_2 \\ M_3 & -L_3 \end{bmatrix}^T K(j\omega)^{\sim} d\omega, \quad (8.18)$$

where

$$K(s) := \begin{bmatrix} sI - \hat{A}_2 & 0 \\ 0 & sI + \hat{A}_3^T \end{bmatrix}^{-1}.$$

Since Φ is a contraction, the integrand of (8.18) is positive. Therefore, $P_0 > 0$. Thus, we have established that $P_0 > 0$ is necessary for the solvability of the model-matching problem.

The above results are summarized as follows :

THEOREM 8.1 There exists a stable Q satisfying (8.3) if and only if the solution P_0 of (8.10) is positive definite. In that case, Φ and Q are parameterized as (8.16)(8.11) and (8.17)(8.14), respectively, where S is an arbitrary contraction in $\mathbf{B}\,\mathbf{H}^{\infty}_{mxr}$.

Finally, we shall discuss the application of Theorem 8.1 to the H^{∞} control problem. Assume that the plant P(s) described in (4.13) has a state-space form

$$P(s) = \{A, \begin{bmatrix} B_1 & B_2 \end{bmatrix}, \begin{bmatrix} C_1 \\ C_2 \end{bmatrix}, \begin{bmatrix} D_{11} & D_{12} \\ D_{21} & 0 \end{bmatrix}\}, \tag{8.19}$$

with $A \in \mathbf{R}_{nxn}$, $D_{11} \in \mathbf{R}_{mxr}$, $D_{12} \in \mathbf{R}_{rxr}$. We put $D_{22} = 0$ without loss of generality. Let $F \in \mathbf{R}_{rxn}$ and $H \in \mathbf{R}_{nxr}$ be such that

$$A_F := A + B_2F, \qquad A_H := A + HC_2 \tag{8.20}$$

are both stable. It is well-known that the matrices T_1, T_2 and T_3 in (4.18) are given by

$$\begin{bmatrix} T_1 & T_2 \\ T_3 & 0 \end{bmatrix} = \{\begin{bmatrix} A_F & -B_2F \\ 0 & A_H \end{bmatrix}, \begin{bmatrix} B_1 & B_2 \\ B_1 + HD_{21} & 0 \end{bmatrix},$$
$$\begin{bmatrix} C_1 + D_{12}F & -D_{12}F \\ 0 & C_2 \end{bmatrix}, \begin{bmatrix} D_{11} & D_{12} \\ D_{21} & 0 \end{bmatrix}\}. \tag{8.21}$$

The assumption (A_3) implies that both D_{12}^{-1} and D_{21}^{-1} exist and

$$\hat{A}_2 = A_F - B_2D_{12}^{-1}(C_1 + D_{12}F) = A - B_2D_{12}^{-1}C_1 \tag{8.22a}$$

$$\hat{A}_3 = A_H - (B_1 + HD_{21})D_{21}^{-1}C_2 = A - B_1D_{21}^{-1}C_2 \tag{8.22b}$$

are anti-stable. From (8.3), it follows that

$$L_2 = B_2 D_{21}^{-1}, \qquad L_3^T = D_{21}^{-1} C_2. \tag{8.23}$$

It is obvious that $R_2 = [-I \;\; 0]$ and $R_3 = -[I \;\; I]^T$ satisfy (8.4). Therefore, from (8.6), it follows that

$$M_2 = B_2 D_{12}^{-1} D_{11} - B_1, \qquad M_3^T = D_{11} D_{21}^{-1} C_2 - C_1 \tag{8.24}$$

The solution P_0 of (8.10) is represented as

$$P_0 = \begin{bmatrix} P_{11} & I \\ I & P_{22} \end{bmatrix} \tag{8.25}$$

where P_{11} and P_{22} are the solution of Lyapunov-type equations

$$(A - B_2 D_{12}^{-1} C_1) P_{11} + P_{11} (A - B_2 D_{12}^{-1} C_1)^T$$
$$= B_2 D_{12}^{-1} D_{12}^{-T} B_2^T - (B_1 + B_2 D_{12}^{-1} D_{11})(B_1 + B_2 D_{12}^{-1} D_{11})^T \tag{8.26a}$$
$$(A - B_1 D_{21}^{-1} C_2)^T P_{22} + P_{22} (A - B_1 D_{21}^{-1} C_2)$$
$$= C_2^T D_{21}^{-T} D_{21}^{-1} C_2 - (C_1 - D_{11} D_{21}^{-1} C_2)^T (C_1 - D_{11} D_{21}^{-1} C_2) \tag{8.26b}$$

The system $\Pi_2(s)$ in (8.14) is calculated to be

$$\Pi_2(s) = \left\{ \begin{bmatrix} -\hat{A}_2^T & 0 \\ A_{21} & A_H \end{bmatrix}, \begin{bmatrix} 0 & 0 \\ 0 & B_1 + HD_{21} \end{bmatrix} + \begin{bmatrix} I & 0 \\ 0 & -I \end{bmatrix} P_0^{-1} \begin{bmatrix} L_2 & -M_2 \\ M_3 & -L_3 \end{bmatrix}, \right.$$
$$\left. \begin{bmatrix} C_{11} & -F \\ -D_3 M_2^T & C_3 \end{bmatrix}, \begin{bmatrix} -D_{12}^{-1} & D_{12}^{-1} D_{11} \\ 0 & D_{21} \end{bmatrix} \right\},$$

where $A_{21} = -(B_1 + HD_{21})(B_1 - B_2 D_{12}^{-1} D_{11})^T$, $C_{11} = D_{12}^{-1}((C_1 + D_{12}F)P_{11} + L_2^T - D_{11}^T M_2^T)$. .Based on the above representation, the controller $K(s)$ achieving the norm bound $\|\Phi\|_\infty < 1$ is given by

$$K(s) = V_{11} + V_{12} S (I - V_{22} S)^{-1} V_{21}$$

$$V(s) = \{A, B, C, D\}$$

$$A = -\hat{A}_2^T - [I \;\; 0] P_0^{-1} \begin{bmatrix} M_2 \\ L_3 \end{bmatrix} (M_2^T - L_3^T P_{11}) \quad B = [I \;\; 0] P_0^{-1} \begin{bmatrix} -M_2 & L_2 \\ -L_3 & M_3 \end{bmatrix} \begin{bmatrix} D_{21}^{-1} & 0 \\ 0 & I \end{bmatrix}$$

$$C = \begin{bmatrix} D_{12}^{-1} & 0 \\ 0 & I \end{bmatrix} \left\{ \begin{bmatrix} L_2^T \\ M_2^T \end{bmatrix} - \begin{bmatrix} M_3^T \\ L_3^T \end{bmatrix} P_{11} \right\}, \qquad D = \begin{bmatrix} D_{12}^{-1} D_{11} D_{21}^{-1} & -D_{12}^{-1} \\ D_{21}^{-1} & 0 \end{bmatrix}$$

The extension to the case where T_2^{-1} and/or T_3^{-1} are no longer anti-stable is found in [22]

9 . CONCLUSION

It has been shown that the classical function-theoretic treatment of the interpolation problem is transferred to the purely algebraic theory of conjugation in the state space. It is important to notice that, if we limit our scope to the systems with rational transfer functions, then the algebraic aspect always dominates which sometimes enables us to discard heavy and advanced mathematical tools. It is not our intension to deemphasize the role of mathematics. On the contrary, as is seen from the short history of H^∞ control theory, mathematics plays a vital role at the initial stage of the problem formulation where the solvability is the most important issue. At the same time, it is of supreme importance for engineering theory to have a clear, elementary and self-contained framework of exposition, in order to penetrate into the modern technology. The theory of conjugation stated in this article is a state-space representation of the classical interpolation theory which gives the most elementary framework for H^∞ control theory for systems with rational transfer functions.

From the space limitation, the more advanced theory of conjugation cannot be exposed which deals with the four-block problem in the most general way; neither can the discrete-time case. Here, we only mention some literature to appear along this line [25] [27]. Also the other applications of classical interpolation in system theory, specially in signal processing, have not been discussed [6][17].

ACKNOWLEDGEMENT

Professor Jan Willems has been one of the greatest leaders in system theory for many years, not only in terms of his numerous outstanding and original achievements, but also in terms of his attractive humanity. It is the author's great pleasure to be able to contribute to his birthday present in such a novel way. The author is grateful to Profs. Schumacher and Nijmeijer for giving him the possibility of making a contribution to this important volume.

REFERENCES

[1] L. V. Ahlfors, *Complex Analysis* , ,McGraw-Hill, 1979.

[2] J. A. Ball, "Nevanlinna-Pick interpolation: Generalizations and applications," Proc. of Special Year in Operator Theory, Indiana Universyity1985-1986.

[3] J. A. Ball and J. W. Helton, "A Beurling-Lax theorem for the Lie group U(m,n) which contains most classical interpolation theory, "J. Operator Theory, vol.9, pp. 107-142, 1983

[4] B. C. Chang and J. B. Pearson, "Optimal disturbance reduction in linear multivariable systems, " IEEE Trans. Automat. Contr., vol AC-29, pp. 880-887, 1984.

[5] P. Delsarte, Y. Genin and Y. Kamp, "The Nevanlinna-Pick problem for matrix-valued functions," SIAM J. of Appl. Math., vol. 36, pp. 47-61, 1979.

[6] P. Delsarte, Y. Genin and Y. Kamp, "On the role of the Nevanlinna-Pick problem in circuit and system theory," Circuit Th. and Appl., vol.9, pp.177-187, 1981.

[7] P. Dewilde and H. Dym, "Lossless chain scattering matrices and optimal linear prediction : The vector case," Circuit Theory Appl., vol. 9, pp. 135-175, 1981.

[8] J. C. Doyle, "Synthesis of robust controllers and filters, " Proc. IEEE Conf. Decision and Control, San Antonio, pp. 109-114, 1983.

[9] H. Dym, "J-contractive matrix functions, reproducing kernel Hilbert spaces and interpolation,' Monograph, Dept. of Theoretical Math., The Weizmann Inst. Science, 1988.

[10] R. M. Fano, "Theoretical limitations on the broadband matching of arbitary impedances," J. Franklin Inst., vol. 249, pp. 57-83, 1960.

[11] I. P. Fedcina, "Solvability criteria of the Nevanlinna-Pick tangent problem," Mat. Issled. Kinshinev (in Russian) yp : 4, pp. 213-227, 1972.

[12] B. A. Francis and G. Zames, "On H^∞ -optimal sensitivity theory for SISO feedback systems," IEEE Trans. Automat. Contr., vol. AC-29, pp. 880-887, 1984.

[13] B. A. Francis, J. W. Helton and G. Zames, "H^∞ optimal feedback controllers for linear multivariable systems, " IEEE Trans. Automat. Contr., vol. AC-29, pp. 888-900, 1984.

[14] B. A. Francis, *A Course in H^∞ Control Theory*, Springer Verlag, New York, 1987.

[15] J. S. Freudenberg and D. P. Looze, *Frequency Domain Properties of Scalar and Multivariable Feedback Systems*, Springer, New york 1988.

[16] Y. Genin, P. van Dooren and T. Kailath, "On Σ-lossless transfer functions and related questions," Linear Algebra and Appl., vol. 50 pp. 251-275, 1983.

[17] T. T. Georgiou and P.P. Khargonekar, "Spectral factorization and Nevanlinna-Pick interpolation," SIAM J. of Control & Optimiz., vol.25, pp.754-766, 1987.

[18] J. W. Helton, "Broadbanding : Gain equalization directly from data," IEEE Trans. Circuit and Systems, vol. CAS-28, pp. 1125-1137, 1981.

[19] J. W. Helton, "Non-Euclidean functional analysis and electronics, " Bull. Amer. Math. Soc., vol. 7, pp. 1-64, 1982.

[20] F. Itakura and S. Saito, "Digital filtering techniques for speech analysis and synthesis," in Proc. 7th Int. Cong., Acoust., Budapest, Paper 25-c-1, pp. 261-264, 1971.

[21] V. E. Katsnel'son "Methods of J-theory in continuous interpolation problems of analysis," Harihov, translated by T. Ando, 1982.

[22] R. Kawatani and H. Kimura, "Synthesis of reduced-order H^∞ controller," to appear in Int. J. Control.

[23] P.P. Khargonekar and A. Tannenbaum, "Non-Euclidian metrics and the robust stabilization of systems with parameter uncertainty," IEEE Trans. Automat. Control, vol. AC-30, pp.1005-1013, 1985.

[24] H. Kimura, "Robust stabilizability for a class of transfer functions, " IEEE Trans. Automat. Contr., vol. AC-29, pp. 788-793, 1984.

[25] H. Kimura, "Directional interpolation approach to H^∞-optimization and robust stabilization," ibid., vol. AC-32, pp. 1085-1093, 1987.

[26] H. Kimura, "Directional interpolation in the state space," Systems and Control Letters, vol. 10, pp. 317-324, 1988.

[27] H. Kimura, "Conjugation, interpolation and model-matching in H^∞," Int. J. Control, vol. 49, pp. 269-307, 1989.

[28] H. Kimura and R. Kawatani, "Synthesis of H^∞ controller based on conjugation, " Proc. IEEE Conf. on Decision and Control, Austin, pp. 7-13, 1988.

[29] H. Kimura, "Conjugation of Hamilton systems and model-matching in H^∞," under preparation.

[30] D. J. Limebeer and B. D. O. Anderson,"An interpolation theory approach to H^∞ controller degree bounds," Linear Algebra and its Appl., vol. 98, pp.347-386, 1988.

[31] K. Z. Liu and T. Mita, "Conjugation and H^∞ control of discrete-time systems, " to appear in Int. J. Control.

[32] A. G. J. MacFarlane (ed.), *Frequency-Response Methods in Control Systems*, IEEE Press, NY., 1979.

[33] D. Q. Mayne, "The design of linear multivariable systems," Automatica, vol. 9, pp. 201-207, 1973.

[34] R. Nevanlinna, "Über beschränkte Funktionen die in gegebenen Punkten vorgeschreibene Funktionswerte bewirkt werden," Ann. Acad. Sci. Fenn., Ser A, vol. 13, pp. 1-71, 1919.

[35] G. Pick, "Über die beschränkungen analytischer Funktionen, welche durch vorgegebene Funktionswerte bewirkt werden," Math. Ann.,vol. 77, pp. 7-23, 1916.

[36] V. P. Potapov, "The multiplicative structure of J-contractive matrix functions, " Amer. Math. Soc. Transl., vol. 15, pp. 131-243, 1960.

[37] A. C. M. Ran, "State space formulas for a model matching problem, " Systems and Control Letters, vol. 12, pp. 17-21, 1989.

[38] H. H. Rosenbrock, *Computer-Aided Control System Design,* Academic Press, 1974.

[39] D.Sarason, "Generalized interpolation in H^∞, " Trans. Amer. Math. Soc., vol. 127, pp. 180-203, 1967.

[40] I. Schur, "Uber die Potenzreihen, die im Inneren des Einheitskreises beschrankt sind, " 1; 2, J. Reine Angew. Math., vol. 147, pp. 205-232, 1917; vol. 148, pp. 122-145, 1918.

[41] A. Tannenbaum,"Modified Nevanlinna-Pick interpolation and feedback stabilization of linear plants with uncertainty in the gain factor," Int. J. Control, vol. 36, pp. 331-336, 1982.

[42] M. Vidyasagar and H. Kimura, "Robust controllers for uncertain linear multivariable systems, " Automatica, vol. 22, pp. 85-94, 1986.

[43] D. C. Youla, "A new theory of broadband matching, " IEEE Trans. on Circuit Th., vol. CT-11, pp. 30-50, 1964.

[44] D. C. Youla and M. Saito, "Interpolation with positive real functions," J of Franklin Inst., vol. 284, pp.77-108, 1967.

[45] G. Zames, "Feedback and optimal sensitivity; model reference transformations, multiplicative seminorms, and approximate inverses, " IEEE Trans. Automat. Control, vol. AC-23, pp. 301-320, 1981.

[46] G. Zames and B. A. Francis, "Feedback, minimax sensitivity and optimal robustness," IEEE Trans. Automat. Contr., vol. AC-28, pp.585-601, 1983.

Generalized State-Space Systems and Proper Stable Matrix Fractions

V. Kučera

Czechoslovak Academy of Sciences
Institute of Information Theory and Automation
182 08 Prague 8, Czechoslovakia

Abstract

The concept of proper stable rational matrix fraction is applied here to the study of linear systems. The most natural class of systems to consider in this context are generalized state-space systems. These systems will be defined and their basic properties reviewed. The key properties of internal properness and stability will be discussed in detail.

The main problem which is addressed here is that of stabilization by feedback. For a given generalized state-space linear system the family of all generalized state-space linear controllers that make the closed loop system internally proper and stable will be characterized in parametric form. The significance of this result will then be illustrated on the design of specific control systems.

Historical Background

The first attempts to characterize the family of feedback controllers that stabilize a given system can be traced back to the early seventies. Kučera (1974a) solved the problem for single-input single-output linear discrete-time systems through all stable solutions of a Diophantine equation. The extension of this idea to multi-input multi-output systems appeared in Kučera (1974b; 1975; 1979). Independently Youla, Jabr and Bongiorno (1976) obtained a characterization of continuous-time stabilizing controllers in terms of a stable rational matrix parameter. This was an explicit way to express all solutions of the Diophantine equation involved.

To obtain these results, the transfer function of the system was expressed in terms of polynomial matrix fractions. Desoer and co-workers (1980) formulated the problem in a more general algebraic setting and showed that the type of fractions employed should be matched with the ultimate requirement on the resulting feedback system. This idea was further elaborated by Vidyasagar (1985) and led to replacing polynomial fractions by stable rational fractions. These two fractions are of course related, see Antsaklis (1986), but the use of adequate fractions greatly simplifies the analysis.

This fractional approach has proved particularly useful in studying the generalized state-space linear systems. In these systems we are concerned not only with stability but also with properness of the system to be designed; this is to avoid both unstable and impulsive behaviour. Therefore the proper and stable rational fractions were used by Kučera (1984; 1986) to delineate all such controllers in parametric form.

Generalized State-Space Linear Systems

The linear system (E,F,G,H) governed by the equations

$$E\dot{x}(t) = Fx(t) + Gu(t), \quad t \geq 0 \qquad (1)$$
$$y(t) = Hx(t)$$

is called a generalized state-space system. Such systems are also called implicit, singular, descriptor or semi-state systems, see Rosenbrock (1974), Luenberger (1977), Verghese (1978), Campbell (1980), Verghese, Lévy and Kailath (1981) and Lewis (1986). The matrices E, F, G and H are real, of respective size $n \times n$, $n \times n$, $n \times q$ and $p \times n$, and u denotes the input, y the output and x the (generalized) state of the system.

It is to be noted that E and F in (1) are square matrices. The implicit systems whose E and F matrices are rectangular are treated by Bernhard (1982) and Grimm (1988). Feedback for such systems, however, is not yet well understood.

If the polynomial matrix $sE - F$ is nonsingular the system (1) is said to be regular. In this case unique solutions of (1) are obtained for all $x(0-)$ and $u(t)$. The $p \times q$ matrix

$$T(s) = H(sE - F)^{-1}G \tag{2}$$

is the transfer function of system (1).

The free response $x(t)$, $t \geq 0$ of a regular system (1) may exhibit exponential modes associated with the finite poles of $(sE - F)^{-1}$ and impulsive modes associated with the infinite poles of $(sE - F)^{-1}$. The transfer function (2) may be any rational matrix, possibly improper or unstable.

From the modelling point of view, any interconnection of integrators, differentiators and scalors can be described by equations (1) and, conversely, equations (1) always represent such an interconnection. As an example, a pure integrator I can be described by (1) with

$$E = 1, \quad F = 0, \quad G = 1, \quad H = 1$$

a pure differentiator D by

$$E = \begin{bmatrix} 0 & 1 \\ 0 & 0 \end{bmatrix}, \quad F = \begin{bmatrix} 1 & 0 \\ 0 & 1 \end{bmatrix}, \quad G = \begin{bmatrix} 0 \\ -1 \end{bmatrix}, \quad H = [1 \quad 0]$$

and a scalor k by

$$E = 0, \quad F = 1, \quad G = -1, \quad H = k.$$

The order of system (1) is n but the number of dynamical elements (i.e., integrators and differentiators) may be lower.

The number of integrators equals the number of the finite poles of $(sE - F)^{-1}$ given by deg det $(sE - F)$. The total number of the dynamical elements whose initial conditions are independent is equal to the total number of the poles of $(sE - F)^{-1}$ given by rank E. For example, the system described by

$$E = \begin{bmatrix} 0 & 1 \\ 0 & 1 \end{bmatrix}, \quad F = \begin{bmatrix} 1 & 0 \\ 0 & 0 \end{bmatrix}, \quad G = \begin{bmatrix} 0 \\ 1 \end{bmatrix}, \quad H = [1 \quad 0]$$

and shown in Fig. 1 has order two while having one exponential mode and no impulsive mode. Note that $x_2(0-)$ is a joint initial condition for I and D.

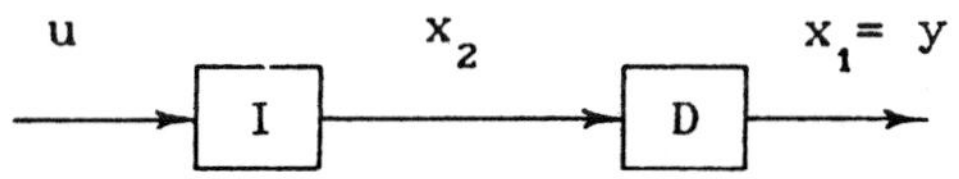

Fig. 1 A system which has less modes than dynamical elements

We say that system (1) is reachable if

$$\text{rank } [sE - F \quad G] = n$$

for every complex s and

$$\text{rank } [E \quad G] = n.$$

We say that system (1) is observable if

$$\text{rank} \begin{bmatrix} sE - F \\ H \end{bmatrix} = n$$

for every complex s and

$$\text{rank} \begin{bmatrix} E \\ H \end{bmatrix} = n.$$

A dynamical interpretation of these notions can be found in Cobb (1984).

A generalized Kalman decomposition will bring system (1) to the form that displays the reachable and observable part, the reachable but unobservable part, the observable but unreachable part, and the unreachable and unobservable part. The transfer function of a regular system (1) depends only on the reachable and observable part. Hence only the regular systems that are reachable and observable are completely characterized by their transfer function.

On the other hand, given any rational $p \times q$ matrix $T(s)$, there exists a realization (E,F,G,H) satisfying (2). Among all realizations there is some in which the matrices E and F have their smallest size n, and such a realization is reachable and observable. For details see Conte and Perdon (1982) and Grimm (1988).

Properness and Stability

The behaviour of system (1) at $t = 0$ and $t \to \infty$ is of considerable importance. We say that a regular system (1) is (internally) proper if the rational matrix $(sE - F)^{-1}$ is proper, i.e., has no infinite poles. We say that a regular system (1) is (internally) stable if the rational matrix $(sE - F)^{-1}$ is stable, i.e., has no finite poles in the closed right half-plane $\mathrm{Re}\ s \geq 0$.

This definition of properness was introduced by Kučera (1984). The free response $x(t)$, $t \geq 0$ of a proper system (1) comprises no impulsive modes at $t = 0$ for every initial condition $x(0-)$. The definition of stability is standard: the free response $x(t)$, $t \geq 0$ of a stable system (1) tends to the origin as $t \to \infty$ for every $x(0-)$.

The above notions of properness and stability reflect internal properties of the system and are to be strictly distinguished from the properness and stability as viewed by an external observer. The latter simply amounts to the properness and stability of the system transfer function (2). An example of a regular system which is externally but not internally proper and stable is given by

$$E = \begin{bmatrix} 1 & 0 & 0 \\ 0 & 0 & 1 \\ 0 & 0 & 0 \end{bmatrix}, \quad F = \begin{bmatrix} 0 & 1 & 0 \\ 0 & 1 & 0 \\ 0 & 0 & 1 \end{bmatrix}, \quad G = \begin{bmatrix} 0 \\ 0 \\ -1 \end{bmatrix}, \quad H = [1 \ 0 \ 0]$$

and visualized in Fig. 2. Observe that the system has one

unreachable mode at $s = 0$ and one unobservable mode at $s = \infty$.

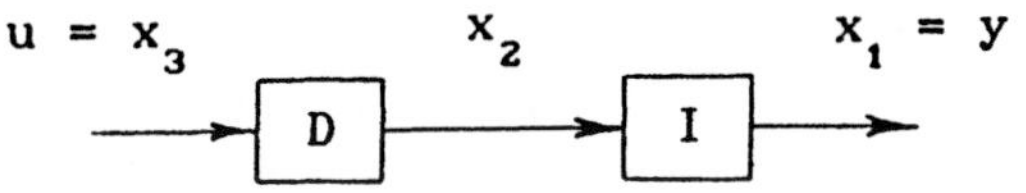

Fig. 2 An externally but not internally proper and stable system

Proper and Stable Matrix Fractions

The properness and stability of system (1) is most effectively studied by means of proper and stable matrix fractions. It is well known that rational functions which are proper and stable form a ring whose field of fractions is isomorphic with the field of rational functions. As a result, every real rational matrix $T(s)$ can be factorized as

$$T(s) = A_1^{-1}(s)\, B_1(s) \tag{3}$$

where $A_1(s)$ and $B_1(s)$ are proper stable rational matrices, and also as

$$T(s) = B_2(s)\, A_2^{-1}(s) \tag{4}$$

where $A_2(s)$ and $B_2(s)$ are proper stable rational matrices as well. We speak of <u>proper stable matrix fractions</u>: (3) is a left one, (4) is a right one.

The units of the ring of proper stable rational functions are sometimes called biproper and bistable rational functions. Similarly a proper stable rational matrix whose inverse exists and is also proper and stable will be termed <u>biproper and bistable.</u>

When $A_1(s)$ and $B_1(s)$ in (3) are relatively left prime (over the ring of proper stable rational functions), they are in a sense unique (Vidyasagar, 1985). More precisely, if $A_1(s)$, $B_1(s)$ and $A_1'(s)$, $B_1'(s)$ are two pairs of relatively left prime, proper stable rational matrices such that

$$T(s) = A_1^{-1}(s)\ B_1(s) = A_1'^{-1}(s)\ B_1'(s)$$

then

$$A_1'(s) = U_1(s)\ A_1(s), \qquad B_1'(s) = U_1(s)\ B_1(s)$$

for a biproper and bistable rational matrix $U_1(s)$. A similar result holds when $A_2(s)$ and $B_2(s)$ in (4) are relatively right prime.

The relatively prime matrix fractions (3) and (4) are naturally closely related. In particular, $\det A_1(s)$ equals $\det A_2(s)$ up to multiplication by a biproper and bistable rational function.

Feedback Systems

Let us now consider two generalized state-space systems

$$\begin{aligned} E_1\dot{x}_1(t) &= F_1x_1(t) + G_1u_1(t), \qquad t \geq 0 \\ y_1(t) &= H_1x_1(t) \end{aligned} \tag{5}$$

where E_1 and F_1 are $n_1 \times n_1$, G_1 is $n_1 \times q$, H_1 is $p \times n_1$ and

$$\begin{aligned} E_2\dot{x}_2(t) &= F_2x_2(t) + G_2u_2(t), \qquad t \geq 0 \\ y_2(t) &= H_2x_2(t) \end{aligned} \tag{6}$$

where E_2 and F_2 are $n_2 \times n_2$, G_2 is $n_2 \times p$ and H_2 is $q \times n_2$. We connect them according to

$$\begin{aligned} u_1(t) &= v_2(t) - y_2(t) \\ u_2(t) &= v_1(t) + y_1(t) \end{aligned} \tag{7}$$

where v_1 and v_2 are external inputs. The resulting closed loop system is shown in Fig. 3.

Our aim is to study the regularity, properness and stability of this closed loop system. We shall suppose that the component systems (5) and(6) are regular, reachable and observable so that they are completely characterized by their transfer functions

$$T_1(s) = H_1(sE_1 - F_1)^{-1}G_1$$

and

$$T_2(s) = H_2(sE_2 - F_2)^{-1}G_2.$$

The special structure of the closed loop system implied by (7) then makes it possible to study its regularity, properness and stability by means of $T_1(s)$ and $T_2(s)$.

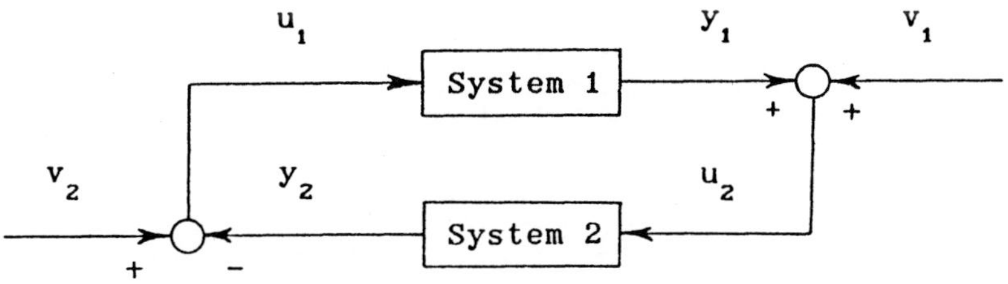

Fig. 3 A prototype of feedback system

To this effect we write $T_1(s)$ in terms of proper stable matrix fractions,

$$T_1(s) = A_1^{-1}(s)\, B_1(s) = B_2(s)\, A_2^{-1}(s) \tag{8}$$

where $A_1(s)$, $B_1(s)$ are relatively left prime while $A_2(s)$, $B_2(s)$ are relatively right prime. Similarly we write $T_2(s)$ in terms of proper stable matrix fractions as

$$T_2(s) = P_1^{-1}(s)\, Q_1(s) = Q_2(s)\, P_2^{-1}(s), \tag{9}$$

where $P_1(s)$, $Q_1(s)$ are relatively left prime while $P_2(s)$, $Q_2(s)$ are relatively right prime. Then we have the following result first announced by Kučera (1984).

<u>Theorem 1</u>. Let (5) and (6) be regular, reachable and observable systems giving rise to the transfer functions (8) and (9), respectively. Then the closed loop system defined by (5), (6) and (7) is regular, proper and stable if and only if the rational matrix

$$A_1(s)\, P_2(s) + B_1(s)\, Q_2(s) \tag{10}$$

is biproper and bistable or, equivalently, the rational matrix

$$P_1(s)\ A_2(s) + Q_1(s)\ B_2(s) \tag{11}$$

is biproper and bistable.

Proof: To prove the necessity, let the closed loop system be regular. We consider the transfer function

$$T_{11}(s)=\begin{bmatrix}A_2(s)\\B_2(s)\end{bmatrix} [P_1(s)A_2(s) + Q_1(s)B_2(s)]^{-1}[-Q_1(s) \qquad P_1(s)]$$

which relates the inputs v_1, v_2 and the outputs u_1, y_1 of the closed loop system, along with the transfer function

$$T_{22}(s)=\begin{bmatrix}P_2(s)\\Q_2(s)\end{bmatrix} [A_1(s)P_2(s) + B_1(s)Q_2(s)]^{-1}[A_1(s) \qquad B_1(s)]$$

which relates the inputs v_1, v_2 and the outputs u_2, y_2 of the closed loop system. If the closed loop system is proper and stable then $T_{11}(s)$ and $T_{22}(s)$ are proper stable rational matrices. Since $P_1(s)$, $Q_1(s)$ are relatively left prime and $A_2(s)$, $B_2(s)$ are relatively right prime, no cancellations are possible in forming $T_{11}(s)$ and $[P_1(s)\ A_2(s) + Q_1(s)\ B_2(s)]^{-1}$ is proper and stable. Since $A_1(s)$, $B_1(s)$ are relatively left prime and $P_2(s)$, $Q_2(s)$ are relatively right prime, no cancellations are possible when forming $T_{22}(s)$, either, and $[A_1(s)\ P_2(s) + B_1(s)\ Q_2(s)]^{-1}$ is proper and stable. Hence the matrices (10) and (11) are biproper and bistable.

To prove the sufficiency, we represent the dynamical action of system (6) on system (5) through the action of the static state feedback

$$\begin{bmatrix}u_1(t)\\u_2(t)\end{bmatrix} = \begin{bmatrix}0 & -H_2\\H_1 & 0\end{bmatrix} \begin{bmatrix}x_1(t)\\x_2(t)\end{bmatrix} + \begin{bmatrix}v_2(t)\\v_1(t)\end{bmatrix}$$

upon the extended system

$$\begin{bmatrix}E_1 & 0\\0 & E_2\end{bmatrix} \begin{bmatrix}\dot{x}_1(t)\\\dot{x}_2(t)\end{bmatrix} = \begin{bmatrix}F_1 & 0\\0 & F_2\end{bmatrix} \begin{bmatrix}x_1(t)\\x_2(t)\end{bmatrix} + \begin{bmatrix}G_1 & 0\\0 & G_2\end{bmatrix} \begin{bmatrix}u_1(t)\\u_2(t)\end{bmatrix} .$$

When we denote

$$E = \begin{bmatrix} E_1 & 0 \\ 0 & E_1 \end{bmatrix}, \quad F = \begin{bmatrix} F_1 & 0 \\ 0 & F_2 \end{bmatrix}, \quad G = \begin{bmatrix} G_1 & 0 \\ 0 & G_2 \end{bmatrix}, \quad K = \begin{bmatrix} 0 & -H_2 \\ H_1 & 0 \end{bmatrix}$$

the regularity, properness and stability of the closed loop system are coded in the polynomial matrix $sE - F - GK$.

As (5) and (6) are observable systems there exist proper stable rational matrices $B_{2s}(s)$ and $Q_{2s}(s)$ such that

$$(sE_1 - F_1)^{-1}G_1 = B_{2s}(s)\ A_2^{-1}(s)$$

$$(sE_2 - F_2)^{-1}G_2 = Q_{2s}(s)\ P_2^{-1}(s) \tag{12}$$

and (Hautus and Heymann, 1978)

$$B_2(s) = H_1B_{2s}(s), \quad Q_2(s) = H_2Q_{2s}(s). \tag{13}$$

As (5) and (6) are reachable systems, the proper stable rational matrices

$$A(s) = \begin{bmatrix} A_2(s) & 0 \\ 0 & P_2(s) \end{bmatrix}, \qquad B_s(s) = \begin{bmatrix} B_{2s}(s) & 0 \\ 0 & Q_{2s}(s) \end{bmatrix} \tag{14}$$

are relatively right prime. Hence we can write

$$\begin{bmatrix} A\ (s) \\ B_s(s) \end{bmatrix} = \begin{bmatrix} \bar{A}\ (s) \\ \bar{B}_s(s) \end{bmatrix} D^{-1}(s) \tag{15}$$

where $\bar{A}(s)$, $\bar{B}(s)$ are relatively right prime polynomial matrices and

$$\begin{bmatrix} \bar{A}\ (s) \\ \bar{B}_s(s) \end{bmatrix}$$

is column reduced with column degrees $d_1, d_2, \ldots, d_{p+q}$, say, while $D(s)$ is a polynomial matrix having column degrees $d_1, d_2, \ldots, d_{p+q}$ and $D^{-1}(s)$ is stable.

It follows from (12) through (14) that

$$(sE - F)\ B_s(s) = GA(s).$$

By subtracting $GKB_s(s)$, we obtain

$$(sE - F - GK)\ B_s(s) = G\ [A(s) - KB_s(s)]. \tag{16}$$

This relation will enable us to study $sE - F - GK$ through $A(s) - KB_s(s)$.

Now let either (10) or (11) be a biproper and bistable rational matrix, say $U_1(s)$ or $U_2(s)$, respectively. Then it follows from (8) and (9) that

$$\begin{bmatrix} P_1(s) & -Q_1(s) \\ B_1(s) & A_1(s) \end{bmatrix} \begin{bmatrix} A_2(s) & Q_2(s) \\ -B_2(s) & P_2(s) \end{bmatrix} = \begin{bmatrix} U_1(s) & 0 \\ 0 & U_2(s) \end{bmatrix}$$

is biproper and bistable. Applying (13), we obtain

$$A(s) - KB_s(s) = \begin{bmatrix} A_2(s) & Q_2(s) \\ -B_2(s) & P_2(s) \end{bmatrix}$$

so that $A(s) - KB_s(s)$ is biproper and bistable.

The factorization (15) gives

$$A(s) - KB_s(s) = [\bar{A}(s) - K\bar{B}_s(s)]\, D^{-1}(s) \tag{17}$$

and relation (16) can be rewritten in terms of polynomial matrices,

$$(sE - F - GK)\, \bar{B}_s(s) = G\, [\bar{A}(s) - K\bar{B}_s(s)]. \tag{18}$$

We first observe that $\bar{A}(s) - K\bar{B}_s(s)$ is nonsingular. Then

$$X(s) = \bar{B}_s(s)\, [\bar{A}(s) - K\bar{B}_s(s)]^{-1}$$

is a rational solution of the equation

$$(sE - F - GK)\, X(s) = G.$$

Hence, over the field of rational functions,

$$\begin{aligned} \text{rank } (sE - F - GK) &= \text{rank } [sE - F - GK \quad G] \\ &= \text{rank } [sE - F \quad G] \begin{bmatrix} I_{n_1+n_2} & 0 \\ -K & I_{p+q} \end{bmatrix} \\ &= \text{rank } [sE - F \quad G] \\ &= n_1 + n_2 \end{aligned}$$

so that $sE - F - GK$ is nonsingular. This implies that the closed loop system is regular.

Then (18) can be given the form

$$(sE - F - GK)^{-1}G = \bar{B}_s(s)\,[\bar{A}(s) - K\bar{B}_s(s)]^{-1}.$$

As (5) and (6) are reachable systems, the matrices $sE - F - GK$ and G are relatively left prime. As $\bar{A}(s)$ and $\bar{B}_s(s)$ are relatively right prime, so are the matrices $\bar{A}(s) - K\bar{B}_s(s)$ and $\bar{B}_s(s)$. As a result,

$$\det\,[\bar{A}(s) - K\bar{B}_s(s)] = c\,\det\,(sE - F - GK)$$

where c is a nonzero real constant. Consequently, by (17),

$$\det\,[A(s) - KB_s(s)] = c\,\frac{\det\,(sE - F - GK)}{\det D}\,. \tag{19}$$

Since $A(s) - KB(s)$ is biproper, (19) yields

$$\deg\det\,(sE - F - GK) = \deg\det D(s) = \sum_{i=1}^{p+q} d_i\,.$$

For reachable systems (5) and (6)

$$\sum_{i=1}^{p+q} d_i = \operatorname{rank} E,$$

see Kučera and Zagalak (1988), so that the closed loop system has no impulsive modes and is therefore proper. Furthermore, since $A(s) - KB_s(s)$ is bistable, $\det\,(sE - F - GK)$ is stable by (19) and the closed loop system is stable. ///

To illustrate Theorem 1, we check the feedback system shown in Fig. 4 for internal properness and stability. The first

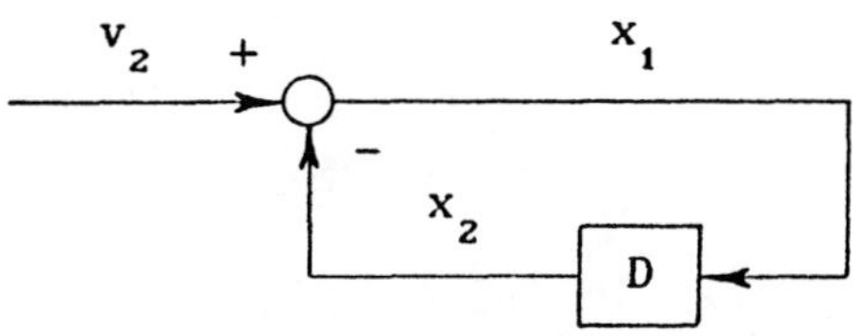

Fig. 4 An example of the feedback system

subsystem is a scalor described by

$$E_1 = 0, \quad F_1 = 1, \quad G_1 = -1, \quad H_1 = 1$$

and the second one is a pure differentiator given by

$$E_2 = \begin{bmatrix} 0 & 1 \\ 0 & 0 \end{bmatrix}, \quad F_2 = \begin{bmatrix} 1 & 0 \\ 0 & 1 \end{bmatrix}, \quad G_2 = \begin{bmatrix} 0 \\ -1 \end{bmatrix}, \quad H_2 = [1 \ 0].$$

Hence

$$A_1(s) = A_2(s) = 1, \quad B_1(s) = B_2(s) = 1$$

$$P_1(s) = P_2(s) = \frac{1}{s+a}, \quad Q_1(s) = Q_2(s) = \frac{s}{s+a}$$

for an arbitrary positive real constant a and

$$A_1(s)\ P_2(s) + B_1(s)\ Q_2(s) = \frac{s+1}{s+a},$$

a biproper and bistable rational function. Hence the system is proper and stable.

Admissible Controllers

When feedback systems are used for control purposes, the subsystem (5) is usually given and called the plant while the second one (6) is a controller to be found so that the overall control system behaves in a desired manner. One usually requires that the control system be internally proper and stable, thus avoiding impulsive and unstable exponential modes.

Given the plant, any controller that renders the resulting closed loop system regular, proper and stable will be called admissible. The family of admissible controllers will now be described following Kučera (1984; 1986).

Theorem 2. Let (5) be a regular, reachable and observable plant that gives rise to the transfer function (8). Let $P_1(s)$, $Q_1(s)$ be an arbitrary proper stable rational solution of the equation

$$P_1(s)\ A_2(s) + Q_1(s)\ B_2(s) = I_q \tag{20}$$

such that $P_1(s)$ is nonsingular. Then

$$P_1^{-1}(s)\ Q_1(s) \tag{21}$$

range over all biproper and bistable rational matrices, say $U_1(s)$ in (20) or $U_2(s)$ in (22). However,

$$\bar{P}_1(s) = U_1(s)\, P_1(s), \qquad \bar{Q}_1(s) = U_1(s)\, Q_1(s),$$

and

$$\bar{P}_2(s) = P_2(s)\, U_2(s), \qquad \bar{Q}_2(s) = Q_2(s)\, U_2(s),$$

define just alternative relatively prime matrix fractions (21) and (23),

$$P_1^{-1}(s)\, Q_1(s) = \bar{P}_1^{-1}(s)\, \bar{Q}_1(s)$$

and

$$Q_2(s)\, P_2^{-1}(s) = \bar{Q}_2(s)\, \bar{P}_2^{-1}(s). \qquad ///$$

If $P_1'(s)$, $Q_1'(s)$ is a particular solution of (20) and $P_2'(s)$, $Q_2'(s)$ is a particular solution of (22) then the general solutions of these equations read (Kučera, 1979)

$$P_1(s) = P_1'(s) + B_1(s)\, V_1(s)$$
$$Q_1(s) = Q_1'(s) - A_1(s)\, V_1(s)$$

and

$$P_2(s) = P_2'(s) + V_2(s)\, B_2(s)$$
$$Q_2(s) = Q_2'(s) - V_2(s)\, A_2(s)$$

where $V_1(s)$ and $V_2(s)$ range over proper stable rational matrices of appropriate size. Therefore we can characterize the family of transfer functions $T_2(s)$ of all admissible controllers for the given plant (5) in terms of one (proper stable rational) <u>parameter</u> matrix $V_1(s)$ or $V_2(s)$ as follows

$$\begin{aligned} T_2(s) &= [P_1'(s) + B_1(s)\, V_1(s)]^{-1}\, [Q_1'(s) - A_1(s)\, V_1(s)] \\ &= [Q_2'(s) - v_2(s)\, A_2(s)]\, [P_2'(s) + V_2(s)\, B_2(s)]^{-1}. \end{aligned} \tag{24}$$

Finally we note that Theorem 2 specifies just the transfer functions of all admissible controllers and not their generalized state-space realizations. In fact any reachable and observable realization (6) of the transfer function (21) or (23) is an admissible controller.

is the transfer function of an admissible controller (6) for the plant (5). ALso let $P_2(s)$, $Q_2(s)$ be an arbitrary proper stable rational solution of the equation

$$A_1(s)\, P_2(s) + B_1(s)\, Q_2(s) = I_p \tag{22}$$

such that $P_2(s)$ is nonsingular. Then

$$Q_2(s)\, P_2^{-1}(s) \tag{23}$$

is the transfer function of an admissible controller (6) for the plant (5). Furthermore, the transfer functions of all admissible controllers (6) for the plant (5) are generated in this way.

Proof: Since $A_2(s)$ and $B_2(s)$ are relatively right prime (over the ring of proper stable rational functions), equation (20) has a solution $P_1(s)$, $Q_1(s)$ in the class of proper stable rational matrices. We shall show that there exists a solution with $P_1(s)$ nonsingular. To this end, let $P_1'(s)$, $Q_1'(s)$ be any solution of (20). Then

$$P_1(s) = P_1'(s) + B_1(s)\, V$$

$$Q_1(s) = Q_1'(s) - A_1(s)\, V$$

is also a solution of (20) for any real constant $q \times p$ matrix V. Since both $A_1(s)$ and $A_2(s)$ are nonsingular, there exists a real s_o such that the constant matrices $A_1(s_o)$ and $A_2(s_o)$ are nonsingular. Put

$$V = A_1^{-1}(s_o)\, Q_1'(s_o).$$

Then $Q_1(s_o) = 0$ so that $P_1(s_o) = A_2^{-1}(s_o)$ by (20). Hence $P_1(s)$ is nonsingular. The solvability of equation (22) can be established along the same lines.

Since the identity matrices are biproper and bistable, it follows from Theorem 1 that any solution of (20) or (22) with $P_1(s)$ and $P_2(s)$ nonsingular defines via (21) or (23) the transfer function of an admissible controller for the given plant. We shall show that the transfer functions of all admissible controllers are generated in this way. Indeed the whole family is generated by solutions $\bar{P}_1(s)$, $\bar{Q}_1(s)$ of equation (20) or $\bar{P}_2(s)$, $\bar{Q}_2(s)$ of equation (22) whose right-hand sides

A Structured Plant

The plant (5) to be controlled is often structured in the following way:

$$E_1\dot{x}_1(s) = F_1x_1(t) + G_1u_1(t) + G_1'u_1'(t), \quad t \geq 0$$
$$y_1(t) = H_1x_1(t)$$
$$y_1'(t) = H_1'x_1(t) \qquad (25)$$

where E_1 and F_1 are $n_1 \times n_1$ matrices while G_1 is $n_1 \times q$, G_1' is $n_1 \times q'$ and H_1 is $p \times n_1$, H_1' is $p' \times n_1$. We interpret x_1 as the state, u_1 as the control input, u'_1 as the disturbance input, y_1 as the measured output and y_1' as the controlled output. The closed loop system defined by (25) and (6),(7) is then shown in Fig. 5.

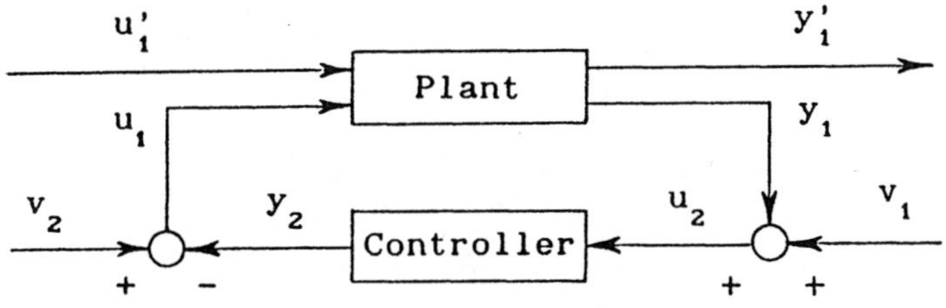

Fig. 5 A feedback system with structured plant

We shall analyze the internal properness and stability of this closed loop system. The transfer function of (25) is given by

$$T_1(s) = \begin{bmatrix} H_1 \\ H_1' \end{bmatrix} (sE_1 - F_1)^{-1}[G_1 \quad G_1'].$$

Let the proper stable matrix fractions defined in (8) be partitioned as follows

$$T_1(s) = \begin{bmatrix} A_1^{11}(s) & 0 \\ A_1^{21}(s) & A_1^{22}(s) \end{bmatrix}^{-1} \begin{bmatrix} B_1^{11}(s) & B_1^{12}(s) \\ B_1^{21}(s) & B_1^{22}(s) \end{bmatrix}$$

$$= \begin{bmatrix} B_2^{11}(s) & B_2^{12}(s) \\ B_2^{21}(s) & B_2^{22}(s) \end{bmatrix} \begin{bmatrix} A_2^{11}(s) & A_2^{12}(s) \\ 0 & A_2^{22}(s) \end{bmatrix}^{-1} \tag{26}$$

where $A_1^{11}(s)$ is $p \times p$, $A_1^{22}(s)$ is $p' \times p'$ and $A_2^{11}(s)$ is $q \times q$, $A_2^{22}(s)$ is $q' \times q'$. The zero blocks can always be achieved by biproper and bistable transformations. When the transfer function of (6),

$$T_2(s) = H_2(sE - F_2)^{-1}G_2,$$

is augmented compatibly with $T_1(s)$, the proper stable matrix fractions defined in (9) take the form

$$\begin{bmatrix} T_2(s) & 0 \\ 0 & 0 \end{bmatrix} = \begin{bmatrix} P_1(s) & 0 \\ 0 & I_{p'} \end{bmatrix}^{-1} \begin{bmatrix} Q_1(s) & 0 \\ 0 & 0 \end{bmatrix}$$

$$= \begin{bmatrix} Q_2(s) & 0 \\ 0 & 0 \end{bmatrix} \begin{bmatrix} P_2(s) & 0 \\ 0 & I_{q'} \end{bmatrix}^{-1} \tag{27}$$

and one has the following corollary of Theorem 1.

<u>Corollary</u>. Let $\left(E_1,\ F_1,\ [G_1\ \ G_1'],\ \begin{bmatrix} H_1 \\ H_1' \end{bmatrix}\right)$ and $(E_2,\ F_2,\ G_2,\ H_2)$ be regular, reachable and observable systems that give rise to the transfer functions (8) and (9) partitioned as shown in (26) and (27). Then the closed loop system (25), (6) and (7) is regular, proper and stable if and only if the rational matrices

$$A_1^{11}(s)\ P_2(s) + B_1^{11}(s)\ Q_2(s),\quad A_1^{22}(s)$$

are biproper and bistable or, equivalently, the rational matrices

$$P_1(s)\ A_2^{11}(s) + Q_1(s)\ B_2^{11}(s),\quad A_2^{22}(s)$$

are biproper and bistable. ///

To illustrate this result, consider a plant (25) with transfer function

$$T_1(s) = \frac{1}{a(s)} \begin{bmatrix} b(s) & c(s) \\ d(s) & e(s) \end{bmatrix}$$

where $a(s)$, $b(s)$, $c(s)$, $d(s)$ and $e(s)$ form a quintuple of relatively prime proper and stable rational functions. Let $g_1(s)$ be any greatest common divisor of $a(s)$, $b(s)$, $c(s)$ and write

$$a(s) = a_1(s)\, g_1(s),\ b(s) = b_1(s)\, g_1(s),\ c(s) = c_1(s)\, g_1(s).$$

Let $g_2(s)$ be any greatest common divisor of $a(s)$, $b(s)$, $d(s)$ and write

$$a(s) = a_2(s)\, g_2(s),\ b(s) = b_2(s)\, g_2(s),\ d(s) = d_2(s)\, g_2(s).$$

Let $g(s)$ be any greatest common divisor of $a(s)$ and $b(s)\, e(s) - c(s)\, d(s)$ and write $a(s) = f(s)\, g(s)$. Then

$$A_1^{11}(s) = a_1(s),\ B_1^{11}(s) = b_1(s),\ A_1^{22}(s) = f(s)\, g_1(s)$$

$$A_2^{11}(s) = a_2(s),\ B_2^{11}(s) = b_2(s),\ A_2^{22}(s) = f(s)\, g_2(s)$$

so that an admissible controller (6) exists if and only if (i) $a(s)$ and $b(s)$ are relatively prime and (ii) $a(s)$ divides $b(s)\, e(s) - c(s)\, d(s)$ in the ring of proper stable rational functions.

Model Matching

We shall now illustrate the use of Theorem 1 and Theorem 2 in the analysis and synthesis of control systems. Consider a plant described by

$$\begin{aligned} E_1\dot{x}_1(t) &= F_1x_1(t) + G_1u(t), \quad t \geq 0 \\ y(t) &= H_1x_1(t) \\ z(t) &= H_1'x_1(t) \end{aligned} \tag{28}$$

where E_1 and F_1 are $n_1 \times n_1$ matrices, G_1 is $n_1 \times q$, H_1 is $p \times n_1$ and H_1' is $m \times n_1$. It is assumed that $(E_1, F_1, G_1, \begin{bmatrix} H_1 \\ H_1' \end{bmatrix})$ is reachable and observable. The problem of <u>model matching</u> with internal properness and stability consists in finding an admissible controller of the form

$$E_2\dot{x}_2(t) = F_2x_2(t) + G_2y(t) + G_2'v(t), \quad t \geq 0$$
$$u(t) = H_2x_2(t) \tag{29}$$

where E_2 and F_2 are $n_2 \times n_2$ matrices, G_2 is $n_2 \times p$, G_2' is $n_2 \times r$ and H_2 is $q \times n_2$, such that the transfer function from v to z in the closed loop system (28), (29) shown in Fig. 6 equals a given (proper stable rational) model matrix M(s).

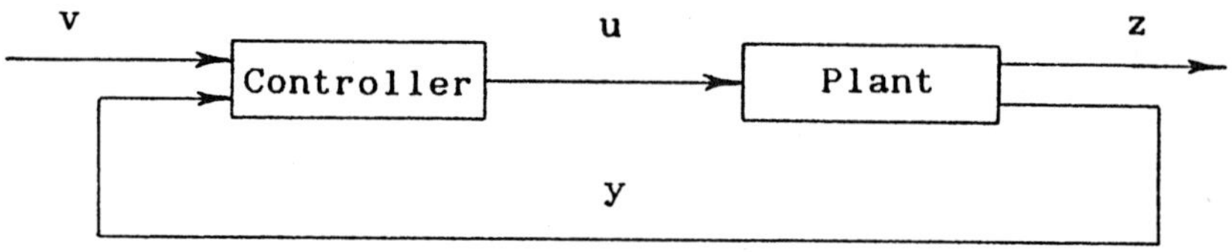

Fig. 6 Block diagram of model matching system

We define the matrix fractions

$$H_1(sE_1 - F_1)^{-1}G_1 = B_2(s)\,A_2^{-1}(s)$$
$$H_1'(sE_1 - F_1)^{-1}G_1 = C_2(s)\,A_2^{-1}(s) \tag{30}$$

where $A_2(s)$, $B_2(s)$ and $C_2(s)$ are proper stable rational matrices such that $A_2(s)$, $\begin{bmatrix} B_2(s) \\ C_2(s) \end{bmatrix}$ are relatively right prime. Similarly we define

$$-H_2(sE_2 - F_2)^{-1}G_2 = P_1^{-1}(s)\,Q_1(s) \tag{31}$$
$$H_2(sE_2 - F_2)^{-1}G_2' = P_1^{-1}(s)\,R_1(s)$$

for some proper stable rational matrices $P_1(s)$, $Q_1(s)$ and $R_1(s)$.

<u>Theorem 3</u>. The model matching problem is solvable if and only if

(i) $A_2(s)$ and $B_2(s)$ are relatively right prime

(ii) $C_2(s)$ is a left divisor of $M(s)$

over the ring of proper stable rational functions.

Any and all admissible controllers possess the transfer functions (31) where $P_1(s)$, $Q_1(s)$ and $R_1(s)$ form any proper stable rational solution triple of the equations

$$P_1(s)\ A_2(s) + Q_1(s)\ B_2(s) = I_q \tag{32}$$

$$C_2(s)\ R_1(s) = M(s) \tag{33}$$

such that $P_1(s)$ is nonsingular.

Proof: Let the closed loop system (28), (29) be proper and stable. By Theorem 1, the matrix $P_1(s)\ A_2(s) + Q_1(s)\ B_2(s)$ is biproper and bistable. Hence (i) follows. Let the transfer function relating v and z equal M(s). Then

$$M(s) = C_2(s)\ [P_1(s)\ A_2(s) + Q_1(s)\ B_2(s)]^{-1} R_1(s)$$

and $[P_1(s)\ A_2(s) + Q_1(s)\ B_2(s)]^{-1} R_1(s)$ is proper and stable. This implies (ii).

Conversely let (i) hold. Then there exist proper stable rational matrices $P_1(s)$, $Q_1(s)$ with $P_1(s)$ nonsingular such that (32) holds. Condition (ii) then entails the existence of a proper stable rational matrix $R_1(s)$ such that (33) is verified. Any triple $P_1(s)$, $Q_1(s)$ and $R_1(s)$ satisfying (32) and (33) defines a controller, via (29), that effects the model matching. By Theorem 2, any and all these controllers are admissible when appropriately realized. ///

This theorem is taken from Kučera (1986) and extends to generalized state-space systems the plethora of trasfer-function results available in the literature on model matching of ordinary state-space systems. Among others, see Wolovich (1974), Morse (1975), Pernebo (1981), and Malabre and Kučera (1984).

Output Regulation

We shall demonstrate the use of Theorem 1 and Theorem 2 on another example. Consider a plant described by

$$E_1\dot{x}_1(t) = F_1x_1(t) + G_1u(t), \quad t \geq 0 \tag{34}$$
$$y(t) = H_1x_1(t)$$
$$z(t) = H_1'x_1(t)$$

where E_1 and F_1 are $n_1 \times n_1$ matrices, G_1 is $n_1 \times q$, H_1 is $p \times n_1$ and H_1' is $m \times n_1$. It is assumed that $(E_1, F_1, G_1, \begin{bmatrix} H_1 \\ H_1' \end{bmatrix})$ is reachable and observable. Consider also a reference generator

$$E_2\dot{x}_2(t) = F_2x_2(t) + G_2v(t), \quad t \geq 0 \tag{35}$$
$$w(t) = H_2x_2(t)$$

where E_2 and F_2 are $n_2 \times n_2$, matrices, G_2 is $n_2 \times r$ and H_2 is $m \times n_2$. The problem of <u>output regulation</u> with internal properness and stability consists in finding an admissible controller of the form

$$E_3\dot{x}_3(t) = F_3x_3(t) + G_3y(t) + G_3'v(t), \quad t \geq 0$$
$$u(t) = H_3x_3(t) \tag{36}$$

such that the transfer function from v to $w - z$ in the closed loop system (34), (36) shown in Fig. 7 is proper and stable.

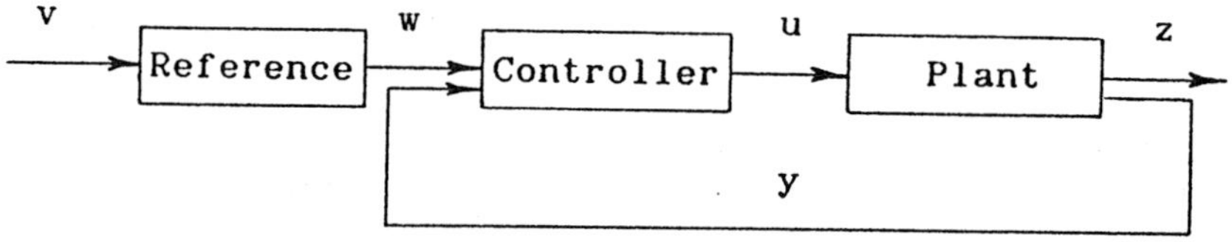

Fig. 7 Block diagram for output regulation system

Denoting $e(t) = w(t) - z(t)$ the regulation error, the above formulation means that $e(t)$, $t \geq 0$ is to be free of impulsive and unstable modes for every initial conditions $x_1(0-)$, $x_2(0-)$ and $x_3(0-)$. In particular $z(t)$ will asymptotically follow $w(t)$.

We define the matrix fractions

$$H_1(sE_1 - F_1)^{-1}G_1 = B_2(s)\, A_2^{-1}(s) \tag{37}$$

$$H_1'(sE_1 - F_1)^{-1}G_1 = C_2(s)\, A_2^{-1}(s)$$

$$H_2(sE_2 - F_2)^{-1}G_2 = F_1^{-1}(s)\, G_1(s)$$

where $A_2(s)$, $B_2(s)$, $C_2(s)$ are $F_1(s)$, $G_1(s)$ are proper stable rational matrices such that $F_1(s)$, $G_1(s)$ are relatively left prime and $A_2(s)$, $\begin{bmatrix} B_2(s) \\ C_2(s) \end{bmatrix}$ are relatively right prime. Similarly define

$$-H_3(sE_3 - F_3)^{-1}G_3 = P_1^{-1}(s)\, Q_1(s) \tag{38}$$

$$H_3(sE_3 - F_3)^{-1}G_3' = P_1^{-1}(s)\, R_1(s)$$

for some proper stable rational matrices $P_1(s)$, $Q_1(s)$ and $R_1(s)$.

<u>Theorem 4</u>. The output regulation problem is solvable if and only if

(i) $A_2(s)$ and $B_2(s)$ are relatively right prime

(ii) $C_2(s)$ and $F_1(s)$ are internally skew prime

over the ring of proper stable rational functions.

Any and all admissible controllers possess the transfer functions (38) where $P_1(s)$, $Q_1(s)$, $R_1(s)$ along with $S_2(s)$ form any proper stable rational solution quadruple of the equations

$$P_1(s)\, A_2(s) + Q_1(s)\, B_2(s) = I_q \tag{39}$$

$$C_2(s)\, R_1(s) + S_2(s)\, F_1(s) = I_m \tag{40}$$

such that $P_1(s)$ is nonsingular.

Proof: Let the closed loop system (34), (36) be proper and stable. By Theorem 1, the matrix $P_1(s)\, A_2(s) + Q_1(s)\, B_2(s)$ is biproper and bistable. Hence (i) follows. Let the transfer function from v to e be proper and stable. It is given by

$$e(s) = \{I_p - C_2(s)\, [P_1(s)\, A_2(s) + Q_1(s)\, B_2(s)]^{-1} \times$$
$$\times\; R_1(s)\}\; F_1^{-1}(s)\; G_1(s)\; v(s)$$

where

$$X_1(s) = [P_1(s)\, A_2(s) + Q_1(s)\, B_2(s)]^{-1} R_1(s)$$

is proper and stable. Since $F_1(s)$ and $G_1(s)$ are relatively left prime, we conclude that $F_1(s)$ is a right divisor of $I_p - C_2(s)\, X_1(s)$, i.e., there exists a proper stable rational matrix $Y_2(s)$ such that

$$C_2(s)\, X_1(s) + Y_2(s)\, F_1(s) = I_p.$$

This proves (ii) using the terminology of Wolovich (1978).

Conversely let (i) hold. Then there exist proper stable rational matrices $P_1(s)$ and $Q_1(s)$ with $P_1(s)$ nonsingular such that (39) holds. Condition (ii) implies (Wolovich, 1978) the existence of proper stable rational matrices $R_1(s)$, $S_2(s)$ such that (40) is verified. Then

$$e(s) = S_2(s)\, v(s)$$

in view of (39), (40) and it is proper and stable. Hence any $P_1(s)$, $Q_1(s)$ and $R_1(s)$ satisfying (39) and (40) define, via (38), a controller that effects the output regulation. By Theorem 2, any and all these controllers are admissible when appropriately realized. ///

This theorem is taken from Kučera (1986) and extends to generalized state-space systems the transfer-function results of many researchers, including Wonham and Pearson (1974), Francis (1977), Bengtsson (1977), Wolovich and Ferreira (1979), Pernebo (1981) and Francis and Vidyasagar (1983).

Although the results of Theorems 1 through 4 are stated for reachable and observable systems, which are completely characterized by their transfer functions, these results hold true for all systems whose impulsive and unstable modes are reachable and observable.

References

Antsaklis, P.J. (1986). Proper stable transfer matrix factorizations and internal system descriptions. IEEE Trans. Automat. Contr., AC-31, 634-638.

Bengtsson, G. (1977). Output regulation and internal models: A frequency domain approach. Automatica, 13, 335-345.

Bernhard, P. (1982). On singular implicit linear dynamical systems. SIAM J. Contr. Optimiz., 20, 612-633.

Cambell, S. L. (1980). Singular Systems of Differential Equations. Pitman, San Francisco.

Cobb, J. D. (1984). Controllability, observability, and duality in singular systems. IEEE Trans. Automat. Contr., AC-29, 1076-1082.

Conte, G. and A. Perdon (1982). Generalized state-space realizations of non-proper rational transfer functions. Syst. Contr. Lett., 1,4.

Desoer, C.A., R. W. Liu, J. Murray and R. Saeks (1980). Feedback system design: The fractional representation approach to analysis and synthesis. IEEE Trans. Automat. Contr., AC-25, 399-412.

Francis, B.A. (1977). The multivariable servomechanism problem from the input-output viewpoint. IEEE Trans. Automat. Contr., AC-22, 322-328.

Francis, B.A. and M. Vidyasagar (1983). Algebraic and topological aspects of the regulator problem for lumped linear systems. Automatica, 19, 87-90.

Grimm, J. (1988). Realization and canonicity for implicit systems. SIAM J. Contr. Optimiz., 26, 1331-1347.

Hautus, M. L. J. and M. Heymann (1978). Linear feedback, an algebraic approach. SIAM J. Contr. Optimiz., 16, 83-105.

Kučera, V. (1974a). Closed-loop stability of discrete linear single-variable systems. Kybernetika, 10, 146-171.

Kučera, V. (1974b). Algebraic theory of discrete optimal control for multivariable systems. Kybernetika, vols. 10-12, pp. 1-240. Published in instalments.

Kučera, V. (1975). Stability of discrete linear feedback systems. Preprints 6th IFAC World Congress, Boston, vol. 1, paper 44.1.

Kučera, V. (1979). Discrete Linear Control: The Polynomial Equation Approach. Wiley, Chichester.

Kučera, V. (1984). Design of internally proper and stable systems. Preprints 9th IFAC World Congress, Budapest, vol. VIII, pp. 94-98.

Kučera, V. (1986). Internal properness and stability in linear systems. Kybernetika, 22, 1-18.

Kučera, V. and P. Zagalak (1988). Fundamental theorem of state feedback for singular systems. Automatica, 24, 653-658.

Lewis, F.L. (1986). A survey of linear singular systems. J. Circuits, Syst., Signal Proc., 5, 3-36.

Luenberger, D.G. (1977). Dynamic equations in descriptor form. IEEE Trans. Automat. Contr., AC-22, 312-321.

Malabre, M. and V. Kučera (1984). Infinite structure and exact model matching problem: A geometric approach. IEEE Trans. Automat. Contr., AC-29, 266-268.

Morse, A.S. (1975). System invariants under feedback and cascade control. Proc. Internat. Symp. Mathematical System Theory, Udine, Italy. Springer-Verlag, New York, 61-74.

Pernebo, L. (1981). An algebraic theory for the design of controllers for linear multivariable systems. IEEE Trans. Automat. Contr., AC-26, 171-194.

Rosenbrock, H.H.(1974). Structural properties of linear dynamical systems. Int. J. Control, 20, 191-202.

Verghese, G. C. (1978). Infinite-frequency behaviour in generalized dynamical systems. Ph. D. Thesis, Dept. of Electrical Engineering, Stanford University.

Vidyasagar, M. (1985). Control System Synthesis: A Factorization Approach. M.I.T. Press, Cambridge, MA.

Wolovich, W.A. (1974). Linear Multivariable Systems. Springer-Verlag, New York.

Wolovich, W.A. (1978). Skew-prime polynomial matrices. IEEE Trans. Automat. Contr., AC-23, 880-887.

Wolovich, W.A. and P. Ferreira (1979). Output regulation and tracking in linear multivariable systems. IEEE Trans. Automat. Contr., AC-24, 460-465.

Wonham, W.M. and J. B. Pearson (1974). Regulation and internal stabilization in linear multivariable systems. SIAM J. Contr. Optimiz., 12, 5-18.

Youla, D.C., H.A. Jabr and J.J. Bongiorno (1976). Modern Wiener-Hopf design of optimal controllers, Part II. IEEE Trans. Automat. Contr., AC-21, 319-338.

Scattering Theory, Unitary Dilations and Gaussian Processes

S. K. Mitter

Department of Electrical Engineering and Computer Science and
Laboratory for Information and Decision Systems
Massachusetts Institute of Technology, Cambridge, MA 02139, U. S. A.
and
Scuola Normale Superiore, Pisa, Italy

Y. B. Avniel

Department of Electrical and Computer Engineering
Drexel University, Philadelphia, PA 19104, U. S. A.

I. INTRODUCTION

The themes of modelling and representation of linear deterministic and stochastic systems have dominated much of systems theory in the last thirty years. In the critical period of this development attention was focussed on the reconciliation between the input-output (external) and state space (internal) points of view of systems. A central result in this development is the statement that a minimal (in the sense of dimension) state space realization of linear finite-dimensional system is unique (up to isomorphism) and corresponds to one which is both controllable and observable. In recent years Jan Willems has forcefully argued that this input-output-state space view is narrow and inadequate to deal with models of dynamical systems arising out of physics, econometrics as well as stochastic processes (notably Markov processes). For one thing, there is no natural identification of what is an input and what is an output in these systems. For another, the need to fix the initial state as an equilibrium state in

(1) This research has been supported by the Air Force Office of Scientific Research under Grant AFOSR-85-0227 and of the Army Research Office under Grant ARO-DAAL03-86-k-017 through the Center for Intelligent Control Systems.

conventional realization theory is unnatural and leads to conceptual difficulties. For a detailed exposition of this work see [23] .

The other important theme in systems theory is one of optimization and approximation. Models of systems obtained from physical principles or data are often of high dimension. When these models are used for prediction and feedback control it is necessary to obtain approximate models of much lower order so that the algorithms for prediction and/or control are computationally tractable. There is an important question here as to what is the appropriate representation on which the approximation (reduction) process should be carried out. Zames [24] has argued that this approximation should be done on an input-output basis and an internal representation of the approximate input-output map could then be obtained for the purpose of prediction and control. This argument rests on the notion that two systems may be near each other in an input-output sense (for example in L^∞-topology) and yet may drastically differ dimensionally in their internal representations. The two processes of approximation and internal representation do not, in general, commute and working with the input-output representation for the needs of approximation is a more stable operation.

There is an apparent contradiction here since we have just argued that the input-output view so prevalent in the early days of systems theory is not an appropriate one. Fortunately, Scattering Theory as developed by Adamjan and Arov [1] and Lax and Phillips [15] and the theory of Abstract Hankel Operators comes to the rescue here. The work of Adamjan, Arov and Krein (for example, [2]) and Ball and Helton [6] provide a mathematical framework for dealing with representation and approximation issues in a Hilbert space setting in a rigorous manner. In the systems theory context, Scattering Theory for Gaussian Processes was investigated in the doctoral thesis of Y. Avniel at M.I.T. [cf. 5] and Scattering Theory and Approximation of Linear Systems has been investigated by Willems himself in his own framework for dynamical Systems [22]. A state space view of Hankel approximation has been provided by Glover [13] in an important paper.

Intimately connected with the theory of Scattering is the theory of minimal unitary dilations of contraction semigroups contracting strongly to zero. Indeed, the theorem of Nagy asserts that every contraction semigroup contracting strongly to zero has a (unique) minimal unitary dilation. This theorem has a physical interpretation of coupling a dissipative system to a heat bath so that the resulting composite system is conservative [cf. 17]. The dual of this question, namely, how certain observables of an infinite-dimensional conservative system can exhibit dissipative behavior has been investigated by Picci by using the theory of stochastic realization [cf. 20].

This semi-expository paper consists of two parts. In the first part we describe in an essentially self-contained manner the Scattering Theory associated with stationary Gaussian processes. This is done in discrete time to avoid certain technical difficulties. The new contribution in this part of this paper is the result which states that for completely non-deterministic stationary Gaussian processes the spectral density can be recovered (up to unitary isomorphism) from the Hankel operator induced by the scattering function associated with the process. This result to some extent justifies using this Hankel operator for model reduction in stochastic systems. We then show the relationship of this scattering view to the theory of Unitary Dilations and Markovian representations. These latter ideas are all contained in the work of Lindquist and Picci [cf.18] and Foias and Frahzo [10]. The exposition serves to show that if we restrict ourselves to a ℓ^2-theory then representation questions for stochastic (and deterministic) systems are nothing else but a version of Scattering Theory à la Adamjan, Arov and Krein. It is worth mentioning that the Scattering Function plays an important role in the parametrization of the unit ball of the quotient space L^∞/H^∞ and the corresponding extension problem for Hankel operators.

The second part of the paper is concerned with the theory of minimal unitary dilations of contraction semi-groups, its relation to positive definite functions on a group and the theory of open systems [8]. We also present a new construction of a unitary dilation of contraction semi-groups which makes evident the coupling to white noise (heat bath) which is implicit in the construction of the dilation.

2.1 NOTATION

Z stands for the set of integers, $\delta(n)$ for the indicator function of $\{0\} \subset Z$, C the complex numbers, and for $a \in C$, $\bar{a}$ denotes the complex conjugate of a. For a matrix $A = (a_{ij})_{i,j=1}^{p}$ we denote by A^* the Hermitian conjugate of $A : A^* = (b_{ij})_{i,j=1}^{p}$

$b_{ij} = + \bar{a}_{ji}$, and by A' its transposition. For a family of subsets $\{M_j\}_j$ of Hilbert space H, we denote by $\bigvee_j M_j$ the smallest closed linear manifold (subspace) that includes each M_j, and by $\bigwedge_j M_j$ the largest subspace contained in each of them

(their intersection). $\overline{M}_j$ denotes the closure of M_j in H. For subspaces M, N, of H, $M \ominus N$ denotes the orthogonal complement of N in M. For a countable family $\{M_j\}$ of mutually orthogonal subspaces: $M_i \perp M_j$ $i \neq j$, we let $\sum_j \oplus M_j$ be their orthogonal

sum. P_M stands for the orthogonal projection of H onto the subspace M. For a bounded linear operator $A : H_1 \to H_2$ of Hilbert space H_1 into H_2, we denote by [A] the matrix of A with respect to specified orthonormal bases in H_1, H_2.

$A^* : H_1 \to H_2$ denotes the adjoint of A. A|M stands for the restriction of A to the subspace $M \subset H_1$. $B(H_1, H_2)$ denotes the Banach space of all bounded linear operators from H_1 into H_2 with $B(H) = B(H,H)$.

By $\ell_2(-\infty,\infty; N)$ we denote the usual Hilbert space of sequences $\{hj\}_{j=-\infty}^{\infty}$ with values in (the Hilbert space) N for which $\sum_j \|h_j\|_N^2 < \infty$. $\ell_2(0,\infty; N)$, $\ell_2(-\infty, -1; N)$ are seen naturally as subspaces of $\ell_2(-\infty,\infty; N)$. L_2, L^∞ will denote respectively the Lebesgue spaces on the circle $T = \{e^{i\lambda} : \lambda \in [-\infty, \infty]\}$ (with respect to the normalized Lebesgue measure $\frac{d\lambda}{2\pi}$ of square integrable, essentially bounded complex valued functions). Each function can be viewed as defined on $[-\pi, \pi]$. Similarly for the spaces $L_2(C^p)$, $L_\infty(C^p)$ of functions f taking values in (C^p) for which $\|f(\pm)\|_{C^p} \in L_p$, $\|f(\pm)\|_{C^p} \in L_\infty$ respectively. $L_\infty(B(C^p))$ is defined analogously for weakly measurable, $B(C^p)$ valued functions f for which ess. sup $\{\| f(e^{i\lambda}) \|_{B(C^p)} : \lambda \in [-\pi,\pi]\} < \infty\}$. $H_2^{\pm}$ are the subspaces of L_2 defined by

$$H_2^+ = \{f \in L_2 : \frac{1}{2\pi} \int_{-\pi}^{\pi} f(e^{i\lambda})e^{-in\lambda}) \quad d\lambda = 0, n = -1, -2, \dots \}$$

$$H_2^- = \{f \in L_2 : \frac{1}{2\pi} \int_{-\pi}^{\pi} f(e^{i\lambda})e^{-in\lambda} \quad d\lambda = 0, n = 0, 1, 2, \dots \},$$

and we have the orthogonal decomposition $L_2 = H_2^+ \oplus H_2^-$. Each $f \in H_2^+$ having a Fourier series

$$f(e^{i\lambda}) \sim \sum_0^\infty a_n e^{in\lambda}$$

generates the function

$$g(z) = \sum_0^\infty a_n z^n$$

belonging to the Hardy class H_2 of functions g(z) holomorphic in $|z| < 1$ and such that

$$\| g \|_{H_2} = \sup_{0<r<1} \left[\frac{1}{2\pi} \int_{-\pi}^{\pi} | g(re^{i\lambda})|^2 d\lambda \right]^{1/2} < \infty .$$

Moreover, the (a.e. existing) radial limit $g(e^{i\lambda})$ of g(z) equals $f(e^{i\lambda})$ a.e. and $\| f \|_{L_2} = \| g \|_{H_2}$. The function g(z) is seen as the analytic extension of $f \in H_2^+$ to the unit disc $|z| < 1$ and is denoted by f(z). We identify H_2^+ and H_2 and denote them commonly by H_2. Using the conjugation with respect to the unit circle ($z \to \frac{1}{\bar{z}}$) by the reflection principles for $f \in H_2 \subset L_2$ the function $\bar{f}$ defined by $\bar{f}(e^{i\lambda}) = \overline{f(e^{i\lambda})}$ has an analytic extension to $|z| > 1$: $f(\frac{1}{\bar{z}})$ which we again denote by $\bar{f}$. The space $\bar{H}_2 = f \in L_2 : \bar{f} \in H_2$ is the space of functions $f \in L_2$ having an analytic extension to the exterior of the disc $|z| < 1$ and we have

$$\| f \|_{L_2} = \sup_{\rho>1} \left[\frac{1}{2\pi} \int_{-\pi}^{\pi} | f(\rho e^{i\lambda})|^2 d\lambda \right]^{1/2} .$$

$f \in \bar{H}_2$ are called conjugate analytic.

Analogously for the Banach space L_∞ we have the subspaces $H_\infty = H_\infty^+ \subset L_\infty$ of functions $f \in L_\infty$ having an analytic extension f(z) to $|z| < 1$ with

$$\| f \|_{L_\infty} = \sup_{|z|<1} | f(z)| = \| f \|_{H_\infty} .$$

Similarly, for the Hilbert space $L_2(C^p)$ we have the subspaces $H_2^+(C^p) = H_2(C^p)$, $H_2^-(C^p)$ with the orthogonal decomposition $L_2(C^p) = H_2(C^p) \oplus H_2^-(C^p)$. In $L_\infty(B(C^p))$, again $H_\infty(B(C^p))$ is defined as the subspace of functions in $L_\infty(B(C^p))$ whose negatively indexed (matrix valued) Fourier coefficients vanish. For $\theta \in H_\infty(B(C^p))$

the function θ^* defined by $\theta^*(e^{i\lambda}) = [\theta(e^{i\lambda})]^*$ is identified with its analytic extension $\theta^*(\frac{1}{\bar{z}}) = [\theta(\frac{1}{\bar{z}})]^*$ to $|z| > 1$.

A function $f \in H$ is called *inner* if $|f(e^{i\lambda})| = 1$ a.e. . Similarly for $\theta \in H_\infty(B(C^p))$ if $\theta(e^{i\lambda})$ is unitary a.e. . $f \in H_2$ is called *outer* if $\bigvee_{n\geq 0} \{\chi^n f\} = H_2$ where χ denotes the function on T defined by $\chi(e^{i\lambda}) = e^{i\lambda}$. For $\varphi \in L_\infty(B(C^p))$ the Toeplitz operator $T_\varphi : H_2(C^p) \to H_2(C^p)$ whose matrix is block Toeplitz with respect to the standard basis $\{e^{ik\lambda}e_1, e^{ik\lambda}e_2, ..., e^{ik\lambda}e_p\}_{k\geq 0}$, $\{e_1, e_2, ..., e_p\}$ being the standard basis in C^p, is defined by $T_\phi f = \pi_+(\phi f)$ where π_+ is the Riesz projection of $L_2(C^p)$ onto $H_2(C^p)$. H_ϕ will denote the Hankel operator [with block Hankel matrix with respect to the standard bases in $H_2(C^p)$, $H_2^-(C^p)$], $H_\phi : H_2(C^p) \to H_2^-(C^p)$ defined by $T_\phi f = \pi_-(\phi f)$, π_- being the Riesz projection of $L_2(C^p)$ onto $H_2^-(C^p)$. The con-vention we employ regarding a Henkel operator as acting from $H_2(C^p)$ into $H_2^-(C^p)$ is not in accordance with the one employed in systems theory, where we act on $H_2^-(C^p)$ into $H_2(C^p)$: $H_\phi f = \pi_+(\phi f)$. It, however, conforms to that employed by Adamjan-Arov-Krein and enables us to use their results without modification, as well as to refer to them.

2.2 SCATTERING THEORY

Let H be a complex separable Hilbert space and let U be a unitary operator on H. A subspace D_+ is said to be *outgoing* for (U,H) if it satisfies

(i) $UD_+ \subset D_+$

(2.2.1)+ (ii) $\bigwedge_{-\infty}^{\infty} U^n D_+ = \{0\}$

(iii) $\bigvee_{-\infty}^{\infty} U^n D_+ = H$.

A subspace D_- for which

(i) $U^* D_- \subset D_-$

(2.2.1)$_-$ (ii) $\bigwedge_{-\infty}^{\infty} U^n D_- = \{0\}$

(iii) $\bigvee_{-\infty}^{\infty} U^n D_- = H$

is said to be *incoming* for (H,U).

2.2.1 DEFINITION. A quadruple (U,H,D_+,D_-) satisfying (2.2.1) is said to be a *scattering system.*

2.2.2 THEOREM (Translation Representation Theorem [19, Th. II.1.1]). *Let* (U,H,D_+) *be outgoing. Then there exists a Hilbert space* N_+ *and a unitary map* r_+ *of* H *onto* $\ell_2(-\infty,\infty;N_+)$ *such that*

(i) $r_+ [D_+] = \ell_2 (0,\infty; N_+)$,

(2.2.2)

(ii) $U_+ = r_+ U r_+^{-1}$

is the right shift operator on $\ell_2 (-\infty,\infty; N_+)$. This representation is unique up to automorphism of N_+ .

Proof (Standard (cf. [15, p. 77]). We give the proof to establish various quantities introduced later. By $(2.2.1)_+$ - (ii) the operator $U|D_+$ is an isometry having no unitary. By Wold's decomposition theorem [19, Th. I.1.1] we may write uniquely

$$(2.2.3)\qquad D_+ = \sum_{n=0}^{\infty} \oplus\, U^n N_+ \quad , \qquad N_+ = D_+ \ominus UD_+ \quad .$$

Since for any $m > 0$

$$U^{-m} D_+ = U^{-m}[D_+ \ominus U^m D_+) \oplus U^m D_+] =$$

$$U^{-m}\,[(\sum_{k=0}^{m-1} \oplus\, U^k N_+) \oplus U^m D_+] =$$

$$(\sum_{k=-1}^{-m} \oplus\, U^k N_+) \oplus D_+) \quad ,$$

we obtain by $(2.2.1 - iii)_+$ that

$$(2.2.4)\qquad H = \sum_{-\infty}^{\infty} \oplus\, U^n N_+ \qquad .$$

It follows that for arbitrary $h \in H$

$$h = \sum_{-\infty}^{\infty} \oplus\, U^n P_{N_+} U^{-n} h, \quad \| h \|_H^2 =$$

$$\sum_{-\infty}^{\infty} \| P_{N_+} U^{-n} h \|_H^2 .$$

Hence the map

$$r_+ : H \to \ell_2(-\infty,\infty; N_+)$$

defined by

$$(2.2.5)\qquad r_+ h = \{P_{N_+} U^{-n} h\}_{n=-\infty}^{\infty}$$

is isometric. Since for $\{h_n\}_{-\infty}^{\infty} \in \ell_2(-\infty,\infty; N_+)$, $h = \sum_{-\infty}^{\infty} U^n h_n \in H$, the map r_+ is onto and thus unitary. By (2.2.3) we obtain (i). From (2.2.5)

$$r_+Uh = P_{N_+} \{U^{-(n-1)} h\}_{n=-\infty}^{\infty} = U_+ (r_+h) \quad ,$$

and (ii) follows. By (2.2.4) U is a bilateral shift of multiplicity equal to dim N_+ and the uniqueness follows.

2.2.3 DEFINITION. The representation $(U_+, \ell_2(0,\infty; N_+), \ell_2(-\infty,\infty; N_+))$ is called an *outgoing transition representation.*

For (U,H,D_-) incoming we similarly obtain

(2.2.6) $$D_- = \sum_{n=-\infty}^{0} \oplus U^n N_- \quad , \qquad N_- = D_- \ominus U^* D_-$$

and

(2.2.7) $$H = \sum_{-\infty}^{\infty} \oplus U^n N_- \quad .$$

For the corresponding map r_- of H onto $\ell_2(-\infty,\infty; N_-)$ we define

(2.2.8) $$r_- h = \{P_{N_-} U^{-(n+1)} h\}_{n=-\infty}^{\infty} \quad .$$

Thus

(i) $r_-[D_-] = \ell_2(-\infty, -1; N_-)$,

(ii) $U_- = r_- U r_-^{-1}$

is the right shift on $\ell_2(-\infty,\infty; N_-)$. The representation $(U_-, \ell_2(-\infty, -1; N_-)$, $\ell_2(-\infty,\infty; N_-)$ is called an *incoming translation representation.*

2.2.4 DEFINITION ([1] , [15]). The operator

$$S = r_-, r_+^{-1} : \ell_2(-\infty,\infty;\ (C^p)) \to \ell_2(-\infty,\infty;\ (C^p))$$

is called the *abstract scattering operator.*

Clearly S is unitary. Denoting by V the right shift on $\ell_2(-\infty,\infty;(C^p))$, we readily obtain by the translation representation theorem

$$\text{(2.2.9)} \qquad SV = r_- r_+^{-1} V = r_- U r_+^{-1} = VS .$$

Let $F : \ell_2(-\infty,\infty;\ (C^p)) \to L_2(C^p)$ be the Fourier transform operator. The unitary operator

$$FSF^{-1} : L^2(C^p) \to L^2(C^p)$$

thus commutes by (2.2.9) with L_χ , the operator of multiplication by χ. It follows [19] that FSF^{-1} is a Laurent operator $L_S \in L_\infty(B(C^p))$, such that

$S(e^{i\lambda})$ a.e. is a unitary map on C^p.

2.2.5 DEFINITION ([1], [15]). S is called the *scattering matrix.*

It is clear from the translation representation theorem that S is determined to within right and left multiplication by unitary transformations on C^p.

The unitary maps $F_- = Fr_-$, $F_+ = Fr_+$ are called the incoming and outgoing spectral representation. We have the following:

a) $F_-(D_- = H^2(C^p)$.

(2.2.9) b) $F_+(D_+) = SH^2(C^p)$.

c) $F_-(Uh) = \chi Fh,\ h \in H$.

Moreover, the operator

(2.2.10) $P_- P_+ : D_+ \to D_-$ is unitarily equivalent to the Hankel operator H_S,

where $P_\pm = P_{D_\pm}$, and the operator

(2.2.11) $PD_-^\perp P_+ : D_+ \to D_-^\perp$ is equivalent to the Toeplitz operator T_S.

2.3 COMPUTATION OF THE SCATTERING FUNCTION FOR REGULAR, MAXIMAL RANK, STATIONARY GAUSSIAN SEQUENCES

Let (Ω, A, P) be a fixed probability space and let

$$\{\underline{y}(n) : n \in Z\} \quad , \underline{y}(n) = \begin{pmatrix} y_1(n) \\ y_2(n) \\ \cdot \\ \cdot \\ \cdot \\ y_p(n) \end{pmatrix}$$

be a centered stationary process with $y_j(n) \in \mathcal{L}_2(\Omega, \mathcal{A}, P)$ $j=1, \ldots, p$. Let $f_{\underline{y}\underline{y}}(\lambda) = (f_{kj}(\lambda))_{k,j=1}^p$, $\lambda \in [-\pi, \pi]$ be its spectral density satisfying

(2.3.1) $$\frac{1}{2\pi} \int_{-\pi}^{\pi} \log \det f_{\underline{y}\underline{y}}(\lambda)\, d\lambda > -\infty$$

i.e., the process is *regular* and of *maximal rank.* Let

$$H = H\underline{y} = \bigvee_{n \in Z} \{y_1(n), y_2(n), \ldots, y_p(n)\} \subset \mathcal{L}_2(\Omega, \mathcal{A}, P)$$

be the space spanned by the process and let U be the unitary shift operator on H associated with the $\underline{y}$ process [21, p. 14]:

$$Uy_j(n) = y_j(n+1) \qquad j = 1, \ldots, p, \qquad n \in Z \quad .$$

We consider the *past* and *future* of $\{\underline{y}(n)\}_{-\infty}^{\infty}$ defined by

$$D_- = H_{\underline{y}}^- (0) = \bigvee_{k\leq 0} \{y_1(k), \ldots, y_p(k)\} \quad ,$$

$$D_+ = H_{\underline{y}}^+ (0) = \bigvee_{k\geq 0} \{y_1(k), \ldots, y_p(k)\} \quad .$$

By (2.3.1) it follows [21, Th. II.6.1]

$$\bigwedge_{-\infty}^{\infty} U^n D_- = \{0\} = \bigwedge_{-\infty}^{\infty} U^n D_+ .$$

We readily obtain that (U, H, D_+, D_-) is a scattering system.

Now let (U, H, D_+, D_-) be the scattering system associated with the regular maximal rank $\underline{y}$ process. The subspace $N_- = D_- \ominus U^* D_-$ $(N_- = D_- \ominus U^* D_-)$ is the *forward (backward)* innovation subspace at n = 0. Since for a scattering system $(U,H, D_+ D_-)$ we have

$$\dim N_- = \text{multiplicity } U = \dim N_+ \quad ,$$

we can arrange the maps r_+ to be onto $\ell^2(-\infty,\infty; (C^p))$.

We next compute the scattering matrix S for the $\underline{y}$ process. Let $\{v_1(0),\ldots,v_p(0)\}$ be an orthonormal basis for N_-. Let $v_j(n) = U^n v_j(0)$ and define

$$\underline{v}(n) = \begin{pmatrix} v_1(n) \\ \cdot \\ \cdot \\ \cdot \\ v_p(n) \end{pmatrix} , \qquad n \in Z .$$

By (2.2.4) the process $\{v(n)\}_{-\infty}^{\infty}$ is a (centered) white noise process with covariance $R_{vv}(n) = \delta(n) I_{C^p}$ constituting the forward innovation process for the $\underline{y}$ process. It is determined up to a choice of basis in N_-. By (2.2.6) we may write

$$\underline{y}(0) = \sum_{-\infty}^{\infty} A(k)\underline{v}(k) \qquad A(k) = (a_{ij}(k))_{i,j=1}^{p} \qquad A(k) = [0],\ k > 0\ .$$

(Wold's representation). It follows from (2.2.8)

$$(2.3.2) \qquad r_{-}y_j(0) = \left\{ \sum_{m=1}^{p} \alpha_{jm}(k+1)v_m(0) \right\}_{k=-\infty}^{\infty}$$

Identifying N_- with (C^p) we readily obtain the representation

$$r_{-}y_j(0) = \begin{pmatrix} \alpha_{j1}(k+1) \\ \cdots\cdots \\ \cdots\cdots \\ \alpha_{jp}(k+1) \end{pmatrix}_{k=-\infty}^{\infty} .$$

Consider the function

$$\Lambda(z) = \sum_{-\infty}^{\infty} A'(k)z^k \quad .$$

Since

$$\sum_{k=-\infty}^{\infty} \sum_{i,j=1}^{p} |\alpha_{ij}(k)|^2 \le \sum_{j=1}^{p} \|y_j(0)\|_H^2 ,$$

$\Lambda(z)$ is analytic in $|z| > 1$. For $\Lambda(z)$ we have,

$$\frac{1}{2\pi} \Lambda^*(z)\Lambda(z) = f_{\underline{y}\underline{y}}(\lambda)$$

By the incoming properties

$$H_2^-(C^p) = \bigvee_{n<0} \{e^{in\lambda}\Lambda(e^{i\lambda})\underline{a} : \underline{a} \in (C^p)\}$$

i.e., Λ is *conjugate outer* [14, p. 121). Thus

$$(2.3.3) \qquad (Fr_{-}y_1(0), \ldots, Fr_{-}y_p(0)) = \bar{\chi}\Lambda \quad .$$

Since the translates (in $H_{\underline{y}}$) of $y_1(0),\ldots,y_p(0)$ and their linear combinations are dense in $H_{\underline{y}}$. Fr_- is determined by the above expression.

We now consider the outgoing representation. Let $\varepsilon_1(0),\ldots,\varepsilon_p(0)$ be an orthonormal basis in N^+. We similarly obtain

$$\underline{y}(0) = \sum_{-\infty}^{\infty} B(k)\underline{\varepsilon}(k) \qquad B(k) = (\beta_{ij}(k))_{i,j=1}^{p}, \qquad B(k) = [0],\ k<0 .$$

This representation constitutes the representation of y(0) in terms of the backward innovation process $\{\underline{\varepsilon}(n)\}_{-\infty}^{\infty}$, $\underline{\varepsilon}(n) = \begin{pmatrix} \varepsilon_1(n) \\ \cdot \\ \cdot \\ \cdot \\ \varepsilon_p(n) \end{pmatrix}$. We define

$$\Gamma(z) = \sum_{0}^{\infty} B'(k) z^k$$

which is analytic in $|z| < 1$. In a similar fashion we obtain by direct computation

$$\frac{1}{2\pi}\Gamma^*(z)\Gamma(z) = f_{\underline{y}\underline{y}}(\lambda), \qquad z = e^{i\lambda}$$

with Γ being outer. Also

(2.3.4) $\quad (Fr_+ y_1(0), \ldots Fr_+ y_p(0)) = \Gamma$.

Combining (2.3.3) with (2.3.4), we obtain

$$S\Gamma = \bar{\chi}\Lambda$$

and thus

$$S = \bar{\chi}\Lambda\Gamma^{-1} .$$

One easily verifies that $S(e^{i\lambda})$ is unitary a.e. $\lambda \in [-\pi,\pi]$. We thus obtain

2.3.1 THEOREM. *For a regular maximal rank process* $\{\underline{y}(n)\}$ *we have*

$$S = \bar{\chi}\Lambda\Gamma^{-1}$$

where S *is determined up to left and right multiplication by constant unitary matrices.*

For the case p = 1 we have

2.3.2 COROLLARY. *For a regular process* $\{\underline{y}(n)\}_{-\infty}^{\infty}$

$$S = \bar{\chi}\,\frac{\bar{\Gamma}}{\Gamma} \quad ,$$

and S *is determined up to multiplication by a constant of unit modulus.*

Proof. The outer function $\bar{\Lambda}$ satisfies $|\bar{\Lambda}| = |\Gamma|$ on T and thus $\bar{\Lambda} = \gamma\Gamma$ a.e. where γ is a constant of unit modulus.

2.3.3 REMARK. The scattering matrix S was defined by outer and conjugate outer factors of the density $f_{\underline{y}\underline{y}}$. Since those are determined up to left multiplication by a constant unitary matrix, we may wish to make a canonical choice (which amounts to choosing specific orthonormal bases in N_+, N_-) in the following fashion: For $\Gamma(0)$ we consider its polar decomposition $\Gamma(0) = KP$ (K unitary, $P > 0$) and define $\Gamma_1(z) = K^{-1}\Gamma(z)$. For Γ_1 we have $\Gamma_1(0) > 0$. This Γ_1 is unique. Similarly for Λ. In this way, the density $f_{\underline{y}\underline{y}}$ will have a unique S associated with it. From the viewpoint of seeing S as the phase function associated with f_{yy} , this may be appealing.

2.4 COMPLETELY NON-DETERMINISTIC STATIONARY SEQUENCES, THEIR SCATTERING FUNCTIONS, AND INDUCED HANKEL OPERATORS

The scattering function of stationary sequences plays the role of the Heisenberg S-matrix in quantum mechanics. The physics of quantum systems is believed to be contained in the S-matrix and this object can in principle be determined experimentally. A natural question then is whether the scattering function of stationary Gaussian sequences, which measures the interaction between the past and future of the process, determines the spectral density of the process. To answer this question we introduce the class of completely non-deterministic processes.

2.4.1 DEFINITION [7]. The process y is said to be completely non-deterministic if

$$(2.4.1) \qquad H^-_{\underline{y}}(0) \cap H^+_{\underline{y}}(1) = \{0\}.$$

This condition states that no value in $H^+_{\underline{y}}(1)$ can be predicted without error based on the information $H^-_{\underline{y}}(0)$. This condition is more restrictive than regularity (cf.

BLOOMFIELD-JEWELL-HAYASHI, loc.cit., for an example of a regular process which is completely non-deterministic).

2.4.2 THEOREM. *The scattering matrix* S *determines the spectral density* $f_{\underline{y}\underline{y}}$ *up to the form* $K^* f_{\underline{y}\underline{y}} K$ *where* K *is a constant* $p \times p$ *non-singular matrix iff* $\underline{y}$ *is completely non-deterministic .*

2.4.3 REMARK. For p=1, this result was obtained by Levinson and McKean [16].

2.4.4 LEMMA. *The scattering matrix S determines the density* $f_{\underline{y}\underline{y}}(\lambda)$ *up to the form*

(2.4.2) $\quad K^* f_{\underline{y}\underline{y}}(\lambda) K$

where K *is a constant pxp non-singular matrix, iff*

(2.4.3) $\quad \dim \operatorname{Ker} T_S = p$.

Proof. First note that for any representation of S

$$S = \bar{\chi} Y X^{-1}$$

with the columns of X in $H_2(C^p)$ and those of $\bar{\chi}Y$ in $H_2^-(C^p)$, the columns of X belong to Ker T_S. Moreover (on T) F_+

$$Y^*Y = (SX)^*SX = X^*X \quad .$$

Assume (2.4.3) holds. It thus follows that

(2.4.4) $\quad X(e^{i\lambda}) = \Gamma(e^{i\lambda})K$

where K is a pxp full rank constant matrix. Thus,

$$\frac{1}{2\pi} X^*(z)\, X(z) = \frac{1}{2\pi} K^*\Gamma^*(z)\Gamma(z)K = K^* f_{\underline{y}\underline{y}}(\lambda)\, K \qquad z = e^{i\lambda}$$

proving the 'if' part.

Now assume (2.4.3) not to hold, i.e., dim Ker $T_S > p$. We can thus find a pxp matrix $X(e^{i\lambda})$ of full rank a.e. λ such that the columns of X belong to Ker T_S and (2.4.4) does not hold. If we define

$$Y = \bar{\chi} S X$$

then the columns of $\bar{\chi}Y$ are in $H_2^-(C^p)$ and $S = \bar{\chi} Y X^{-1}$ with $Y^*Y = X^*X$. The result follows.

We next characterise condition (2.4.3) on a process level.

2.4.5 LEMMA. *We have*

$$F_+^* [\text{Ker } T_S] = H_{\underline{y}}^-(0) \wedge H_{\underline{y}}^+(0) \quad .$$

Proof. Let $0 \neq f \in$ Ker T_S . From the well-known identity $H_S^* H_S + T_S^* T_S = 1$, it follows that

$$H_S^* H_S f = f$$

i.e.,

$$\| H_S f \| = \| f \| .$$

From (2.2.10) and (2.2.11) we obtain for $\xi = F_+^* f \in H_y^+(0)$

$$\| P_- \xi \| = \| \xi \| ,$$

and $\xi \in H_{\underline{y}}^-(0)$. Thus

$$F_+^* [\text{Ker } T_S] \subset H_{\underline{y}}^-(0) \wedge H_{\underline{y}}^+(0) .$$

Now let $\xi \in H_{\underline{y}}^-(0) \wedge H_{\underline{y}}^+(0)$. It follows from (2.2.10) and (2.2.11)

$$H_S(F_+ \xi) = F_- \xi .$$

Let $f = F_+ \xi \in H_2(C^p)$. We obtain

$$\|H_S f\| = \|F_- \xi\| = \|\xi\| = \|F_+ \xi\| = \|f\|$$

and

$$H_S^* H_S f = f .$$

Thus $f \in \operatorname{Ker} T_S$ which implies

$$F_+ [H_{\underline{y}}^- (0) \wedge H_{\underline{y}}^+ (0)] \subset \operatorname{Ker} T_S .$$

By the unitarity of F_+

$$\dim \operatorname{Ker} T_S = \dim H_{\underline{y}}^- (0) \wedge H_{\underline{y}}^+ (0) ,$$

and since $\underline{y}$ is regular and of full rank we readily conclude

$$\dim H_{\underline{y}}^- (0) \wedge H_{\underline{y}}^+ (0) = p \quad \text{iff} \quad \dim H_{\underline{y}}^- (0) \wedge H_{\underline{y}}^+ (1) = 0 .$$

Proof of Theorem 2.4.2. Combine Lemmas 2.4.4. and 2.4.5.

The converse question, namely, when is a function $S \in L_\infty(B(C^p)$ the scattering matrix of some full rank, p-dimensional completely non-deterministic process is of interest.

We first observe that any $S \in L_\infty(B(C^p))$ which is unitary valued a.c. on T is the scattering matrix of the canonical scattering system [1]

$$U = L\chi , H = L_2(C^p) , D_+ = SH_2(C^p) , D_- = H_2^-(C^p) .$$

The above question amounts to characterizing all scattering systems (U, H, D_+, D_-) for which there exists a set $\{\varepsilon_1, \ldots, \varepsilon_p\}$ of linearly independent vectors such that

$$H = \overline{\operatorname{span} \{U^n \xi_j : j=1, \ldots, p, n=0, \pm 1, \ldots\}}$$

$$D_+ = \overline{\text{span}\ \{U^n \xi_j : j=1, \ldots, p,\ n \geq 0\}}$$

$$D_- = \overline{\text{span}\ \{U^n \xi_j : j=1, \ldots, p,\ n \leq 0\}}$$

and such that any other linearity independent set satisfying the above is of cardinality p. The corresponding process will be $\{\underline{\xi}(n)\}_{-\infty}^{\infty}$ where $\underline{\xi}(0) = \begin{pmatrix} \xi_1 \\ \cdot \\ \cdot \\ \cdot \\ \xi_p \end{pmatrix}$,

$$\underline{\xi}(n) = \begin{pmatrix} U^n \xi_1 \\ \cdot \\ \cdot \\ \cdot \\ U^n \xi_p \end{pmatrix},$$ and the spectral density is obtained by

$f_{\underline{\xi\xi}}(\lambda) = \left(\dfrac{d(E_\lambda \xi_i, \xi_j)_H}{d\lambda} \right)_{i,j=1,\ldots,p}$ $\{E_\lambda : \lambda \in [-\pi, \pi]\}$ being the resolution of the identity for U. The answer is given in the following.

2.4.6 THEOREM. *Let* $S \in L_\infty(B(C^p))$ *be such that*

(i) $S(e^{i\lambda})$ *is* a.e. λ *a unitary map on* (C^p) ,

(ii) $\dim \operatorname{Ker} T_S = p$.

Then there exists a p-dimensional full rank completely non-deterministic process $\underline{y}$ whose scattering matrix is S .

Proof. Let $\Gamma_1, \Gamma_2, \ldots, \Gamma_p$ span the kernel of T_S and define

$$\Gamma = [\,\Gamma_1 \mid \Gamma_2 \mid \ldots \mid \Gamma_p\,] .$$

Let

$$\Lambda = S\Gamma .$$

Since $\Lambda_j = (S\Gamma_j) + \pi_+ (S\Gamma_j) = \pi_-(S\Gamma_j)$, $j=1,\dots, p$, the columns of $\Lambda = [\Lambda_1 | \Lambda_2 | \dots | \Lambda_p]$ are in $H_2^-(C^p)$ and by (i)

$$\Lambda^*(z)\Lambda(z) = \Gamma^*(z)\Gamma(z) \qquad z = e^{i\lambda} .$$

If we define

$$f_{\underline{y}\,\underline{y}}(\lambda) = \frac{1}{2\pi} \Gamma^*(e^{i\lambda})\, \Gamma(e^{i\lambda})$$

the theorem follows provided we show that Γ is outer and $\chi\Lambda$ conjugate outer. Let $U = L_\chi$ and define :

$$\hat{D}_- = \bigvee_{n\le -1} \{\chi^n\Lambda_1, \dots, \Lambda_p\} \subset H_2^-((C^p)) ,\ \hat{D}_+ = \bigvee_{n\ge 0} \{\chi^n S\Gamma_1, \dots, \chi^n S\Gamma_p\} \subset SH_2(C^p) .$$

Let

(2.4.5) $$\hat{H} = (\bigvee_{n\in Z} U^n \hat{D}_-) \vee (\bigvee_{n\in Z} U^n \hat{D}_+) .$$

It is easily verified that $(2.2.1)_\pm$ - (i), (ii) holds for $(U,D_\pm)$. In [1] Adamjan-Arov show [1, Th. 2.5] that a quadruple (U,H,D_+,D_-) satisfying $(2.2.1)_\pm$ - (i), (ii) and (2.4.5) has a scattering matrix S which is unitary valued a.e. on T iff

$$\bigvee_{n\in Z} U^n D_+ = H = \bigvee_{n\in Z} U^n D_-$$

and, moreover, from their generalized functional model [1, Th. 2.1] we need have

$$D_- = H_2^-(C^p), \ D_+ = \hat{S}\, H_2(C^p) .$$

A straightforward computation gives

$$\hat{S} = S$$

and the result follows.

From the theorem of Nehari [3],and its vector generalization we know that for a bounded Hankel operator H with symbol $\varphi \in L_\infty$, there exists a function $\varphi_\mu \in L^\infty$ such that

$$\|H_\varphi\| = \|\varphi_\mu\|_\infty.$$

The function φ_μ is called a minifunction for H_φ. In general φ_μ is not unique. We however have

2.4.7 THEOREM. *The Hankel operator* H_S *determines* S *uniquely. Indeed,* S *is its unique minifunction.*

Proof. From Lemma 2.4.4 we note that $f \in \operatorname{Ker} T_S$ is an eigenvector of $H_S^* H_S$ corresponding to the eigenvalue $\|H_S\| = 1$. Since $S = \chi \Lambda \Gamma^{-1}$ every column of Γ belongs to this kernel. Thus, the projection of the above eigenspace on the first coordinate in $l_2(0,\infty;(C^p))$ spans $\Gamma(0)$. Now observe that for $\Gamma(0)$ we have, because of its outer property in $H2(B(C^p))$,

$$\log \frac{|\det \Gamma(0)|}{(2\pi)^{p/2}} = \frac{1}{4\pi} \int_{-\pi}^{\pi} \log \det f_{\underline{yy}}(\lambda)\, d\lambda > -\infty ,$$

so that $\Gamma(0)$ is of full rank. We conclude that the aforementioned projection is *onto* the firtst coordinate space. According to a result of Adamjan-Arov-Krein [2, Corollary 3.1] for Hankel operator H_ϕ to have a unique minifunction, it is sufficient that the projection of the eigenspace of $H_\phi^* H_\phi$ corresponding to $\|H_\phi\|$ on the first coordinate space be onto. The result follows.

2.4.8 REMARK. It is of interest to observe that since for a completely non-deterministic process, the eigenvectors of $H_S^* H_S$ corresponding to the eigenvalue $\|H_S\|$ are only the columns of Γ, the projection of this eigenspace on the first coordinate is not only onto, but also 1-1. In [2, Sec. 2] it is shown that for any Hankel operator $H : H_2(C^p) \to H_2^-(C^p)$ satisfying this condition its unique minifunction is of the form ρS, $\rho = \|H\|$.

Thus, up to a constant multiple $\rho > 0$ all minifunctions of such Hankel operators are in 1-1 correspondence with regular, full rank, completely nondeterministic processes.

The case of Rational Functions. Let $\{y(n)\}_{-\infty}^{\infty}$ have rational density

$$f_{\underline{yy}}(\lambda) = \frac{|P(z)|^2}{|Q(z)|^2} \qquad z = e^{i\lambda},$$

where the polynomials P,Q have no zeros in $|z| < 1$ and are relatively prime. Since $f_{\underline{yy}} \in L_1$, the polynomial Q has its zeros in $|z| > 1$. Write

$$P = P_1 P_2$$

where P_1 of degree k has its zeros on T and P_2 in $|z| > 1$. For $P_1(z) = \prod_{j=1}^{k} (z-\alpha_j)$, $\{\alpha_j\}_{j=1}^{k} \subset T$ we have

$$\frac{\overline{P_1(e^{i\lambda})}}{P_1(e^{i\lambda})} = e^{-ik\lambda}(-1)^k \prod_{j=1}^{k} \bar{\alpha}_j .$$

Thus

$$(2.4.6) \qquad S = \gamma\bar{\chi}^{k+1} \frac{\bar{\psi}_c}{\psi_c} \qquad \psi_c = \frac{P_2}{Q}, \qquad \gamma = (-1)^k \prod_{j=1}^{k} \bar{\alpha}_j ,$$

where $|\gamma| = 1$ and ψ_c is outer. In [3] Adamjan-Arov-Krein show that (2.4.6) is the general form of unimodular minifunctions and that in this case k+1 is the dimension of the eigenspace corresponding to the singular value $1 = \|H_S\|$. *We conclude that a regular process with rational spectral density is completely nondeterministic iff it has no zeros on T.*

2.5 MARKOV PROCESSES AND UNITARY DILATIONS

In this section we show how a Markov structure is intrinsically associated with unitary dilations and the resulting Scattering theory of Lax and Phillips. The results in this section are due to Lindquist and Picci [18] and Foias and Frahzo

[10]. In some sense however the results of this section are essentially contained in Adamjan-Arov [1]. This is demonstrated in this section.

In a Hilbert space setting, a centered stationary process $\{\underline{x}(n)\}_{-\infty}^{\infty}$ is said to be Markov if for all $n \geq s$

$$P_{H_2^-(s)} P_{X(s)} h \qquad h \in H_2^+(n) \quad ,$$

where $X(s) = \text{span}\{x_j(s): j=1,\ldots, m\}$. In our setting, all stationary processes will be generated by the shift U (on $H_{\underline{y}}$) associated with the $\underline{y}$ process. Thus, for a stationary process $\{\underline{x}(n)\}_{-\infty}^{\infty}$ (in $H_{\underline{y}}$) we will have $\underline{x}(n) = U^n \underline{x}(0)$. It readily follows from above that one can define the notion of a Markov subspace $X \subset H_{\underline{y}}$ (for U) if X satisfies (see [18]) $\underline{y}$

$$(2.5.1) \qquad P_{\mathop{\vee}\limits_{-\infty}^{s} U^m X} U^n x = P_{U^s X} U^n x \quad , \qquad n \geq s,\ x \in X \quad .$$

Thus X is a Markov subspace (for U) iff the process $\{U^n X\}$ has the (weak) Markov property. In what follows a Markov process $\{U^n X\}$ will invariably arise in this fashion.

Markov subspaces $X \subset H_{\underline{y}}$ which are *representations* for the process $\underline{y}$, i.e., for which

$$\{y_1(0), \ldots, y_p(0)\} \subset X \ ,$$

satisfy

$$H_{\underline{y}} = \mathop{\vee}\limits_{-\infty}^{\infty} U^n X \quad ,$$

and are said to be of full range. There is a direct relationship between Markov processes of full range and unitary dilations (see also [10]). Recall [19] that a unitary operator U on a Hilbert space H is said to be the minimal unitary (power) dilation of a contraction A on $X \subset H$ if

$$A^n = P_X U^n | X \qquad n \geq 0 \text{ and } H = \mathop{\vee}\limits_{-\infty}^{\infty} U^n X \text{ (minimality).}$$

2.5.1 PROPOSITION. $X \subset H_{\underline{y}}$ *is a Markov subspace of full range iff* U *(on* $H_{\underline{y}}$*) is the minimal unitary (power) dilation of the state operator*

$$A = P_X U | X : X \to X$$

Proof. From (2.5.1) we obtain for $x, x' \in X$ and $m,n \geq 0$

$$(U^{-m}x, U^n x') = (U^{-m}x, P_X U^n x).$$

Denoting $A(n) = P_X U^n | X$, we obtain

$$(x, A(m+n)x') = (x, U^{m+n}x') = (U^{-m}x, U^n x') = (U^{-m}x, P_X U^n x')$$

$$= (x, P_X U^{-m} P_X U^n x') = (x, A(m)A(n)x') .$$

We infer that $A(m+n) = A(m)A(n)$ and

$$A(n) = A^n(1) = A^n .$$

Since X is of full range, we conclude that U in $H_{\underline{y}}$ is the minimal unitary dilation of A (in X). This proves the 'only if' part. The 'if' part follows by reversing the argument.

Having made the connection between a Markov process $\{U^n K\}$ and the dilation property characterizing it, the work of Adamjan-Arov [1] on the duality between dilation theory and the scattering operator model is directly applied. First, note that the process $\{U^n X\}$ is regular, i.e., satisfies

$$\bigwedge_{n \geq 0} \bigvee_{k \leq n} U^k X = \{0\} = \bigwedge_{n \geq 0} \bigvee_{k \geq n} U^k X$$

iff

$$A^n \to 0 \quad , \quad A^{*n} \to 0 \quad (n \to \infty) .$$

Second, those Markov processes which in addition to being regular represent $\underline{y}$ (and are thus of full range) correspond to scattering systems according to a result of Adamjan-Arov [1, Th. 3.4] :

2.5.2 THEOREM. *Let* $X \subset H_{\underline{y}}$ *be a regular Markov subspace of full range. Then* $H_{\underline{y}}$ *decomposes and, moreover, uniquely into the orthogonal sum*

$$H_{\underline{y}} = D_- \oplus X \oplus D_+,$$

where $(U, H_{\underline{y}}, D_+, D_-)$ *is a scattering system.*

2.5.3 DEFINITION. A scattering system $(U,H,D_+,D_-)_X$ for which

$$D_- \subset D_+$$

is called a Lax-Phillips (L-P) scattering system.

Let $\{U^n X\}$ be an arbitrary regular Markov process of full range, and $(U, H_{\underline{y}}, D_+, D_-)_X$ its associated L-P scattering system. Let $\Theta_X(e^{i\lambda})$ be the corresponding scattering matrix. For the induced incoming spectral representation F^-_X we obtain

$$F^-_X [D_-] = H^-_2(C^p) \quad , \quad F^-_X [D_-] = \Theta_X H^2(C^p) .$$

Since $D_+ \perp D_-$

$$\Theta_X \in H_\infty(C^p)) .$$

To each regular full range Markov process there is thus associated an inner function Θ_X, which is the scattering matrix of the corresponding L-P system $(U, H_{\underline{y}}, D_+, D_-)_X$. From [1, Th. 3.3] it follows that the scattering matrices Θ associated with *regular full range Markov processes* are precisely the inner functions $\Theta \in H_\infty B(C^p))$ which are purely contractive [19, p. 188], i.e.; for which

$$\|\Theta(0)\| < 1 \quad .$$

2.6 FACTORIZATION OF THE SCATTERING MATRIX AND MODELLING

The 1-1 correspondence

$$X \leftrightarrow (U,\ H_{\underline{y}}\ ,\ D_+,\ D_-)_X$$

enables us to translate the *realization problem of finding all regular Markovian representaions* for y to a covering problem in $L_2(C^p)$ via the outgoing spectral representation for $(U,\ H_{\underline{y}}\ ,D_+,D_-)$. Let $X \subset H_{\underline{y}}$ be a regular Markov subspace reprsenting y, $(U,\ H_{\underline{y}}\ ,D_+,D_-)_X$ its L-P scattering system, ΘX its scattering matrix, and F^+_X the corresponding outgoing spectral representation. Since $\{y_1(0),...,y_p)0)\} \subset X$ it follows

(2.6.1) $$F^+_X\,[H^-_y(0)] \in F^+_X\left[\bigvee_{k\le n} U^n X\right] = F^+_X\,[D_- \oplus X] = H^-_2(C^p)\ ,$$

and $F^+_X\,[H^-_y(0)]$ is a full range left shift invariant subspace of $H^-_2(C^p)$. Let V be the corresponding inner function obtained from the Beurling-Lax theorem, i.e.,

$$F^+_X\,[H^-_y(0)] = V^* H^-_2(C^p)\ .$$

Be the unitary of F^+_X

$$F^+_X(y(0)) = \bar{\chi} V^* \Lambda\ .$$

Thus

$$\{y_1(0)\ ,...,\ y_p(0)\} \subset X$$

translate under F_X^+ to the equivalent condition

$$(2.6.2) \qquad \bar{\chi}V^*\Lambda \in H_2^-(C^p) \ominus \Theta^* H_2^-(C^p) \, .$$

Conversely, if Θ, V are inner functions for which (2.6.2) holds, the mapping

$$y(0) \rightarrow \bar{\chi}V^*\Lambda$$

induces in a natural fashion a spectral representation $F_{\Theta,V}$ for which

$$X = F_{\Theta,V}^{-1} \, [H_2^-(C^p) \ominus \Theta^* H_2^-(C^p)\,]$$

is a Markovian representation for y, and its corresponding L-P scattering system has its scattering matrix Θ_X coinciding with Θ. We have therefore proved the following

2.6.1 THEOREM. *Finding all models (realizations) of* y *is equivalent to finding all inner functions* Θ_1 *such that*

$$\bar{\chi}V^*\Lambda \in H_2^-(C^p) \ominus \Theta_1^* H_2^-(C^p)$$

for some inner function V. *Each model corresponds to a pair* (Θ_1, V).

2.6.2 COROLLARY. *All regular Markovian representations of* y *are paramterized by precisely those inner functions* Θ_1 *for which*

$$(2.6.3) \qquad V^*S = \Theta_1^*\Theta_2 \qquad \Theta_2 \quad H_\infty(B(C^p)\,)$$

for some inner function V.

Proof. (2.6.2) holds iff $\Theta_1\bar{\chi}\Lambda \in H_2(C^p)$ iff $\Theta_1 V^* S\Gamma \in H_2(C^p)$. Since Γ is outer the latter holds iff $\Theta V^* S \in H_\infty(B(C^p))$, i.e., iff (6.3) holds.

2.6.3 COROLLARY. *All regular Markov subspaces* $X \subset H_y^-(0)$ *representing* y *are parametrized by those and only those inner functions* Θ_1 *for which*

$$(2.6.4) \qquad S = \Theta_1^* \Theta_2 \qquad \Theta_2 \in H_\infty(B(C^p)).$$

Proof. $X \in H_y^-(0)$ implies $D_+ \oplus X = \bigvee_{n \le 0} U^n X \subset H_y^-(0)$ combining with (6.1) we conclude that V is a constant unitary matrix.

The possibility of writing the scattering matrix X in the form (2.6.4) has an interpretation on a process level. By the Beurling-Lax theorem, (2.6.4) holds iff (the invariant subspace for the left shift):

$$H_2^-(C^p) \ominus (\text{range } H_S) \text{ is of full range (for } \chi)$$

which is equivalent to

$$(2.6.5) \qquad H_y^-(0) \ominus P_{H_y^-(0)} H_y^+(0) \qquad \text{is of full range (for U).}$$

A^n $L_\infty(B(C^p))$ function satisfying (2.6.4) is called [9] *strictly non-cyclic* , and the corresponding process - having a strictly non-cyclic scattering matrix - is called strictly monocyclic. We thus obtain [18, Lemma 7.3 and Th. 7.6]:

2.6.4 COROLLARY. *Let*

$$S = Q_1^* Q_2 = P_2 P_1^*$$

be respectively the left, right coprime factorization of S. *Then all minimal regular Markov subspaces representing* y*are parametrized by those and only those inner functions* Θ_1 *such that*

$$(2.6.6.) \qquad V^* S = \Theta_1^* \Theta_2$$

where Θ_1, Θ_2 *are left coprime and* V *is an arbitrary left divisor of* P_2. *Moreover we have*

$$\det \Theta_1 = \det Q_1 .$$

Proof. Combine Corollary (2.6.2) with [14, Lemma III. 5-8].

The general Fuhrmann degree theory for strictly non-cyclic functions [12, Ch. iii.5] now arises naturally - S playing a central role. All regular Markovian subspaces $X \subset H_{\underline{y}}$ representing y are parametrized by an inner function $\Theta_1 \in H_\infty(B(C^p))$ will be of degree

$$d(\Theta_1) = \det \Theta_1 ,$$

an inner function in H_∞, and

$$d(Q_1) \text{ divides } d(\Theta_1).$$

Thus, the degree of the minimal subspace is the lowest, in the sense that $d(Q_1)$ is the weakest among the degrees of all other regular Markovian subspaces representing y. Applying [12,Th. II. 14.11] we infer that two minimal regular Markov subspaces representing y are quasi-similar.

2.6.5 COROLLARY. *For a* $p = 1$ *dimensional process* y *the minimal Markovian representation* of y *is parametrized by the inner function* q_1 *for which*

$$s = \bar{q}_1 q_2$$

is a coprime factorization.

III. UNITARY DILATIONS OF IRREVERSIBLE EVOLUTIONS

In the previous section we have shown how we can associate a Markov semigroup with a Gaussian process via the associated Scattering System. In this section, we consider the dual view of associating a Unitary Dilation with a Markov semigroup.

The evolution of a Hamiltonian (conservative) system is reversible while the evolution of real physical system is not. The real system returns to a state of thermal equilibrium at a temperature determined by its surroundings. This is the physical interpretation of the Poincaré Recurrence Theorem. We may argue that the construction of a unitary dilation of a Markov semigroup is the abstract interpretation of coupling a physical system of a finite number of degrees of freedom to a heat bath thereby producing a Hamiltonian system of infinite number of degrees of freedom. The ideas of this section are due to Ford, Kac, Mazur [11], Lewis and Thomas [17], Evans and Lewis [9] and the first author.

We undertake the development of this section in continuous time since this is its natural setting.

3.1 NOTATION AND PRELIMINARIES

Let H be a separable Hilbert space with scalar product, $< \cdot , \cdot >_H$ and norm $\| \cdot \|_H$. When there is no confusion the subscript H will be dropped. **R** denotes the set of real numbers , $\mathbf{R}^+$ the non-negative real numbers and $\mathbf{R}^-$ the nonpositive real numbers. For $I \subset \mathbf{R}$, an interval, $L^2(I)$ denotes the space of Borel measurable square-integrable complex-valued functions and $L^2(I;H)$ the space of Borel-measurable H-valued square integrable functions. $W^{1,2}(\mathbf{R};H)$ denotes the Sobolev space of H-valued functions and $W^{1,2}(I;H)$ the Sobolev space obtained by restriction of $W^{1,2}(\mathbf{R};H)$. Let $(T_t)_{t\in \mathbf{R}^+}$ denote a one-parameter, strongly continuous, contractive semigroup on H, with $T_0 = I$, and let $-A$ be the infinitesimal generator of $(T_t)_{t\in \mathbf{R}^+}$. We know $T_t := e^{-tA}$ is a contractive semi-group iff, $\forall\, x \in D(-A)$, $\exists\, x^* \in H$ with $\|x^*\| = 1$, $(x^*, x) = \|x\|$ and $\mathrm{Re}\, <x^*, Ax> \geq 0$.

A semigroup of contractions $(T_t)_{t\in \mathbf{R}^+}$ is said to contract strongly to zero if $\forall\, h \in H$, we have $\lim_{t\to\infty} \|T_t h\| = 0$.

3.2 MINIMAL UNITARY DILATION OF A CONTRACTIVE SEMIGROUP

3.2.1 DEFINITION. Given a contractive semi-group $(T_t)_{t\in \mathbf{R}^+}$ on H, we say that a strongly continuous one-parameter group $(U_t)_{t\in \mathbf{R}}$ on $\mathcal{H}$ is a unitary dilation of (T_t, H) if there exists an isometry $i: H \to \mathcal{H}$ such that the following diagram commutes:

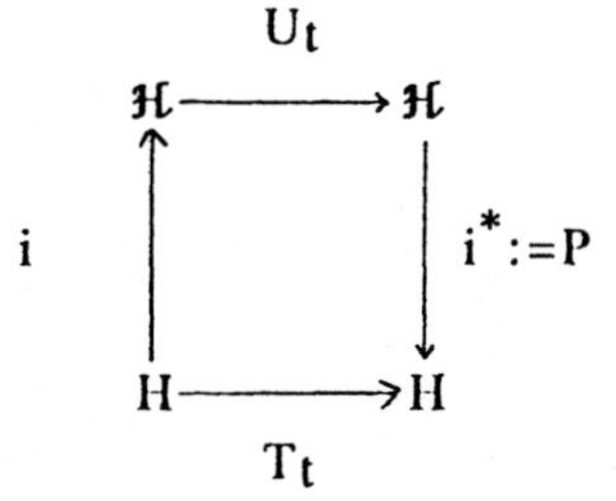

The unitary dilation is said to be minimal if

$$\mathcal{H} = \overline{\cup \left\{ U_t(iH) \mid t \in \mathbf{R} \right\}} \ .$$

3.2.2 THEOREM [LAX-PHILLIPS]. *Let* (T_t, H) *be a strongly continuous, contractive, semigroup contracting strongly to zero. Then there exists a unitary dilation* $(U_t, \mathcal{H})$. *The dilation* $(U_t, \mathcal{H})$ *has the canonical representation with* $\mathcal{H} = L^2(\mathbf{R}; N)$, N *a Hilbert space and* $(U_t | t \in \mathbf{R})$ *being the unitary group of right translations on* $L^2(\mathbf{R}; N)$:

(3.2.1) $$(U_t f)(s) = f(s-t).$$

Proof. Since $(T_t)_{t \in R^+}$ is a contraction,

(3.2.2) $$Q(h) < Ah, h > + < h, Ah > \geq 0 \ \forall \ h \in D(A).$$

Let $N_0 = \mathrm{Ker}[Q(h)]$ and let P be the canonical projection of D(A) onto the quotient space $D(A)/N_0$. On $D(A)/N_0$ there exists a scalar product $< \cdot \, , \cdot >_A$ such that

(3.2.3) $$< Ph, Pk >_A = < Ah, k > + < k, Ah > \ , \ \forall \ h,k \in D(A).$$

Let N denote the Hilbert space completion of $D(A)/N_0$ with respect to the norm induced by (3.2.3). Therefore

(3.2.4) $$\int_{-t}^{0} \|PT_{-s}h\|_A^2 ds = \|h\|^2 - \|T_t h\|^2, \ \forall \ h \in D(A) \ t \geq 0.$$

If we let $t \to \infty$, since T_t contracts strongly to zero, there exists an isometric embedding $i: H \to L^2(\mathbf{R}; N)$, such that on D(A),

$$(ih)(s) = PT_{-s}h \ , \ \forall \ s \leq 0.$$

Regarding $L^2(\mathbf{R}^-; N)$ as a subspace of $L^2(\mathbf{R}; N)$, we have for $\forall \ h \in D(A)$ and $t \geq 0$

$$(U_t ih)(s) = \begin{cases} PT_{t-s}h & s \leq t \\ 0 & s > t \end{cases}$$

$$= (iT_t h)(s) + \eta_t(s)$$

where $\eta_t(s) \in L^2(\mathbf{R}^+; N) \subset i(H)^{\perp}$. Hence, $\forall \ t \leq 0$

$T_t = i^* U_t i$, and therefore U_t is a unitary dilation of T_t on $\mathcal{H} = L^2(\mathbf{R}; N)$.

The unitary dilation we have constructed is in fact minimal. This is done by constructing a linear stochastic differential equation involving an operator-valued Brownian motion. We first introduce positive definite kernels and consider their decomposition.

3.2.3 DEFINITION. A map $K: \mathbf{R} \times \mathbf{R} \to B(H)$ is said to be a kernel. The set of all such maps is denoted by $K(\mathbf{R}; H)$. A kernel K is said to be positive definite if $\forall\ h_1, \ldots, h_n$ in H and $x_1, \ldots, x_n$ in $\mathbf{R}$.

$$\text{(3.2.5)} \qquad \sum_{i,j=1}^{n} < K(x_i, x_j)h_j, h_i > \geq 0.$$

3.2.4 DEFINITION. Let $K \in K(\mathbf{R}; H)$. Let H' be a Hilbert space and let $V: \mathbf{R} \to B(H;H')$ be such that $K(x,y) = V(x)^* V(y)$. Then V is said to be a Kolmogoroff decomposition of K. This decomposition is minimal if $H' = \cup\{V(x)h | x \in \mathbf{R}, h \in H\}$. One can prove that every positive definite kernel has a minimal Kolmogoroff decomposition. This is done with the aid of the reproducing kernel Hilbert space associated with K.

With the notation of Theorem 3.2.2, let us introduce an operator-valued Brownian motion as follows:

Let $W: \mathbf{R} \to B(N; H)$ be the map given by

$$\text{(3.2.6)} \qquad (W_t \eta)(s) = \begin{cases} \chi_{[0,t]}(s)\eta, & t \geq 0 \\ -\chi_{[t,0]}(s)\eta, & t < 0, \end{cases}$$

where $\eta \in N$ and $\chi(.)$ is the characteristic function.

Consider the positive-definite kernel:

$(s,t) \to (s \wedge t)I_N$, where I denotes the identity operator. Then

$$(s \wedge t)I_N = W_t^* W_s.$$

In the sequel we denote by $(D(A), |.|)$ the Hilbert space $D(A)$, with the graph norm.

3.2.5 THEOREM. *Let* $(U_t, \mathcal{H})$ *be the dilation of* (T_t, H) *given in Theorem* 3.2.2. *Then there exists a bounded linear operator* B: (D(A), |.|) $\rightarrow$ N *and an operator-valued Brownian motion* $W_t : R \rightarrow B(N; M)$, *where*

$$M = \cup\{W_s\eta \mid s \in \mathbf{R}, \eta \in N\} \text{ and } W_t \textit{ satisfies } (3.2.6) \textit{ such that}$$

$$(3.2.7) \qquad (U_t i - U_s i)h = -\int_s^t U_r i A h dr + (W_t - W_s)Bh, \ \forall\ h \in D(A).$$

Proof. The proof is constructed by verifying equation (3.2.7) for $h \in D(A^2)$ and then by density for $h \in D(A)$. For $h \in D(A^2)$ one can show that a solution is given by

$$U_t i h = e^{-A(t-s)} U_s i h + \int_s^t W(dr) B e^{-A(t-r)} h$$

where the last term is a Wiener integral, which can be defined by an integration by parts formula. The fact that U_t is a minimal unitary dilation follows from the fact that w_t is a minimal Kolmogoroff decomposition.

3.2.6 REMARK. The stationary solution of the equation is given by

$$U_t i h = \int_{-\infty}^t W(ds) B e^{-A(t-s)} h.$$

We may verify that this U_t defines a regular stationary Gaussian process and there is a Lax-Phillips structure associated with it. We may also obtain an ordinary stochastic differential equation for the Markov semigroup attached to this Lax-Phillips system.

3.2.7 A NEW REPRESENTATION OF THE DILATION.

Let us assume that the semigroup $(T_t)_{t \in R^+}$ on H is self-adjoint with generator -A. Then A is a positive self-adjoint operator which we assume to be injective. In this case. one can show that there exists a minimal unitary dilation $(U_t, \mathcal{H})$, where $\mathcal{H} = H \oplus L^2(\mathbf{R}; H)$.

Let us write a vector $\varphi \in L^2(\mathbf{R}; H)$ as $\varphi = \varphi^+ + \varphi^-$ with $\varphi^+ \in L^2(\mathbf{R}^+; H)$ and $\varphi^- \in L^2(\mathbf{R}^-; H)$. Then one can write the unitary dilation for $t \in \mathbf{R}^+$ as

$$(3.2.8) \qquad U_t = \begin{pmatrix} T_t & \mathcal{A}_t \\ B_t & S_t + C_t \end{pmatrix}, \text{ where}$$

$\mathcal{A}_t$: $L^2(\mathbf{R}; D(A^{1/2})) \to H$: $\varphi \to (2A)^{1/2} \int_0^t T_{t-s}\varphi(s)ds$

B_t: $D(A^{1/2}) \to L^2(\mathbf{R}; H)$: $h \to (B_t h)(s) = \chi_{[0,t]}(s)(-2A)^{1/2}T_{t-s}h$

C_t: $L^2(\mathbf{R}; D(A)) \to L^2(\mathbf{R}; H)$: $\varphi \to (C_t\varphi)(s) = \chi_{[0,t]}(s)(-2A)^{1/2}\mathcal{A}_{t-s}\varphi$

S_t: $L^2(\mathbf{R}; H) \to L^2(\mathbf{R}; H)$: $\varphi \to (S_t\varphi)(s) = \varphi(s-t)$

$\mathcal{A}_t$, B_t, C_t are densely defined contractions.

Moreover writing $U_t = e^{it\mathcal{K}}$, on physical grounds the Hamiltonian $\mathcal{K}$ can be written as

(3.2.9) $$\mathcal{K} = \mathcal{K}_s \oplus \mathcal{K}_c \oplus \mathcal{K}_R$$

$\mathcal{K}_R$ the Hamiltonian of the reservoir is the generator of the shift of Brownian motion. $\mathcal{K}_s$, the Hamiltonian of the system is zero. $\mathcal{K}_c$, the Hamiltonian of the coupling is of the form

(3.2.10) $$\mathcal{K}_c = \begin{pmatrix} 0 & -iC^* \\ iC & 0 \end{pmatrix}$$

where C: $D(A^{1/2}) \to L^2(\mathbf{R}; H)$: $h \to \delta_0 \otimes (2A)^{1/2}h$, δ_0 being the Dirac mass (this is formal and needs to be justified) and

$$C^*: L^2(\mathbf{R}; H) \to D(A^{1/2}): \varphi \to (2A)^{1/2}\varphi(o)$$

(this is also formal).

We now give some indications on how the new construction is arrived at. Since $(T_t)_{t\in \mathbf{R}+}$ is contractive, the quadratic form $F(x,x) = (Ax,x) + (x,Ax) = \lim_{\varepsilon\downarrow 0} \frac{\|x\|^2-\|T_\varepsilon x\|^2}{\varepsilon} \le 0$ for $x \in D(A)$. We claim that there exists an operator $C : D(A) \to H_1$ where H_1 is a Hilbert space equal to $\overline{CD(A)}$, such that $\|Cx\|^2 = F(x,x)$ $\forall$ $x\in D(A)$. In a similar manner there exists a pair (C', H_2) for $t\in \mathbf{R}_-$, such that $\|C'\|^2 = (A^*x,x) + (x,A^*x)$ $\forall$ $x\in D(A^*)$. The operators C and C' are to be thought of as coupling operators. In the self-adjoint case $C = (2A)^{1/2}$.

The idea is to construct the dilation on the space $\mathcal{H} = L^2(\mathbf{R}_-;H_1) \oplus H \oplus L^2(\mathbf{R}_+; H_2)$. Now since (T_t, H) is a Markov semigroup, we must have $(U_s \circ P_0^\perp \circ U_t) \circ i(\xi)$ is orthogonal to $i(H)$ where $i : H \to \mathcal{H}$ is the injection $\xi \to \begin{pmatrix} 0 \\ \xi \\ 0 \end{pmatrix}$, $P_0 : \mathcal{H} \to H$ is the orthogonal projection and $P_0^{\perp}$ is the orthogonal projection on the orthogonal complement of H.

This suggests picturing the unitary dilation as follows:

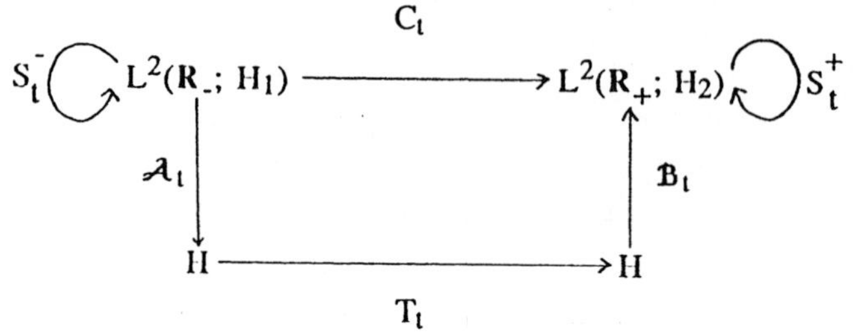

The operators $\mathcal{A}_t^*$ and $\mathcal{B}_t$ "couple" (T_t, H) to $(S_t^-, L^2(R_-; H_1))$ and $(S_t^+, L^2(R_+; H_2))$ where S_t^- and S_t^+ are right shifts. For example $\mathcal{B}_t$ is given by

$$(B_t\xi)(s) = \begin{cases} -C+T_{t-s}\xi & \text{if } s \in [0,t] \\ 0 & \text{if } s \notin [0,t]) \end{cases}, \xi \in D(A) \quad .$$

It can be shown that $\mathcal{B}_t$ is a contraction. In the self-adjoint case there is a simplification and it is enough to couple H to $L^2(\mathbf{R}; H)$ and we try to give an intuitive justification of (3.2.8).

In a physical setting the shifts will correspond to the random behaviour of the heat bath and will be the flow of Brownian motion. We expect the coupling between the system and the heat bath to be instantaneous and this coupling will take place via the coupling operator $(2A)_{1/2}$. For $t \geq 0$, we therefore expect that a vector $\begin{pmatrix} x \\ 0 \end{pmatrix} \in \begin{pmatrix} D(A) \\ L^2(\mathbf{R}_+;H) \end{pmatrix}$ to be transformed into $\begin{pmatrix} -Axdt \\ db_t \otimes (-(2A)^{1/2}x) \end{pmatrix}$ in time dt, where b_t denotes standard Brownian motion and $db_t \otimes (-(2A)^{1/2}x)$ is an element of $L^2(\mathbf{R}^+; H) = L^2(\mathbf{R}^+) \otimes H$ (tensor product). The second component $db_t \otimes (-(2A)^{1/2}x)$ in integrated form is essentially $\mathcal{B}_t$ in (3.2.8).

Finally, we can explain the form of of K_s, K_R and K_c on physical grounds. Since the time evolution on H is self-adjoint, it does not contain a unitary part and we expect H_s to be zero. The fact that K_c should be of the form (3.2.10) follows from the same argument given above.

REFERENCES

1. Adamjan, V.M. and Arov, D.Z.: On unitary couplings of semi-unitary operators, Amer. Math. Soc. Transl. 95(2)(1970), 75-129.
2. Adamjan, V.M., Arov, D.Z., and Krein, M.G.: Infinite Hankel block matrices and related extension problems, Amer. Math. Soc. Transl. 3(2)(1978), 133-156.
3. Adamjan, V.M., Arov, D.Z., and Krein, M.G.: Infinite Hankel matrices and generalized Caratheodory-Fejer and Riesz problems, Functional Anal. Appl. 2. (1968), 1-18.
4. Adamjan, V.M., Arov, D.Z., and Krein, M.G.: Analytic properties of Schmidt pairs for a Hankel operator and the generalized Schur-Takagi problem, Mat. Sb. 86(128)(1971), 34-75; Math. U.S.S.R. Sb. (1971), 31-73.
5. Avniel, Y.: Realization and approximation of stationary stochastic processes, Report LIDS-TH-1440, Laboratory for Information and Decision System, MIT, Cambridge, MA., February 1985.
6. Ball,J. and Helton, W.: Interpolation problems of Pick-Nevanlinna and Loewner types for meromorphic matrix functions: parametrization of the set of all solutions, Integral Equations and Operator Theory, 9, 1986, 155-203.
7. Bloomfield, P.B., Jewell, N.P., and Hayashi, E.: Characterizations of completely nondeterministic stochastic processes, Pacific J. of Math. 107 (1983), 307-317.
8. Davies, E.B.: Quantum Theory of Open Systems, Academic Press, New York, 1976.
9. Evans, D.E. and Lewis,J.T.: Dilations of irreversible evolutions in algebraic quantum theory, Comm. of Dublin Institute for Advanced Studies, No. 24, 1977.
10. Foias, C. and Frazho, E.: A note on unitary dilation theory and state spaces, Acta Sci. Math. 45 (1983), 165-175.
11. Ford, G.W., Kac, M., and Mazur, P.: Statistical mechanics of assemblies of coupled oscillators, J. of Math. Physics, 6, 1965, 504-515.
12. Fuhrmann, P.A.: Linear Systems and Operators in Hilbert Space. McGraw-Hill, New York, 1981.
13. Glover, K.: All optimal Hankel-norm approximations of linear multi-variable systems and their L^∞-error bounds, Int.J. of Control, 39, 1984, 1115-1193.
14. Helson, H.: Lectures on Invariant Subspaces. Academic Press, New York, 1964.

15. Lax, P.D. and Phillips, R.S.: Scattering Theory. Academic Press, New York, 1967.

16. Levinson, N. and McKean, H.P.: Weighted trigonometrical approximations on $\mathbf{R}^1$ with applications to the Germ field of stationary Gaussian Noise, Acta Math., 112, 1964, 99-143.

17. Lewis, J.T. and Thomas, L.C.: How to make a heat bath, Functional Integration, ed. A.M.Arthurs, Oxford, Clarendon Press 1974.

18. Lindquist, A. and Picci, G.: Realization theory for multivariate stationary Gaussian processes, SIAM J. Control and Optimization 23 (1985), 809-857.

19. Sz-Nagy, B. and Foias, C.: Harmonic Analysis of Operators on Hilbert Space. Amsterdam, North-Holland, 1970.

20. Picci, G.: Application of Stochastic Realization Theory to a Fundamental Problem of Statistical Physics, in Modelling, Identification and Robust Control, (eds.: C.I. Byrnes and A. Lindquist), Elsevier Science Publishers B.V. (North-Holland), 1986.

21. Rozanov, Y.A.: Stationary Random Processes. Holden-Day, San Francisco, 1963.

22. Willems, J.C. and Heij, C.: Scattering Theory and Approximation of Linear Systems, in Proceedings of the 7th International Symposium on the Mathematical Theory of Networks and Systems MTNS-85, June 10-14, 1985, Stockholm, North-Holalnd, Amsterdam, 1985.

23. Willems, J.C.: Models for Dynamics, to appear in Dynamics Reported.

24. Zames, G.: Private Communication.

On the Theory of Nonlinear Control Systems

H. Nijmeijer

Dept. of Applied Mathematics, University of Twente
P.O. Box 217, 7500 AE Enschede, the Netherlands

Abstract A review of some of the recent advances in nonlinear control theory is presented. A central theme in the discussion is the notion of state feedback and its use for altering the structural properties of a system. The relations between various synthesis problems, controlled invariance and a general decomposition problem are investigated.

1. Introduction

One of the most important new developments in nonlinear system theory in the last decade has been, without any doubt, the introduction of invariant and controlled invariant distributions, [20,21,25]. This (differential) geometric approach, as it is often called, provides a mathematically elegant and effective approach for solving various synthesis problems. In this regard, we mention the Disturbance Decoupling Problem and the Input-Output Decoupling Problem. Moreover, this approach provides a very clear and sophisticated theoretical picture for understanding nonlinear control systems, and therefore is valuable in the interpretation of classical system theoretic concepts like observability, accessibility (controllability) and invertibility.

Invariant and controlled invariant distributions in nonlinear system theory play the same role as - and in fact generalize - invariant and controlled invariant subspaces in linear system theory. For linear systems, the crucial features of such subspaces are well known; see, for instance, the geometric approach of Wonham [51] and the work of Basile and Marro [3]. In this approach the essential tools stem from linear algebra and involve linear mappings, subspaces and so on.

For general nonlinear systems, i.e. a system having its dynamics of the form $\dot{x} = f(x,u)$, various characterizations of controlled invariant distributions have been obtained, cf. [23,37,38], though certainly many open mathematical problems still remain.

From a control theoretic point of view the aforementioned concepts deeply involve the notion of state feedback. It is certainly one of the essential contributions of the past decade in nonlinear control that a more advanced theory of state variable feedback has been initiated; see for numerous examples, the conference proceedings [16,9]. We note that the most common type of feedback studied is static, though some interesting results with dynamic state feedback have also appeared recently. Another important topic in the theory of nonlinear control systems is that of decomposition of a system. The idea is to study how a nonlinear system can be decomposed, in a nontrivial manner, as the interconnection of lower dimensional systems, thereby reducing the control complexity of the original system. A basic contribution on the (cascade) decomposition of a nonlinear system is due to the Krener cf. [30], which is also reminiscent of the later introduced notion of controlled invariance. However, the analysis of system decomposition is still far from its completion.

The purpose of this paper is twofold. Firstly the main results about controlled invariance and its relations with static and dynamic state feedback problems are reviewed in sections 2 and 3. Secondly, in section 4, we formulate the general decomposition problem for a nonlinear system and discuss the interrelations with the dynamic state feedback problems of sections 2 and 3.

Acknowledgement This paper expresses the author's personal view on present day nonlinear control theory, and may not reflect the opinion of others. Since the time I was a graduate student, I have appreciated the continuing enthousiasm and inspiration from Jan C. Willems. His fiftieth birthday serves as an excellent occasion to thank him with this paper. I also want to express my gratitude to Arjan van der Schaft for the many discussions we have had on this and other papers.

2. Controlled invariance

2.1 Linear systems

First, we briefly review parts of the standard geometric theory for linear systems, cf. [3,51]. Consider the linear system:

$$\Sigma \quad : \quad \dot{x} = Ax + Bu \tag{2.1}$$

with $x \in \mathcal{X} = \mathbb{R}^n$, $u \in \mathcal{U} = \mathbb{R}^m$ and A and B matrices of appropriate sizes. A subspace $\mathcal{V} \subset \mathcal{X}$ is called **controlled invariant** if there exists a linear static state feedback

$$u = Fx + I_m v, \tag{2.2}$$

where $v \in \mathcal{U}$ is a new input signal, such that $\mathcal{V}$ is invariant for the closed-loop dynamics

$$\Sigma_F \quad : \quad \dot{x} = (A+BF)x + B\,v; \tag{2.3}$$

that is,

$$(A+BF)\,\mathcal{V} \subset \mathcal{V} \tag{2.4}$$

This is equivalent to the requirement

$$A\mathcal{V} \subset \mathcal{V} + \text{im } B \tag{2.5}$$

Choosing a basis $\{e_1, \ldots, e_k, e_{k+1}, \ldots, e_n\}$ for $\mathcal{X}$, such that $\{e_1, \ldots, e_k\}$ forms a basis for $\mathcal{V}$, the system Σ_F can be written as

$$\frac{d}{dt}\begin{pmatrix} x_1 \\ x_2 \end{pmatrix} = \begin{pmatrix} (A+BF)_{11} & (A+BF)_{12} \\ 0 & (A+BF)_{22} \end{pmatrix}\begin{pmatrix} x_1 \\ x_2 \end{pmatrix} + \begin{pmatrix} B_1 \\ B_2 \end{pmatrix} v \tag{2.6}$$

where x_1 and x_2 are k- and (n-k)- dimensional vectors respectively. Clearly, (2.6) yields a linear system "modulo $\mathcal{V}$" as

$$\dot{x}_2 = (A+BF)_{22}x_2 + B_2 v \tag{2.7}$$

In a more abstract way, the system (2.7) is a linear system, to be denoted as Σ_F, on the quotient space $\overline{\mathcal{X}} = \mathcal{X} \pmod{\mathcal{V}}$. Namely, letting $\Pi : \mathcal{X} \longrightarrow \overline{\mathcal{X}}$ be the projection along $\mathcal{V}$, (2.4) implies the existence of a linear mapping $\overline{A + BF} : \bar{\mathcal{X}} \longrightarrow \bar{\mathcal{X}}$ such that $\Pi(A+BF) = (\overline{A+BF})\Pi$; the quotient system $\bar{\Sigma}_F$ is defined as

$$\bar{\Sigma}_F \ : \dot{\bar{x}} = (\overline{A+BF})\bar{x} + \bar{B}\, v \tag{2.8}$$

where $\bar{B} = \Pi B$. Equation (2.7) is therefore nothing else than a matrix representation of (2.8). This explains how to obtain $\bar{\Sigma}_F$ from Σ, given the condition (2.4). To recover the system Σ from the quotient system $\bar{\Sigma}_F$ we need to add to (2.7) the dynamics

$$\dot{x}_1 = A_{11}x_1 + A_{12}x_2 + B_1\, u \tag{2.9a}$$

together with the "inverse" feedback law

$$v = -F_1x_1 - F_2x_2 + I_m\, u \tag{2.9b}$$

where we have written $F = [F_1 \vdots F_2]$ in the obvious way.
Note that the equations (2.9a,b) constitute, in fact, a dynamic state feedback. For an arbitrary system $\bar{\Sigma} : \dot{\bar{x}} = A\bar{x} + \bar{B}v$ a dynamic state feedback or precompensator is defined as

$$\mathcal{P} : \begin{cases} \dot{z} = Pz + Q\bar{x} + Ru \\ v = Sz + T\bar{x} + Uu \end{cases} \tag{2.10}$$

where $z \in \mathcal{Z} = \mathbb{R}^{\nu}$, $v \in \mathcal{U}$, and P, Q, R, S, T and U are matrices of appropriate sizes. The integer ν is the dimension of the precompensator $\mathcal{P}$. The system $\bar{\Sigma}$ in closed loop with the dynamic feedback $\mathcal{P}$ will be denoted as $\bar{\Sigma}\cdot\mathcal{P}$. The preceding analysis shows that the precompensator $\mathcal{P}$ defined via (2.9a,b) applied to the system (2.7) precisely reproduces the original dynamics (2.1). This may be represented as in Diagram 1.

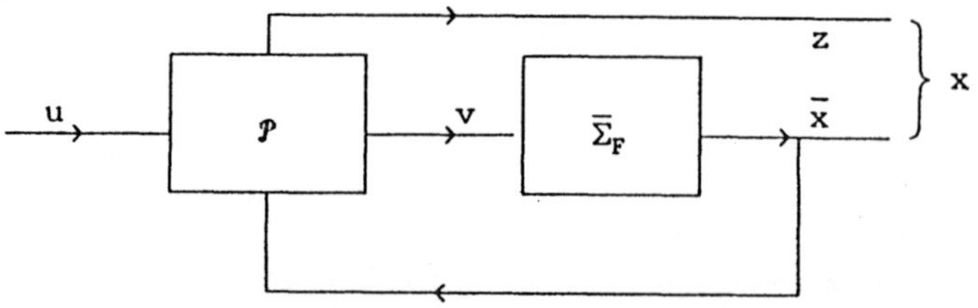

Diagram 1 : $\Sigma = \bar{\Sigma}\cdot\mathcal{P}$

In geometric linear system theory, decoupling problems are often phrased in terms of controlled invariant subspaces. For instance, in the Disturbance Decoupling Problem (DDP), one considers the system

$$\begin{cases} \dot{x} = Ax + Bu + Eq \\ y = Cx \end{cases} \tag{2.11}$$

where x and u are as before and $q \in \mathbb{Q} = \mathbb{R}^{\ell}$ and $y \in \mathbb{R}^{p}$ are the disturbances and outputs respectively. In the DDP, one searches for a feedback law (2.2) which isolates the disturbances from the outputs; the problem is solvable if, and only if, there exists a controlled invariant subspace $\mathcal{V}$ satisfying.

$$\text{im } E \subset \mathcal{V} \subset \ker C. \tag{2.12}$$

Provided (2.12) holds, a solution of the DDP is given by a feedback that satisfies (2.4). Notice that (2.12) in Diagram 1 implies that the disturbances only enter via $\mathcal{P}$ and that the output only depends upon $\bar{x}$.

Remark Usually one solves the DDP by searching for a feedback law $u = Fx$, rendering a solution without a reference input. As the solvability conditions are the same in both cases, we prefer to consider the class of feedbacks (2.3), because the new input v may be used to achieve further controller design goals.

2.2 Nonlinear systems

We now review some essentials from the differential geometric approach to nonlinear systems.

A nonlinear input-state system, or shortly control system Σ is a 3-tuple $\Sigma(M,B,f)$ where M is a manifold, B is a fiber bundle over M with projection $\pi : B \longrightarrow M$ and f is a smooth mapping such that Diagram 2 commutes, where π_M denotes the natural projection of TM on M.

$$\begin{array}{ccc} B & \xrightarrow{\quad f \quad} & TM \\ & \searrow^{\pi} \quad \swarrow_{\pi_M} & \\ & M & \end{array}$$

Diagram 2: A nonlinear system $\Sigma(M,B,f)$

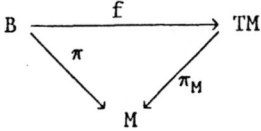

In this definition, M is considered as the state space of the system while the fibers of B represent the state-dependent input spaces; see [6,49,46,37] for the motivation of this definition. Because $\pi : B \longrightarrow M$ is a fiber bundle, we can locally identify B as the Cartesian product of the state space M and the fiber space U. Choosing local coordinates x for M and (x,u) for B where u denotes the coordinates for the fibers, this definition locally reduces to the equation

$$\dot{x} = f(x,u), \tag{2.13}$$

where we have abused notation by letting $(x,u) \longmapsto (x, f(x,u))$. In this framework, static state feedback can be defined as a bundle isomorphism $\alpha : B \longrightarrow B$, i.e., for each x in M, α maps the fiber $\pi^{-1}(x)$ diffeomorphically into itself. With the same abuse of notation as before, the feedback α is locally described as

$$u = \alpha(x,v). \tag{2.14}$$

Next, we introduce the notion of controlled invariance for a system $\Sigma(M,B,f)$. In what follows we use some standard terminology coming from differential geometry [4]. Let D be a **regular** distribution on M, i.e. D is smooth, constant dimensional and involutive; that is, for each x in M, D(x) is a linear subspace of T_xM depending smoothly on x, of constant dimension - say k - and for each pair of vector fields X_1 and X_2 on M with $X_1, X_2 \in D$ the Lie bracket $[X_1, X_2]$ belongs to D.

Let Σ be a nonlinear control system which is locally described as in (2.13). We say that the regular distribution D is **controlled invariant** if there exists a static state feedback α, which is locally given as in (2.14), such that the closed loop dynamics

$$\dot{x} = f(x,\alpha(x,v)) = \tilde{f}(x,v) \tag{2.15}$$

satisfies

$$[\tilde{f}(\cdot,v), D] \subset D \text{ for every constant } v \in U, \tag{2.16}$$

where $O \times U$ is a local trivialization of $\pi : B \longrightarrow M$. Here (2.16) means that $[\tilde{f}(,v), X] \in D$ for each vector field X belonging to D. The condition (2.16) implies, similarly to the linear case, see (2.6), a sort of local decomposition. This can be seen as follows. Locally around each point $x^0 \in M$ there exist local coordinates $x = (x^1,\ldots,x^n)^T$ such that (Frobenius, cf [4]).

$$D = \operatorname{span}\left\{\frac{\partial}{\partial x^1},\ldots\ldots,\frac{\partial}{\partial x^k}\right\}. \tag{2.17}$$

We obtain from (2.16) that the system (2.15) in these coordinates is of the form

$$\begin{cases} \dot{x}_1 = f_1(x_1, x_2, v) \\ \dot{x}_2 = f_2(x_2, v) \end{cases} \tag{2.18}$$

where $x_1 = (x^1,\ldots,x^k)^T$ and $x_2 = (x^{k+1},\ldots, x^n)^T$. At this point we emphasize that finding a suitable feedback α can be understood as the selection of properly chosen fiber respecting coordinates for the fiber bundle $\pi : B \longrightarrow M$.

A necessary and sufficient conditon for the local existence of a state feedback (2.14) defined around an arbitrary point in B yielding the invariance of the distribution D for the system (2.15) is that (see [38])

$$f_*(\pi_*^{-1}(D)) \subset \dot{D} + f_*(\Delta_0), \tag{2.19}$$

provided that the distribution $f_*(\Delta_0) \cap \dot{D}$ has constant dimension. Here, $f_* : TB \longrightarrow T(TM)$ is the mapping defined as $f_*(b,\xi) = (f(b), Df_b(\xi))$, $\pi_* : TB \longrightarrow TM$ is defined similarly, $\dot{D}$ is a regular distribution on TM which in the coordinates $(x^1,\ldots,x^n, \dot{x}^1,\ldots,\dot{x}^n)$ for TM has the form (compare with (2.17))

$$\dot{D} = \operatorname{span}\left\{\frac{\partial}{\partial x^1},\ldots,\frac{\partial}{\partial x^k},\frac{\partial}{\partial \dot{x}^1},\ldots,\frac{\partial}{\partial \dot{x}^k}\right\}$$

and finally Δ_0 is the distribution on B given by $\Delta_0 = \{X \in TB \mid \pi_* X = 0\}$. In general, (2.19) only assures the local existence of a static state feedback α; to guarantee that such an α can be globally defined, further assumptions are needed (cf. [8,23,37]).

Remark For an affine nonlinear control system

$$\dot{x} = f(x) + \sum_{i=1}^{m} g_i(x)u_i , \tag{2.20}$$

the condition (2.19) takes the more familiar form

$$\begin{cases} [f, D] \subset D + \operatorname{span}\{g_1,\ldots,g_m\} \\ [g_i, D] \subset D + \operatorname{span}\{g_1,\ldots,g_m\}, \ i = 1,\ldots m \end{cases} \tag{2.21}$$

provided that the distribution $D \cap \operatorname{span}\{g_1,\ldots,g_m\}$ has constant dimension.

Henceforth we consider a system which satisfies the condition (2.19) and so a suitably locally defined state feedback α brings the system into the form (2.18). As in the linear case, the local controlled invariance of the distribution D induces locally a nonlinear system on the "manifold" M(mod D), namely, see (2.18),

$$\bar{\Sigma}_\alpha : \dot{x}_2 = f_2(x_2, v) \tag{2.22}$$

(Note that M(mod D) locally forms a neighborhood in $\mathbb{R}^{n-k}$, but in general M(mod D) is not a Hausdorff manifold). This explains how we locally obtain from the system Σ given in (2.13), via a feedback (2.14), the quotient system $\bar{\Sigma}_\alpha$ given in (2.22). To recover (2.13) again from (2.22), we need to add the dynamics

$$\dot{x}_1 = f_1(x_1, x_2, u) \tag{2.23a}$$

together with the "inverse" feedback law

$$v = \alpha^{-1}(x_1, x_2, u) \ , \tag{2.23b}$$

where for each $x = (x_1, x_2)^T$, $\alpha^{-1}(x,\cdot)$ is the inverse of the mapping $\alpha(x,\cdot)$. We observe that the equations (2.23a,b) define (locally) a particular dynamic state feedback. In general, for a system locally described by $\dot{\bar{x}} = f(\bar{x}, v)$ a dynamic state feedback is given as

$$\mathcal{P} \quad \begin{cases} \dot{z} = p(z, \bar{x}, u) \\ v = \alpha(z, \bar{x}, u) \end{cases} \tag{2.24}$$

where the state of the precompensator z belongs to an open neighbourhood $Z \subset \mathbb{R}^\nu$. The system $\bar{\Sigma}$ together with the dynamic state feedback $\mathcal{P}$ will be denoted as $\bar{\Sigma}\cdot\mathcal{P}$. The foregoing analysis yields that the precompensator $\mathcal{P}$ defined via (2.23a,b) applied to (2.22) precisely reproduces (2.13) and thus Diagram 1 is locally valid in this nonlinear situation too.

Regular distributions that are (locally) controlled invariant play an essential role in nonlinear synthesis problems. As an example we briefly discuss the local nonlinear Disturbance Decoupling Problem.

Let

$$\begin{cases} \dot{x} = f(x, u, q) \\ y = h(x) \end{cases} \tag{2.25}$$

be a nonlinear control system in local coordinates with disturbances q and outputs y. Let π be the projection $\pi : (x,u,q) \longmapsto x$ and $\tilde{\pi}$ the projection $\tilde{\pi} : (x,u,q) \longmapsto (x,u)$. Then there locally exists a static state feedback $u = \alpha(x,v)$ which isolates the disturbances from the outputs if, and only if, there is a regular distribution D on M satisfying

$$f_* (\pi_*^{-1}(D)) \subset \dot{D} + f_* (\pi_*^{-1}(0)) \tag{2.26a}$$

$$f_* (\tilde{\pi}_*^{-1}(0)) \subset \dot{D} \tag{2.26b}$$

$$D \subset \ker dh \tag{2.26c}$$

provided the distributions $f_* (\pi_*^{-1}(0)) \cap \dot{D}$ and $f_* (\tilde{\pi}_*^{-1}(0))$ have constant dimension, see [38]. Notice that (2.26a) precisely yields, for each constant q, the condition (2.19). Again, as in the linear case, (see (2.12)), the equations (2.26a,b,c) imply that in the corresponding flow Diagram 1 of the nonlinear system (2.25), the disturbances enter via $\mathcal{P}$ and the outputs y only depend on the state $\bar{x} = x_2$.

3. Static and dynamic state feedback

Consider again a linear or nonlinear system Σ of the form (2.1), respectively (2.13). In various controller design problems one faces the question of adding control loops to the system Σ such that the closed loop system satisfies a set of prescribed design goals. The aforementioned Disturbance Decoupling Problem forms a, perhaps naive but simple, illustration of such a synthesis problem. Of course the control loops to be added are not randomly chosen but usually depend on the observations of the system Σ. To simplify our discussion we henceforth assume that the state x of Σ is available for controller design - in general this may not be the case and one has to resort to more restrictive alternatives, as for instance output feedback. The control loops now depend on the state x and we arrive at a situation which is depicted in Diagram 3.

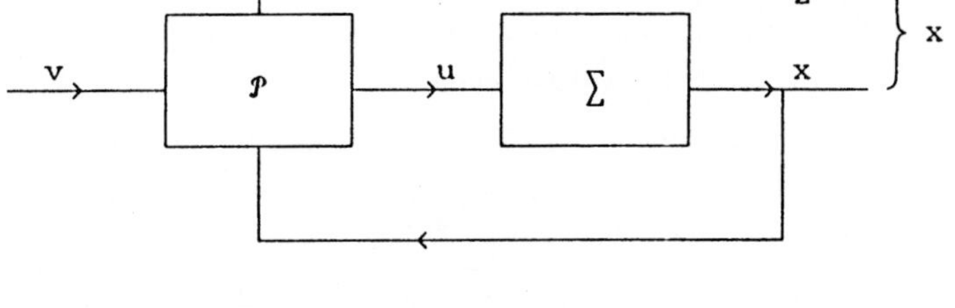

Diagram 3: Closed loop system $\Sigma \cdot \mathcal{P}$

As in section 2, $\mathcal{P}$ is assumed to be another system (in analogy with Σ either linear or nonlinear) where the inputs consists of a new set of reference inputs v together with the state x of Σ, and the control u appears as the output of $\mathcal{P}$. Thus, $\mathcal{P}$ typically is of the form (2.10) or (2.14) and forms a dynamic state feedback. Notice that it also includes the static state of feedback (2.2) (respectivily (2.14)).

Remark 3.1 The closed loop system in Diagram 3 has the same structure as the system depicted in Diagram 1. However a closer inspection shows the obvious differences: in Diagram 1 the system Σ appears as the precompensated system of a system $\bar{\Sigma}$ whereas in the latter case the possible closed loop systems $\Sigma \cdot \mathcal{P}$ are considered.

The importance of the feedback structure given above cannot be overemphasized. In essence almost all controller design problems can be formulated as: Given the system Σ, find, if possible, a precompensator $\mathcal{P}$ such that the closed loop system $\Sigma \cdot \mathcal{P}$ has some prescribed properties. Depending on the nature of the problem further requirements on the precompensator $\mathcal{P}$ may be imposed. In the last decades nonlinear control theory has focused on typical synthesis problems as described below. (Note that only a sample of such problems will be discussed here; various others have also been addressed in the literature).

3.1 The feedback stabilization problem

One of the most widely studied problems in control theory is the feedback stabilization problem. That is given the system Σ described as in (2.13) with an equilibrium point $f(x_0, u_0) = 0$, the question is whether or not a static state feedback $u = \alpha(x)$ with $u_0 = \alpha(x_0)$ exists, which renders the closed loop system (locally) asymptotically stable. And if it does exist, can it be assumed to have certain smoothness properties, such as C^k, C^∞ or analytic, see also [28]. Thus in our setting the precompensator $\mathcal{P}$ is static, with reference inputs v set equal to zero.

Many results around the local stabilization problem have appeared through the last decades, but in its full generality the question is still far from being solved. As we will not pursue this problem here, we refer the reader to [2] and the references therein. We also refer to the recent work of Byrnes and Isidori, where a slightly different approach to the global stabilization problem is advocated; see [7] and the references therein.

3.2 The feedback linearization problem

Consider the nonlinear system Σ locally described as in (2.13). Then the problem is to find, if possible, a precompensator $\mathcal{P}$ such that in suitable local coordinates the system $\Sigma\cdot\mathcal{P}$ is a controllable linear system. For static state feedbacks, i.e. $\mathcal{P}$ described as in (2.14), this question has been solved completely. In fact, this has been done in [27] and [21] for affine nonlinear systems (2.20) - thereby extending the results of [5] and [29] - and then for general systems the solution was given in [45].

To describe this solution we introduce with (2.13) the so called "extended" system.

$$\begin{cases} \dot{x} = f(x,u) \\ \dot{u} = \tilde{u} \end{cases} \tag{3.1}$$

which is an affine nonlinear system since the controls $\tilde{u}$ appear linearly. Now, the nonlinear system (2.13) is locally feedback linearizable into a controllable linear system if and only if the corresponding "extended" system (3.1) is. It is therefore enough to give the solution of the feedback linearization problem for an affine control system (2.20). With the system (2.20) we introduce the distributions Δ_i as follows. Let Δ_0 = span $\{g_1,\ldots,g_m\}$ and define recursively $\Delta_k = [f,\Delta_{k-1}]$, $k = 1,2\ldots$. Then the system (2.20) is locally feedback linearizable around a point x_0 with $f(x_0) = 0$ if and only if the distributions Δ_k, $k = 0,1,\ldots\ldots, n-1$ are regular and dim (Δ_{n-1}) = dim M. Observe that the aforementioned conditions on the distributions Δ_k are purely geometric. One can also formulate them in an algebraic manner, see e.g. [19] and this brings the feedback linearization problem in contact with the classical (Cartan) differential geometric theory of normal forms of Pfaffian systems, see [19] and [17].

So far we have discussed the local linearization of a nonlinear system via a static state feedback. An interesting extension of this question is when dynamic state feedback is also allowed. Some preliminary results on this idea may be found in [26,33], but a complete solution is still far away.

3.3 Model Matching Problems

Given the system Σ as in (2.13), together with an output mapping, say $y = h(x)$, find, if possible, a dynamic precompensator $\mathcal{P}$ such that $\mathcal{P}\cdot\Sigma$ matches a given reference model Σ_M described as

$$\Sigma_M \begin{cases} \dot{x}_m = f_m(x_m, v) \\ y_m = h_m(x_m), \end{cases} \tag{3.2}$$

Note that the linear model matching problem, i.e. Σ, Σ_M and $\mathcal{P}$ are linear, was solved in the early seventies, cf. [35,36]; see also [32]. In the last decade, several versions of the nonlinear problem have been studied.

In a first problem of this type (see [24]) it is required that the input-output behaviour of $\mathcal{P}\cdot\Sigma$ matches the input-output behaviour of a linear model Σ_M (thus (3.2) forms a linear system).

In other words, the Volterra kernels of $\mathcal{P}\cdot\Sigma$, $V^k_{\mathcal{P}\cdot\Sigma}$, should coincide with the Volterra kernels V^k_M of a linear system Σ_M, $k = 1,2,\ldots$, and thus satisfy V^1_M is independent of x and $V^k_M = 0$ for $k > 1$.

Notice that the matching of input-output behaviour between $\mathcal{P}\cdot\Sigma$ and Σ_M does not impose conditions on the 0-order Volterra kernels; the autonomous behavior may differ in principle.

Sufficient - but not necessary - conditions for the solvability for this matching of a prescribed linear input-output behaviour are given in [24]. As a matter of fact in [24] the problem is solved for those systems Σ which do have a linear input-output behaviour after applying a suitable static state feedback, and this class of systems is identified in the same reference. Note that this forms a weaker linearization problem then the one discussed in section 3.2.

A second, less restrictive, model matching problem has been studied in [12]. In this case no further assumptions on the model Σ_M are made and the question is to search for conditions on Σ which guarantee the existence of a precompensator $\mathcal{P}$ such that the input-output behaviours of $\mathcal{P}\cdot\Sigma$ and Σ_M are identical. So far, only sufficient conditions for the solvability of this matching problem are known and these are based on the reduction of the problem to a Disturbance Decoupling Problem with Measurements, that is a Disturbance Decoupling Problem in which the disturbances are measured (compare with the exposition on the DDP in section 2).

3.4 Decoupling Problems

A large class of synthesis problems essentially involve the isolation (or decoupling) of a subset of the inputs from the output. So far we have encountered the Disturbance Decoupling Problem with or without Measurements as a typical example.

Another example of a problem of this type fitting in the general formulation is the input-output decoupling problem. Consider again the nonlinear system

Σ locally described by (2.13) together with a set of output mappings $y_1 = h_1(x), \ldots, y_m = h_m(x)$. Here the number of output-blocks equals the number of inputs.

In the input-output decoupling problem one searches for a precompensator $\mathcal{P}$ such that the closed loop system $\mathcal{P}\cdot\Sigma$ with the above outputs $y_1, \ldots, y_m$ is input-output decoupled, i.e. the i-th input v_i does not influence the j-th output y_j, $j \neq i$ and v_i "controls" the i-th output y_i, $i = 1, \ldots, m$. This decoupling problem has received a lot of attention in the literature and several versions of the problem have been solved. In particular the situation where it is required that $\mathcal{P}$ is a static state feedback, a local solution involving the notion of controlled invariant distributions has been obtained for affine nonlinear systems, [40,41], see also [45] for a treatment of the problem for general nonlinear systems. For affine nonlinear systems with scalar outputs also the problem with a dynamic precompensator has been solved, see [10,42]. The general block decoupling problem with a dynamic state feedback is solved in [11,18,34]. Stability issues are now being addressed for decoupling problems, see [22] for the noninteracting control problem and [48] for the disturbance decoupling problem.

Remark 3.2 A promising recent approach is given by Fliess in [13] where differential algebraic tools are used in the study of various nonlinear synthesis problems. Without doubt these methods will complement and enrich throughout the next decade the differential geometric approach to nonlinear control theory that has been described here.

4. Decomposition of systems

In the previous section, (see Remark 3.1), we observed the similarity between Diagrams 1 and 3. The purpose of this section is to pursue the structure of Diagram 1 in more detail. Basically the configuration of Diagram 1 shows that the given system Σ, either linear or nonlinear, appears as the closed loop system $\Sigma = \bar{\Sigma}\cdot\mathcal{P}$ for a certain system $\bar{\Sigma}$ and a precompensator $\mathcal{P}$, both also linear or nonlinear. Of course, in general, $\bar{\Sigma}$ and $\mathcal{P}$ are unknown and the problem is: under which conditions does such a factorization of Σ exist. This can be viewed as a Decomposition Problem for Σ. The main motivation for looking at this problem is that when Σ allows a factorization as in Diagram 1, then Σ is built up from the simpler systems $\bar{\Sigma}$ and $\mathcal{P}$. Here a "simpler system" is used as a synonym for a system having a smaller state space dimension (see also [30]). This can be useful in the control of the more complex system Σ. The decomposition of a system as a cascade of simpler systems already appeared in the Krohn-Rhodes theory for finite automata, cf. [1,31]. A system Σ is said to admit a cascade decomposition if Σ decomposes as in Diagram 4.

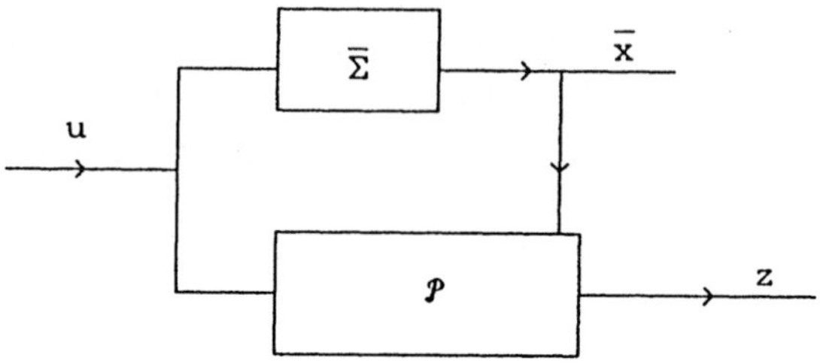

Diagram 4: Cascade decomposition of $\bar{\Sigma}$ and $\mathcal{P}$

Note that a cascade decomposition indeed is a special type of the decomposition described by Diagram 1; namely, take in the precompensator $\mathcal{P}$ given by (2.10) or (2.24) the linking map $v = u$. Cascade decompositions for nonlinear systems have been studied in Lie-algebraic terms in [30,44] and from a differential geometric viewpoint in [14,15] and [23,39] where in the latter papers the most explicit relation with (controlled) invariant distributions is given. Also other decomposition-types such as parallel- and series-decompositions, have been studied, see e.g. [44].

On the other hand the decomposition problem as stated here has not yet been investigated in its full generality. Clearly as was shown in section 2, one way in which a (local) decomposition as depicted in Diagram 1 appears, is when we deal with controlled invariant subspaces or distributions. This is however a rather limited way in which such a factorization appears. In the sequel, we will indicate one other possible way of decomposing affine nonlinear systems.

Consider the affine nonlinear system with a one dimensional input

$$\dot{x} = f(x) + g(x)\, u \tag{4.1}$$

Assume we work around a point x_0 with $g(x_0) \neq 0$. Then there exist local coordinates $x = (x^1,\dots,x^n)^T$ such that $g(x) = \partial/\partial x^n$, (Frobenius' Theorem, see section 2). In these coordinates, the system (4.1) takes the form

$$\begin{cases} \dot{x}^1 = f_1(x^1,\ldots, x^n) \\ \vdots \\ \dot{x}^{n-1} = f_{n-1}(x^1,\ldots, x^n) \\ \dot{x}^n = f_n(x^1,\ldots, x^n) + u \end{cases} \tag{4.2}$$

Applying the static state feedback $u = -f_n(x^1,\ldots, x^n) + \tilde{u}$ yields the system

$$\begin{cases} \dot{x}^1 = f_1(x^1,\ldots, x^n) \\ \vdots \\ \dot{x}^{n-1} = f_{n-1}(x^1,\ldots, x^n) \\ \dot{x}^n = \tilde{u} \end{cases} \tag{4.3}$$

A closer inspection of the last equation of (4.3) shows that the variable x^n appears as an arbitrary smooth function taking values in a neighbourhood of x_0^n, and therefore can be interpreted as an input function. In this way we have obtained a decomposition of the system (4.1) into the system $\bar{\Sigma}$ locally given by

$$\bar{\Sigma} \begin{cases} \dot{x}^1 = f_1(x^1,\ldots, x^{n-1}, v) \\ \vdots \\ \dot{x}^{n-1} = f_{n-1}(x^1,\ldots, x^{n-1}, v) \end{cases} \tag{4.4}$$

and the precompensator $\mathcal{P}$ as given by

$$\mathcal{P} \begin{cases} \dot{x}^n = f_n(x^1,\ldots, x^{n-1}, x^n) + u \\ v = x^n \end{cases} \tag{4.5}$$

Note that this procedure applied to a single input linear system $\Sigma : \dot{x} = Ax + bu$ yields that the subspace im b can be factored out. For such a linear system, the above procedure can be iterated several times, and so it follows that all subspaces $\mathrm{im}\, b + \ldots + A^k \mathrm{im}\, b$ can be factored out, thereby yielding that any almost controlled invariant subspace of $\dot{x} = Ax + bu$ can be factored out; see also [50,47] for the definition and further properties on almost controlled invariance. Obviously, writing the system (4.1) as the system (4.4) together with the precompensator (4.5) is only one way of decomposing (4.1). For a further decomposition of (4.1) one needs a more general approach for dealing with not necessarily affine nonlinear systems as automatically appear in (4.4). One such result for the system (4.1) is as follows, cf. [43]. Define the involutive distribution D as the involutive closure of the vector fields $g, \mathrm{ad}_f g, \ldots, \mathrm{ad}_f^k g$, and assume this distribution is regular, i.e. has constant dimension. Then, if

$$\text{rank } [f,D] \ (\text{mod } D) \ \le 1, \tag{4.6}$$

the distribution D can be "factored out", thereby locally yielding a nonlinear system $\bar{\Sigma}$ on the state space $\mathcal{X}$ (mod D), together with a precompensator $\mathcal{P}$ of dimension equal to the dimension of D. Note that the requirement (4.6) is in the case of a linear system automatically satisfied for distributions (= subspaces) D of the form im b + A im b + ...+ A^kim b. The above results on single input nonlinear systems can be viewed as a first step towards the Decomposition Problem for nonlinear system. Today the general Decomposition Problem is usually not addressed. Further research in this direction is certainly needed. The relation (or duality) with the problems of section 3 can perhaps be exploited fruitfully.

References

[1] M.A. Arbib, **The algebraic theory of machines, languages and semigroups**, Academic Press New York (1968).

[2] A. Bacciotti, "The local stabilizability problem for nonlinear systems",IMA Jnl. Math. Control & Inform. 5, pp. 27-39, (1988).

[3] G. Basile & G. Marro, "Controlled and conditioned invariant subspaces in linear system theory", Jnl. Optimiz. Theory Appl. 3, pp. 306-315, (1969).

[4] W.M. Boothby, **An introduction to differential manifolds and Riemannian geometry**, Academic Press, New York, (1975).

[5] R.W. Brockett, "Feedback invariants for nonlinear systems", Proc. VIIth IFAC Worlds Congress, Helsinki, pp. 1115-1120, (1978).

[6] R.W. Brockett, "Global descriptions of nonlinear control problems, vector bundles and nonlinear control theory", CBMS Conference Notes, (1979).

[7] C.I. Byrnes & A. Isidori, "Heuristics for nonlinear control", in **Modelling and Adaptive Control**, C.I. Byrnes & A. Kurzhanski (Eds.), LNCIS 105, Springer, Berlin, pp. 48-70, (1988).

[8] C.I. Byrnes & A.J. Krener, "On the existence of globally (f,g)-invariant distributions", in **Differential Geometric Control Theory**, R.W. Brockett, R.S. Millman & H.J. Sussmann (Eds.), Birkhäuser, Boston, pp. 209-225, (1983).

[9] C.I. Byrnes & A. Lindquist (Eds.), **Theory and applications of nonlinear control systems**, North Holland, Dordrecht, (1986).

[10 J. Descusse & C.H. Moog, "Decoupling with dynamic compensation for strong affine nonlinear systems", Int. Jnl. Control 42, pp. 1387-1398, (1985).

[11] M. Di Benedetto, J.W. Grizzle & C.H. Moog, "Rank invariants of nonlinear systems", SIAM Jnl. Control & Optimiz., to appear, (1989).

[12] M. Di Benedetto & A. Isidori, "The matching of nonlinear models via dynamic state feedback", SIAM Jnl. Control & Optimiz. 24, pp. 1063-1075, (1986).

[13] M. Fliess, "Automatique et corps différentiels", Forum. Math. 1, to appear, (1989).

[14] M. Fliess, "Décompositions en cascades des systèmes automatiques et feuilletages invariants", Bull. Soc. Math. France **113**, pp. 285-293, (1985).

[15] M. Fliess, "Cascade decompositions of nonlinear systems, foliations and ideals of transitive Lie algebras", Systems & Control Lett. 5, pp. 263-265, (1985).

[16] M. Fliess & M. Hazewinkel (Eds.), **Algebraic and geometric methods in nonlinear control theory**, Reidel, Dordrecht, (1986).

[17] R.B. Gardner, "Differential geometric methods interfacing control theory", in **Differential Geometric Control Theory**, R.W. Brockett, R.S. Millman & H.J. Sussmann (Eds.), Birkhäuser, Boston, pp. 117-180, (1983).

[18] J.W. Grizzle, M. Di Benedetto & C.H. Moog, "Computing the differential output rank of a nonlinear system", Proc. 26th CDC, Los Angeles, pp. 142-145, (1987).

[19] R. Hermann, "Perturbation and linearization of nonlinear control systems", in **Systems, Information and Control**, vol. II, L.R. Hunt & C.F. Martin (Eds.), Math. Sci. Press., pp. 195-238, (1984).

[20] R.M. Hirschhorn, "(A,B)-invariant distributions and disturbance decoupling of nonlinear systems", SIAM Jnl. Control & Optimiz. 19, pp. 1-19, (1981).

[21] L.R. Hunt, R. Su & G. Meijer, "Design for multi-input nonlinear systems", in **Differential Geometric Control Theory**, R.W. Brockett, R.S. Millmann & H.J. Sussmann (Eds.), Birkhäuser, Boston, pp. 268-298, (1983).

[22] A. Isidori & J.W. Grizzle, "Fixed modes and nonlinear noninteracting control with stability", IEEE Trans. Autom. Control 33, pp. 907-914, (1988).

[23] A. Isidori, A.J. Krener, C. Gori-Giorgi & S. Monaco, "Nonlinear decoupling via feedback: a differential geometric approach", IEEE Trans. Automat. Control 26, pp. 331-345, (1981).

[24] A. Isidori, "The matching of a prescribed linear input-output behaviour in a nonlinear system", IEEE Trans. Automat. Control, 30, pp. 258-265, (1985).

[25] A. Isidori, **Nonlinear control systems: an introduction**, LNCIS 72, Springer-Verlag, Berlin, (1985).

[26] A. Isidori, A. de Luca & F. Nicolo, "Control of robot arm with elastic joints via nonlinear dynamic feedback", Proc. 24th CDC, Ft. Lauderdale (Fl), pp. 1671-1679, (1985).

[27] B. Jakubczyk & W. Respondek, "On linearization of control systems", Bull. Acad. Polon. Sci. (Math.), **28**, pp. 517-522, (1980).

[28] M. Kawski, "Stabilization of nonlinear systems in the plane", Syst. Control Lett. 12, pp. 169-176, (1989).

[29] W. Korobov, "Controllability, stability of some nonlinear systems", Differencialnyje Uravnienje **9**, pp. 466-469, (1973).

[30] A.J. Krener, "A decomposition theory for differentiable systems", SIAM Jnl. Control & Optimiz. **15**, pp. 813-829, (1977).

[31] K.B. Krohn & J.L. Rhodes, "Algebraic theory of machines: I. The main decomposition Theorem", Trans. Amer. Math. Soc. **116**, pp. 450-464, (1965).

[32] M. Malabre, "Structure à l'infini des triplets invariants. Application à la poursuite parfaite de modèle", in **Analysis and Optimization of Systems**, A. Bensoussan & J.L. Lions (Eds.), LNCIS 44, Springer Verlag, Berlin (1982).

[33] R. Marino, B. Charlet & J. Levine, "Two Sufficient conditions for dynamic feedback linearization of nonlinear systems", in **Analysis and Optimization of Systems**, A. Bensoussan & J.L. Lions (Eds.) LNCIS 111, Springer-Verlag, Berlin, pp. 181-191, (1988).

[34] C.H. Moog & J.W. Grizzle, "Découplage non linéaire vu de l'algèbre linéaire", C.R. Acad. Sci. Paris, t.307, Série I, pp. 497-500, (1988).

[35] B.C. Moore & L.M. Silverman, "Model matching by state feedback and dynamic compensations", IEEE Trans. Automat. Contr. **17**, pp. 491-497, (1972).

[36] A.S. Morse, "Structure and design of linear model following systems", IEEE Trans. Automat. Contr. **18**, pp. 346-354, (1973).

[37] H. Nijmeijer & A.J. van der Schaft, "Controlled invariance for nonlinear systems", IEEE Trans. Automat. Contr. **27**, pp. 907-914, (1982).

[38] H. Nijmeijer & A.J. van der Schaft, "The disturbance decoupling problem for nonlinear control systems". IEEE Trans. Automat. Contr. **28**, pp. 621-623, (1983).

[39] H. Nijmeijer & A.J. van der Schaft, "Partial symmetries for nonlinear systems", Math. Syst. Theory **18**, pp. 79-96, (1985).

[40] H. Nijmeijer & J.M. Schumacher, "Zeros at infinity for affine nonlinear control systems", IEEE Trans. Automat. Contr. **30**, pp. 566-573, (1985).

[41] H. Nijmeijer & J.M. Schumacher, "The regular local noninteracting control problem for nonlinear control systems", SIAM Jnl. Contr. & Optimiz. **24**, pp. 1232-1245, (1986).

[42] H. Nijmeijer & W. Respondek, "Dynamic input-output decoupling of nonlinear control systems", IEEE Trans. Automat. Contr. **33**, pp. 1065-1070, (1988).

[43] H. Nijmeijer, "Preliminaries on almost controlled invariance of single input nonlinear systems", Proc. 27th CDC, Austin (TX) pp. 932-933, (1988).

[44] W. Respondek, "On decomposition of nonlinear control systems", Syst. Contr. Lett. **1**, pp. 301-308, (1982).

[45] A.J. van der Schaft, "Linearization and input-output decoupling for general nonlinear systems", Syst. Contr. Lett. **5**, pp. 27-33, (1984).

[46] A.J. van der Schaft, **System theoretic description of physical systems**, CWI Tract **3**, Amsterdam, (1984).

[47] H.L. Trentelman, **Almost invariant subspaces and high gain feedback**, CWI Tract **29**, Amsterdam, (1986).

[48] L. van der Wegen & H. Nijmeijer, "The local disturbance decoupling problem with stability for nonlinear systems", Syst. Control Lett. 12, pp. 139-149, (1989).

[49] J.C. Willems, "System theoretic models for the analysis of physical systems", Ricerche di Automatica **10**, pp. 71-106, (1979).

[50] J.C. Willems, "Almost invariant subspaces: an approach to high gain feedback design - part I, almost controlled invariant subspaces", IEEE Trans. Automat. Contr. **26**, pp. 235-252, (1981).

[51] W.M. Wonham, **Linear multivariable control: a geometric approach**, Springer-Verlag, Berlin, (1979).

Aggregation of Linear Systems in a Completely Deterministic Framework

G. Picci

Universitá di Padova
Dipartimento di Elettronica e Informatica
Via Gradenigo 6/A, 35131 Padova, Italy
and
LADSEB-CNR, Padova, Italy

ABSTRACT: Although Professor WILLEMS'exhortation to theologians and cosmologists to get into infinite dimensional Systems Theory could hardly be dissented with, it is suggested in this paper that the study of non complete Behaviours may make a quite enlightening exercise also for finite-dimensional minded System Theorists.

1. INTRODUCTION

In this paper we study a particular representation or "modelling" problem for Dynamical Systems that we would like to name *Deterministic Aggregation*. In very general (and somewhat imprecise) terms the problem can be described as follows.

One would like to generate the output trajectories, $t \to y(t)$, of a "large" autonomous dynamical system like

$$\begin{aligned} \dot{z}(t) &= F(z(t)) \\ & \qquad\qquad z(t) \in M \subset \mathbb{R}^N \\ y(t) &= H(z(t)) \end{aligned} \tag{1.1}$$

by means of a smaller dimensional system *with inputs*, say

$$\begin{aligned} \dot{x}(t) &= f(x(t)) + g(x(t))u(t) \\ y(t) &= h(x(t)) \end{aligned} \tag{1.2}$$

where $x(t) \in X \subset \mathbb{R}^n$. The crucial requirements are that $n<N$ (aggregation) and that u be *locally free*, i.e. an "input" variable, in the sense of WILLEMS [1986, 1988]. Of course, to make the problem interesting, the system (1.1) should be assumed *irreducible* in the sense that no lower dimensional autonomous system can generate the same output signal.

If for every initial state $z(0) \in M$ there are pairs $(x(0), u(\cdot))$ such that the output of the large system (1.1) with initial state z(0) and the output of the system (1.2) with initial state x(0) and input $u(\cdot)$, *are the same function of time*, we say that (1.1) is *aggregable* and that (1.2) is an *aggregation* of (1.1).

Our motivation for considering aggregation problems of the type roughly defined above originally came from Statistical Mechanics. Some common grounds and motivations probably also exist with the notion of aggregation considered in Dynamic Economic Theory. See AOKI [1976, 1980].

In PICCI [1988, 1989] a stochastic version of the idea of aggregation is introduced and aggregability is studied in detail for Linear Hamiltonian Systems. In the stochastic frame, at some stage the large system is "randomized" by introducing an invariant probability measure for the state evolution flow. The output $y(\cdot)$ becomes then a stationary stochastic process defined on the phase space M. One has then to search for a smaller dimensional *stochastic* Dynamical System like (1.2) where now u is *white noise*, producing an output process stochastically equivalent in some sense to the output process of (1.1). The problem is essentially phrased as a *Stochastic Realization Problem.*

The particular case of Linear Hamiltonian Systems is especially instructive since it can be analyzed in detail and from several different points of view. In particular, since the natural invariant measures for Linear Hamiltonian Systems are *Gaussian*, the aggregation problem can be rephrased as a stochastic Realization Problem for *Gaussian processes.*

The deterministic problem that we shall consider here will still deal with Linear Hamiltonian Systems. In fact we shall eventually make contact with (and complement) the "deterministic" approach tried at the beginning of our previous paper PICCI [1986].

In this paper we shall use the language and the general ideas of WILLEMS' Theory of Dynamical Systems. In this framework the problem can be formulated very naturally as a *realization problem* for a certain concretely defined *Behaviour*. In this respect, a non-secondary goal of this paper will also be to clarify, by means of an example, some of the relations existing between WILLEMS' Theory and the ideas of *Stochastic Modelling* formulated in Stochastic Realization Theory. As noticed a while ago (compare LINDQUIST-PICCI [1985]) there are some non superficial points of contacts of the two Theories. A surprise, for some, may come from the discovery that the point at which they really come to overlap is the (linear) infinite dimensional "non complete" case. This case, which unfortunately is largely yet to be worked out, has a very rich structure and potentially covers interesting applications.

A reason why infinite dimensional Systems and random processes are so strictly related objects may be traced back (at least for the linear-Gaussian case) to the equivalence proven in PICCI [1988b].

2 - LINEAR HAMILTONIAN SYSTEMS

We shall formulate the aggregation problem in a linear time-invariant setting. The structure of candidate aggregate models will be defined first. In a linear-time-invariant framework the natural candidates are linear finite dimensional models of the type

$$\begin{aligned} \dot{x} &= Ax + Bu \\ y &= Cx \end{aligned} \qquad (S)$$

where x(t) belongs to some n-dimensional vector space X, y is the m-dimensional output (or "external") variable and u, the latent "driving" input, ranges on a vector space $\mathcal{U}$ of p-dimensional vector functions *closed under concatenation*. In particular $\mathcal{U}$ contains all finite linear combinations of indicator functions. The natural topology to be given to $\mathcal{U}$ will be discussed later. For the moment it will just be assumed to contain all measurable functions $u: \mathbb{R} \to \mathbb{R}^p$ for which (S) has a global solution for each initial state $x \in X$. This $\mathcal{U}$ is clearly a linear vector space closed under concatenation.

We shall only consider models (S) that are *irreducible* in the sense that the *representation map* Σ_S: $X \times \mathcal{U} \to (\mathbb{R}^m)^{\mathbb{R}}$ of the system, defined as

$$\Sigma_S(x,u)(t) = C\,e^{At}x + \int_0^t C\,e^{A(t-s)}\,Bu(s)ds \qquad (2.1)$$

is *injective* on the product space $X \times \mathcal{U}$. Note that Σ_S is a linear shift-invariant, everywhere defined, map. Comparing with WILLEMS [1983, p. 588] it is seen that the range space of Σ_S, $\mathcal{B}:=\mathcal{R}(\Sigma_S)$ is just the *Behaviour* of the system (S).

Obviously a necessary condition for irreducibility is that (A,C) be an observable pair. It is also immediate to check that if (S) is irreducible *and* p>0 then its Behaviour $\mathcal{B}$ is an *infinite dimensional* vector space. The obvious consequence is,

PROPOSITION 2.1

A necessary condition for an irreducible linear autonomous system

$$\begin{cases} \dot{z} = Fz \\ y = Hz \end{cases} \qquad z(t) \in \mathbb{R}^N \qquad (2.2)$$

to admit aggregate models of the type (S) *is that* $N=\infty$. *In other words, the system* (2.2) *can be aggregable only if it is infinite dimensional.*

Proof: Let N be finite, then the Behaviour of (2.2) is N-dimensional. It can therefore be described by an irreducible model of the type (S) only if p=0. But then both models are autonomous and observable and so necessarily n=N. So no aggregation is possible.

□

We shall abstain from commenting about possible "physical" justifications and/or interpretations of this condition. The interested reader may find some speculations in PICCI [1989].

Due to the previous observation we are forced to consider a slightly different question than the one we originally had in mind, namely what interesting classes of (infinite dimensional) linear autonomous systems are aggregable.

A candidate is defined below.

DEFINITION 2.1

A Linear Hamiltonian System is a triple $\{H, \Phi(t), h^*\}$ *where* H *(the phase space) is a real separable Hilbert space,* $\{\Phi(t)\}_{t\in\mathbb{R}}$ *(the Hamiltonian flow) is a strongly continuous group of unitary operators on* H *and* $h^*: H \to \mathbb{R}^m$ *is an m-tuple of linear-bounded functionals (the observables of the system)*

$$h_k^*(\cdot) = \langle h_k, \cdot\rangle, \qquad h_k \in H, \quad k = 1,\dots,m \;.$$

The state or "phase" of the system, $z(t) \in H$, *evolves in time according to* $z(t) = \Phi(t)z(0)$ *or, symbolically, is a solution of*

$$\dot{z}(t) = F\, z(t) \tag{2.3}$$

where F *is the generator of* $\Phi(t)$ *(a real, self-adjoint, densely defined operator). Each scalar observable produces an output signal,*

$$y_k(t) = \langle h_k, z(t)\rangle \qquad t \in \mathbb{R} \quad , \quad k=1,\dots,m \;.$$

The (column) output vector $y(t)=\mathrm{col}[y_1(t),\dots,y_m(t)] \in \mathbb{R}^m$ *is the external variable of the system. It will be written as*

$$y(t) = \langle h, z(t)\rangle = \langle h, \Phi(t)z(0)\rangle \tag{2.4}$$

where $\langle h, z\rangle := \mathrm{col}[\langle h_1, z\rangle,\dots,\langle h_m, z\rangle]$, $z\in H$.

A Linear Hamiltonian System is "nonsingular" if there are no nontrivial fixed vectors for $\Phi(t)$, *i.e.* $\Phi(t)x = x,\ \forall\, t, \Rightarrow x=0$. *(This is the same as* F *having no zero eigenvalues).*

A Linear Hamiltonian System is "irreducible" if,

$$\overline{\mathrm{span}}\,\{\Phi(t)h_k;\ k=1,\dots,m,\ t \in \mathbb{R}\} = H \tag{2.5}$$

i.e. $h_1,\dots,h_m$ *are generators for the Hilbert space* H *with respect to* $\Phi(t)$. *This is equivalent to the map* $z \to \langle h, \Phi(\cdot)z\rangle$ *being injective and, in turn, to the non existence of a proper* $\Phi(t)$*-invariant subspace* $H_o \subset H$ *such that the Linear Hamiltonian System* $\{H_o, \Phi_o(t), h_o^*\}$ *is equivalent (same behaviour) to* $\{H, \Phi(t), h^*\}$. *Here* $\Phi_o(t)$ *and* h_o^* *are the restrictions of* $\Phi(t)$ *and* h^* *to the invariant subspace* H_o.

The terminology may seem arbitrary but is not. As reported in PICCI [1988], it can be shown that the phase space of any nonsingular Linear Hamiltonian System has a natural *symplectic structure* i.e. it can always be split into the orthogonal direct sum of two isomorphic real Hilbert spaces, $H=H_p \oplus H_q$ and that there exists a *symplectic* bounded operator J on H, i.e. an operator satisfying $J^2 = -I$, $J^* = -J$, having the matrix representation

$$J = \begin{bmatrix} 0 & -E^* \\ E & 0 \end{bmatrix},$$

where $E: =J|_{H_p} : H_p \to H_q$ is unitary.

There is an injective positive self adjoint operator V (the *potential* operator) on H for which the orthogonal direct sum $H_p \oplus H_q$ is reducing. V commutes with E (and E^*) i.e. $VE = EV$ and the generator of $\Phi(t)$ can be expressed as

$$F = \begin{bmatrix} 0 & -E^* V \\ EV & 0 \end{bmatrix}.$$

The "differential equation" (2.3) for $z(t) = \mathrm{col}(p(t),q(t))$ then has the canonical structure

$$\dot{p}(t) = -E^* V\, q(t), \qquad \dot{q}(t) = EV\, p(t) \tag{2.6}$$

It is not hard to see that this is the general form which the canonical equations of a system with a general *nonsingular* quadratic Hamiltonian, transform into under the linear transformation which normalizes the quadratic Hamiltonian to the "*energy norm*" $1/2\ \|z\|^2$.

The coding of infinite dimensional linear Hamiltonian models in the terms of Def. 2.1 is at the roots of LAX-PHILLIPS scattering theory for hyperbolic systems. More examples are given in LEWIS and MAASSEN [1984].

There are plenty of interesting examples of infinite dimensional Linear Hamiltonian System for which aggregation is indeed a deep, relevant problem. We may add, on the mathematical side, that evolution models governed by a unitary group can always be seen as *dilations* of a very broad family of possible linear infinite dimensional autonomous systems (see FUHRMANN [1981] for details on this). So the theory could really be given a much wider scope than the one we restrict to in this paper.

3 - ON THE BEHAVIOUR SPACE OF A LINEAR HAMILTONIAN SYSTEM

The *Behaviour space* of a Linear Hamiltonian System is

$$\mathcal{H}: = \{y_z(\cdot) = \langle h, \Phi(\cdot)z\rangle; \ z \in H\} \tag{3.1}$$

where we use $y_z(\cdot)$ to denote the dependence of the output trajectory on the initial state z. Note that $\mathcal{H}$ is made of $\mathbb{R}^m$-valued continuous functions and has a natural real vector space structure with addition of $\mathbb{R}^m$-valued vector functions and multiplication by real scalars defined in the obvious way. Henceforth we shall assume that the Linear Hamiltonian System $\{H, \Phi(t), h^*\}$ is non singular and *irreducible*. This in particular means that the finite sums

$$z = \Sigma\, \alpha_k^T\, \Phi(t_k)h \qquad \alpha_k \in \mathbb{R}^m \tag{3.2}$$

are dense in H.

We shall introduce in $\mathcal{H}$ the inner product

$$\langle y_z(\cdot), y_w(\cdot)\rangle_{\mathcal{H}}: = \langle z,w\rangle \tag{3.3}$$

Since T: $z \to y_z(\cdot)$ is linear and injective by irreducibility, (3.3) is well defined and makes $\mathcal{H}$ into a real Hilbert space. By means of the T map then we obtain a system unitarily equivalent to $\{H, \Phi(t), h^*\}$, denoted $\{\mathcal{H}, S(t), ev_o\}$, where S(t) is the left translation group on time functions, $(S(t)f)(s) = f(s+t)$, and $ev_o(f)=f(0)$, is the evaluation map at zero.

$$\begin{array}{ccccc} H & \xrightarrow{\Phi(t)} & H & \xrightarrow{h^*} & \mathbb{R}^m \\ T\downarrow & & \downarrow T & \nearrow_{ev_o} & \\ \mathcal{H} & \xrightarrow[S(t)]{} & \mathcal{H} & & \end{array} \tag{3.4}$$

We define the *covariance matrix* $\Lambda(\cdot)$ of the Linear Hamiltonian System $\{H, \Phi(t), h^*\}$ as

$$\Lambda_{kj}(t): = \langle h_k, \Phi(t)h_j\rangle \qquad k,j = 1,\dots,m\ . \tag{3.5}$$

Using a self explanatory vector notation, we may express $\Lambda(\cdot)$ directly also as

$$\Lambda(t) = \langle h, \Phi(t)h^T \rangle .$$

LEMMA 3.1

The Behaviour space of an irreducible Linear Hamiltonian System is generated by the translates of the columns of its covariance matrix i.e.

$$\mathcal{H} = \overline{\text{span}} \{S(t)\Lambda(\cdot)\alpha \quad ; \quad \alpha \in \mathbb{R}^m, \ t \in \mathbb{R}\} \tag{3.6}$$

where the closure is with respect to the metric induced by the scalar product $\langle \cdot, \cdot \rangle_{\mathcal{H}}$. *Moreover* $\mathcal{H}$ *is a reproducing kernel Hilbert space with reproducing kernel* $\Lambda(\cdot)$.

Proof:

Under the unitary map T the generators $h_1, \ldots, h_m$ for H are mapped into $y_{h_1}(\cdot), \ldots, y_{h_m}(\cdot)$, that is into the columns of the covariance matrix $\Lambda(\cdot)$ (compare (3.5)). Then the columns are generators for $\mathcal{H}$ with respect to the translation group. The reproducing kernel property follows from $T[\Phi^*(t)h_j] = \Lambda_j(\cdot - t)$ (j-th column) and therefore,

$$y_z(t) = \langle \Phi^*(t)h, z \rangle = \text{col}\{\langle \Phi^*(t)h_1, z \rangle \ldots \langle \Phi^*(t)h_m, z \rangle\}$$

$$= \text{col}\ \{\langle \Lambda_1(\cdot - t), y_z \rangle_{\mathcal{H}}, \ldots, \langle \Lambda_m(\cdot - t), y_z \rangle_{\mathcal{H}}\} \tag{3.7}$$

$$: = \langle \Lambda(\cdot - t), y_z \rangle_{\mathcal{H}}$$

□

Since Λ is a continuous positive definite kernel it has a Bochner representation

$$\Lambda(t) = \int_{-\infty}^{+\infty} e^{i\lambda t}\, dM(i\lambda) \tag{3.8}$$

where M is a finite positive Hermitian matrix valued measure, the *spectral distribution measure of the system*, defined on the Borel sets of the imaginary axis. It is well known that M contains all spectral information about the unitary group $\Phi(t)$ on H. In fact, by the spectral representation theorem, $\Phi(t)$ acting on H is unitarily equivalent to multiplication by $e^{i\lambda t}$ on the Hilbert space $L^2_m(\mathbb{I}, dM)$ of m-dimensional measurable functions on the imaginary axis, square integrable with respect to the matrix measure M (see e.g. FUHRMANN [1981]). It is customary to write the function f_z corresponding to $z \in H$ as a *row* vector function so that the inner product in $L^2_m(\mathbb{I}, dM)$ ($\equiv L^2(dM)$) is written as

$$<f_{z_1}, f_{z_2}>_{L^2(dM)} = \int_{-\infty}^{+\infty} f_{z_1}(-i\lambda)dM(i\lambda)f^T_{z_2}(i\lambda) \quad . \tag{3.9}$$

We now come to an important result which will be instrumental in connecting aggregation to "spectral factorization", as explained in the the next section.

PROPOSITION 3.2

Every element $y_z \in \mathcal{H}$ *has a spectral representation in* $L^2(dM)$ *of the form*

$$y_z(t) = \int_{-\infty}^{+\infty} e^{i\lambda}dM(i\lambda)\hat{f}^T_z(i\lambda) \tag{3.10}$$

where $\hat{f}_z$ *is the spectral representation of the initial phase* z *in* $L^2(dM)$. *The inner product* $<y_{z_1}, y_{z_2}>_{\mathcal{H}}$ *is equal to the right hand member in* (3.9).

Proof. The proof of (3.10) follows from the formula

$$y_z(t) = <\Phi^*(t)h, \ z>$$

after expressing the inner product in the right hand side by means of the representatives in $L^2(dM)$.

□

For finite dimensional Linear Hamiltonian Systems $M(i\lambda)$ has just a finite number of jumps at points $\pm i\omega_k$ k=1,...,N/2. On the opposite side, suppose that $\Phi(t)$ has *absolutely continuous spectrum*. Then there is an L^1-valued nonnegative matrix Ψ such that

$$M(\Lambda) = \int_\Delta \Psi(i\lambda)d\lambda$$

for every Borel set $\Delta \subset \mathbb{I}$. It is not hard to imagine that in this case the representation (3.10) may be made to look like the Fourier transform of a convolution. In spite of this encouraging appearance an additional difficulty pops up in this case however, as $\mathcal{H}$ behaves like a truly infinite dimensional function space, in particular it is *not finitely determined* or "complete" in the sense of WILLEMS [1986, 1988]. This fact may be feared to be in a sharp contrast with the famous equivalence: ‚Completeness ⇔ realizability by a finite dimensional linear model of the type (S), so much advertised in WILLEMS'work. (Although we are not aware of any explicit statement of this kind in the continuous-time case). It will be clear in the next section that in the present setting the difficulty is indeed present and so we shall have to do something about it at the very *problem formulation* level.

Let $\mathcal{V}$ be a vector space of $\mathbb{R}^q$-valued time functions defined on $\mathbb{R}$. We introduce the subsets $\mathcal{V}_a^+$ and $\mathcal{V}_a^-$ of *forward*, and *backward truncations* of elements $f \in \mathcal{V}$, say

$$\mathcal{V}_a^+ = \{f_a^+ : [a, +\infty] \to \mathbb{R}^q \quad ; \quad f_a^+(t) = f(t),\ t\geq 0 \quad f \in \mathcal{V}\}$$

and similarly for $\mathcal{V}_a^-$.

DEFINITION 3.3

The linear finite-dimensional model

$$\dot{x}_+ = A_+x_+ + B_+u \qquad y = C_+x_+ \tag{S_+}$$

is a forward realization *of* $\mathcal{H}$ *if, for every finite* $a \in \mathbb{R}$ *the range space of the "forward" representation map,*

$$\Sigma_a^+ : X \times \mathcal{U}_a^+ \to (\mathbb{R}^m)^{[a,+\infty)}$$

where,

$$\Sigma_a^+ (x(a), u)(t) = C_+e^{A_+(t-a)}x(a) + \int_a^t C_+e^{A_+(t-\sigma)}B_+u(\sigma)d\sigma, \qquad t\geq a$$

is $\mathcal{H}_a^+$. *Dually, we say that*

$$\dot{x}_- = A_-x_- + B_-u\ , \qquad y = C_-x_- \tag{S_-}$$

is a backward realization *of* $\mathcal{H}$ *if for every finite* a *the range space of the "backward" representation map,*

$$\Sigma_a^-: X \times \mathcal{U}_a^- \to (\mathbb{R}^m)^{[-\infty, a]}$$

where,

$$\Sigma_a^-(x(a),u)(t) = C_- e^{A_-(t-a)} x(a) + \int_a^t C_- e^{A_-(t-\sigma)} B_- u(\sigma) A\sigma, \qquad t \le a$$

is the whole of $\mathcal{H}_a^-$.

Clearly a forward or backward realization of a *complete* system is just a bonafide realizations in the sense defined in the previous section. This is so because of the very definition of a complete (=finitely determined) subspace. An incomplete behaviour may instead admit forward or backward realization although *not* admitting finite dimensional realizations.

REMARK

In the above Definition there is one technical point which is not spelled out in sufficient detail. Of course the question here is to generate $\mathcal{H}$ as an (infinite dimensional) *Hilbert space*. In order to do this, $\Sigma_a^\pm$ should be viewed as Hilbert space operators and, $\mathcal{U}$ (and X) should also be given a Hilbert space structure. (In fact this should be done also for the appropriate truncation spaces $\mathcal{H}_a^\pm$, $\mathcal{U}_a^\pm$etc.). The equalities $\mathcal{H}_a^\pm = \mathcal{R}(\Sigma_a^\pm)$ are then understood to mean that $\Sigma_a^\pm$ are bounded linear operators acting on the appropriate Hilbert spaces $X \times \mathcal{U}_a^\pm$ with closed range, equal to $\mathcal{H}_a^\pm$. This observation brings us back to the question of the natural topology that should be given to $\mathcal{U}$.

It seems difficult to give a general answer to this question. It will however be shown in the next section that for *aggregable Linear Hamiltonian Systems* (in the sense that $\mathcal{H}$ admits a finite dimensional forward or backward realization of dimension strictly smaller than N = dim $\mathcal{H}$) the input space $\mathcal{U}$ *must* have the Lebesgue space $L^2(\mathbb{R};\mathbb{R}^p)$ structure.

The study of the relations existing between forward and backward realizations of $\mathcal{H}$ will have to be delayed until some more precise information on the spectrum of A_+ an A_- will be available. See Theorem 4.7.

4. NECESSARY CONDITIONS FOR AGGREGABILITY

In this section we study necessary conditions for aggregability of a Linear Hamiltonian System. The class of candidate aggregate models is the class of finite dimensional linear systems (S) defined at the beginning of sect. 2. *Aggregation* will however be understood in the weakened sense discussed at the end of the previous section, namely we shall say that *an (irreducible)* Linear Hamiltonian System *is aggregable if there is a forward* (S_+) *or a backward* (S_-) *realization of its behaviour space* $\mathcal{H}$, *of dimension strictly smaller than* N(=dim$\mathcal{H}$). Of course, since N=∞ has already been shown to be a necessary condition for aggregability we just require realizability by a *finite dimensional* forward or backward model.

LEMMA 4.1

Let (S) *be an irreducible (forward or backward) realization of the Behaviour space* $\mathcal{H}$ *of an infinite dimensional Linear Hamiltonian System. Consider the decomposition of the state space of* (S), $X=X_i\oplus X_\pm$, *where* X_i *is the purely imaginary (generalized) eigenspace of* A *and let* $x_i=\mathrm{col}(x_i,x_\pm)$ *be a representation of* x *relative to the above direct sum decomposition. Then,*

i) *The purely imaginary eigenspace* X_i *is uncontrollable. This is equivalent to* $P^{X_i}B = 0$ (P^{X_i}*=projection onto* X_i *along* $X_\pm$) *or to* $\mathcal{R}(B) \subset X_\pm$.

ii) (S) *has a direct sum decomposition,*

$$\dot{x}_i = A_i x_i$$

$$\dot{x}_\pm = A_\pm x_\pm + B_\pm u$$

$$y = C_i x_i + C_\pm x_\pm$$

where $A_i=A|_{X_i}$, $A_\pm=A|_{X_\pm}$, (A_i,C_i) *is observable and* $(A_\pm,C_\pm,B_\pm)$ *is a minimal triple, in particular* $(A_\pm,B_\pm)$ *is controllable. By irreducibility* $B_\pm$ *has also full column rank* p.

iii) $\mathcal{H}$ *admits a shift invariant reducing subspace* $\mathcal{H}_i$ *which is realized by the subsystem* $\dot{x}_i=A_i x_i$, $y=C_i x_i$, *of the same dimension* $n_i=\dim X_i$. *Every trajectory* $y \in \mathcal{H}_i$ *is a linear combination of* n_i *simple real harmonic oscillations.*

Proof:

Note that each trajectory in $\mathcal{H}$ is an uniformly bounded function as $|y_k(t)|\le\|h_k\|\ \|z\|$ by (2.4).

i) Because of observability each imaginary mode shows in the output and therefore it cannot (have Jordan blocks of length greater than one and) be controllable, for otherwise it would be possible to find a suitable input $u \in \mathcal{U}$ producing an unbounded output. We stress here that such an input u can be found even in $\mathcal{U}=L^2(\mathbb{R};\mathbb{R}^p)^{(*)}$.

ii) Recall that the eigenmodes of $A_\pm$ diverge as t goes to $+\infty$ or to $-\infty$. Assume $X_\pm$ is strictly larger than the controllable subspace for $(A_\pm, B_\pm)$. Then there are initial states which produce state trajectories (evolving on subspaces) unaffected by the inputs. By observability they show in the output and produce unbounded output trajectories. Hence the subsystem $(C_\pm, A_\pm, B_\pm)$ must be controllable.

iii) The eigenvalues of A_i must also be eigenvalues of F. The corresponding eigenspace H_i is reducing for $\Phi(t)$ by elementary spectral theory, so the same holds for $\mathcal{H} = TH_i$.

□

By the above result nothing really interesting (from the aggregation point of view) is attached to the imaginary eigenspace of a realization of $\mathcal{H}$. The purely oscillatory component, $\dot{x}_i=A_ix_i$, $y=C_ix_i$, of (S) describes a shift invariant subspace $\mathcal{H}_i\subset \mathcal{H}$ of the same dimension n_i. In fact (assuming irreducibility) this subsystem is isomorphic to a purely oscillatory component, $\dot{z}_i= F_iz_i$, $y=h^*(z_i)$, $F_i=F|H_i$, of the original Linear Hamiltonian System.

For this reason in the following we shall feel free to assume, unless otherwise explicitly stated, that *the oscillatory subsystem has been removed from any realization* (S), tacitly assuming that the same operation was done also on the given Behaviour space $\mathcal{H}$.

(*) Professor E. Fornasini has been very helpful with this, showing to me how to construct such an L^2 input.

Recall now the notions of *incoming* and *outgoing* subspaces for a unitary group $U(t)$ on a Hilbert space H. A subspace $S \subset H$ is *incoming* for $U(t)$ if letting $S_t := U(t)S$, one has $U^*(t)\, S \subset S$ for $t \geq 0$ (i.e. S_t is *increasing* in time) and

$$\bigvee_{t \in \mathbb{R}} S_t = H\,, \qquad \bigcap_{t \in \mathbb{R}} S_t = \{0\}\ . \tag{4.1}$$

Dually, $\bar{S} \subset H$ is *outgoing* for $U(t)$ if $U(t)\bar{S} \subset \bar{S}$ (i.e. $\bar{S}_t$ is decreasing for t increasing) and

$$\bigvee_{t \in \mathbb{R}} \bar{S}_t = H\,, \qquad \bigcap_{t \in \mathbb{R}} \bar{S}_t = \{0\} \tag{4.2}$$

(the wedge means closed vector sum). Note that the intersection conditions in (4.1) and (4.2) really only regard the tail behaviour of $\{S_t\}$ at $t=-\infty$ and of $\{\bar{S}_t\}$ at $t=+\infty$.

THEOREM 4.2

Assume that the Behaviour space $\mathcal{H}$ *of a Linear Hamiltonian System has a forward realization* (S_+), *i.e.* $\forall\ a \in \mathbb{R}$, $\forall\ z \in H\ \exists\ x(a) \in X$ *and* $u \in \mathcal{U}_a^+$, *such that*

$$y_z(t) = C_+ e^{A_+(t-a)} x(a) + \int_a^t C_+ e^{A_+(t-s)} B_+ u(s)ds \tag{4.3}$$

for all $t \geq a$.

Then the translation group $S(t)$ *on* $\mathcal{H}$ *has an outgoing subspace.*

Proof:

Define

$$S_a := \{y \in \mathcal{H};\ \exists\ x \in X \text{ such that } y(t) = C_+ e^{A_+(t-a)} x\ ,\ \text{for } t \geq a\} \tag{4.4}$$

By assumption S_a is a subspace of $\mathcal{H}$. (Its elements are obtained by setting u=0 in (4.3)).

The family $\{S_a\}_{a \in \mathbb{R}}$ is defined by letting $S_a = S^*(a)S_o$. It is easy to see that $S_o \subseteq S_a$ for $a \geq 0$, for if $y \in S_o$ then,

$$y(t) = C_+ e^{A_+(t-a)} x(a), \quad x(a) = C_+ e^{A_+ a} x(0)$$

for all $t \geq a$. Moreover by assumption each truncated space S_a^- contains $\mathcal{H}_a^-$ and so for every $y \in \mathcal{H}$, each segment $y_{[-\infty,a]}$ belongs, for any $a \in \mathbb{R}$, to the backward a-truncation of the space

$$S_\infty := \bigvee_{a \in \mathbb{R}} S_a$$

This means that $S_\infty = \mathcal{H}$. Finally,

$$S_{-\infty} := \bigcap_a S_a$$

consists of those functions $y \in \mathcal{H}$ such that

$$y(t) = C_+ e^{A_+(t-a)} x(a) \quad \text{for} \quad t \geq a \,,$$

for every $a \in \mathbb{R}$. This means that all $y(\cdot) \in S_{-\infty}$ are of the form $t \to C_+ e^{A_+ t} x$, *on the whole time axis.* Now, we may without loss of generality assume (S_+) irreducible. Setting u=0 in (4.3) and invoking observability of (C_+, A_+) we reach the conclusion that A_+ *must be stable,* i.e.

$$\mathrm{Re}\, \lambda(A_+) < 0 \tag{4.5}$$

(recall that we factored out the imaginary eigenspace of (S_+)). But then all nonzero element of $S_{-\infty}$ would have to be unbounded as $t \to -\infty$ contradicting uniform boundedness of each $y \in \mathcal{H}$. Therefore $S_{-\infty} = \{0\}$.

The above proves that S_o is incoming for the adjoint $S^*(t)$ of the translation group. Now define $\hat{S}_a := S_{-a}$, $a \in \mathbb{R}$. Then $\hat{S}_a = S(a)\hat{S}_o$, $a \in \mathbb{R}$ and $\hat{S}_o$ is outgoing for S(t).

□

REMARK

A totally analogous statement holds in case $\mathcal{H}$ has a *backward* realization (S_-). The stability of A_+, implied by the forward representation property, is now replaced by *antistability* of A_-, i.e.

$$\mathrm{Re}\, \lambda(A_-) > 0 \tag{4.6}$$

Then, by arguing exactly as in the proof of Theorem 4.2 it can be shown that

$$\mathcal{T}_a := \{y \in \mathcal{H};\ \exists x \in X \text{ such that } y(t) = C_- e^{A_-(t+a)} x, \text{ for } t \le a\} \tag{4.7}$$

form an increasing family of subspaces which fills up the whole of $\mathcal{H}$ as $a \to +\infty$ and shrinks to the zero function when $a \to -\infty$. In other words, *the translation group* $S(t)$ *admits an incoming subspace.*

□

Theorem 4.4 describes a key structural condition to be satisfied by aggregable Linear Hamiltonian Systems. This condition has a number of system-theoretical implications which we shall explore in the following. In order to do this we first need a representation Lemma.

LEMMA 4.3

Let $\Sigma^+ : L^2_q(\mathbb{R}) \to \mathcal{H}$ *be a linear bounded operator commuting with translation i.e.*

$$\Sigma^+ S(t) f = S(t)\Sigma^+ f \qquad \forall f \in L^2_q(\mathbb{R})$$

Assume Σ^+ *maps* $L^2_q(-\infty, a)$, *considered as a subspace of* $L^2_q(\mathbb{R})$, *onto* S_a, $\forall a \in \mathbb{R}$. *Then* Σ^+ *restricted to* $L^2_q(-\infty, a)$ *has the representation,*

$$(\Sigma^+ f)(t) = \int_{-\infty}^{a} W_+(t-s) f(s)\,ds, \qquad t \in \mathbb{R} \tag{4.8}$$

where $W_+(t) = C_+ e^{A_+ t} N_+$ *for* $t \ge 0$ *and is zero for* $t < 0$.

A dual result holds for any translation invariant operator $\Sigma^- : L^2_q(\mathbb{R}) \to \mathcal{H}$ *mapping* $L^2_q(a, +\infty)$ *onto the incoming subspace* $\mathcal{T}_a$, $\forall a \in \mathbb{R}$.

Proof:

Pick $f \in L^2_q(\mathbb{R}_-)$, then $y(t) = (\Sigma^+ f)(t)$, for $t \ge 0$, can be written as $C_+ e^{A_+ t} x(0)$ for a unique $x(0)$. There is then a well defined linear map $f \to x(0) \in X$ from $L^2_q(\mathbb{R}_-)$ onto the n-dimensional linear space X, which (assuming $X = \mathbb{R}^n$ without loss of generality) must necessarily have the form

$$x(0) = \int_{-\infty}^{0} Q(s) f(s)\,ds$$

for some nxq matrix Q with rows in $L^2_q(\mathbb{R}_-)$. Now S(-a)u is mapped into $S(-a)y \in S_a$ so that

$$y(t-a) = C_+ e^{A_+(t-a)} \int_{-\infty}^{a} Q(a,s)f(s-a)ds, \qquad t \geq a .$$

But S(t-a) also belongs to S_o and therefore

$$y(t-a) = C_+ e^{A_+(t-a)} e^{A_+ a} x(0) \qquad t \geq a .$$

□

Comparing the two expressions we get the conclusion.

Now, by a fundamental result in Analysis, the existence of incoming, resp. outgoing, subspaces S_a, $\bar{S}_a$, for $S^*(t)$ on $\mathcal{H}$ is equivalent to the existence of *unitary* operators (called *translation representations*),

$$\Sigma^+ : L^2_q(\mathbb{R}) \to \mathcal{H} \quad \text{and} \quad \Sigma^- : L^2_q(\mathbb{R}) \to \mathcal{H}$$

commuting with translation and also inducing unitary maps,

$$\Sigma^+ : L^2_q(-\infty, a) \to S_a, \qquad \Sigma^- : L^2_q(a, +\infty) \to \bar{S}_a$$

for all $a \in \mathbb{R}$. (Here we have chosen to work with the adjoint "left" translation $S^*(t)$, as it seems to be easier to visualize). The $S^*(t)$-outgoing subspace $\bar{S}_a$ is $\mathcal{T}_{-a}$ with $\mathcal{T}_a$ defined by (4.7). The number q is the *multiplicity* of S(t) on $\mathcal{H}$ (i.e. the minimal number of generating vectors for $\mathcal{H}$). We have $q \leq m$ as the m representatives $h_1, \ldots, h_m$ of the read-out map h^* are a (generally non minimal) set of generators for $\mathcal{H}$. Compare (3.4).

Using Lemma 4.3 we immediately obtain a representation of the type (4.8) for Σ^+ and a similar backward representation for Σ^-,

$$(\Sigma^- f)(t) = \int_a^{+\infty} W_-(t-s)f(s)ds, \qquad f \in L^2_q(a, +\infty) \tag{4.9}$$

where $W_-(t) = C_- e^{A_- t} N_-$ for $t \leq 0$ and zero for $t > 0$. Since both Σ^+ and Σ^- are injective on L^2_q the matrices N_+, N_- both have full column rank q. Now the translation representations Σ^+, Σ^-, are just the input-output maps of two linear systems

$$\begin{aligned} \dot{x}_+ &= A_+x_+ + N_+f \\ y &= C_+x_+ \end{aligned} \qquad (\hat{S}_+)$$

and

$$\begin{aligned} \dot{x}_- &= A_-x_- + N_-f \\ y &= C_-x_- \end{aligned} \qquad (\hat{S}_-)$$

both acting on the input space $L^2_q(\mathbb{R})$. It follows from unitarity that $(\hat{S}_+)$ and $(\hat{S}_-)$ are *irreducible* forward and backward realizations of $\mathcal{H}$. Moreover, as they have the same state spaces as the original realizations (S_+) and (S_-), it must hold that $B_+=N_+Q_+$, $B_-=N_-Q_-$ where $Q_\pm$ are qxq nonsingular matrices. So for an irreducible realization the number of inputs p must be equal to q = multiplicity of S(t). Actually more is true.

COROLLARY 4.4

There is just one choice of the input space $\mathcal{U}$ *(modulo isomorphisms of the input alphabet) by which any irreducible forward (or backward) realization becomes a one to one norm preserving map onto* $\mathcal{H}$ *(i.e. a Hilbert space isomorphism). This choice is* $\mathcal{U} = L^2_q(\mathbb{R})$, *where* q *is the multiplicity of* S(t).

Proof:

The corollary is just a restatement of the uniqueness of the translation representation. There is a technical detail on how one should naturally make (say) the forward representation, $\hat{\Sigma}^+_a : X \times L^2(a,+\infty) \to \mathcal{H}^+_a$ into a unitary map starting from a unitary $\Sigma^+ : L^2_q(\mathbb{R}) \to \mathcal{H}$. This involves introducing a suitable norm on X, but we shall skip the details.

□

From now on we shall fix $\mathcal{U} = L^2_p(\mathbb{R})$. For irreducible systems p = q, and every irreducible realization can be *identified* with a translation representation of $\mathcal{H}$. It is worth remarking at this point (although this fact has implicitly already been used) that, because of the translation representation theorem there is no need to think of (say) a forward realization of $\mathcal{H}$ as a representation holding only on half lines $[a,+\infty]$. In fact the *entire trajectories* of $\mathcal{H}$ are generated by the input-output map $u \to \Sigma_u^+$, $u \in L_q^2(\mathbb{R})$, of the realization. This is equivalent to saying that the system (S_+) *is started at the "boundary state"* $x(-\infty)=0$ at $t = -\infty$. Similarly, any backward realization generates $\mathcal{H}$ starting from the *boundary state* $x(+\infty)=0$.

All the preparatory work done so far eventually leads to a very explicit and neat characterization of aggregability.

THEOREM 4.5

If a Linear Hamiltonian System is aggregable then its spectral distribution measure M *is absolutely continuous. In fact, let* $\Psi(i\lambda)$ *be the* m×m *spectral density matrix of* M *and* $\hat{W}_+(s) = C_+(sI-A_+)^{-1}B_+$ *be the transfer function of a forward realization. Then* $\hat{W}_+$ *is an analytic (in* $\mathbb{R}e>0$*) spectral factor of* Ψ *i.e.*

$$\Psi(i\lambda) = \hat{W}_+(i\lambda)\hat{W}_+(-i\lambda)^T . \qquad (SF_+)$$

Similarly, the transfer function $\hat{W}_-(s) = C_-(sI-A_-)^{-1}B_-$ *of a backward realization is a coanalytic (analytic in* $\mathbb{R}e<0$*) spectral factor, i.e.*

$$\Psi(i\lambda) = \hat{W}_-(i\lambda)\hat{W}_-(-i\lambda)^T . \qquad (SF_-)$$

In particular, Ψ *must be a rational function of* $i\lambda$.

Proof:

Let (S_+) be an irreducible forward realization of $\mathcal{H}$. The translation representation corresponding to (S_+) can be composed with the unitary map T^{-1} of (3.4), yielding a (unitary) translation representation $T_+ : L_q^2(\mathbb{R}) \to$ $\to H$. If we rename $W_+(-s) := V_+(s)$, the equality

$$< \Phi^*(t)h,\ z > = y_z(t) = \int_{-\infty}^{+\infty} W_+(t-s)u(s)ds$$

can be rewritten as

$$< \Phi^*(t)h,\ z > = < S^*(t)V_+,\ u >_{L^2_q(\mathbb{R})}$$

for all corresponding pairs (u,z) under T_+. This clearly implies that under T_+, h_k *corresponds to the k-th row* $W_{+,k}(-\cdot)^T$ of $W_+(-\cdot)$. Note incidentally, that $W_{+,k}(-\cdot)$ is in $L^2_q(\mathbb{R})$. Hence the k-th column of the covariance matrix of the system has the expression

$$\Lambda_k(t) = \int_{-\infty}^{+\infty} W_+(t-s)W_{+k}(-s)^T ds$$

for k=1,...,m. This relation is equivalent to (SF_+). A totally analogous argument also works for (SF_-).

□

As we see, the existence of aggregate models depends ultimately on the spectrum of $\Phi(t)$ which has to be of *Lebesgue type* (plus perhaps a finite number of imaginary eigenvalues to take into account also the oscillatory component which we decided to ignore in order to simplify the exposition) and on the observables $<h_k,\cdot>$. An interesting question in this respect would be to "design" the observables of a Linear Hamiltonian System (known to have an evolution operator $\Phi(t)$ of finite multiplicity and with "essentially" Lebesgue Spectrum as described above) in such a way to obtain rationality of the spectrum.

Let us very explicitly point out that the condition of rationality of the Spectrum stated in the Theorem is also *sufficient* for aggregability. This has been shown (using a slightly different language) in PICCI [1986] and we shall not repeat the argument here. More then that, *any full rank rational analytic solution of* (SF_+) and, respectively, *any full rank rational coanalytic solution of* (SF_-), *provides a forward and, respectively, a backward representation of* $\mathcal{H}$. This is immediate from the spectral representation (3.10), which, given any such solution $\hat{W}$ of (say) (SF_+), can be rewritten as

$$y_z(t) = \int_{-\infty}^{+\infty} e^{i\lambda t}\, \hat{W}(i\lambda)\hat{u}(i\lambda)d\lambda \qquad (4.10)$$

where

$$\hat{u}\ (i\lambda):=\hat{W}(-i\lambda)^T f_z^T(i\lambda) \tag{4.11}$$

This u is just the driving input function corresponding to the initial phase z of the Linear Hamiltonian System. Note that by irreducibility, the functions

$$f_z(i\lambda) = \Sigma\, \alpha_k^T\, e^{i\lambda t_k}, \qquad \alpha_k \in \mathbb{R}^m$$

span $L^2(dM)$ (compare (3.2) and therefore the inputs given by (4.11) span exactly L_q^2. Here, of course, the number of inputs is determined by the full rank condition on $\hat{W}$. Recall that q = rank W = rank Ψ = multiplicity of $S(t)$.

COROLLARY 4.6

The transfer functions $\hat{W}_+$ *of the irreducible forward realizations of* $\mathcal{H}$ *(defined modulo multiplication from the right by a constant orthogonal* qxq *matrix) are the analytic spectral factors of* Ψ *which have minimal McMillan degree (also called minimal stable spectral factors). A completely analogous statement holds for the transfer functions* $\hat{W}_-$ *of the irreducible backward models.*

Proof:

We only need to prove the equivalence between minimality of $\hat{W}_+$ as a full rank spectral factor and irreducibility of the corresponding realization. Suppose $\hat{W}_+$ is non minimal, then it has a *coprime factorization* $\hat{W}_+=\hat{W}_*Q$ with Q a nonconstant rational inner matrix and $\hat{W}_*$ outer. "Coprime" means that the McMillan degrees of the factors add up to the MwMillan degree of $\hat{W}_+$. Now it easy to check that for any minimal realization of Q, say,

$$\begin{aligned} \dot{x}_Q &= A_Q x_Q + B_Q u \\ y_Q &= C_Q x_Q + u \end{aligned} \qquad \dim x_Q(t) = n_Q \qquad (Q)$$

for every $x=x_Q(0)$ there is a (feedback) input $u(\cdot)$ which makes $y_Q(t)=0$, $\forall t \in \mathbb{R}$ (i.e. $\mathcal{V}^* = \mathbb{R}^{nQ}$). This implies that the representation map $\Sigma_Q: X_Q \times \mathcal{U} \to (\mathbb{R}^q)^{\mathbb{R}}$ of the (minimal) system (Q) is certainly *not injective*. A minimal realization of $\hat{W}$ has state space $X = X_* \oplus X_Q$ and hence if $n_Q>0$ it cannot be irreducible since for initial states $x=col(0,x_Q)$ there is always an input function producing identically zero output.

□

We shall finally bring out the relation between forward and backward realizations. There is a natural pairing of forward-backward realizations of $\mathcal{H}$ which in a certain sense describes how a given (irreducible) realization is modified under "time reversal". In algebraic terms, this correspondence ties together solutions $(\hat{W}_+,\hat{W}_-)$ of (SF_+) and (SF_-) having the same zero structure.

THEOREM 4.7

Let (S_+) *be an irreducible forward realization of* $\mathcal{H}$. *There is a companion backward realization* (S_-) *describing the same output trajectories in the backward direction, i.e. if* y *is described as:*

$$y_a^+ = \Sigma_a^+ (x,u) \qquad \forall\, a \in \mathbb{R}$$

then there is a corresponding input v *of* (S_-) *such that,*

$$y_a^- = \Sigma_a^- (x,v) \qquad \forall\, a \in \mathbb{R} .$$

(S_-) *is starting at* t=a *from the same initial state as* (S_+), $x_-(a) = x_+(a)$. *The realization* (S_-) *has parameters*

$$A_- = -PA_+^TP^{-1} , \qquad B_- = B_+ \qquad C_- = C_+$$

where P *is the unique simmetric solution of the Lyapunov equation*

$$A_+P + PA_+^T + B_+B_+^T = 0$$

The input v *driving the backward systems (and producing the same output trajectory as the forward system) is given by*

$$v(t) = u(t) - B^TP^{-1}x_+(t) \quad .$$

□

5. EPILOGUE

As the picture is starting to become quite recognizable it is probably now a very good time to end the story. The striking similarity between the present setup and the Linear-Gaussian Stochastic realization framework as exposed e.g. in LINDQUIST PICCI [1985] could be pushed to the extreme details, at the risk of boring the reader to death. Instead of doing this we should just like to point out, very briefly, some unexpected "deterministic" versions of phenomena which a priori might be tought to be peculiar only of "stochastic" modelling.

One such thing is *irreversibility* of the aggregate description, which we would like to define as the capability of a model of describing the given behaviour $\mathcal{H}$ only in one direction of time. A *forward* model, starting in the state $x(t_0)$ only describes the trajectories of $\mathcal{H}$ for $t \geq t_0$. If we want to look backward (i.e. "reverse time") and describe the trajectories of $\mathcal{H}$ for $t \leq t_0$ we need to use the companion *backward* model which is definitely *different* (A_+ cannot be similar to $-A_+^T$) from the forward one.

This irreversibility at the aggregate level, contrasts sharply with the reversibility of the "microscopic" Hamiltonian system (where instead $-F^* = F$). In a *linear* context, the present theory "explains" a famous and fascinating paradox of Statistical Mechanics which seems to have been puzzling physicists since the end of last century. WILLEMS' notion of *completeness* has been instrumental in this explaination.

REFERENCES

AOKI M. [1976]. On fluctuations in microscopic states of a large system. *Directions in Large Scale Systems*, Y.C. Ho and S.K. Mitter eds. Plenum Press New York.

[1980]. Dynamics and control of a system composed of a large number of similar subsystems. *Dynamic Optimization and Mathematical Economics*. Pan Tai Liu ed. Plenum Press New York.

FUHRMANN P.A. [1981]. *Linear Systems and Operators in Hilbert Spaces*. Mc Graw Hill New York.

LAX P.D., PHILLIPS R.S. [1967]. *Scattering Theory*. Academic Press, New York

LINDQUIST A., PICCI G. [1985]. Realization theory for multivariable stationary Gaussian processes. *SIAM J. Control Optim.* *23*, 809-857.

LEWIS J.T., MAASSEN H. [1984]. Hamiltonian models of classical and quantum stochastic processes. *Quantum Probability and Applications to the Quantum Theory of Irreversible Processes.* L. Accardi, A. Frigerio and V. Gorini eds. Springer L.N. in Mathematics, 1055, Springer Verlag.

PICCI G. [1986]. Applications of stochastic realization theory to a fundamental problem of statistical physics. *Modelling Identification and Robust Control*, C.I. Byrnes and A. Lindquist eds. North Holland.
[1988a]. Hamiltonian representation of stationary processes. *Operator Theory Advances and Applications*, 35, pp. 193-215.
[1988b]. Stochastic aggregation. *Linear Circuits Systems and Signal Processing, Theory and Applications*, C.I. Byrnes, C. Martin, R. Saeks eds. North Holland.
[1989]. Stochastic aggregation of dynamical systems. Submitted for publication.

WILLEMS J.C. [1983]. Input-output and state space representations of finite-dimensional linear time-invariant systems. *Lin. Alg. Appl.*, 50, 581-608.
[1986]. From time series to linear systems. Part I. Finite Dimensional Linear Time Invariant Systems. *Automatica*, 22, 561-580.
[1988]. Models for Dynamics. *Dynamics Reported*, 2, Wiley and Teubner.

Linear System Representations

J. M. Schumacher

Centre for Mathematics and Computer Science (CWI)
Kruislaan 413, 1098 SJ Amsterdam, the Netherlands
and
Department of Economics, Tilburg University
P. O. Box 90153, 5000 LE Tilburg, the Netherlands

1. Introduction

The theory of system representations is concerned with the various ways in which a 'system' (a dynamical relation between several variables) can be described in mathematical terms. This paper will concentrate on the class of linear, time-invariant, deterministic, finite-dimensional systems, for which there exists indeed a variety of representations. The study of system representations is of interest for two reasons, which correspond to two different points of view. First of all, even when representation types (or 'model classes') are mathematically equivalent, the ease with which a particular problem is handled may be quite representation-dependent. Also, it may happen for instance that a problem is best understood theoretically in one representation, but that another representation is most useful for the numerical solution. Thus, one should be able to switch from one representation to another. The study of the corresponding transformations belongs to representation theory. The second reason for interest in system representations is connected with the modeling problem. Often, a model for a physical system is built up by writing down equations for the components and for the connection constraints. In this way, one obtains a system representation. It may be useful, though, to rewrite the equations; the derivation of the Euler-Lagrange equations of mechanics could be cited as an example. Again, we have here a problem of transformation between system representations.

Interest in the theory of system representations has been stimulated in recent years by a series of papers by J. C. Willems [64, 68-70, 72, 73]. In this work, the 'modeling' point of view has been emphasized. As noted by Willems, even such raw data as an observed time series can already be taken as a system representation, and the identification problem then becomes a problem of transformation of representations. In this paper, we shall concentrate on representations by equations rather than by measured data. A survey of system representations and transformations will be presented in the spirit of [71]. We shall use the notion of 'external equivalence', again following

Willems.

The next section contains a brief historical survey of system representations in connection with control theory, centered on the description of linear, finite-dimensional, deterministic systems. After that, we shall attempt to give an up-to-date account of the results concerning the representation of this class under external equivalence. Section 4 will be devoted to an application of the theory to the idea of a factor system, and the paper will be closed with conclusions and research perspectives.

2. System representations: a historical sketch

The birth of mathematical control theory is often dated 1868, the year of the publication of J.C. Maxwell's paper "On Governors" [42]. In this paper, Maxwell deals with a number of contrivances that in his time were in use to regulate the operation of steam engines. Maxwell uses second order equations to describe the motions of the engine itself and the regulators. He takes the coupling of the different parts into account and linearizes to obtain a coupled set of second-order linear differential equations. As an example, the following equations appear for a steam engine regulated by a combination of Thomson's governor with Jenkin's governor (in Maxwell's notation):

$$A\frac{d^2\theta}{dt^2} + X\frac{d\theta}{dt} + K\frac{d\phi}{dt} + T\phi + J\psi = P - R$$

$$B\frac{d^2\phi}{dt^2} + Y\frac{d\phi}{dt} - K\frac{d\theta}{dt} = 0$$

$$C\frac{d^2\psi}{dt^2} + Z\frac{d\psi}{dt} - T\phi = 0.$$

Here, $P - R$ denotes the effective driving torque. The main variable is θ, which represents the deviation of the main shaft angle from its nominal value. The variables ϕ and ψ correspond to the two governors. Maxwell then writes the general solution for θ, which, by the standard theory of ordinary differential equations, involves a linear combination of exponential functions. These exponential functions are determined by the roots of a polynomial equation that can be derived readily from the given system. Maxwell writes n for the unknown, and obtains a fifth-degree equation by setting

$$\begin{vmatrix} An^2 + Xn & Kn + T & J \\ -K & Bn + Y & 0 \\ 0 & -T & Cn^2 + Zn \end{vmatrix} = 0$$

(a factor n has been cancelled right away in the second row). He is then confronted with the problem of determining conditions on the coefficients under which all solutions of this equation are located in the left half of the complex plane. This, of course, led to the work of Routh on conditions for the stability of polynomials of arbitrary degree. We see that Maxwell's fifth-order equation arises from the application of a fourth-order controller to a second-order system, and that the conditions for stability are given by him in terms of the zeros of a polynomial matrix that is obtained directly from a standard modeling procedure.

Maxwell used second-order differential equations, but it gradually became standard in the nineteenth century to write differential equations in first-order form. The fact that a higher-order differential equation in one variable may be replaced by a first-order equation in several variables was actually already known in Cauchy's time. The Lagrangian equations of mechanics were later put into a suitable first-order form by Hamilton; towards the end of the century, Poincaré and Lyapunov used first-order vector representations systematically. Naturally, therefore, representations of this type (called *state* representations later on) have dominated control-theoretical work that was done in close connection with the theory of ordinary differential equations. This concerned mainly linear stability theory at first, but later, in the first decades of the twentieth century, attention shifted to nonlinear problems. This line of research was held up high especially in the USSR (see for instance the survey by Minorsky in [45]).

The work in connection with differential equations had a natural tendency to emphasize closed-loop systems, obtained by combining a given system with a given controller. Indeed, for such systems one may readily apply the powerful methods from the theory of ordinary differential equations and allied disciplines, such as the theory of differential-difference (delay) equations. The analysis by Maxwell, as briefly described above, is an example of this approach. The closed-loop point of view is quite satisfactory for many problems in mechanical engineering. To the communications engineer, however, it is more natural to use an open-loop point of view, in which a system is viewed as an operation that acts on an input signal and produces an output signal. This 'operational' point of view called for a representation which would express the output signal as the result of some operator acting on the input signal. Such a representation is provided, at least for linear systems, by the convolution integral. However, competing representations were soon to appear. Indeed, the use of complex quantities for the representation of complex signals, the Fourier and Laplace transforms, and Heaviside's Operational Calculus were all in principle available by the turn of the century. The value of these techniques was gradually recognized among electrical engineers, be it certainly not without resistance (see for instance [46]). From the mathematical point of view, the use of operational methods led to the introduction of techniques quite different from the ones usually found in the theory of differential equations. Applications of complex function theory were limited at first to partial fraction expansions and computation of integrals, but the appearance of the Nyquist criterion [47] made engineers realize that full-fledged function theory was a natural tool to use in the analysis of linear systems [10, p. 9]. Function-theoretic tools, in particular Cauchy's theorem, were used extensively by Bode in his book [9], which incorporated the celebrated Bode gain-phase relation and the minimum phase concept. The development of the root locus method by Evans in 1948 [18] firmly established the view of the transfer function as a function defined on the complex plane rather than just on the real frequency axis. For a more extensive discussion of the development of frequency-domain methods, we refer to [40]. We will not at all review the developments in the area of stochastic systems. In connection with what just has been said, however, it is interesting to quote Wiener on some of the differences between his own work and that of Kolmogorov:

> ... my work, unlike the explicitly published work of Kolmogoroff, concerns the instrumentation which is necessary to realize the theory of prediction in automatic apparatus for shooting ahead of an airplane. This engineering bias leads me to emphasize more than does Kolmogoroff the problem of prediction in terms of linear operators in the scale of frequency, rather than in similar operators on the scale of time. [63, p. 308]

While the communication engineers developed their own methods, work on the ODE-type approach to control systems was still continuing, in particular in the Soviet Union. During the Second World War, a research centre was formed in Kazan where work on applied problems was done by outstanding mathematicians such as L. S. Pontryagin, who had already acquired fame because of his pre-war contributions to topological algebra. After the war, research efforts in control theory continued at various mathematical institutes in the USSR. One important research direction centered around 'Aizerman's conjecture' [1], a nonlinear generalization of the Nyquist criterion. This problem called for a representation of systems with an explicitly appearing input variable, unlike the setting that was mainly used before in the 'ODE' framework. Systems with one input were studied first, in line with the original work of Nyquist, but the extension to several inputs was a natural one. For instance, Letov [35] considered in 1953 the following system (in original notation):

$$\dot{\eta}_k = \sum_{\alpha=1}^{n} b_{k\alpha}\eta_\alpha + n_{k1}\xi_1 + n_{k2}\xi_2 \qquad (k = 1, \cdots, n),$$

$$\dot{\xi}_1 = f_1(\sigma_1), \quad \sigma_1 = \sum_{\alpha=1}^{n} p_{1\alpha}\eta_\alpha - r_{11}\xi_1 - r_{12}\xi_2\,,$$

$$\dot{\xi}_2 = f_2(\sigma_2), \quad \sigma_2 = \sum_{\alpha=1}^{n} p_{2\alpha}\eta_\alpha - r_{21}\xi_1 - r_{22}\xi_2\,.$$

We recognize the first equation (with hindsight, perhaps) as a linear state equation with two inputs.

The early fifties saw the rise of modern *optimal control theory*. One of the first problems to be studied was time-optimal control. In some applications, it is natural to consider control strategies in which one switches between full power in one direction and full power in the reverse direction. This motivated a study of differential equations with discontinuous forcing terms by D. W. Bushaw at Princeton University [11]. Bushaw noted that the switching instant could be optimized to obtain a transfer from one state to another in minimal time. Subsequently, J. P. LaSalle observed that 'bang-bang' policies would be optimal among all possible control policies which lead from a given state to another. LaSalle used a nonlinear formulation, but later on Bellman *et al.* considered linear systems [7]. In this paper, Bellman and his co-authors required invertibility of the input matrix (as we would now call it), so in particular they let the number of inputs be equal to the number of states. In independent work, Gamkrelidze [25] considered shortest time problems for linear systems with n states and r inputs. He writes the following state equation [25, p. 451]:

$$\dot{\mathbf{x}} = A\mathbf{x} + \mathbf{b}_1 u^1 + \cdots + \mathbf{b}_r u^r$$

which is practically the formula $\dot{x} = Ax + Bu$ that has become ubiquitous in control theory.

By the end of the fifties, the time had come for an amplification of the notion of 'state' far beyond its meaning as the vector that appears when dynamic equations are written in a first-order form. This was due to the role that this concept had to play in Bellman's dynamic programming method, but also to developments in the theory of automata (finite state machines; Nerode equivalence).

In control theory, the announcement by Pontryagin of his Maximum Principle at the International Mathematical Congress in Edinburgh in 1958 had a tremendous impact on research in optimal control. Bang-bang control problems, in which one seeks to steer from one state to another, naturally led to the formulation of the concept of *controllability* by R. E. Kalman. This concept, and the dual notion of observability, turned out to play a crucial role in what Kalman called the *realization problem*:

> Given an (experimentally observed) impulse response matrix, how can we identify the linear dynamical system which generated it? [30, p. 153]

The word 'realization' is used here in a sense that is different from the traditional usage in electrical engineering. There, one would look for realization of a given driving-point impedance as an actual or idealized electrical circuit (cf. also the use of the term 'realize' in the quotation from Wiener given above). Although Kalman did advertize the state space realization as a 'blueprint' which could serve as a basis for implementation in an analog network [28], this connection was hardly emphasized in subsequent research.

In the newly founded SIAM Journal on Control, E. Gilbert argued that the transfer representation was misleading and could lead to erroneous results. His point was that unobservable and/or uncontrollable states could be created by system composition:

> Thus transfer-function matrices may satisfactorily represent all the dynamic modes of the subsystems but fail to represent all those of the composite system. Furthermore, the loss of hidden response modes is not easy to detect because of the complexity of the transfer-function matrices and matrix algebra. [27, p. 140]

To develop linear control theory from the state space point of view, it had to be shown that the familiar concepts from the frequency domain could be translated to state space terms. For this, the new realization theory was an indispensible tool. Gilbert [27] used partial fraction expansion (much in the tradition of Heaviside, one might say) to obtain a state space realization for a transfer matrix having only simple poles. This method can be extended to the general situation (not necessarily simple poles), but then becomes somewhat involved (see [50]). A more elegant realization algorithm was published by Kalman and B. L. Ho in 1966 [28]. The algorithm was based on a new parametrization of the transfer matrix — new at least to control theory: in 1894, A. A. Markov had already used essentially the same parametrization for a study of continued fractions [41]. The 'Markov parameters' are the first (matrix) coefficients in the power series development around infinity of a proper rational matrix.

For a while, 'realization theory' was, at least to the system theorist, practically equivalent to the determination of a state space representation from a transfer matrix given through its Markov parameters. The seventies, however, brought a renewed interest in polynomial representations. An important impetus for this development came from the appearence of Rosenbrock's book [51] on multivariable systems. In this work, Rosenbrock considered input/output systems given in the form

$$T(s)\xi = U(s)u$$
$$y = V(s)\xi + W(s)u$$

where all matrices are polynomial. Great emphasis was placed on the study of *equivalence* notions. Rosenbrock found a 'lifting' of Kalman's system equivalence concept to the more general representation displayed above, which he called *strict system equivalence*. It seems safe to say that the systematic development of the theory of system representations, system equivalence and system transformations starts with [51].

From Rosenbrock's system matrix, the transfer matrix is represented as $V(s)T^{-1}(s)U(s) + W(s)$, i.e., as a ratio of polynomial matrices. It is not difficult to see that, in fact, every rational matrix can be written in either of the two forms $V(s)T^{-1}(s)$ or $T^{-1}(s)U(s)$, where, moreover, a *coprimeness* condition may be imposed. These *coprime fractional representa-*

tions were used very successfully by Kučera [31, 32] and by Youla *et al.* [79, 80] to give a parametrization of all stabilizing controllers for a given plant. This is an example of a result that appears quite naturally in one representation but would be awkward to derive in some other representations. At the same time, fractional matrix representations were also used in work on infinite-dimensional realization problems done at Harvard University by R. W. Brockett, J. S. Baras, and P. A. Fuhrmann (see for instance [6]). In the infinite-dimensional context, the available mathematical tools strongly suggested to replace polynomials by functions analytic on the unit disk (in the discrete-time case — for continuous-time systems, the class to use would be the set of functions that are analytic on the right half plane). This idea was picked up by researchers in finite-dimensional system theory, who discovered that some difficulties with the Kučera-Youla parametrization could be ironed out by using the ring of rational functions that have no poles in the closed right half plane (including the point at infinity) rather than the ring of polynomials (see for instance [16]). The fractional representation over the ring of proper and stable rational functions was subsequently used extensively in the emerging H^∞-theory, which is in itself an example of an application of function-theoretic techniques to control problems in a way that would probably have been quite beyond the imagination of Nyquist and Bode. On the other hand, H^∞-theory has also relied heavily on state space representations, since the representation in terms of constant matrices makes it possible to use standard numerical software. The cooperation between the two representations was facilitated by the discovery (attributed to D. Aplevich in [62]) that there is an easy way to pass from a state space representation to a fractional representation over the ring of proper stable rational functions. (Fractional representations over the ring of polynomials cannot be obtained in a comparable way from a state space representation.)

Nevertheless, polynomial representations were emphasized again in the mid-seventies when Fuhrmann worked out an elegant procedure to go from a polynomial matrix fraction representation to a state space representation [23]. The discovery of this procedure, now known as *Fuhrmann's realization*, spurred considerable research on the relation between state space concepts, as developed in particular in the 'geometric approach' to linear systems [77], and polynomial or transfer matrix concepts. For an introduction to this, see for example Chapter 1 of [24].

Polynomial matrices, even when less suitable for a number of purposes than stable proper rational matrices, are important in system theory because they arise naturally in modeling. Indeed, a polynomial matrix representation can be written down immediately from a set of linear differential and algebraic equations describing a given system. Maxwell's equations for the controlled steam engine, as given above, may serve as an example. Of course, by the old trick of replacing higher-order derivatives by new variables, it is also possible to obtain a first-order representation. Instead of the Rosenbrock form discussed above, one then gets a representation in the form

$$E\dot{x} = Ax + Bu$$
$$y = Cx + Du,$$

where E, A, B, C, and D are constant matrices. The variable "x" which appears here was called the *descriptor variable* by Luenberger, who was first to make an extensive study of this representation in system theory [37, 38]. Contrary to the standard state space representation, the descriptor form is capable of representing systems having a non-proper transfer function (also called 'non-causal systems' or 'singular systems'). Through the years, the term 'descriptor system' has come to be used almost exclusively for such systems, although this was certainly not Luenberger's original intention — he was trying to emphasize the modeling issue, rather than the question of causality. The descriptor form was used by Verghese [60] to define an equivalence concept which deals neatly with pole/zero cancellations at infinity. This cleared up a problem which had remained unsolved in Rosenbrock's work. Alternative solutions were given later by Anderson, Coppel and Cullen [2] and by Pugh, Hayton and Fretwell [48, 49]. The fact that the notions of equivalence defined by these authors are indeed the same was established by Ferreira [19]. Further comments on descriptor systems will be given in the next section.

In recent years, the study of system representations has been stimulated by the work of J. C. Willems. There are several important points where his approach is different from other approaches discussed above. First of all, Willems uses an intrinsic definition of system equivalence (i.e., one that does not depend on a specific representation). He does this by defining a 'system' simply as 'a family of trajectories of given variables' (such as the port voltages and currents of an electrical network, or forces and displacements in a mechanical system). The given variables which appear in the definition are also denoted as 'external variables', to distinguish them from 'internal variables' which are possibly used as auxiliary quantities in a description of the system. The external variables may consist of what are usually called 'inputs' and 'outputs', but, as shown in section 4 of this paper, other interpretations can sometimes also be useful. The family of trajectories is also referred to as a 'behavior' $\mathcal{B}$.

In this approach, there is some flexibility associated with the choice of the function space to which the trajectories that make up the system are supposed to belong. In the study of differential equations, one normally uses function spaces that allow for exponentially growing solutions (such as C^{∞}, or the space of distributions). In the context of system theory, however, it also makes sense to consider for instance only those trajectories that are square integrable. Different choices of function spaces lead, in this way, to different notions of system; put in another way, they lead to different equivalence relations on system descriptions. More on this will be said below. Willems has shown [68] that, if one interprets 'external variables' as 'inputs and outputs' and uses the classical function spaces alluded to above, the equivalence relation that emerges is in fact different from the equivalence relations that were mentioned above.

It should be noted that the definition of a 'system' as a family of trajectories is not new. Compare, for instance, McMillan's definition of a $2n$-pole:

> The constraints imposed by a general $2n$-pole N on voltages and currents are completely described by the totality of pairs $[\underline{v}, \underline{k}]$ which N admits. We shall *define* a general $2n$-pole, therefore, as
> (i) a collection of n oriented ideal branches, as in 4.11, and
> (ii) a list of pairs $[\underline{v}, \underline{k}]$ of voltages and currents admitted in these branches.
> [44, p. 228]

(The oriented ideal branches in §4.11 of McMillan's paper serve just to define the pairing of the terminals.) In recent work in system theory, the equivalence notion as used by Willems has in fact occurred in several places; see [4, p. 513] ('external equivalence') and [8, p. 92] ('input-output equivalence'). Nevertheless, there is no doubt that the consequences of the acceptance of this intrinsic definition of what a system is have been explored to the fullest in the work of Jan Willems.

3. A ROAD MAP OF REPRESENTATIONS

In this section, we shall review the available representations for a specific class of systems, viz., the class of finite-dimensional, deterministic, time-invariant, real, linear systems in continuous time, without further special structure. (The addition 'without further special structure' refers to the fact that we shall not consider special properties that arise, for instance, for systems defined on a symplectic space.) This is the class that has served as sort of a standard in system theory during the last three decades, except that *causality* is often imposed as an additional requirement. This condition was not included in the list above for two reasons. First of all, we are sometimes interested in external variables that are not to be considered as 'inputs' and 'outputs' (cf. Section 4 of this paper, for instance), and in such cases causality need not be a relevant issue. Secondly, even when we do distinguish inputs and outputs, there are no simple ways to tell, at a general level of representation such as Rosenbrock's system matrix, whether a given system is causal or not [51, p. 51]. Imposing causality as a constraint on such general linear system representations would therefore be awkward.

3.1 Notions of equivalence

When discussing system representations, we will have to specify under which circumstances we shall say that two representations are equivalent in the sense that they correspond to the same system. There are three main options. There is the notion of *strong equivalence*, which boils down to Kalman's concept of equivalence for causal input/output systems in standard state space form. Definitions of this equivalence (by specification of a list of allowed transformations) were given at the level of descriptor systems by Verghese [60] and by Pugh *et al.* [48, 49], and by Anderson *et al.* at the level of the Rosenbrock system matrix [2]. Secondly, for every class of representations that have a given input/output structure and that define a transfer matrix, one has the notion of *transfer equivalence* according to which two representations are equivalent if and only if they define the same transfer matrix. Finally, if one considers representations that define a family of trajectories of the external variables (an 'external behavior' in the sense of [68]), then there is the notion of *external equivalence* according to which two representations are equivalent if and only if they induce the same external behavior.

As noted before, external equivalence can in fact be understood in various ways, depending on the choice of a function space for the trajectories, and on the choice of external variables. There is also some freedom that arises from the interpretation of the external variables. For example, if we allow only permutation transformations on the external variables, this means that these variables are interpreted as quantities which each have there own meaning and are measured on a fixed scale. On the other hand, if we allow general invertible linear transformations, then the implication is that the vector of external variables is understood as an element of a general linear space. It goes without saying that, depending on the problem one has at hand, some of the external variables can be interpreted in one way and others in another way. (The same might be said about the choice of a function space.) The term 'external equivalence' will be used for what might be called the 'classical' interpretation: the function space is such that exponentially growing solutions are admitted (we shall use C^∞ to make life a little bit easier), and only permutation operations will be allowed on the external variables. We call this the 'classical' form because it would seem that the notion of equivalence that is used (often implicitly) in treatments of ordinary differential equations is of this type. If one uses an L_2-space rather than a C^∞-space as a trajectory space, then (cf. [74]) the corresponding notion of external equivalence turns out to be an extension of transfer equivalence, in the sense that it coincides with transfer equivalence on the class of systems that define a transfer matrix. Suppose now that one has a system of equations in the form

$$\sigma x = Ax + Bu \tag{3.1}$$

$$y = Cx + Du. \tag{3.2}$$

One might propose to take u, y, and x as external variables following C^{∞}-trajectories, to interpret u and y in a 'classical' sense, and to interpret x as a variable in a general linear space. The resulting concept of equivalence is Kalman's equivalence. It may be suspected that a similar re-interpretation in terms of external equivalence is also possible for strong equivalence.

To keep the presentation manageable, we shall consider transformations under 'classical' external equivalence. For other types of equivalence, the picture will be different but similar. We will discuss special representations for systems equipped with an i/o structure, but the particular representations that are available only for causal systems will be omitted.

3.2 A catalog of representations

We start by listing a number of representations. A number of basic types will be distinguished that are different by appearance; within these, we distinguish subtypes that do not differ notationally but that are subject to more or less severe constraints.

The most unspecific type of representations we shall take into consideration is the AR/MA class. An AR/MA representation is specified by two polynomial matrices $P(s)$ and $Q(s)$, which determine the external behavior consisting of all trajectories w of the external variables for which there exists a trajectory ξ of the internal variables such that

$$\begin{aligned} P(\sigma)\xi &= 0 \\ w &= Q(\sigma)\xi. \end{aligned} \tag{3.3}$$

In the continuous-time interpretation we use here, σ stands for d/dt. The class is called AR/MA because of the discrete-time interpretation in which σ is the shift: in this case, (3.3) implies that the external variables are expressed as a moving average of the internal variables, which themselves satisfy an autoregressive equation. Every representation of this kind can trivially be rewritten as a 'system with auxiliary variables' [68] (later also called an 'ARMA' representation by Willems [73]), which is defined by an equation of the form

$$P'(\sigma)\xi + Q'(\sigma)w = 0; \tag{3.4}$$

simply take

$$P'(s) = \begin{bmatrix} P(s) \\ Q(s) \end{bmatrix}, \quad Q'(s) = \begin{bmatrix} 0 \\ -I \end{bmatrix}. \tag{3.5}$$

On the other hand, it is also easy to write an AR/MA representation for a system with auxiliary variables, by extending the space of internal variables and writing

$$P(s) = [P'(s) \quad Q'(s)], \qquad Q(s) = [0 \quad I]. \tag{3.6}$$

We see that the AR/MA representation is, in general, less parsimonious in the use of internal variables than the representation as a system with auxiliary variables. Since we are looking for an unspecific representation, this might be construed as an argument against the representation in the form (3.4). Actually, when dealing with systems described by *partial* differential equations, one easily runs into clear-cut cases in which an AR/MA representation appears much more naturally than a representation with auxiliary variables as in (3.4).

For systems with an i/o structure, another general representation is RSM (Rosenbrock system matrix [51]). An RSM representation is specified by four polynomial matrices $T(s)$, $U(s)$, $V(s)$, $W(s)$, where $T(s)$ is square and invertible. The external behavior defined by an RSM representation consists of the set of all input trajectories u and output trajectories y for which there exists an internal-variable trajectory ξ such that the following equations hold:

$$\begin{aligned} T(\sigma)\xi &= U(\sigma)u \\ y &= V(\sigma)\xi + W(\sigma)u. \end{aligned} \tag{3.7}$$

The third polynomial representation we shall consider is the AR representation [69]. An AR representation is specified by a single polynomial matrix $R(s)$, which should have as many columns as there are external variables. The external behavior it defines is simply the set of all external-variable trajectories w satisfying

$$R(\sigma)w = 0. \tag{3.8}$$

We shall always require $R(s)$ to have full row rank; this simply means that the equations specified by the rows of $R(s)$ are independent. An AR representation given by $R(s)$ will be said to be *minimal* if the sum of the row degrees of $R(s)$ is minimal in the set of all AR representations of the same system. One can show (see for instance [69, Thm.6]) that a matrix $R(s)$ is minimal in this sense if and only if it is row proper. The class of minimal AR representations will be denoted by $\mathrm{AR}_{\mathrm{min}}$. If the external variable is partitioned into inputs and outputs, the defining matrix $R(s)$ of an AR representation will be divided into two blocks $R_1(s)$ and $R_2(s)$, which correspond to outputs and inputs respectively. If $R_1(s)$ is square and nonsingular, the representation so obtained will be called an LMF representation ('left matrix fraction').

By introducing new internal variables, it is easy to transform an AR/MA representation to a first-order form

$$\begin{aligned} \sigma G\xi &= F\xi \\ w &= H\xi \end{aligned} \tag{3.9}$$

(F, G, and H are constant matrices). This representation, specified by the three matrices F, G, and H, will be called the *pencil* representation ([33]; cf. also [4, 56]), and the corresponding class of representations will be denoted by P. To be complete, one should also indicate the spaces on which the various mappings are defined, and so we shall sometimes also give a P representation as a six-tuple $(F, G, H; Z, X, W)$ where F and G are mappings from the 'internal variable space' Z to the 'equation space' X, and H maps Z into the external variable space W. A descending chain of subclasses can be formed by putting more and more strict requirements on the triple (F, G, H). If G is surjective, the corresponding class will be denoted by P_{dv}, because this class is closely related to the DV representations that will be discussed below. The class of representations which in addition satisfy the condition that $[G^{\mathrm{T}} \;\; H^{\mathrm{T}}]^{\mathrm{T}}$ is injective will be denoted by P_{io}; in a representation of this type, one can easily see which partitionings of the external variables into inputs and outputs will lead to a causal i/o structure (cf. [33], Lemma 5.1 and Lemma 6.1). Finally, pencil representations that also satisfy the requirement that $[sG^{\mathrm{T}} - F^{\mathrm{T}} \;\; H^{\mathrm{T}}]^{\mathrm{T}}$ has full column rank for all $s \in \mathbb{C}$ form a class that will be denoted by $\mathrm{P}_{\mathrm{min}}$. It has been shown in [33] (Prop. 1.1) that a pencil representation is minimal under external equivalence if and only if it belongs to $\mathrm{P}_{\mathrm{min}}$.

Next in our collection of representations is the DV (driving-variable) representation [4, 68, 69, 73], which, as already mentioned, is closely related to the P_{dv} class. A DV representation is specified by four constant matrices A, B, C', and D', which determine an external behavior by the equations

$$\begin{aligned} \sigma\xi &= A\xi + B\eta \\ w &= C'\xi + D'\eta \end{aligned} \tag{3.10}$$

(ξ and η are auxiliary variables). The class of DV representations for which the matrix D' is injective will be denoted by $\mathrm{DV}_{\mathrm{io}}$. If also the requirement is imposed that the 'system pencil'

$$\begin{bmatrix} sI - A & B \\ C' & D' \end{bmatrix}$$

has full column rank for all s, then we obtain a class of representations that will be denoted by $\mathrm{DV}_{\mathrm{min}}$. It has been shown in [68] that a $\mathrm{DV}_{\mathrm{min}}$ representation is minimal in the class of DV representations, in the sense that both the length of ξ and the length of η are minimal.

For input/output behaviors, there are further special representations that may be used. A well-known form is the *descriptor* representation [37, 38]. The class of such representations will be denoted by D. A descriptor representation is specified by five constant matrices E, A, B, C, and D, and determines an input/output behavior by the equations

$$\begin{aligned} \sigma E\xi &= A\xi + Bu \\ y &= C\xi + Du. \end{aligned} \tag{3.11}$$

The domain of the mappings E and A will be denoted by X_{d} (descriptor space), the codomain will be written as X_{e} (equation space).

Quite a few special properties have been used in the literature in connection with this representation (see for instance [5, 13, 36, 52, 61, 78]. We shall use the following conditions. The representation (3.11) is said to be *controllable at infinity* if

$$\operatorname{im} E + \operatorname{im} B + A(\ker E) = X_{\mathrm{e}}. \tag{3.12}$$

It is said to be *reachable at infinity* if

$$\operatorname{im} E + \operatorname{im} B = X_{\mathrm{e}}. \tag{3.13}$$

It is called *observable at infinity in the sense of Verghese* if

$$\ker E \cap \ker C \cap A^{-1}[\operatorname{im} E] = \{0\} \tag{3.14}$$

and *observable at infinity in the sense of Rosenbrock* if

$$\ker E \cap \ker C = \{0\}. \tag{3.15}$$

The representation (3.11) is said to have *no nondynamic variables* if

$$A(\ker E) \subset \operatorname{im} E. \tag{3.16}$$

These are all properties that relate to the point at infinity. We note that, for representations that satisfy (3.16), there is no difference between controllability and reachability at infinity or between the two notions of observability at infinity. In connection with the finite modes, we shall need the following condition: a representation of the form (3.11) is said to have *no finite unobservable modes* if

$$\operatorname{rank}_{\mathbf{R}} \begin{bmatrix} sE - A \\ C \end{bmatrix} = \operatorname{rank}_{\mathbf{R}(s)} \begin{bmatrix} sE - A \\ C \end{bmatrix} \quad \text{for all } s \in \mathbf{C}. \tag{3.17}$$

In principle, a considerable number of descriptor representation types could be formed by taking combinations of the six conditions mentioned above. We shall consider just four types, which together seem to present a reasonable hierarchy. The most unspecific type is the general descriptor form, for which the symbol D has already been introduced. The symbol D_{ri} will be used for the class of descriptor representations that are reachable at infinity. Descriptor representations that have no nondynamic variables and that are both controllable and observable at infinity will be denoted as D_{mi} representations ('minimal at infinity'). Finally, the class of D_{min} representations consists of the D_{mi} representations that have no finite unobservable modes. It is shown in [34] that a descriptor representation is minimal under external equivalence if and only if it belongs to this class.

3.3 The road map

To indicate the connections between the somewhat vast number of representations introduced above, we shall now present a map. The following organizational principles have been applied:

- polynomial representations are on the left, first-order representations on the right;
- representations that do not distinguish between inputs and outputs are in the middle, i/o representations are on the extremes;
- more specific representations are higher up in the diagram than less specific ones.

Moreover, arrows have been used to indicate known transformation procedures (including the trivial ones, which involve no transformation at all, and the very easy ones, such as the transformation from AR to AR/MA). The organization of the diagram is such that arrows going up represent the heaviest computational loads. The result is shown in Fig. 1 below.

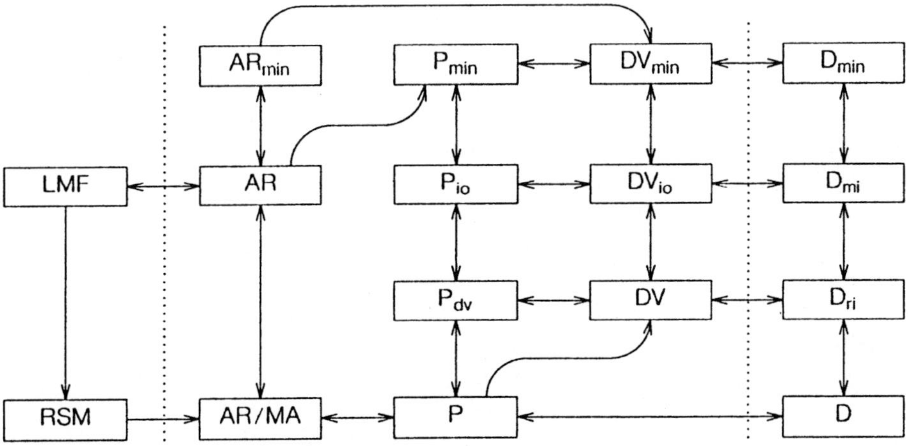

FIGURE 1. Representations and transformations of linear systems.

The arrows going *down* in this diagram all correspond to trivial rewritings or re-interpretations. For instance, an LMF representation is a special case of an RSM representation, obtained by taking $V(s) = I$ and $W(s) = 0$. The connection between LMF and AR is also quite clear. One gets from an RSM representation to an AR/MA representation simply by identifying the inputs with new internal variables. It is quite obvious how to transform the various types of

DV representations to the P representations on the same level, and vice versa. The transformation from AR/MA to P is by the standard trick of replacing higher-order derivatives by new variables. Most of the other transformations require more work, however, and some of the corresponding algorithms will be discussed below.

3.4 Algorithms

We start with the transition from an AR/MA representation to an AR representation. For this, we have the following procedure.

ALGORITHM 1 Let an AR/MA representation be given by $(P(s), Q(s))$. For instance by the algorithm of reduction to Hermite form ([39, pp. 32-33]; see also [29, pp. 375-376] or [12, p. 34]), find a unimodular matrix $U(s)$ such that

$$\begin{bmatrix} U_{11}(s) & U_{12}(s) \\ U_{21}(s) & U_{22}(s) \end{bmatrix} \begin{bmatrix} P(s) \\ Q(s) \end{bmatrix} = \begin{bmatrix} T(s) \\ 0 \end{bmatrix} \tag{3.18}$$

where $T(s)$ has full row rank. Let $R(s)$ be a maximal selection of independent rows from $U_{22}(s)$. Under these conditions, $R(s)$ gives an AR representation that is externally equivalent to the AR/MA representation $(P(s), Q(s))$.

For a proof of this, see [68, Prop. 3.3] or [33, Lemma 4.1]. The algorithm in [68] is actually based on the Smith form; from a computational point of view, this presents a considerable amount of overkill. In the algorithm given above, is is easy to see that $U_{22}(s)$ will automatically have full row rank (so that we simply have $R(s) = U_{22}(s)$) when $P(s)$ has full row rank, which is a natural restriction to impose.

The passage from AR to AR_{min} is just the reduction of a polynomial matrix to row proper form. The standard algorithm to do this is described for instance in [76, pp. 27-29] and in [29, p. 386]. This algorithm essentially requires only operations on constant matrices, and the computational load involved is in general much less than in a transition from AR/MA to AR form.

The steps leading from P to DV, from DV to DV_{io}, and from DV_{io} to DV_{min} are detailed in [56]. These steps can be 'lifted' to the level of P representations, and, in fact, it turns out that they can be derived quite naturally in this context. We shall now explain this in some detail.

First, consider the transition from a general P representations to the P_{dv} representation. From the equation $\sigma G\xi = F\xi$, it follows that any ξ-trajectory satisfying this equation must belong to the subspace $F^{-1}[\text{im}\, G]$. This implies, of course, that $G\xi$ belongs to $GF^{-1}[\text{im}\, G]$. From that fact, it follows that any trajectory $\xi(\cdot)$ satisfying $\sigma G\xi = F\xi$ must actually belong to the subspace $F^{-1}[GF^{-1}[\text{im}\, G]]$, which obviously is contained in $F^{-1}[\text{im}\, G]$. We can go on in this way; a subspace recursion emerges which can be summarized as follows. Let the space on which G and F act be denoted by Z. Define

$$Q^0 = Z \tag{3.19}$$

and

$$Q^{k+1} = F^{-1}GQ^k. \tag{3.20}$$

We have $Q^{k+1} \subset Q^k$ for all k, and so a limit must be reached after finitely many (in fact, at most $\dim Z$) steps. The limit subspace will be denoted by $Q^*(F, G)$ or simply by Q^* if there is no risk of confusion. We arrive at the following algorithm to obtain a P_{dv} representation from a P representation.

ALGORITHM 2 Let $(F, G, H; Z, X, W)$ be a P representation. Compute the subspace Q^* of Z as the limit of the sequence of subspaces defined by (3.19-3.20). Take $\tilde{Z} = Q^*$, $\tilde{X} = GQ^*$, and define $\tilde{F}$, $\tilde{G}$, and $\tilde{H}$ as the restrictions of the respective mappings to $\tilde{Z}$ and $\tilde{X}$. (Note that, by the definition of Q^*, F does indeed map Q^* into GQ^*.) Under these conditions, a P_{dv} representation that is equivalent to the original P representation is given by $(\tilde{F}, \tilde{G}, \tilde{H}; \tilde{Z}, \tilde{X}, W)$.

Next, we consider the transformation from a P_{dv} to a P_{io} representation. Let $(F, G, H; Z, X, W)$ be a P_{dv} representation, and suppose that $[G^T \quad H^T]^T$ is not injective. We can then split up the internal variable space Z as $Z = Z_1 \oplus Z_2$, where $Z_2 = \ker G \cap \ker H$ is nonzero. With respect to this decomposition, write $G = [G_1 \quad 0]$, $F = [F_1 \quad F_2]$, $H = [H_1 \quad 0]$. The equations $\sigma G\xi = F\xi$, $w = H\xi$ then appear in the following form:

$$\sigma G_1\xi_1 = F_1\xi_1 + F_2\xi_2 \tag{3.21}$$

$$w = H_1\xi_1. \tag{3.22}$$

Since there are no restrictions on ξ_2, the above equations are equivalent to

$$\sigma TG_1\xi_1 = TF_1\xi_1 \tag{3.23}$$

$$w = H_1\xi_1 \tag{3.24}$$

where T is any map satisfying $\ker T = \operatorname{im} F_2$. It is natural to let T be surjective, and we see that the above transformation achieves a reduction of the dimension of the internal variable space and perhaps also a reduction of the dimension of the equation space. In more geometric terms, what we have done is the following. Define $S^1 = \ker G \cap \ker H$, and let $Z_1 = Z/S^1$, $X_1 = X/FS^1$. With these definitions, the factor mappings $G_1: Z_1 \to X_1$, $F_1: Z_1 \to X_1$, and $H_1: Z_1 \to W$ are all well-defined, and the representation $(F_1, G_1, H_1; Z_1, X_1, W)$ is equivalent to the original representation.

There is no guarantee that, after this step, the reduced representation is of the P_{io} type, and in general the reduction will have to be repeated a number of times. For instance, the reduction in the second step is determined by the subspace

$$\begin{aligned} \ker G_1 \cap \ker H_1 &= \{z \bmod S^1 \mid Gz \in FS^1 \text{ and } Hz = 0\} \\ &= (G^{-1}FS^1 \cap \ker H) \bmod S^1. \end{aligned} \tag{3.25}$$

The subspace recursion that emerges is the following:

$$S^0 = \{0\} \tag{3.26}$$

$$S^{k+1} = G^{-1}FS^k \cap \ker H. \tag{3.27}$$

We have $S^{k+1} \supset S^k$ at every step, and so after finitely many ($\leqslant \dim Z$) steps a limit is reached. The limit subspace will be denoted by $S^*(F, G, H)$ or simply by S^* if the context is clear. The algorithm to go from a P_{dv} to a P_{io} representation can now be formulated as follows.

ALGORITHM 3 Let $(F, G, H; Z, X, W)$ be a P_{dv} representation. Define the subspace S^* of Z as the limit of the sequence defined by (3.26-3.27). Define $\tilde{Z} = Z/S^*$, $\tilde{X} = Z/FS^*$. With these definitions, the factor mappings $\tilde{G}: \tilde{Z} \to \tilde{X}$, $\tilde{F}: \tilde{Z} \to \tilde{X}$, and $\tilde{H}: \to W$ are well-defined, and $(\tilde{F}, \tilde{G}, \tilde{H}; \tilde{Z}, \tilde{X}, W)$ forms a P_{io} representation that is equivalent to the original P_{dv} representation.

The final transformation in this series is the one that leads from P_{io} to P_{min} representations.

To achieve this reduction, we note that a redundancy in P_{io} descriptions is associated with subspaces N of the internal variable space Z that satisfy the two properties

$$FN \subset GN \tag{3.28}$$

and

$$N \subset \ker H. \tag{3.29}$$

Indeed, if N is a nonzero subspace having these properties, then we can decompose the internal variable space Z and the equation space X in such a way that $H = [H_1 \quad 0]$ and the mappings G and F take the form

$$G = \begin{bmatrix} G_{11} & 0 \\ 0 & G_{22} \end{bmatrix}, \qquad F = \begin{bmatrix} F_{11} & 0 \\ F_{21} & F_{22} \end{bmatrix}. \tag{3.30}$$

Of course, both G_{11} and G_{22} must be surjective. The equations become

$$\sigma G_{11}\xi_1 = F_{11}\xi_1 \tag{3.31}$$

$$\sigma G_{22}\xi_2 = F_{21}\xi_1 + F_{22}\xi_2 \tag{3.32}$$

$$w = H_1\xi_1. \tag{3.33}$$

Because G_{22} is surjective, the second equation can always be satisfied by a suitable choice of ξ_2; therefore, no constraint is imposed on ξ_1. This means that the second equation as well as the variable ξ_2 may be removed without altering the external behavior. Speaking geometrically, this means that we replace Z by $\tilde{Z} = Z/N$ and X by $\tilde{X} = X/GN$, and that the mappings F, G and H are replaced by the respective factor mappings.

The reduction that is obtained in this way increases with N, and so we are interested in the largest element of the set of subspaces satisfying both (3.28) and (3.29). (The fact that this set indeed has a largest element follows from the fact that the set is closed under subspace addition.) Let us denote this largest element by $N^*(F, G, H)$. The question is, how to compute this subspace. The answer to this is provided by the following equality, which expresses perhaps the most basic result in the geometric theory of linear systems:

$$N^*(F, G, H) = Q^*\left(\begin{bmatrix} G \\ 0 \end{bmatrix}, \begin{bmatrix} F \\ H \end{bmatrix}\right). \tag{3.34}$$

Indeed, this gives us an *algorithm* to compute N^*. The proof of (3.34) is not difficult, and may essentially be found in the standard reference [77, p.91]. A considerable amount of translation of terms is needed, though, and the reader may find it easier to construct a direct proof. Rewriting the algorithm (3.19-3.20) a little bit to suit the special form which appears in (3.34), we finally obtain the following algorithm.

ALGORITHM 4 Let $(F, G, H; Z, X, W)$ be a P_{io} representation. Define a sequence of subspaces of Z by

$$N^0 = Z \tag{3.35}$$

$$N^{k+1} = F^{-1}GN^k \cap \ker H. \tag{3.36}$$

Denote the limit subspace by N^*, and define $\tilde{Z} = Z/N^*$, $\tilde{X} = X/GN^*$. With these definitions, the factor mappings $\tilde{G}: \tilde{Z} \to \tilde{X}$, $\tilde{F}: \tilde{Z} \to \tilde{X}$, and $\tilde{H}: \tilde{Z} \to W$ (corresponding to G, F, and H

respectively) are well-defined, and the representation $(\tilde{F}, \tilde{G}, \tilde{H}; \tilde{Z}, \tilde{X}, W)$ is a $\mathsf{P}_{\min}$ representation that is equivalent to the given representation.

It has been proved in [56] (using somewhat different terminology) that this algorithm does indeed lead to a minimal representation.

There is a trivial way to pass from a general pencil representation to a general descriptor representation. If (F, G, H) is a P representation, and $H = [H_y^{\mathrm{T}} \;\; H_u^{\mathrm{T}}]^{\mathrm{T}}$ is the decomposition of H associated with a given partitioning of the external variables into inputs and outputs, then an equivalent D representation is obviously given by

$$\sigma \begin{bmatrix} G \\ 0 \end{bmatrix} \xi = \begin{bmatrix} F \\ H_u \end{bmatrix} \xi + \begin{bmatrix} 0 \\ -I \end{bmatrix} u \tag{3.37}$$

$$y = H_y \xi. \tag{3.38}$$

The main virtue of this transformation is that it doesn't require computation. A transformation that does a better job at preserving minimality properties is given by the following algorithm.

ALGORITHM 5 Let $(F, G, H; Z, X, W)$ be a pencil representation, and let an i/o structure be given, so that $H = [H_y^{\mathrm{T}} \;\; H_u^{\mathrm{T}}]^{\mathrm{T}}$. Decompose the internal variable space Z as $Z_0 \oplus Z_1 \oplus Z_2$ where $Z_1 = \ker G \cap \ker H_u$, and $Z_1 \oplus Z_2 = \ker G$. Accordingly, write

$$G = [G_0 \;\; 0 \;\; 0], \qquad F = [F_0 \;\; F_1 \;\; F_2], \tag{3.39}$$

$$H_y = [H_{y0} \;\; H_{y1} \;\; H_{y2}], \qquad H_u = [H_{u0} \;\; 0 \;\; H_{u2}]. \tag{3.40}$$

The matrix H_{u2} has full column rank, and by renumbering the u-variables if necessary, we can write

$$H_{u0} = \begin{bmatrix} H_{10} \\ H_{20} \end{bmatrix}, \qquad H_{u2} = \begin{bmatrix} H_{12} \\ H_{22} \end{bmatrix} \tag{3.41}$$

where H_{22} is invertible (or empty, if $\ker G \subset \ker H_u$). Define descriptor parameters by

$$E = \begin{bmatrix} G_0 & 0 \\ 0 & 0 \end{bmatrix}, \qquad A = \begin{bmatrix} F_0 - F_2 H_{22}^{-1} H_{20} & F_1 \\ H_{10} - H_{12} H_{22}^{-1} H_{20} & 0 \end{bmatrix}, \qquad B = \begin{bmatrix} 0 & F_2 H_{22}^{-1} \\ -I & H_{12} H_{22}^{-1} \end{bmatrix},$$

$$C = [H_{y0} - H_{y2} H_{22}^{-1} H_{20} \;\; H_{y1}], \qquad D = [0 \;\; H_{y2} H_{22}^{-1} H_{20}]. \tag{3.42}$$

These parameters define a D representation without nondynamic variables that is externally equivalent to the original P representation. Moreover, if the given representation is of the P_{dv} (P_{io}, $\mathsf{P}_{\min}$) type, then the obtained representation is of the D_{ri} (D_{mi}, $\mathsf{D}_{\min}$) type.

The proofs of the statements above are given in [34]. At the 'P_{dv}' level and higher, it might be said that the algorithm in fact uses the driving-variable representation as an intermediate step, so that the DV representations fit into the picture as shown in Fig. 1. The converse transformation is obtained as follows.

ALGORITHM 6 Let a D_{ri} representation be given by (E, A, B, C, D) (so that $[E \;\; B]$ is surjective). Choose coordinates in such a way that

$$E = \begin{bmatrix} I & 0 \\ 0 & 0 \end{bmatrix}, \qquad A = \begin{bmatrix} A_{11} & A_{12} \\ A_{21} & A_{22} \end{bmatrix}, \qquad B = \begin{bmatrix} B_{11} & B_{12} \\ B_{21} & B_{22} \end{bmatrix},$$

$$C = [C_1 \quad C_2], \qquad D = [D_1 \quad D_2] \tag{3.43}$$

where B_{22} is invertible. Define matrices A', B', C', and D' by

$$\begin{bmatrix} sI - A' & -B' \\ C' & D' \end{bmatrix} =$$

$$= \begin{bmatrix} I & 0 & 0 & B_{12}B_{22}^{-1} \\ 0 & I & 0 & -D_2B_{22}^{-1} \\ 0 & 0 & I & 0 \\ 0 & 0 & 0 & B_{22}^{-1} \end{bmatrix} \begin{bmatrix} sI - A_{11} & -A_{12} & -B_{11} \\ C_1 & C_2 & D_1 \\ 0 & 0 & I \\ A_{21} & A_{22} & B_{21} \end{bmatrix}. \tag{3.44}$$

The DV representation given by the four-tuple (A', B', C', D') is externally equivalent to the given D representation. Moreover, if the given descriptor representation is in the D_{mi} (D_{min}) class, the resulting driving-variable representation is in the DV_{io} (DV_{min}) class.

For a proof of these statements, see again [34]. More refined statements could be made; for instance, it is clear that to obtain a DV_{io} representation from the algorithm above, it is sufficient that the D representation we start with is reachable at infinity and observable in the sense of Verghese.

The corresponding reduction to minimal form in the 'DV' branch can be thought of as a reformulation of the above in special coordinates. The details have been worked out in [56]. The reductions take a somewhat different form at the 'D' level. Verghese [60] already gave a simple algorithm to remove nondynamic variables. It has been shown in [33] (Lemma 7.3 and Lemma 7.4) how to reduce a given descriptor representation in case it does not satisfy either one of the conditions '$[E \quad B]$ surjective' or '$[E^T \quad C^T]^T$ injective'. Clearly, by repeating these reduction steps if necessary, it is always possible to arrive at a situation in which these conditions are satisfied. The final passage to D_{min} comes down to removing the finite unobservable modes. This might for instance be done via reduction to the Weierstrass canonical form of the pencil $sE - A$ [26] followed by an application of the well-known procedure to remove unobservable modes in standard state space systems.

Finally, we come to the transformation from AR to P_{min}. This is essentially the Fuhrmann realization [23, 24]. In [33], the transformation is given the following form.

ALGORITHM 7 Let an AR representation be specified by $R(s)$. Consider the following spaces of rational vector functions in a formal parameter λ (π_- denotes projection onto the proper rational functions, W is the space of external variables, k is the number of rows of $R(s)$):

$$X^R = \{w(\lambda) \in \lambda^{-1}W[[\lambda^{-1}]] \mid \pi_- R(\lambda)w(\lambda) = 0\} \tag{3.45}$$

$$X_R = \{p(\lambda) \in \mathbb{R}^k[\lambda] \mid \exists w(\lambda) \in \lambda^{-1}W[[\lambda^{-1}]] \text{ s.t. } p(\lambda) = R(\lambda)w(\lambda)\} \tag{3.46}$$

$$N^R = \{w(\lambda) \in \lambda^{-1}W[[\lambda^{-1}]] \mid R(\lambda)w(\lambda) = 0\}. \tag{3.47}$$

The following mappings (G and F from $X^R/\lambda^{-1}N^R$ to X_R, H from $X^R/\lambda^{-1}N^R$ to W) are well-defined:

$$G: w(\lambda) \bmod \lambda^{-1}N^R \mapsto R(\lambda)w(\lambda) \tag{3.48}$$

$$F: w(\lambda) \bmod \lambda^{-1}N^R \mapsto R(\lambda)\pi_-(\lambda w(\lambda)) \tag{3.49}$$

$$H : w(\lambda) \bmod \lambda^{-1} N^R \mapsto w_{-1}. \tag{3.50}$$

With these definitions, (F, G, H) is a minimal pencil representation that is externally equivalent to the AR representation given by $R(s)$.

This version differs from Fuhrmann's original one in two respects. First, the resulting representation is given in pencil form rather than in standard state space form, so that it becomes possible to consider noncausal systems. (The Fuhrmann realization has been used before in a noncausal context [14, 75], but only by separating finite and infinite frequencies, and under the assumption that a transfer matrix exists.) Secondly, the procedure is presented as one under external equivalence, rather than as one under transfer equivalence.

The transformation algorithm given above is abstract, and may be used very well in theoretical considerations. However, a more computational form can also be given (see [33, §8]). This requires the given representation to be in $\mathrm{AR}_{\min}$ form, and produces a representation in $\mathrm{DV}_{\min}$ form, which explains the arrow between the corresponding boxes in our map of linear system representations.

4. The factor system

In [67], J. C. Willems has pointed out that there is a close connection between the notion of an 'almost controlled invariant subspace' and that of a 'factor system'. Before discussing the connection, let us briefly recall what these two notions mean. To define the factor system, following the development in [65], let first X be a finite-dimensional vector space over $\mathbb{R}$. Also, let A be a linear mapping from X into itself, and let B be a linear mapping ranging in X. The *smooth system* $\Sigma(A, B)$ on X determined by A and B is the following set of C^∞-functions from $\mathbb{R}$ into X:

$$\Sigma(A, B) = \{x(\cdot) \in C^\infty(\mathbb{R}; X) \mid \dot{x}(t) - Ax(t) \in \operatorname{im} B \text{ for all } t\}. \tag{4.1}$$

Let Σ be a smooth system on X and let K be a subspace of X. Consider the following set of trajectories on the factor space X/K:

$$\Sigma/K := \{x(\cdot) \bmod K \mid x(\cdot) \in \Sigma\}. \tag{4.2}$$

If this set of trajectories is a smooth system on X/K, then Σ/K is called the *factor system* determined by Σ and K.

The notion of an almost controlled invariant subspace can be defined in the same context. So let us assume that a state space X, a state mapping A, and an input mapping B have been given. A subspace K is said to be *almost controlled invariant* [66] if for every $\epsilon > 0$ and for every x_0 in K there exists a trajectory $x(\cdot)$ in $\Sigma(A, B)$ such that $x(0) = x_0$ and $\operatorname{dist}(x(t), K) \leqslant \epsilon$ for all $t \geqslant 0$. This concept has many applications in control theory, of which some are reviewed in the contribution by J. L. Willems to this volume.

Given a smooth system $\Sigma(A, B)$, one would of course like to know under what conditions on K the set Σ/K is a factor system. It is claimed in [67] (Theorem A) that this will hold if and only if K is almost controlled invariant. In the cited paper, only a sketchy proof is provided for the 'if' part of this statement, and the 'only if' part is given without proof. Later on, a detailed proof of the 'if' part has been provided in [59], but a complete proof of the reverse implication is still lacking in the literature. Our goal in this section is to provide a short proof of Theorem A of [67], using a result in [56]. This proof is essentially based on manipulation of representations.

In the previous section, algorithms were presented for the removal of redundancies in pencil representations. In these algorithms, certain subspace recursions played a key role. We will also need these recursions below, as well as some related recursions which we will introduce now. To the sequence of subspaces S^k defined by (3.26-3.27), another sequence $\hat{S}^k$ can be related by

$$\hat{S}^k = G^{-1}FS^k. \tag{4.3}$$

From (3.26-3.27), we see that this sequence might also be defined by the recursion

$$\hat{S}^0 = \ker G \tag{4.4}$$

$$\hat{S}^{k+1} = G^{-1}F[\hat{S}^k \cap \ker H]. \tag{4.5}$$

Denoting $\lim \hat{S}^k$ by $\hat{S}^*$, we also see from the definitions that $S^* = \hat{S}^* \cap \ker H$. It is furthermore useful to introduce two subspace recursions that do not take place in the 'internal variable space' Z but in the 'equation space' X. The first of these is obtained if we define

$$V^k = GN^k. \tag{4.6}$$

The corresponding recursion is

$$V^0 = X \tag{4.7}$$

$$V^{k+1} = G[F^{-1}V^k \cap \ker H]. \tag{4.8}$$

Similarly, we define

$$T^k = G\hat{S}^k \ (= FS^k) \tag{4.9}$$

with the corresponding recursion

$$T^0 = \{0\} \tag{4.10}$$

$$T^{k+1} = F[G^{-1}T^k \cap \ker H]. \tag{4.11}$$

The limit subspaces resulting from these recursions will be denoted by V^* and T^*, respectively.

The subspaces that have now been introduced play a role in the characterization of some important system invariants in terms of the parameters in a $\mathrm{P_{dv}}$ representation. If $(F, G, H; Z, X, W)$ is a $\mathrm{P_{min}}$ representation of a behavior $\mathcal{B}$, we define the *degree* of this behavior, to be denoted by $\deg(\mathcal{B})$, as $\dim X$. Also, we define the *order* of $\mathcal{B}$, to be denoted by $\mathrm{ord}(\mathcal{B})$, as dim Z. Since a $\mathrm{P_{min}}$ representation is determined up to isomorphisms of the internal variable space and the equation space, the degree and the order are clearly independent of the choice of a particular $\mathrm{P_{min}}$ representation. There are of course many other equivalent characterizations; for instance, the degree is also equal to the sum of the row degrees of the matrix $R(s)$ in any $\mathrm{AR_{min}}$ representation of $\mathcal{B}$, and to the dimension of the state space in any minimal state space representation of any causal input-output behavior that can be obtained from $\mathcal{B}$ by partitioning the external variables in inputs and outputs. (For a catalog of such results, see [69], Thm. 6.)

From the fact that the internal variable space in a $\mathrm{P_{min}}$ representation is obtained from the internal variable space in a given $\mathrm{P_{dv}}$ representation by successively factoring out the subspaces S^* and N^*, it might be suspected that the degree is given in terms of a $\mathrm{P_{dv}}$ representation by $\mathrm{codim}(N^* + S^*)$. It has been established in [56] (Thm. 4.1) that this is indeed the case. The relevant result may be summarized, with some rephrasing, as follows.

PROPOSITION 4.1 *Let a behavior $\mathcal{B}$ be given by a $\mathrm{P_{dv}}$ representation $(F, G, H; Z, X, W)$. Define subspaces S^*, N^*, and $\hat{S}^*$ of Z, and subspaces V^* and T^* of X by the recursions (3.26-3.27), (3.35-3.36), (4.4-4.5), (4.7-4.8), and (4.10-4.11) respectively. We then have the following equalities:*

$$\deg(\mathcal{B}) = \operatorname{codim}(N^* + \hat{S}^*) = \operatorname{codim}(V^* + T^*) \tag{4.12}$$

$$\operatorname{ord}(\mathcal{B}) = \operatorname{codim}(N^* + S^*). \tag{4.13}$$

In case $\ker H$ *contains* $\ker G$*, an alternative formula for the order is*

$$\operatorname{ord}(\mathcal{B}) = \operatorname{codim}(V^* + (T^* \cap G(\ker H))). \tag{4.14}$$

Our next concern is to characterize a 'smooth system' in terms of system invariants. This is described in the following lemma.

LEMMA 4.2 *A linear time-invariant behavior* $\mathcal{B}$ *with external variable* w *has a representation in the form*

$$\sigma x = Ax + Bu \tag{4.15}$$

$$w = x \tag{4.16}$$

if and only if $\mathcal{B}$ *has no static constraints (i. e. for all* $w_0 \in W$ *there exists a* $w \in \mathcal{B}$ *such that* $w(0) = w_0$*), and* $\dim W$ *is equal to* $\operatorname{ord}(\mathcal{B})$.

PROOF Consider the 'if' part first. If $\dim W$ equals $\operatorname{ord}(\mathcal{B})$, then there exists a $\mathrm{P}_{\min}$ representation

$$\sigma G\xi = F\xi \tag{4.17}$$

$$w = H\xi \tag{4.18}$$

in which the matrix H is square. From the requirement that $\mathcal{B}$ has no static constraints, it follows that H must be nonsingular. Let G^+ denote a right inverse of G, and let $\tilde{F}$ be a mapping satisfying $\operatorname{im}\tilde{F} = \ker G$. The equation (4.17) is then equivalent to

$$\sigma\xi = G^+ F\xi + \tilde{F}\eta \tag{4.19}$$

where η is a new internal variable. Using a nonsingular transformation of the ξ-variable, we can replace H by the identity mapping, and then the desired form is reached.

For the 'only if' part, we first note that the behavior defined by (4.15-4.16) has no static constraints. To determine the order of the behavior represented by (4.15-4.16), we have to take into account the fact that this representation is not minimal. Let T be any surjective mapping such that $\ker T = \operatorname{im} B$; then (4.15) is equivalent to

$$\sigma Tx = TAx. \tag{4.20}$$

Moreover, the representation (4.20-4.16) is minimal and we see that $\dim W$ is equal to $\operatorname{ord}(\mathcal{B})$, as claimed.

To obtain the main result of this section we combine the above characterization of smooth systems, the result that gives the order in terms of a P_{dv} representation, and a characterization of almost controlled invariant subspaces in terms of subspace recursions, taken from [66].

THEOREM 4.3 *Let a smooth system* $\Sigma(A, B; X)$ *be given, and let* K *be a subspace of* X*. Under these conditions, the set of trajectories of* Σ *modulo* K*,* Σ/K*, is a smooth system if and only if* K *is almost controlled invariant.*

PROOF Let $C: X \to X/K$ be the factor mapping. Obviously, a representation of the behavior Σ/K is given by

$$\sigma x = Ax + Bu \tag{4.21}$$

$$w = Cx \tag{4.22}$$

and so we have to find the conditions on K under which this is a smooth system. First of all, note that the behavior Σ / K can have no static constraints because otherwise the original system Σ would also have static constraints, which we know is not the case. Therefore, from the above lemma and the proposition we see that Σ / K is a smooth system if and only if

$$\dim X/K = \operatorname{codim}(V^* + (T^* \cap G(\ker H))) \tag{4.23}$$

where everything is taken with respect to the parameters

$$G = [I \quad 0], \qquad F = [A \quad B], \qquad H = [C \quad 0]. \tag{4.24}$$

(Note that indeed $\ker H$ contains $\ker G$, so that the above formula applies.) Rewriting the V^*- and T^*-algorithms for the above special values of the $\mathrm{P_{dv}}$ parameters while keeping in mind that $\ker C = K$, we obtain

$$V^0 = X \tag{4.25}$$

$$V^{k+1} = K \cap A^{-1}(V^k + \operatorname{im} B) \tag{4.26}$$

and

$$T^0 = \{0\} \tag{4.27}$$

$$T^{k+1} = A[T^k \cap K] + \operatorname{im} B. \tag{4.28}$$

The algorithm (4.25-4.26) is recognized as the *invariant subspace algorithm* [77, p.91]. If we define $\hat{T}^k = T^k \cap K$, then the associated recursion is

$$\hat{T}^0 = \{0\} \tag{4.29}$$

$$\hat{T}^{k+1} = K \cap (A\hat{T}^k + \operatorname{im} B) \tag{4.30}$$

and this is recognized as the *controllability subspace algorithm* [77, p. 107], also known as the *almost controllability subspace algorithm* [66]. Noting that $G[\ker H] = \ker C = K$, we see that we always have

$$K \supset V^* + (T^* \cap K) \tag{4.31}$$

so that the condition (4.23) may be rewritten as

$$K = V^* + (T^* \cap K) = V^* + \hat{T}^*. \tag{4.32}$$

But this is exactly the condition given in [66] for a subspace K to be almost controlled invariant with respect to (A, B).

5. Conclusions

It should be emphasized that our 'road map' of system representations covers only a small area in the large field of representation theory. We have only been looking at the 'classical' form of external equivalence, thereby excluding representations such as the matrix fractional form over the ring of proper and stable rational functions, which is one of the main tools in the latest developments in control theory [21, 43]. Also, there are many other classes of systems for which representation theory leads to useful results. This of course includes the generalizations to nonlinear and infinite-dimensional systems, but important new aspects also arise if one considers systems with particular properties. A simple example is provided by the case of linear systems with a Hamiltonian or a gradient structure, such as appear in the modeling of mechanical structures and electrical networks. The problem of setting up state equations for such systems, starting from (higher-order) differential equations and algebraic constraint equations, is in fact a classical one. For a treatment following lines as presented here, see [57]. Of course, the Hamiltonian structure is important in the nonlinear context as well, and the problem of dealing with systems with mixed differential and algebraic equations comes up naturally for instance in setting up models for robots. For general nonlinear systems, the relations between systems of higher-order differential equations on the one hand and the standard state space form on the other have been widely discussed; an early reference is [22], and [15, 20, 55, 58] provide a sample of recent contributions. It has been shown in [54], a nonlinear system of algebraic and differential equations in a DV-type form can be reduced to a minimal representation in standard state space form if and only if certain integrability conditions are satisfied. In the nonlinear case, the partitioning of external variables into inputs and outputs to obtain a causal i/o structure is, in general, a local construction. This could be one of the reasons for interest in a nonlinear version of the pencil form. Such a nonlinear pencil form might be specified by giving a submanifold of the tangent bundle of a manifold of internal variables, plus a mapping from that manifold to the manifold of external variables.

Representation theory for *stochastic* systems is a very well developed subject. The richer structure of stochastic systems allows for a variety of representations, some of which are discussed in the contribution by J. H. van Schuppen to the present volume. However, it seems that not so much study has been made of questions concerning nonminimal representations, such as sometimes appear in modeling problems. As an example, consider an electrical network with linear elements containing some noisy resistors. Writing down network equations in the usual way, one could write down a representation in the form

$$G\dot{\xi} = F\xi + J\eta \tag{5.1}$$

$$w = H\xi \tag{5.2}$$

where η is 'white noise', and w represents the port variables. It requires proof to show that this can be rewritten in the standard form

$$\dot{x} = Ax + Bu + Nv \tag{5.3}$$

$$y = Cx + Du + Mv \tag{5.4}$$

where v is white noise, and w has been partitioned into inputs u and outputs y. Representation of stochastic systems is also the subject of debate in econometric circles (see for instance [3, 17]).

Some aspects of the representation of infinite-dimensional linear systems are discussed in the contribution of R. F. Curtain to this volume. A great deal of effort has been spent by the infinite-dimensional systems community on trying to fit into the standard (A, B, C, D) framework equations like the following one (the normalized string equation with forces and displacements at both ends as external variables):

$$\frac{\partial^2}{\partial t^2}\phi(x, t) = \frac{\partial^2}{\partial x^2}\phi(x, t) \tag{5.5}$$

$$w(t) = \begin{bmatrix} \phi(0, t) \\ \phi(1, t) \\ \phi'(0, t) \\ \phi'(1, t) \end{bmatrix}. \tag{5.6}$$

(The variable x is used here as the spatial variable, and the prime denotes differentiation with respect to x.) Such an equation would fit more naturally into representations of the pencil type. This advantage doesn't come without a price, however; whereas standard semigroup theory is available for writing down solutions of the equations in (A, B, C, D) form, another route will have to be taken for systems in pencil form. Nevertheless, it would seem to be worth the effort to pursue this direction. It should be noted that a representation which easily incorporates equations like the string equation above has been proposed by D. Salamon under the name 'boundary control systems' [53]; however, this class was introduced by Salamon for specific purposes, and the restrictions he imposes are consequently more severe than one would like to see in a pencil representation.

The theory of system representations can be viewed as a theory of modeling. System-theoretic ideas may be applied to modeling problems as well as to control problems, and it may even be that some problems that are now considered as control problems will eventually be looked at rather as representation problems (model matching might fall in this category). In the process, it may be necessary to abandon some conventional wisdom. This paper has been written as a tribute to Jan Willems, one of the best abandoners of conventional thinking that I know.

ACKNOWLEDGEMENT

I would like to thank Margreet Kuijper, Henk Nijmeijer, and Arjan van der Schaft for their comments on an earlier version of this paper.

REFERENCES

1. M. A. AIZERMAN (1949). On a problem concerning the stability in the large of dynamic systems. *Usp. Mat. Nauk. 4*, 187-188. (In Russian.)
2. B. D. O. ANDERSON, W. A. COPPEL, D. J. CULLEN (1985). Strong system equivalence. *J. Austral. Math. Soc. Ser. B 27*, 194-222 (part I), 223-237 (part II).
3. M. AOKI (1988). Nonstationarity, cointegration, and error correction in economic modeling. *J. Econ. Dyn. Contr. 12*, 199-201.
4. J. D. APLEVICH (1981). Time-domain input-output representations of linear systems. *Automatica 17*, 509-522.
5. V. A. ARMENTANO (1984). The pencil $sE - A$ and controllability-observability for generalized linear systems: a geometric approach. *Proc. 23rd IEEE Conf. Dec. Contr.*, IEEE, New York, 1507-1510.
6. J. S. BARAS, R. W. BROCKETT, P. A. FUHRMANN (1974). State space models for infinite-dimensional systems. *IEEE Trans. Automat. Contr. AC-19*, 693-700.

7. R. BELLMAN, I. GLICKSBURG, O. GROSS (1956). On the 'bang-bang' control problem. *Q. Appl. Math. 14*, 11-18.
8. H. BLOMBERG, R. YLINEN (1983). *Algebraic Theory for Multivariable Linear Systems*, Ac. Press, London.
9. H. W. BODE (1945). *Network Analysis and Feedback Amplifier Design*, Van Nostrand, Princeton, NJ.
10. H. W. BODE (1960). Feedback — The history of an idea. *Symposium on Active Networks and Feedback Systems* (Polytechn. Inst. Brooklyn, April 19-21, 1960), Polytechnic Press, New York, 1-17. (Reprinted in R. BELLMAN, R. KALABA (eds.), *Selected Papers on Mathematical Trends in Control Theory*, Dover, New York, 1964.)
11. D. W. BUSHAW (1952). *Differential Equations with a Discontinuous Forcing Term* (Ph. D. thesis), Dept. of Math., Princeton Univ.
12. F. M. CALLIER, C. A. DESOER (1982). *Multivariable Feedback Systems*, Springer, New York.
13. J. D. COBB (1984). Controllability, observability, and duality in singular systems. *IEEE Trans. Automat. Contr. AC-29*, 1076-1082.
14. G. CONTE, A. M. PERDON (1982). Generalized state-space realizations for non-proper rational transfer functions. *Syst. Contr. Lett. 1*, 270-276.
15. P. E. CROUCH, F. LAMNABHI-LAGARRIGUE (1988). State space realizations of nonlinear systems defined by input-output differential equations. A. BENSOUSSAN, J. L. LIONS (eds.). *Analysis and Optimization of System* (Proc. INRIA Conf., Antibes, France, 1988), Lect. Notes Contr. Inf. Sci. 111, Springer, Berlin, 138-149.
16. C. A. DESOER, R. W. LIU, J. MURRAY, R. SAEKS (1980). Feedback system design: The fractional representation approach to analysis and synthesis. *IEEE Trans. Automat. Contr. AC-25*, 399-412.
17. R. F. ENGLE, C. W. J. GRANGER (1987). Co-integration and error correction: Representation, estimation, and testing. *Econometrica 55*, 251-276.
18. W. R. EVANS (1948). Graphical analysis of control systems. *AIEE Trans. 67*, 547-551.
19. P. M. G. FERREIRA (1987). On system equivalence. *IEEE Trans. Automat. Contr. AC-32*, 619-621.
20. M. FLIESS (1988). Nonlinear control theory and differential algebra. C. I. BYRNES, A. KURZHANSKI (eds.). *Modelling and Adaptive Control* (Proc. IIASA Conf., Sopron, Hungary, 1986), Lect. Notes Contr. Inf. Sci. 105, Springer, Berlin, 134-145.
21. B. A. FRANCIS (1987). *A Course in H_∞ Control Theory*, Lect. Notes Contr. Inf. Sci. 88, Springer, Berlin.
22. M. I. FREEDMAN, J. C. WILLEMS (1978). Smooth representations of systems with differentiated inputs. *IEEE Trans. Automat. Contr. 23*, 16-22.
23. P. A. FUHRMANN (1976). Algebraic system theory: an analyst's point of view. *J. Franklin Inst. 301*, 521-540.
24. P. A. FUHRMANN (1981). *Linear Systems and Operators in Hilbert Space*, McGraw-Hill, New York.
25. R. V. GAMKRELIDZE (1958). Theory of time-optimal processes for linear systems. *Izvestia Akad. Nauk SSSR 22*, 449-474. (In Russian.)
26. F. R. GANTMACHER (1959). *Matrix Theory (Part II)*, Chelsea, New York. (Russian original: 1954.)
27. E. GILBERT (1963). Controllability and observability in multivariable control systems. *SIAM J. Control 1*, 128-151.

28. B. L. Ho, R. E. KALMAN (1966). Effective construction of linear, state-variable models from input/output functions. *Regelungstechnik 14*, 545-548.
29. T. KAILATH (1980). *Linear Systems*, Prentice-Hall, Englewood Cliffs, N. J.
30. R. E. KALMAN (1963). Mathematical description of linear systems. *SIAM J. Control 1*, 152-192.
31. V. KUČERA (1974). Algebraic theory of discrete optimal control for multivariable systems. *Kybernetika 10-12*, 1-240. (Published in installments.)
32. V. KUČERA (1975). Algebraic approach to discrete linear control. *IEEE Trans. Automat. Contr. 20*, 116-120.
33. M. KUIJPER, J. M. SCHUMACHER (1989). *Realization of autoregressive equations in pencil and descriptor form*, Report BS-R8903, CWI, Amsterdam.
34. M. KUIJPER, J. M. SCHUMACHER (1989). *Transformations under external equivalence for descriptor systems.* (In preparation.)
35. A. M. LETOV (1953). Stability of control systems with two actuators. *Prikl. Mat. Mekh. 4.*
36. F. L. LEWIS (1986). A survey of linear singular systems. *Circuits Systems Signal Process. 5*, 3-36.
37. D. G. LUENBERGER (1977). Dynamic equations in descriptor form. *IEEE Trans. Automat. Contr. AC-22*, 312-321.
38. D. G. LUENBERGER (1978). Time-invariant descriptor systems. *Automatica 14*, 473-480.
39. C. C. MACDUFFEE (1956). *The Theory of Matrices*, Chelsea, New York. (Reprint of original, 1933.)
40. A. G. J. MACFARLANE (1979). The development of frequency-response methods in automatic control. *IEEE Trans. Automat. Contr. AC-24*, 250-265. (Reprinted in: A. G. J. MACFARLANE (ed.), *Frequency-Response Methods in Control Systems*, IEEE Press, New York, 1979, pp. 1-16.)
41. A. A. MARKOV (1894). On functions obtained by converting series into continued fractions. *Zap. Petersburg Akad. Nauk.* (In Russian. Reprinted in: A. A. MARKOV, *Collected Works*, Moscow, 1948, pp. 78-105.)
42. J. C. MAXWELL (1868). On Governors. *Proc. Royal Soc. London 16*, 270-283.
43. D. MCFARLANE, K. GLOVER (to appear). *Robust Controller Design Using Normalized Coprime Factor Plant Descriptions.*
44. B. MCMILLAN (1952). Introduction to formal realizability theory. *Bell Syst. Tech. J. 31*, 217-279 (part I), 541-600 (part II).
45. N. MINORSKY (1947). *Introduction to Nonlinear Mechanics*, Edwards Bros., Ann Arbor, Mich. (Originally published 1944-1946 as Restricted Reports by the David W. Taylor Model Basin, U. S. Navy.)
46. P. J. NAHIN (1988). *Oliver Heaviside: Sage in Solitude*, IEEE Press, New York.
47. H. NYQUIST (1932). Regeneration theory. *Bell Syst. Techn. J. 11*, 126-147. (Reprinted in R. BELLMAN, R. KALABA (eds.), *Selected Papers on Mathematical Trends in Control Theory*, Dover, New York, 1964, and in A. G. J. MACFARLANE (ed.), *Frequency-Response Methods in Control Systems*, IEEE Press, New York, 1979.)
48. A. C. PUGH, G. E. HAYTON, P. FRETWELL (1983). Some transformations of matrix equivalence arising from linear systems theory. *Proc. Automat. Contr. Conf.* (San Francisco, CA).
49. A. C. PUGH, G. E. HAYTON, P. FRETWELL (1987). Transformations of matrix pencils and implications in linear systems theory. *Int. J. Contr. 45*, 529-548.
50. H. H. ROSENBROCK (1968). Computation of minimal representations of a rational transfer-

function matrix. *Proc. IEE 115*, 325-327.

51. H. H. ROSENBROCK (1970). *State Space and Multivariable Theory*, Wiley, New York.
52. H. H. ROSENBROCK (1974). Structural properties of linear dynamical systems. *Int. J. Control 20*, 191-202.
53. D. SALAMON (1987). Infinite dimensional systems with unbounded control and observation: a functional analytic approach. *Trans. AMS 300*, 383-431.
54. A. J. VAN DER SCHAFT (1987). On realization of nonlinear systems described by higher-order differential equations. *Math. Syst. Th. 19*, 239-275. (Correction: *Math. Syst. Th. 20* (1987), 305-306.)
55. A. J. VAN DER SCHAFT (1989). Representing a nonlinear state space system as a set of higher-order differential equations in the inputs and outputs. *Syst. Contr. Lett. 12*, 151-160.
56. J. M. SCHUMACHER (1988). Transformations of linear systems under external equivalence. *Lin. Alg. Appl. 102*, 1-34.
57. J. M. SCHUMACHER (1988). *State representations of linear systems with output constraints*, Report OS-R8807, CWI, Amsterdam. (To appear in *MCSS*.)
58. E. D. SONTAG (1988). Bilinear realizability is equivalent to existence of a singular affine I/O equation. *Syst. Contr. Lett. 11*, 181-187.
59. H. L. TRENTELMAN (1986). *Almost Invariant Subspaces and High Gain Feedback*, CWI Tract 29, CWI, Amsterdam.
60. G. C. VERGHESE (1978). *Infinite Frequency Behavior in Generalized Dynamical Systems* (Ph. D. thesis), Information Systems Lab., Stanford Univ.
61. G. C. VERGHESE, B. LÉVY, T. KAILATH (1981). A generalized state space for singular systems. *IEEE Trans. Automat. Contr. AC-26*, 811-831.
62. M. VIDYASAGAR (1984). The graph metric for unstable plants and robustness estimates for feedback stability. *IEEE Trans. Automat. Contr. AC-29*, 403-418.
63. N. WIENER (1950). Comprehensive view of prediction theory. *Proc. Int. Congr. Math., Cambridge, Mass., 1950*, vol. 2, 308-321. (Reprinted in: P. MASANI (ed.), *Norbert Wiener. Collected Works with Commentaries*, MIT Press, Cambridge, Mass., 1981, 109-122.)
64. J. C. WILLEMS (1979). System theoretic models for the analysis of physical systems. *Ricerche di Automatica 10*, 71-106.
65. J. C. WILLEMS (1980). Topological classification and structural stability of linear systems. *J. Diff. Eq. 35*, 306-318.
66. J. C. WILLEMS (1980). Almost A (mod B)-invariant subspaces. *Astérisque 75-76*, 239-248.
67. J. C. WILLEMS (1981). Almost invariant subspaces: An approach to high gain feedback design. Part I: Almost controlled invariant subspaces. *IEEE Trans. Automat. Contr. AC-26*, 235-252.
68. J. C. WILLEMS (1983). Input-output and state-space representations of finite-dimensional linear time-invariant systems. *Lin. Alg. Appl. 50*, 581-608.
69. J. C. WILLEMS (1986). From time series to linear system. Part I: Finite dimensional linear time invariant systems. *Automatica 22*, 561-580.
70. J. C. WILLEMS (1986). From time series to linear system. Part II: Exact modelling. *Automatica 22*, 675-694.
71. J. C. WILLEMS (1986). Deducing the input/output and the input/state/output structure from the external behavior. *Proc. 25th IEEE Conf. Dec. Contr.* (Athens, Greece, Dec. 1986), IEEE Press, New York, 1936-1937.
72. J. C. WILLEMS (1987). From time series to linear system. Part III: Approximate modelling.

Automatica 23, 87-115.

73. J.C. WILLEMS (1988). Models for dynamics. U. KIRCHGRABER, H.O. WALTHER (eds.). *Dynamics Reported* (Vol. 2), Wiley/Teubner, 171-269.
74. J.C. WILLEMS, C. HEIJ (1986). Scattering theory and approximation of linear systems. C.I. BYRNES, A. LINDQUIST (eds.). *Modelling, Identification and Robust Control*, North-Holland, Amsterdam, 397-411.
75. H.K. WIMMER (1981). The structure of nonsingular polynomial matrices. *Math. Systems Theory 14*, 367-379.
76. W.A. WOLOVICH (1974). *Linear Multivariable Systems*, Springer, New York.
77. W.M. WONHAM (1979). *Linear Multivariable Control: a Geometric Approach* (2nd ed.), Springer, New York.
78. E.L. YIP, R.F. SINCOVEC (1981). Solvability, controllability, and observability of continuous descriptor systems. *IEEE Trans. Automat. Contr. AC-26*, 702-707.
79. D.C. YOULA, J.J. BONGIORNO, H.A. JABR (1976). Modern Wiener-Hopf design of optimal controllers. Part 1: The single-input-output case. *IEEE Trans. Automat. Contr. 21*, 3-13.
80. D.C. YOULA, J.J. BONGIORNO, H.A. JABR (1976). Modern Wiener-Hopf design of optimal controllers. Part 2: The multivariable case. *IEEE Trans. Automat. Contr. 21*, 319-338.

Optimal Control

H. J. Sussmann

Department of Mathematics, Rutgers University
New Brunswick, NJ 08903, U. S. A.

§1. Introduction .

Optimal Control Theory (OCT), in its modern sense, began in earnest in the 1950's, with the formulation and proof of the Pontryagin Maximum Principle (PMP), cf. [P], [Be], [LM]. The PMP is a far-reaching generalization of the Euler-Lagrange equations from the classical Calculus of Variations, and the development of OCT has proceeded, to a large extent, as an outgrowth of the Calculus of Variations, from which it has borrowed many of its themes, problems and methods. More recently, techniques based on ideas from "Geometric Control Theory" have been applied, leading to new and exciting results. This paper is an attempt to survey some of the recent developments in OCT, with a special emphasis on the contribution made by these new techniques.

We will deal exclusively with *finite-dimensional, deterministic* optimal control. In §2 we describe some of the basic problems of OCT. In §3 we discuss results on properties of optimal trajectories, including results on necessary conditions for optimality, and the important particular case of Local Controllability. In §4 we discuss results on optimal synthesis and the value function.

* This work was supported in part by the National Science Foundation under Grant DMS83-01678-01, and by the CAIP Center, Rutgers University, with funds provided by the New Jersey Commission on Science and Technology and by CAIP's industrial members.

§2. The basic problems of optimal control theory .

Throughout this paper, we deal with control systems

$$\dot{x} = f(x,u) \ , \ x \in \Omega \ , \ u \in U \tag{2.1}$$

where Ω is an open subset of $\mathbb{R}^n$, U is a set, and $f : \Omega \times U \to \mathbb{R}^n$ is of class C^1 in x for each fixed u. A *control* for (2.1) is a function $\eta : I \to U$ whose domain $I = \mathrm{Dom}(\eta)$ is an interval of the real line. The control η is *f-admissible* if the time-varying vector field $f_\eta : \Omega \times I \to \mathbb{R}^n$ given by $f_\eta(x,t) = f(x,\eta(t))$ satisfies the conditions of the Carathéodory existence and uniqueness theorem of ordinary differential equations, that is: (i) f_η is jointly measurable in x and t, and (ii) for each compact interval $J \subseteq I$ and each compact $K \subseteq \Omega$ there exists a Lebesgue integrable function $\varphi : J \to \mathbb{R}$ such that

$$\|f_\eta(x,t)\| + \|D_x f_\eta(x,t)\| \leq \varphi(t) \text{ for all } x \in K \, , t \in J \ . \tag{2.2}$$

A *trajectory* for a control $\eta : I \to U$ is an absolutely continuous curve $\gamma : I \to \Omega$ such that $\dot{\gamma}(t) = f(\gamma(t), \eta(t))$ for almost all $t \in I$. The Carathéodory theorem (cf., e.g., [CLe]), implies the local existence and global uniqueness of trajectories, that is: if η is f-admissible, then (i) given $\bar{t} \in I$, $\bar{x} \in \Omega$, there exists $\varepsilon > 0$ with the property that there is a trajectory γ for the restriction of η to the interval $(\bar{t} - \varepsilon, \bar{t} + \varepsilon) \cap I$, such that $\gamma(\bar{t}) = \bar{x}$, and (ii) if γ and δ are trajectories for η such that $\gamma(\bar{t}) = \delta(\bar{t})$ for some $\bar{t}$ in I, then $\gamma \equiv \delta$. If η is f-admissible and γ is a trajectory for η, then (γ, η) is an *f-admissible pair.*

It follows from our definitions that *every piecewise constant control is admissible.* Moreover, the following important *Piecewise Constant Approximation Theorem* (PCAT) holds: if (γ, η) is an admissible pair, $\mathrm{Dom}(\eta)$ is compact, $\bar{t} \in \mathrm{Dom}(\eta)$, $\bar{x} = \gamma(\bar{t})$, $x_j \in \Omega$, $x_j \to \bar{x}$ as $j \to +\infty$, then there exist piecewise constant controls $\eta_j : \mathrm{Dom}(\eta) \to U$ and corresponding trajectories γ_j such that $\gamma_j(\bar{t}) = x_j$ and $\gamma_j \to \gamma$ uniformly as $j \to \infty$. (For a proof, cf. [Su14].)

A control η *starts* at time t_0 if $t_0 \in \mathrm{Dom}(\eta)$ and $\mathrm{Dom}(\eta) \subseteq [t_0, \infty)$. In that case, if γ is a trajectory for η, then the point $x_0 = \gamma(t_0)$ is the *initial point* of γ. The definition of what it means for a control to *end* at a time t_1, and for a point x_1 to be the *terminal point* of a trajectory, is completely analogous. If x_0 and x_1 are the initial and terminal points of a trajectory γ that corresponds to a control η, then we say that γ *goes from* x_0 to x_1, or that η *steers* x_0 to x_1. We want to consider more general terminal conditions, such as the requirement that $\gamma(t) \to 0$ as $t \to \infty$. So we define a *terminal condition on trajectories* (TCT) to be, simply, a set τ of Ω-valued curves with the property that, if two curves γ, δ coincide from a certain time on, then $\gamma \in \tau$ iff $\delta \in \tau$. If τ is a TCT and $\gamma \in \tau$, then we say that γ *satisfies* τ. A particular example of TCT is the *point terminal condition* τ_x, i.e. the set of all curves that end at x. A *target* is a set $\mathcal{T}$ of pairwise disjoint TCT's. A curve γ *reaches the target* $\mathcal{T}$ (or *hits* $\mathcal{T}$, or *ends at* $\mathcal{T}$, or *gets to* $\mathcal{T}$) if γ satisfies some $\tau \in \mathcal{T}$.

We consider an optimal control problem $\mathcal{P} = (\Omega, U, f, L, \mathcal{T}, \varphi)$, given by the specification of (i) a control system (2.1), (ii) a function $L : \Omega \times U \to \mathbb{R}$, called the *Lagrangian,* (iii) a target $\mathcal{T}$, and (iv) a function $\varphi : \mathcal{T} \to \mathbb{R}$. We require that L be of class C^1 in x

for each fixed u. If $\eta : I \to U$ is an admissible control, and $\gamma : I \to \Omega$ is a trajectory for η that satisfies the TCT $\tau \in \mathcal{T}$, then we would like to define the *cost* of (γ, η) by

$$J(\gamma,\eta) = \int_I L(\gamma(t),\eta(t))\,dt + \varphi(\tau) \ . \tag{2.3}$$

However, the function $t \to L(\gamma(t),\eta(t))$ may fail to be integrable (or even measurable) unless some extra technical hypotheses hold. So we need to restrict further the set Dom(J) of those pairs (γ,η) for which $J(\gamma,\eta)$ is defined. For instance, we may take Dom(J) to be the set of those f-admissible (γ,η) such that $\gamma \in \tau$ for some $\tau \in \mathcal{T}$ and the integrand $h(t)$ of (2.3) has a well defined (but possibly infinite) Lebesgue integral (i.e. h is measurable and at least one of the integrals $\int_I h_+$, $\int_I h_-$ is finite, where $h_+ = \max(h,0)$ and $h_- = h_+ - h$). If, as is often the case, the Lagrangian is nonnegative, and the f-admissibility of (γ,η) suffices to imply that the function h is measurable, then this amounts to letting Dom(J) be the set of *all* f-admissible pairs (γ,η). A second possibility is to require η to be *(f,L)-admissible*, i.e. admissible for the *augmented control system* obtained from (2.1) by adding a new variable y with equation $\dot{y} = L(x,u)$. In this case, the integrability of h is guaranteed if I is finite (e.g. if the target $\mathcal{T}$ consists entirely of point TCT's and η has a starting time). If I is allowed to be infinite, then we restrict ourselves even further and require (γ,η) to be such that the integral exists. We shall refer to the above two possibilities as "Formulations I and II of the Optimal Control Problem $\mathcal{P}$."

For $x \in \Omega$ we define $V(x)$ to be the infimum of $J(\gamma,\eta)$ taken over all $(\gamma,\eta) \in$ Dom(J) such that γ starts at x and ends at the target $\mathcal{T}$. (If no such γ exists then of course $V(x) = +\infty$.) The function $V : \Omega \to \mathbb{R} \cup \{-\infty, +\infty\}$ is the *value function*, or *Bellman function*, of the problem $\mathcal{P}$. An *optimal pair* (often simply referred to as an "optimal trajectory") is a pair (γ,η) that has a starting point x and is such that $J(\gamma,\eta) = V(x)$.

We remark that *the choice of formulation* (i.e. the choice of Dom(J) as explained above) *may make a fundamental difference.* This is most dramatically illustrated by the "Lavrentieff phenomenon" (cf., e.g., [BM]). One can exhibit a nonnegative polynomial $L : \mathbb{R}^3 \to \mathbb{R}$ and real numbers x_0, x_1 such that, for the problem of minimizing $\int_0^1 L(x(t),\dot{x}(t),t)\,dt$ among all "curves" $t \to x(t)$ such that $x(0) = x_0$ and $x(1) = x_1$, the solution exists if by "curve" we mean "absolutely continuous function," and also if we mean "Lipschitz function." However, the values V_1, V_2 that correspond to these two interpretations satisfy $V_2 > V_1$. In control terms, this corresponds to the control system $\dot{x} = u$, $\dot{y} = 1$, with an obvious choice of Lagrangian. Requiring $x(\cdot)$ to be Lipschitz amounts to demanding that the control be measurable and bounded, in which case it is clearly (f,L)-admissible. Moreover, it follows easily from the PCAT (applied to the augmented system) that every trajectory $x(\cdot)$ for an (f,L)-admissible control can be approximated by Lipschitz trajectories $x_j(\cdot)$ *in such a way that the costs of the x_j converge to that of x.* Therefore V_2 is precisely the value for Formulation II. On the other hand, the condition that $x(\cdot)$ be absolutely continuous is precisely the requirement that the control be f-admissible, so V_1 is the value for Formulation I.

The choice of formulation is important, in particular, for the problem of *necessary conditions for optimality.* Indeed, the most basic of such conditions is the PMP, and the PMP is valid for Formulation II but may fail for Formulation I. For this reason, *from* *now on it will always be understood that the optimal control problem $\mathcal{P}$ is interpreted in the sense of Formulation II.*

For a control system (2.1) and a target $\mathcal{T}$, define a *presynthesis* to be a family $\Gamma = \{(\gamma_x, \eta_x) : x \in S\}$ of admissible pairs such that, for each $x \in S$, γ_x starts at x and ends at the target. The set S is the *domain* of Γ. Call a presynthesis Γ *memoryless* if, whenever $x \in S$ and $t \in \mathrm{Dom}(\eta_x)$, it follows that $y = \gamma_x(t) \in \mathrm{Dom}(\Gamma)$ and η_y is precisely the restriction of η_x to the interval $\mathrm{Dom}(\eta_x) \cap [t, +\infty)$. A memoryless presynthesis Γ will be referred to as a *partial synthesis.* If the domain of Γ is the largest possible (i.e. the set of *all* points x such that there exists a trajectory that goes from x to the target), then Γ will be called a *synthesis*. If each pair (γ_x, η_x) is optimal for the control problem $\mathcal{P}$, then we call Γ an *optimal synthesis* for $\mathcal{P}$.

The following problems are of interest in Optimal Control Theory:

1. existence of optimal trajectories,
2. characterization of optimal trajectories,
3. properties of optimal trajectories,
4. existence of an optimal synthesis,
5. characterization of optimal syntheses,
6. properties of optimal syntheses,
7. characterization and properties of the value function.

§3. Optimal Trajectories .

The problem of the *existence* of optimal trajectories is very classical, and has been studied in great detail in the 1950's and 1960's, so we choose not to discuss it here, and refer the reader instead to the book [Be] by L.D. Berkovitz for a rather complete account of the results.

Regarding the *characterization* of optimal trajectories, it has been clear for a long time that, except for very special problems such as linear systems with a convex Lagrangian, the question of *sufficient conditions* is essentially hopeless. On the other hand, the search for *necessary conditions for optimality* (NCO's) has been vigorously and fruitfully pursued since the early days of the PMP. Since the PMP is, roughly, a "first order condition," it was natural to look for "high order conditions." Some classical results of this endeavor are described in the papers [GK], [Go], [JS], the book [BJ], and especially H.W. Knobloch's book [Kn], which gives a detailed and mathematically rigorous treatment based on a formalism of asymptotic expansions. More recently, several new techniques have been introduced, many of which are based on the systematic use of *Lie brackets* and *Lie algebras of vector fields.* In many cases, the purpose was to solve some of the other problems in our list. In particular, the *synthesis problem* has led —as explained below— to an interest in specific *properties of optimal trajectories*, and new NCO's had to be developed in order to establish these properties.

To illustrate why Lie brackets come into the picture, let us remark that, in order to obtain necessary conditions for a pair (γ, η) (with $\mathrm{Dom}(\eta)$ assumed to be a compact interval $[a, b]$) to be optimal, the time-honored method is to make a perturbation of η —called a *control variation*— and embed η in a one-parameter family $\{\eta_\varepsilon : 0 \leq \varepsilon \leq \bar{\varepsilon}\}$ of controls with $\mathrm{Dom}(\eta_\varepsilon)) = [a, b_\varepsilon]$. These η_ε will then give rise to trajectories $\gamma_\varepsilon : [a, b_\varepsilon] \to \Omega$ (with $\gamma_\varepsilon(a) = \gamma(a)$). If the variation is smooth in some appropriate sense, then the optimality of (γ, η) will imply that the derivative of $J(\gamma_\varepsilon, \eta_\varepsilon)$ with respect to ε at $\varepsilon = 0$ has to be nonnegative. (We are ignoring here the extra difficulty that the perturbed trajectories must also satisfy the desired target conditions.) If we assume that we are already dealing with the augmented problem, as explained above, then the requirement becomes the condition that $v \cdot \nabla\psi(p) \geq 0$, where $p = \gamma(b)$, ψ is a function on the state space Ω, and v is the *variational vector*

$$v = \frac{d}{d\varepsilon}\bigg|_{\varepsilon=0} \gamma_\varepsilon(b_\varepsilon) \ . \tag{3.1}$$

So the vector v must belong to a particular subset of $\mathbb{R}^n$. When one applies this condition to a reasonably large class of variations, one obtains an NCO. In particular, the PMP is obtained by taking variations of a specially simple kind (cf., e.g. [P]), and more powerful NCO's arise by considering more sophisticated variations.

One is thus naturally led to the study of the set of all possible variational vectors. And it can be seen in a number of ways that *these vectors can be obtained as leading terms of asymptotic expansions that involve Lie brackets.* (This idea, as well as its use for high order NCO's, was proposed by [Bro], [AG2] and [Kr].) For a simple illustration of this situation let us consider a system of the form $\dot{x} = f(x) + ug(x)$, where f and g are C^∞ vector fields and the scalar control u takes values in $U = [-1, 1]$. (This is called a *two vector field system*, because the possible directions of motion at a point p are $X(p)$, $Y(p)$, and their convex combinations, where $X = f - g$ and $Y = f + g$.) Assume moreover that a point p is an *equilibrium*, i.e. that $f(p) = 0$. Then the constant trajectory $\gamma(t) \equiv p$ corresponds to the control $\eta(t) \equiv 0$. One can prove that, if $\eta : [0, T] \to [-1, 1]$ is a general control, and γ_η is the corresponding trajectory with initial condition $\gamma_\eta(0) = p$, then the asymptotic expansion

$$\gamma_\eta(T) \sim \cdots e^{\alpha_3(\eta)b_3} e^{\alpha_2(\eta)b_2} e^{\alpha_1(\eta)b_1} p \tag{3.2}$$

holds, where (i) the b_i are vector fields obtained from certain formal Lie brackets B_i in the two indeterminates F and G by plugging in f for F and g for G, (ii) the B_i are a *P. Hall basis* (cf. [Se], [Vi]) of the free Lie algebra in the indeterminates F and G (for instance, one can take $B_1 = F$, $B_2 = G$, $B_3 = [F, G]$, $B_4 = [F, [F, G]]$, $B_5 = [G, [F, G]]$, $\ldots$), (iii) the $\alpha_i(\eta)$ are *iterated integrals* calculated from η by explicit formulae (for instance, if the P. Hall basis is as indicated above, then $\alpha_1(\eta) = T$, $\alpha_2(\eta) = \int_0^T \eta(t)\,dt$, $\alpha_3(\eta) = \int_0^T t\,\eta(t)\,dt$, $\alpha_4(\eta) = \frac{1}{2}\int_0^T t^2\eta(t)\,dt$, $\alpha_5(\eta) = \int_0^T \int_0^t t\,\eta(t)\,\eta(s)\,ds\,dt$), (iv) the exponential notation stands for the *flow*, i.e. $t \to e^{tV}p$ is the integral curve of the vector field V that goes through p at time $t = 0$, (v) "$\sim$" stands for "asymptotic equality," in the sense that, if P_k denotes the right-hand side of (3.2) with all the B_i of degree $> k$ omitted, then $\gamma_\eta(T) - P_k = O(T^{k+1})$ as $T \to 0$, uniformly in η. (The right-hand side of (3.2) is the *Chen series*. Cf. [Su2] for the asymptotic formula, and [Su11] for the fact that the Chen series is given as an infinite product of exponentials.) If one now manages to choose $\eta_\varepsilon : [0, T_\varepsilon] \to [-1, 1]$ so that all the integrals $\alpha_k(\eta_\varepsilon)$ other than a particular $\alpha_m(\eta_\varepsilon)$ are $o(\varepsilon^{\delta(m)})$ (where $\delta(m)$ is the *degree* of the integral α_m, i.e. the degree of the formal Lie bracket B_m) whereas $\alpha_m(\eta_\varepsilon)$ itself is $\sim c\varepsilon^{\delta(m)}$, with $c \neq 0$, then we will have exhibited a control variation in the direction of $\pm b_m(p)$, the sign being that of c. It is natural to choose $T_\varepsilon = \varepsilon$. The iterated integrals associated to Lie monomials of degree $> \delta(m)$ are then automatically $O(\varepsilon^{\delta(m)+1})$, so the only real difficulty is to make sure that the integrals of degree $i \leq \delta(m)$, whose natural asymptotic behavior is $\sim \varepsilon^i$, are in fact much smaller. As an example, let us exhibit a variation in the direction of the vector $-[g,[f,g]](p)$. We begin by picking a function $\eta : [0,1] \to [-1,1]$ which is L^2-orthogonal to the functions 1, t and t^2 but not $\equiv 0$. (One can take η to be a cubic polynomial.) Then define $\eta_\varepsilon(t) = \eta(\frac{t}{\varepsilon})$, with domain $[0,\varepsilon]$. An easy calculation shows that $\alpha_1(\eta_\varepsilon) = \varepsilon$, $\alpha_2(\eta_\varepsilon) = \alpha_3(\eta_\varepsilon) = \alpha_4(\eta_\varepsilon) = 0$, and $\alpha_5(\eta_\varepsilon) = -c\varepsilon^3$, where $c = \frac{1}{2}\int_0^1(\int_0^t \eta(s)\,ds)^2 dt$, so that $c > 0$. It is clear that the brackets B_i of degree ≤ 3 are precisely B_1, B_2, B_3, B_4 and B_5. Since $e^{\alpha_1(\eta)b_1}p = p$ for all η (because $b_1 = f$ and $f(p) = 0$), and $b_5 = [g,[f,g]]$, we see that $\gamma_{\eta_\varepsilon}(\varepsilon) = p - c\varepsilon^3[g,[f,g]](p) + o(\varepsilon^3)$. So $-[g,[f,g]](p)$is a variational vector.

The preceding example shows how a particular direction can be exhibited as a variational direction by "neutralizing" those directions that in principle might have given rise to larger terms in the asymptotic expansion. Moreover, the example makes it clear that there are "semialgebraic" properties of the iterated integrals that make it possible to neutralize certain vectors but not others. For instance, the fact that $\alpha_5(\eta)$ is negative follows from the equalities $\alpha_2(\eta) = \alpha_3(\eta) = \alpha_4(\eta) = 0$. Therefore the vector $+[g,[f,g]](p)$ cannot be exhibited as a variational vector by this method. On the other hand, it may happen for a particular system that $[g,[f,g]](p)$ is equal to some other vector (e.g. $g(p)$) which is a variational vector for other reasons. In that case, $[g,[f,g]](p)$ will be a variational vector after all. Therefore, in order to understand the variational vectors, one has to understand the semialgebriac properties of iterated integrals that cause certain vectors such as $-[g,[f,g]](p)$ (but not $+[g,[f,g]](p)$) to show up automatically as variational vectors, as well as the interplay between this algebra and the special Lie-algebraic properties of particular systems that permit more general vectors such as $[g,[f,g]](p)$ to be variational in certain cases even though they are not variational in general.

The line of thought outlined above has been actively pursued in recent years, leading to a series of papers on *small-time locally controllable* (STLC) systems. A system Σ is STLC from a point p if for arbitrarily small $T > 0$ the set $\mathcal{R}(\leq T, p)$ (i.e. the reachable set from p in time $\leq T$) contains p in its interior. The study of the STLC condition is important because it is a very interesting special case of the more general question of NCO's. Indeed, a system fails to be STLC from p iff the constant trajectory $\gamma(t) \equiv p$ —which for an equilibrium point of a two vector field system corresponds to the control $\eta(t) \equiv 0$— is a *boundary trajectory*, in the sense that $\gamma(t) \in \partial\mathcal{R}(\leq t, p)$ for all sufficiently small $t > 0$. So a sufficient condition for the STLC property immediately yields a necessary condition for a particular γ to be a boundary trajectory, i.e. an NCO of a special kind.

The search for sufficient conditions for a system to be STLC has been guided by the above ideas. It is at least intuitively clear that, if a suitable definition of "variational vector" is given, then the set $\mathcal{V}(p)$ of all variational vectors at p should be a convex cone, and a system should be STLC from p if $\mathcal{V}(p)$ is the whole space $\mathbb{R}^n$. To make these considerations rigorous one may proceed as in the classical theory of the PMP (cf. [P]), or one may follow the approach proposed by H. Frankowska (cf. [Fra]) and use open mapping theorems for set-valued maps. In either case, one is led to the question of sufficient conditions for $\mathcal{V}(p)$ to be the whole space, and this requires the algebraic analysis outlined above. As expected, Lie brackets arise most naturally. The development of the theory was initiated by H. Hermes in a series of papers ([He1], [He2],[He3]). For the particular case of two vector field systems, Hermes pointed out that it was crucial to distinguish between those brackets with an even number of g's and those with an odd number. Let us assume that we are considering *real analytic systems.* Then we may assume that the "accessibility property" (AP) holds (i.e. that the Lie algebra Λ generated by f and g satisfies $\Lambda(p) = \mathbb{R}^n$), since this is known in any case to be necessary for the reachable set from p to have a nonvoid interior (cf. [SJ]). Hermes singled out the property that $S_k(f,g)(p) = S_{k-1}(f,g)(p)$ for all *even* k and conjectured that, together with the AP, this condition was sufficient for STLC. (Here $S_k(f,g)$ is the linear span of all the brackets of f's and g's with no more than k g's. Notice that the Hermes condition holds if p is an equilibrium and $S_1(f,g)$ already equals $\mathbb{R}^n$. This is precisely the case when the linearization at p is controllable, so the Hermes condition is weaker than the classical condition on the linearization, and the corresponding sufficiency theorem stronger.) In [Su3] it was proved that the Hermes condition is indeed sufficient for the STLC property. This was done by introducing a new technique for handling the algebraic part of the argument, namely, the use of groups of symmetries. Using this technique in a more refined way, a much stronger sufficiency theorem was established in [Su7]. Since then, new and more powerful sufficient conditions have been found by K. Wagner, R.M. Bianchini and G. Stefani, and M. Kawski (cf. [Wa], [BS1], [Ka1], [St1], [St2], and especially the survey article [Ka2] by Kawski). Moreover, the work on the STLC condition has been extended to the study of local controllability about more general —i.e. not necessarily constant— reference trajectories, that is to the general NCO problem that had originally provided the motivation for studying this particular case. Work by A.A. Agrachev and R.V. Gamkrelidze ([AG2]) developed a general formalism for expressing, in terms of Lie brackets, variational vectors arising from perturbations of general controls (cf. also [Su4]). In various other papers (e.g. [AG1], [AG3]), these authors applied their formalism to the derivation of new NCO's. P. Crouch and F. Lamnabhi-Lagarrigue ([CLa]) have pursued the development of the algebra needed to understand the perturbation expansions. Using their work, new NCO's have been obtained by Lamnabhi-Lagarrigue, Bianchini and Stefani (cf. [LLa], [BS2], [LLS], [St3]).

Another set of NCO's were obtained by A. Bressan in [Br1] using methods from nonlinear functional analysis (the Mountain Pass Lemma), and then applied by Bressan himself in his work on synthesis (cf. [Br2]). A third approach to obtaining NCO's has been used by H. Schättler in his synthesis work (cf. [Sc1], [Sc2] and [SSu]). The method here is to use the Campbell-Hausdorff series to compare two controls defined on small time intervals. Suppose, for instance, that we are dealing with the minimum time problem for a two-vector field system as above. Then we may want to study bang-bang controls of the form $\eta^{\tau_1,\ldots,\tau_k}_{t_1,\ldots,t_k}$, i.e. controls obtained by letting $u = -1$ during a time τ_1, then $u = 1$ during a time t_1, then $u = 1$ during a time τ_2, and so on. Such a control steers a point p to a point

$$q = e^{t_k Y} e^{\tau_k X} \ldots e^{t_2 Y} e^{\tau_2 X} e^{t_1 Y} e^{\tau_1 X} p \ . \tag{3.3}$$

If we have two different controls of the above type that steer p to the same point q, then we can use the asymptotic expansion of the right-hand side of (3.3) to compare the total time $\sum t_i + \sum \tau_i$ for both controls, and we may be able to conclude that one of the two cannot be optimal.

Finally, we give an example of an NCO which is of a different type, in that it generalizes the theory of envelopes from the classical Calculus of Variations. Define a *one-parameter field of extremals* (1PFE) to be a family $\{(\gamma_\varepsilon, \eta_\varepsilon, \lambda_\varepsilon) : 0 \leq \varepsilon \leq \bar{\varepsilon}\}$ of triples such that the $(\gamma_\varepsilon, \eta_\varepsilon)$ are extremals, defined on intervals $[a_\varepsilon, b_\varepsilon]$, the $\lambda_\varepsilon : [a_\varepsilon, b_\varepsilon] \to \mathbb{R}^n$ are Pontryagin adjoint variables for the $(\gamma_\varepsilon, \eta_\varepsilon)$, and the a_ε, b_ε, γ_ε and λ_ε "depend smoothly on ε." (Naturally, to make this precise we need a precise definition of smoothness. In [Su12], the particular case of bang-bang extremals is considered, and "smoothness" simply means that the switching times of the controls and the initial conditions for the γ_ε and the λ_ε are continously differentiable functions of ε. A paper developing the general theory under minimal smoothness assumptions is in preparation.) If the a_ε and the $\gamma_\varepsilon(a_\varepsilon)$ do not depend on ε, then we shall say that we have a *fixed initial condition* (FIC) field. Now suppose that we have a Pontryagin extremal (γ, η) defined on an interval $[a, b]$. Define an *envelope* for (γ, η) to be a curve $\delta : [-\bar{\varepsilon}, 0] \to \Omega$, with $\bar{\varepsilon} > 0$ with the property that there is a FIC 1PFE $\{(\gamma_\varepsilon, \eta_\varepsilon, \lambda_\varepsilon) : 0 \leq \varepsilon \leq \bar{\varepsilon}\}$, defined on intervals $[a, b_\varepsilon]$, such that $\delta(-\varepsilon) = \gamma(b_\varepsilon)$ for $0 \leq \varepsilon \leq \bar{\varepsilon}$, and δ is itself a trajectory, corresponding to some control $\theta : [-\bar{\varepsilon}, 0] \to U$, in such a way that $\langle \lambda_\varepsilon(b_\varepsilon), f(\delta(-\varepsilon), \theta(\varepsilon)) \rangle = 0$ for all ε. One then proves that the trajectory obtained by concatenating $\gamma_{\bar{\varepsilon}}$ with δ (with an obvious definition of the controls) has exactly the same cost as (γ, η). In particular, if (γ, η) is optimal then it follows that (δ, θ) is optimal as well. This can be used in many cases to prove that (γ, η) cannot be optimal. For instance, suppose that the terminal point q of γ lies in a region R such that every optimal trajectory in R is bang-bang. Suppose also that we manage to produce an envelope δ which is not bang-bang. Then it follows that (γ, η) is not optimal. This method has been successfully applied in [Su12], [Sc1], [Sc2] and [SSu] to the study of synthesis problems in low dimensions.

We now turn to the problem of *properties* of optimal trajectories. Naturally, there are no sharp boundaries separating this question from that of NCO's since, strictly speaking, an NCO is exactly the same as a statement that an optimal trajectory must have some special property. However, the question that will interest us now is that of proving theorems that say that optimal trajectories must have some interesting "regularity properties." For instance, if it was true that every optimal control is piecewise smooth (assuming now that we are dealing with problems with $U \subseteq \mathbb{R}^m$ and f reasonably smooth in u as well as in x), then this would be an example of the kind of theorem we want. Unfortunately, no such theorem is true in full generality, and two natural questions arise. First, one wishes to know whether some weaker result (e.g. that optimal controls necessarily have countably many discontinuities, or that the set of points of discontinuity has measure zero) is true *in general*. Second, one wishes to know whether strong theorems are true for *interesting special classes* of systems. Two classical examples of the latter situation are (i) the theorem that, for a classical Calculus of Variations problem whose Lagrangian $L(x, \dot{x})$ is of class C^∞ and has an everywhere nonsingular Hessian with respect to $\dot{x}$, all optimal trajectories are of class C^∞, and (ii) the bang-bang theorem from Linear System Theory, according to which, for a linear system $\dot{x} = Ax + Bu$ with a polyhedral control constraint $u \in K = \mathrm{co}(u_1, \ldots, u_k)$, whenever there is a trajectory from a point p to another point q, there is a time-optimal trajectory from p to q which is bang-bang with a finite number N of switchings that can actually be estimated to be $\leq C + DT$, where C, D are constants and T is the optimal time for going from p to q. Notice that the second result does not say that a time-optimal trajectory necessarily has to be bang-bang, since the system may be such that, for instance, one of the state coordinates —say x_1— satisfies $\dot{x}_1 = 1$, in which case *every* trajectory is time-optimal. What the result does say is that "bang-bang trajectories suffice for time-optimality." More generally, let us say that a class **C** of trajectories *suffices* for an optimal control problem $\mathcal{P}$ if, whenever there is an optimal trajectory from a point p to the target $\mathcal{T}$, it follows that there is an optimal trajectory from p to the target which belongs to **C**, and let us say that **C** suffices for a class of problems if it suffices for each problem in the class. (So, for instance, bang-bang trajectories suffice for linear time-optimal control problems with polyhedral constraints.) Notice that this formulation includes the situation when the cost is identically zero and the targets are points, in which case our question is just that of finding classes that are *sufficient for reachability*, i.e. such that whenever it is possible to go from a point p to another point q by means of *some* trajectory, then this can be done by means of a trajectory in **C**.

The problem of finding interesting sufficient classes of trajectories is essentially handled by the use of NCO's, i.e. by assuming that one has an optimal pair (γ, η) and trying to prove —using all available NCO's— that (γ, η) is "special" in some sense. There is, however, a significant twist that gives this line of research a slightly different technical flavor: we are now no longer restricted to working with given (γ, η); we are allowed to modify (γ, η) in order to make it "nicer," as long as we do not change the initial and terminal conditions.

The work on sufficient classes has so far proceeded by a combination of analytic arguments based on the use of the PMP and other NCO's, and geometric reasoning. In many cases (such as the work by Schättler and Bressan mentioned above) it was the need to prove sufficiency theorems that led to the discovery of new NCO's. In most of this work, Lie brackets have played an essential role.

The simplest of the new sufficiency theorems is the result of [Su1], according to which bang-bang trajectories are also sufficient for certain classes of two vector field systems. The specific condition on the systems can easily be understood as a particular consequence of linearity: a two vector field system is linear iff all Lie brackets involving two g's vanish identically, whereas the property of [Su1] says that all brackets with two g's must be expressible in a special way as linear combinations of brackets with only one g (cf. [Su1] for the precise details). It then turns out that this weaker property is all that is required for the bang-bang theorem.

The result of [Su1] only makes use of the PMP. In [Su6], [Su8], [Su9], [Su10], [Su13], [Sc1], [Sc2], [Br2]) a number of other results on sufficient classes were proved for various classes of systems. Remarkably, the result of [Su8], [Su9] and [Su10] shows that, for real analytic two vector field systems in two dimensions, trajectories of piecewise analytic controls (with a *finite* number of points of nonanalyticity), are sufficient for time-optimality. Unfortunately, this result fails, even in two dimensions, as soon as more complicated cost functionals are considered. For instance, for the problem of reaching the origin with minimum cost, subject to $\dot{x} = y$, $\dot{y} = u$, and $|u| \leq 1$, the cost functional being $\int x^2$, it can be shown that the optimal control from any initial condition $p \neq (0,0)$ is bang-bang *with infinitely many switchings.* (This is "Fuller's problem;" cf. [Ma] for a classical treatment, and [Ku] for a an analysis showing that Fuller-like situations appear in most optimal control problems, in sufficiently high dimensions.) One can easily modify Fuller's problem to obtain a *minimum time* problem in $\mathbb{R}^3$ for which similar "pathology" occurs.

One may then ask whether there are any results at all that can be expected to hold in full generality. The answer turns out to be closely related to *real analyticity*. Indeed, it was shown in [Su14] that, if an *arbitrary* measurable function $\eta : [0,T] \to [-1,1]$ is given, then one can construct a two vector field system of class C^∞ in $\mathbb{R}^3$, and points p, q, such that η steers p to q and no other control does. So no class of controls strictly smaller than that of all controls can be sufficient for all C^∞ problems. On the other hand, it turns out that the above construction is not possible if one wants the vector fields to be analytic. As shown in [Su13], the class **RAD** of controls η with the property that η is real analytic on an open dense subset of $\mathrm{Dom}(\eta)$ is sufficient for reachability and for time optimality for all real analytic two vector field problems.

The problem of the general structure and regularity properties of optimal controls, even for the case of two vector field systems (and even in $\mathbb{R}^3$) appears at the moment to be beyond the reach of currently existing methods. (However, substantial progress has been made, in three dimensions, in the understanding of generic cases, cf [Sc1], [Sc2], [Br2].) It is still a mystery where exactly, somewhere in between the very large class **RAD** and the much smaller class of piecewise analytic controls with at most countably many points of nonanalyticity, lies the frontier that characterizes the maximum possible pathology of optimal controls.

§4. Optimal Synthesis .

The *existence problem* for optimal synthesis is closely related to the question of existence of optimal trajectories. Once one knows that an optimal trajectory exists for every initial point p, then an optimal presynthesis is obtained by just choosing one such trajectory for each p, which is certainly possible if one accepts the Axiom of Choice. However, to get a synthesis one has to make the presynthesis *memoryless*. Whether this is possible in general appears at the moment to be an open question.

Regarding the *characterization* of optimal synthesis, the situation is quite remarkably different from what happens for optimal trajectories. Nice *sufficient conditions* can now be proved. Indeed, the question of optimal synthesis is intimately related to the study of the solutions V of the *Hamilton-Jacobi-Bellman (HJB) equation*

$$H(\nabla V(x), x) = 0 \ . \tag{4.1}$$

The function H is the *normal minimized Hamiltonian*, given by $H(\lambda, x) = \tilde{H}(\lambda, 1, x)$, where

$$\tilde{H}(\lambda, \lambda_0, x) = \inf \{\mathcal{H}(\lambda, \lambda_0, x, u) : u \in U\} \ , \tag{4.2}$$

and

$$\mathcal{H}(\lambda, \lambda_0, x, u) = \langle \lambda, f(x,u) \rangle + \lambda_0 L(x,u) \ . \tag{4.3}$$

(Here $\lambda \in \mathbb{R}^n$ and $\lambda_0 \in \mathbb{R}$.)

Now, let Γ be a synthesis and let V_Γ be the corresponding value function, i.e. $V_\Gamma(x) = J(\gamma_x, \eta_x)$. Let V be the value function of the problem under consideration. Then of course Γ is optimal iff $V_\Gamma \equiv V$. So the problem of deciding whether Γ is optimal is precisely that of checking whether $V_\Gamma \equiv V$. Moreover, for Γ to be optimal it is clearly necessary that each trajectory (γ_x, η_x) be a Pontryagin extremal. (Recall that a *Pontryagin extremal* is an (f, L) admissible pair (γ, η) with the property that there exist (i) an absolutely continuous function $t \to \lambda(t)$, defined for $t \in \mathrm{Dom}(\eta)$, and (ii) a constant $\lambda_0 \geq 0$, such that the equalities

$$\dot{\lambda}(t) = -\frac{\partial \mathcal{H}}{\partial x}(\lambda(t), \lambda_0, \gamma(t), \eta(t)) \tag{4.4}$$

and

$$\tilde{H}(\lambda(t), \lambda_0, \gamma(t)) = \mathcal{H}(\lambda(t), \lambda_0, \gamma(t), \eta(t)) = 0 \tag{4.5}$$

hold for almost all $t \in \mathrm{Dom}(\eta)$, and $(\lambda(t), \lambda_0) \neq (0, 0)$ for some —and hence every— value of t. The PMP says that an optimal trajectory must be a Pontryagin extremal.) So an obvious necessary condition for Γ to be optimal is that it be an *extremal synthesis*, i.e. a synthesis all whose trajectories are Pontryagin extremals. We shall use t_x to denote the starting time of the control η_x.

It turns out that, "modulo some technical conditions," the value function of the optimal control problem can be characterized as the unique solution of the HJB equation that satisfies certain boundary conditions. So what we need to know is whether V_Γ is a solution of the HJB equation and whether it satisfies the required boundary conditions. If one compares the HJB equation with the statement of the PMP given above, it is easy to see that the value function V_Γ will satisfy the HJB equation *provided we show that the adjoint variable $\lambda_x(\cdot)$ whose existence is asserted by the PMP can be chosen so that $\nabla V_\Gamma(x) = \lambda(t_x)$, and the corresponding λ_0 can be taken to be equal to* 1. It turns out that —again "modulo technical conditions"— this is indeed possible. So an extremal synthesis is indeed optimal if suitable technical conditions hold.

The preceding paragraph is intendedly vague, because of the repeated reference to unspecified "technical conditions." As written, the paragraph constitutes an accurate reflection of the current state of our knowledge. It is "known" that in some sense the statements are true in general, but no satisfactory way has yet been found to translate them into theorems that are both precise and sufficiently general. We shall refer to the problem of finding such theorems as *the Fundamental Problem of Synthesis Theory (FPST)*. Naturally, the main question is that of finding the appropriate technical conditions that make all the desired results true.

The FPST splits into three parts. *First*, there is the problem of rigorously defining what is meant for a function to be a solution of the HJB equation. The difficulty here is that the value function typically is not everywhere differentiable, so one needs some concept of "weak solution." To illustrate the difficulties involved, consider the simple example of the system $\dot{x} = u$ on the interval $[-1, 1]$, with control constraint $|u| \leq 1$. Assume that we want to reach the boundary of the interval in minimum time. It is easy to see that the value function V is just $V(x) = 1 - |x|$, the HJB equation is $|V'(x)| = 1$, and the boundary condition is $V(-1) = V(1) = 0$. Requiring that a solution of the HJB equation be differentiable everywhere would exclude the function V. If we only require differentiability almost everywhere, or even everywhere except at just one point, then we obtain the desired function V as well as many undesirable functions, such as $-V$.

Second, there is the question of passing from the fact that V_Γ satisfies the HJB equation and the boundary conditions to the conclusion that V_Γ is the value function. It is clear that the value function V satisfies the *Dynamic Programming Inequality (DPI)*

$$V(x) \leq V(y) + \int L(\gamma(t), \eta(t))\, dt \ , \tag{4.6}$$

whenever (γ, η) is a trajectory that goes from x to y and that, conversely, any function that satisfies this inequality plus the boundary conditions must be bounded above by the value function. In particular, if V_Γ satisfies the DPI and the boundary conditions then V_Γ is the value function. So the technical problem here is the passage from the HJB equation to the DPI. Formally, this is done by observing that, the HJB equation implies that $\langle \nabla V(\gamma(t)), \dot{\gamma}(t)\rangle + L(\gamma(t), \eta(t)) \geq 0$, from which the DPI follows by integration. The technical problem is to handle the case when ∇V does not exist everywhere, and therefore the proof by integration is not justified. Actually, all that is used in the formal derivation is the inequality $H(\nabla V(x), x) \geq 0$. Let us call a function that satisfies this inequality in some sense a *subsolution* of HJB. Then the real problem is to find the adequate technical definition of "subsolution" so that it becomes a true theorem that a function is a subsolution iff it satisfies the DPI.

Third, there is the derivation of the HJB equation from the property that Γ is extremal. In this case, there is a formal proof which is just based on differentiating V_Γ at a point x in a direction v. Asssuming, for simplicity, that the target is just a single point, one writes $V_\Gamma(x + hv)$ as an integral, and differentiates with respect to h at $h = 0$. The result is precisely $\langle \lambda(t_x), v\rangle$ and, since v is an arbitrary direction, the desired conclusion follows. One is then left with the technical problem of justifying the differentiation under the integral sign.

V.G. Boltyansky (cf. [Bo]) developed a theory of "regular synthesis" in order to tackle the above problems. Roughly speaking, a regular synthesis in Boltyansky's sense is a synthesis in which the trajectories and the controls depend on x in a "piecewise smooth" way. (The precise definition is quite technical, and we shall not repeat it here.) Boltyansky's result is that a regular synthesis in his sense is indeed optimal. However, it is not hard to give examples of optimal syntheses that are not regular in Boltyansky's sense but are good enough so that Boltyansky's arguments work, possibly after some trivial modifications. So Boltyansky's theory has to be extended and generalized. Examples of such extensions have been proposed in [Bru2], [Su3] and [Su5], but none of these can be regarded as definitive.

The main reason why we do not have a good theory of "regular synthesis" is that we lack a good theory of how "nice" the optimal synthesis has to be for reasonably large classes of problems. Indeed, suppose we could identify a property **P** such that one could prove, for every optimal control problem in a sufficiently large class, that (a) an optimal synthesis with Property **P** exists, and (b) if an extremal synthesis has Property **P** then it is optimal. Then we could regard the FPST as solved: one would consider **P** to be *the* natural property to be required of a synthesis, in that an existence theorem holds, and there is a simple necessary and sufficient condition for optimality, namely, extremality.

So we are naturally led to the question of *properties of optimal syntheses*. The study of this problem began with the work of P. Brunovsky [Bru1], [Bru2], who showed, for certain classes of problems (linear time optimal with a polyhedral control constraint) that a "regular synthesis" (in a sense slightly different from Boltyansky's) exists. Although the class of problems originally studied by Brunovsky was quite limited, his work made a far-reaching contribution, in that it introduced the idea of using the theory of *subanalytic sets* (cf. [Ha], [Hi], [Su15]) in optimal control. As was subsequently noticed (e.g. by Brunovsky himself in [Bru2], and also in [Su3]), the crucial point of his approach is that, thanks to the use of subanalytic sets, the problem of proving existence of a nice synthesis can be reduced to that of proving that certain classes of trajectories are sufficient for optimality. This has led to a number of results on existence of regular synthesis (cf. Bressan [Br2], Schättler [Sc1], [Sc2], Sussmann [Su10]).

The problem of the *characterization of the value function* is closely related to the FPST, except that now we are just given a function V_g and we ask for conditions under which this function will be the value function V. For the synthesis problem the function V_g already comes from a synthesis, so it automatically satisfies $V_g \leq V$. Here, on the other hand, we need conditions that will imply both inequalities $V_g \leq V$ and $V_g \geq V$. An elegant characterization of the value function along these lines is given by the theory of *viscosity solutions* due to M.G. Crandall and P.L. Lions (cf. [CLi], [LS]). A continuous function V is a *viscosity subsolution* of the HJB equation if, whenever φ is a C^1 function defined on a neighborhood of a point x, and such that $V - \varphi$ has a local maximum at x, it follows that $H(\nabla\varphi(x), x) \geq 0$. It is then easy to prove that, if V is continuous, then V is a viscosity subsolution if and only if it satisfies the DPI. Replacing "maximum" by "minimum" and "$H \geq 0$" by "$H \leq 0$" one obtains the definition of viscosity supersolution. Naturally, a viscosity solution is then defined to be a function which is both a subsolution and a supersolution. It then turns out that for large classes of problems the value function can be characterized as the unique viscosity solution of the HJB equation that satisfies the appropriate boundary conditions (cf.[LS]).

REFERENCES

[AG1] Agrachev, A.A. and R.V. Gamkrelidze, "A second-order optimality principle for a time-optimal problem," *Math. Sbornik* **100**, 142 (1976).

[AG2] Agrachev, A.A. and R.V. Gamkrelidze, "The exponential representation of flows and the chronological calculus," *Math. Sbornik* **109**, 149 (1978).

[AG3] Agrachev, A.A. and R.V. Gamkrelidze, "The Morse index and the Maslov index for smooth control systems," *Doklady Akad. Nauk USSR* **287** (1986).

[BM] Ball, J.M. and V.J. Mizel, "One-dimensional variational problems whose minimizers do not satsify the Euler-Lagrange equation," *Arch. Rational Mech. Anal.* **63** (1985), pp. 273-294.

[BJ] Bell, D.J. and D.H. Jacobson, *Singular Optimal Control Problems*, Academic Press (1973).

[Be] Berkovitz, L.D. *Optimal Control Theory*, Springer-Verlag (1974).

[BS1] Bianchini, R.M., and G. Stefani, "Sufficient conditions for local controllability," in *Proceedings 25th IEEE Conference on Decision and Control* (1986).

[BS1] Bianchini, R.M., and G. Stefani, "Local controllability about a reference trajectory," in *Analysis and Optimization of Systems*, A. Bensoussan and J.L. Lions eds., Springer-Verlag Lect. Notes Contr. Inf. Sci. 83 (1986).

[Bo] Boltyansky, V.G., "Sufficient conditions for optimality and the justification of the Dynamic Programming Principle," *SIAM J. Control* 4 (1966), pp. 326-361.

[Br1] Bressan, A., "A high-order test for optimality of bang-bang controls," *SIAM J. Control Opt.* **23** (1985), pp. 38-48.

[Br2] Bressan, A., "The generic optimal stabilizing controls in dimension 3," *SIAM J. Control Opt.* **24** (1986), pp. 177-190.

[Bro] Brockett, R.W., "Lie theory, functional expansions and necessary conditions in optimal control," in *Mathematical Control Theory*, W.A. Coppel ed., Springer-Verlag (1978), pp. 68-76.

[Bru1] Brunovsky, P., "Every normal linear system has a regular synthesis," *Mathematica Slovaca*, **28** (1978), pp. 81-100.

[Bru2] Brunovsky, P., "Existence of regular synthesis for general problems," *J. Diff. Equations*, **38** (1980), pp. 317-343.

[CLe] Coddington, E.A. and N. Levinson, *Theory of Ordinary Differential Equations*, McGraw-Hill (1955).

[CLi] Crandall, M.G. and P.L. Lions, "Viscosity solutions of Hamilton-Jacobi equations," *Trans. Amer. Math. Soc.* **277** (1983), pp. 1-42.

[CLa] Crouch, P.E. and F. Lamnabhi-Lagarrigue, "Algebraic and multiple integral identities," to appear in *Acta Applicandae Mathemathicae.*

[FR] Fleming, W.H. and R.W. Rishel, *Deterministic and Stochastic Optimal Control*, Springer-Verlag (1975).

[Fra] Frankowska, H., "An open mapping principle for set-valued maps," Centre de Recherches Mathématiques, Univ. de Montréal, rapport CRM-1364 (1986).

[GK] Gabasov, V., and F.M. Kirillova, "High-order necessary conditions for optimality," *SIAM J. Control* **10** (1972), pp. 127-168.

[Go] Goh, B.S., "The second variation for the Bolza problem," *SIAM J. Control* 4 (1966), pp. 309-325.

[Ha] Hardt, R.M., "Stratifications of real analytic maps and images," *Inventiones Math.* **28** (1975), pp. 193-208.

[He1] Hermes, H., "On local and global controllability," *SIAM J. Control* **12** (1974), pp. 252-261.

[He2] Hermes, H., "Lie algebras of vector fields and local approximation of attainable sets," *SIAM J. Control Opt.* **16** (1978), pp. 715-727.

[He3] Hermes, H., "Control systems which generate decomposable Lie algebras," *J. Diff. Equations* **44** (1982), pp. 166-187.

[Hi] Hironaka, H., *Subanalytic Sets*, Lect. Notes Istituto Matematico "Leonida Tonelli," Pisa, Italy (1973).

[JS] Jacobson, D.H. and J.L. Speyer, "Necessary and sufficient conditions for optimality for singular control problems: a limit approach," *J. Math. Anal. Appl.* **34** (1971), pp. 239-266.

[Ka1] Kawski, M., "A new necessary condition for local controllability," to appear in *Proceedings Conference on Differential Geometry, San Antonio, Texas, 1988*, Amer. Math. Society Contemporary Mathematics Series.

[Ka2] Kawski, M., "High order small-time local controllability," to appear in *Nonlinear Optimal Control and Controllability*, H.J. Sussmann ed., M. Dekker, Inc.

[KKM] Kelley, H.J., R.E. Kopp and H.G. Moyer, "Singular extremals," in *Topics in Optimization*, G. Leitman ed., Academic Press (1967).

[Kn] Knobloch, H.W., *High Order Necessary Conditions in Optimal Control*, Springer-Verlag (1975).

[Kr] Krener, A.J., "The higher order maximum principle and its application to singular extremals," *SIAM J. Control Opt.* **15** (1977), pp. 256-293.

[Ka2] Kupka, I., "The ubiquity of Fuller's phenomenon," to appear in *Nonlinear Optimal Control and Controllability*, H.J. Sussmann ed., M. Dekker, Inc.

[LLa] Lamnabhi-Lagarrigue, F., "Singular optimal control problems: on the order of a singular arc," *Systems and Control Letters* **9** (1987), pp. 173-182 .

[LLS] Lamnabhi-Lagarrigue, F., and G. Stefani, "Singular optimal control problems: on the necessary conditions for optimality," to appear in *SIAM J. Control Opt.*

[LM] Lee, E.B. and L. Markus, *Foundations of Optimal Control Theory*, J. Wiley, New York (1967).

[LS] Lions, P.L. and T. Souganidis, "Differential games, optimal control and directional derivatives of viscosity solutions of Bellman's and Isaacs' equations," *SIAM J. Control Opt.* **23** (1985), pp. 566-583.

[Ma] Marchal, C., "Chattering arcs and chattering controls," *J. Optim Theory Appl.* **11** (1973), pp. 441-468.

[P] Pontryagin, L.S., V.G. Boltyansky, R.V. Gamkrelidze and E.F. Mischenko, *The Mathematical Theory of Optimal Processes*, J. Wiley (1962).

[Sc1] Schättler, H., "On the local structure of time-optimal bang-bang trajectories in $\mathbb{R}^3$," *SIAM J. Control Opt.* **26** (1988), pp. 186-204.

[Sc2] Schättler, H., "The local structure of time-optimal trajectories in $\mathbb{R}^3$ under generic conditions," *SIAM J. Control Opt.* **26** (1988), pp. 899-918.

[SSu] Schättler, H., and H.J. Sussmann, "On the regularity of optimal controls," *J. Appl. Math. Physics (ZAMP)* **38** (1987), pp. 292-301.

[Se] Serre, J.P., *Lie algebras and Lie groups*, W.A. Benjamin, New York (1975).

[St1] Stefani, G., "Local controllability of nonlinear systems: an example," *Systems and Control Letters* **6** (1985), pp. 123-125.

[St2] Stefani, G., "On the local controllability of a scalar-input control system," in *Theory and Applications of Nonlinear Control systems*, C.I. Byrnes and A. Lindquist eds., Elsevier (1986).

[St3] Stefani, G., "A sufficient condition for extremality," in *Analysis and Optimization of Systems*, A. Bensoussan and J.L. lions eds., Springer-Verlag Lect. Notes Contr. Inf. Sci. 111 (1988), pp. 270-281.

[SJ] Sussmann, H.J. and V. Jurdjevic, "Controllability of nonlinear systems," *J. Diff. Equations* **12** (1972), pp. 95-116.

[Su1] Sussmann, H.J., "A bang-bang theorem with bounds on the number of switchings," *SIAM J. Control Opt.* **17** (1979), pp. 629-651.

[Su2] Sussmann, H.J., "Lie brackets and local controllability: a sufficient condition for scalar input systems," *SIAM J. Control Opt.* **21** (1983), pp. 686-713.

[Su3] Sussmann, H.J., "Lie brackets, real analyticity and geometric control theory," in *Differential Geometric Control Theory*, R.W. Brockett, R.S. Millman and H.J. Sussmann eds., Birkhäuser Boston Inc. (1983), pp. 1-115.

[Su4] Sussmann, H.J., "A Lie-Volterra expansion for nonlinear systems," in *Mathematical Theory of Networks and Systems, Proceedings of the MTNS-83 International Symposium, Beer-Sheva, Israel*, P.A. Fuhrmann Ed., Springer-Verlag (1984), pp. 822-828.

[Su5] Sussmann, H.J., "Lie Brackets and real analyticity in control theory," in *Mathematical Control Theory*, C. Olech ed., Banach Center Publications, Volume 14, PWN-Polish Scientific Publishers, Warsaw, Poland, 1985, pp. 515-542.

[Su6] Sussmann, H.J., "Resolution of singularities and linear time-optimal control," in *Proceedings 23rd IEEE Conference on Decision and Control*, Las Vegas, Nevada (Dec. 1984), pp. 1043-1046.

[Su7] Sussmann, H.J., "A general theorem on local controllability," *SIAM J. Control Opt.* **25** (1987), pp. 158-194.

[Su8] Sussmann, H.J., "The structure of time-optimal trajectories for single-input systems in the plane : the C^∞ nonsingular case," *SIAM J. Control Opt.* **25** (1987), pp. 433-465.

[Su9] Sussmann, H.J., "The structure of time-optimal trajectories for single-input systems in the plane: the general real-analytic case," *SIAM J. Control Opt.* **25** (1987) pp. 868-904.

[Su10] Sussmann, H.J., "Regular synthesis for time-optimal control of single-input real-analytic systems in the plane," *SIAM J. Control Opt.* **25** (1987), pp. 1145-1162.

[Su11] Sussmann, H.J., "A product expansion for the Chen series," in *Theory and Applications of Nonlinear Control Systems*, C. Byrnes and A. Lindquist Eds., North-Holland (1986), pp. 323-335.

[Su12] Sussmann, H.J., "Envelopes, conjugate points and optimal bang-bang extremals," in *Algebraic and Geometric Methods in Nonlinear Control Theory*, M. Fliess and M. Hazewinkel Eds., D. Reidel Publishing Co., Dordrecht, The Netherlands (1986), pp.325-346.

[Su13] Sussmann, H.J., "A weak regularity theorem for real analytic optimal control problems," *Revista Matemática Iberoamericana* **2** (1986), pp. 307-317.

[Su14] Sussmann, H.J., "Recent developments in the regularity theory of optimal trajectories," in *Linear and Nonlinear Mathematical Control Theory*, Rendiconti del Seminario Matematico, Università e Politecnico di Torino, Fascicolo Speciale 1987, pp. 149-182.

[Su15] Sussmann, H.J., "Real analytic desingularization and subanalytic sets: an elementary approach," to appear in *Transactions Amer. Math. Soc.*

[Vi] Viennot, G., *Algèbres de Lie et Monoïdes Libres*, Springer-Verlag (1978).

[Wa] Wagner, K., "Über den Steuerbarkeitsbegriff bei nichtlinearen Kontrollsystemen," *Arch. Math.* **47** (1986), pp. 29-40.

System Theory and Mechanics

A. J. van der Schaft

Dept. of Applied Mathematics, University of Twente
P. O. Box 217, 7500 AE Enschede, the Netherlands

Abstract. This paper discusses a system theoretic approach to mechanics, regarding Hamiltonian systems as conservative "mechanical m-ports". Recent results in the Hamiltonian realization problem are surveyed, and generalizations are being indicated. The potential use for control purposes of the Hamiltonian structure of nonlinear control systems is exemplified.

0. INTRODUCTION

Although mechanical systems always have constituted an important area of application for system and control theory, historically the interplay (in particular on a theoretical level) between system and control theory as an independent scientific discipline and mechanics has been rather weak. Indeed, areas such as electronics or automata theory have had a much larger impact on the development of system and control theory. Especially electrical network theory has enriched the framework of modern system theory from the very beginning with physical concepts such as energy, external power, passivity, reciprocity and duality. In particular the relation of the notion of passivity with (input-output) stability of systems has been one of the first major themes of system theory.

In the beginning of the seventies the notion of a passive (or dissipative) state space system having additional (external and internal) symmetry properties, initially motivated by electrical network theory, was further formalized and extended by Willems [W1,W2], with an eye also towards parts of physics such as thermodynamics, elasticity and mechanics. A systematic comparison of the basic concepts of system theory and of (analytical) mechanics was pioneered by Brockett [B]. In this innovative paper the notions of Hamiltonian and Lagrangian control systems were basically established, and several problems were formulated, which have set the trend for subsequent research in this area.

In the present paper we want to survey some of the main features, in our opinion, of the theory of Hamiltonian control systems as it has evolved over the last fifteen years. First in Section 1 we motivate the definition of Hamiltonian and Lagrangian control systems, very much in the spirit of the paper by Brockett, and indicate some open problems and extensions. The subsequent sections 2,3,4 can be read independently. Section 2 briefly surveys the global coordinate free definition of Hamiltonian control systems using tools from symplectic geometry. In Section 3 we deal with the Hamiltonian realization problem, i.e. the problem of characterizing a Hamiltonian system in terms of its external behavior, which is a first step towards the problem of mechanical synthesis. We mainly survey the results as obtained in [CS], but indicate in the linear case a more general approach which fits into the realization theory as recently developed by Willems [W3,W4]. In Section 4 it is illustrated in three particular cases how the Hamiltonian structure of a system can be profitably used for control purposes, enhancing the robustness of the controller and providing attractive theoretical and computational shortcuts. Finally in Section 5 some concluding remarks are given.

Due to space limitations, several other topics of interest in the interplay between system theory and mechanics are not dealt with here; in particular we like to mention symmetries and conservation laws [S2,S3], infinite dimensional Hamiltonian systems, quantum mechanical control systems [BS,THC,S8], Poisson control systems [San,KM], systems of a gradient nature [C,B], and last but not least, the relations between mechanics and optimal control theory (see also Example 6 in Section 1).

Acknowledgements. It is a great pleasure for me to thank Jan C. Willems at the occasion of his fiftieth birthday for putting me on this track of research, and most of all for conveying his stimulating enthusiasm for scientific research and his neverending curiosity about the concepts of science. I also like to thank Peter Crouch for a very pleasant and fruitful cooperation on the material of Section 3.

1. HAMILTONIAN CONTROL SYSTEMS.

Let us briefly review some basic elements of classical mechanics, see e.g. [G,Wh]. Consider a mechanical system with n degrees of freedom, locally represented by n configuration coordinates $q = (q_1,..,q_n)$. Based on d'Alembert's principle of virtual work one obtains the equations of motion

$$(1.1)\qquad \frac{d}{dt}\left(\frac{\partial T}{\partial \dot{q}_i}\right) - \frac{\partial T}{\partial q_i} = F_i, \qquad i = 1,\dots,n,$$

where $T(q,\dot{q})$ is the kinetic (co-)energy, and F_i are the forces acting on the system. Let us decompose the forces F_i, $i = 1,\dots,n$, into a *conservative* part, i.e. one that is derivable from a potential energy $V(q)$, and a remaining part consisting of *dissipative* and *external* forces, F_i^e, $i = 1,\dots,n$. Defining the *Lagrangian function* $L(q,\dot{q})$ as $T(q,\dot{q}) - V(q)$, one arrives at the celebrated *Euler-Lagrange* equations

$$(1.2)\qquad \frac{d}{dt}\left(\frac{\partial L}{\partial \dot{q}_i}\right) - \frac{\partial L}{\partial q_i} = F_i^e, \qquad i = 1,\dots,n.$$

For $F_i^e = 0$, $i = 1,\dots,n$, i.e. a closed conservative mechanical system, the equations (1.2) also arise from extremizing the action integral $\int L dt$. (This is usually called *Hamilton's principle* [G,Wh].) Alternatively, one could say that the right-hand side of (1.2) *defines* the extra forces F_i^e necessary to let the system evolve according to an *arbitrary* trajectory $q(t)$, see e.g. [T].

From (1.2) one can obtain a *control system* by disregarding dissipative forces and interpreting the external forces F_i^e in (1.2) as *input* or *control* variables. (Notice that this already constitutes a major departure from the classical point of view where F_i^e are usually regarded as *given* functions of $q,\dot{q}$ and/or time t.) More generally, if only *some* degrees of freedom can be directly controlled then one obtains the control system

$$(1.3)\qquad \frac{d}{dt}\left(\frac{\partial L}{\partial \dot{q}_i}\right) - \frac{\partial L}{\partial q_i} = \begin{cases} u_i, & i = 1,\dots,m, \\ 0, & i = m+1,\dots,n, \end{cases}$$

with $u_1,\dots,u_m$ being the (independent) controls. For instance robotic manipulators are of this type, if one neglects dissipation and takes the torques corresponding to the relative joint angle coordinates as inputs, while the remaining, not directly controlled, degrees of freedom may model flexibility.

Even more generally one could consider mechanical control systems described in vector notation as

$$\text{(1.4)} \qquad \frac{d}{dt}\left(\frac{\partial L}{\partial \dot{q}}\right) - \frac{\partial L}{\partial q} = B(q)u, \qquad q \in \mathbb{R}^n,\ u \in \mathbb{R}^m,\ m \leq n,$$

for some $n \times m$ matrix $B(q)$. However, as already argued by Brockett [B], control or input variables in mechanical systems do not necessarily have to appear as external forces (as in (1.3) or (1.4)). This is illustrated by the following simple example (see for a more involved example [B]).

Example 1. Consider a linear mass-spring system attached to a moving frame,

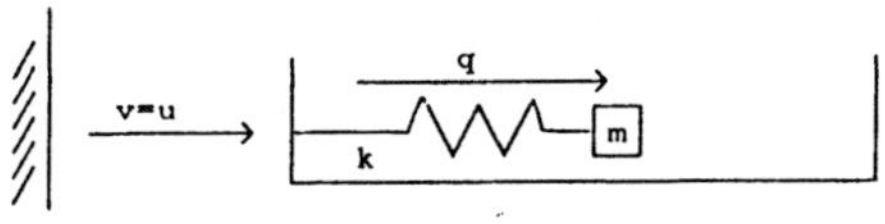

where the input u is the *velocity* of the frame. The Lagrangian $L(q,\dot{q},u) = \frac{1}{2}m(\dot{q}+u)^2 - \frac{1}{2}kq^2$ depends directly on u, and the equations of motion are $\frac{d}{dt}\left(\frac{\partial L}{\partial \dot{q}}(q,\dot{q},u)\right) - \frac{\partial L}{\partial q}(q,\dot{q},u) = m(\ddot{q}+\dot{u}) - kq = 0.$ □

In general, for a mechanical system of n degrees of freedom with a Lagrangian $L(q,\dot{q},u)$ depending directly on control variables u (through the kinetic and/or the potential energy) the equations of motion, in the absence of other forces, are given as

$$\text{(1.5)} \qquad \frac{d}{dt}\left(\frac{\partial L}{\partial \dot{q}_i}(q,\dot{q},u)\right) - \frac{\partial L}{\partial q_i}(q,\dot{q},u) = 0, \quad i = 1,\dots,n.$$

(If u is *given* function of $q,\dot{q},t$, then (1.5) can be interpreted as the Euler-Lagrange equations with zero external forces for the *time-dependent* Lagrangian $L(q,\dot{q},u(q,\dot{q},t))$.)

We notice that equations (1.3) can always be regarded as a special case of (1.5), by taking in (1.5) the control dependent Lagrangian

$$\text{(1.6)} \qquad \tilde{L}(q,\dot{q},u) = L(q,\dot{q}) + \sum_{j=1}^{m} u_j q_j$$

On the other hand, equations (1.4) are a special case of (1.5) *only* in case every j-th row of the transposed matrix $B^T(q)$ is (locally) the *gradient* of some function $H_j(q)$, or, equivalently, if the following integrability conditions are satisfied

$$(1.7) \qquad \frac{\partial B_{ij}}{\partial q_k}(q) = \frac{\partial B_{kj}}{\partial q_i}(q), \qquad i,k = 1,\dots,n,\ j = 1,\dots,m,$$

with $B_{ij}(q)$ the (i,j)-th element of B(q). In this case one recovers (1.4) from (1.5) for the Lagrangian

$$(1.8) \qquad \tilde{L}(q,\dot{q},u) = L(q,\dot{q}) + \sum_{j=1}^{m} u_j H_j(q)$$

Motivated by this we finally define a *Lagrangian control system* as a system in local configuration coordinates q given in the form (1.5), for some Lagrangian $L(q,\dot{q},u)$. We conclude that the main limitations in this definition are that in principle we do not allow for internal *dissipative* forces (we return later to this issue), and secondly that in case the controls are external forces we require the extra integrability condition (1.7) (which is automatically satisfied if, as in (1.4), the controls $u_1,\dots,u_m$ are the generalized forces corresponding to *independent* configuration coordinates). Relaxation of this last condition will involve the use of quasi-coordinates and/or the theory of non-holonomic systems [Wh]; however we will not go into this here.

Let us now pass on to the *Hamiltonian* formulation. For the Lagrangian control system (1.5) we define the generalized momenta in the usual way as

$$(1.9) \qquad p_i = \frac{\partial L}{\partial \dot{q}_i}(q,\dot{q},u), \qquad i = 1,\dots,n.$$

(Notice that p_i may depend on u.) If the $n \times n$ matrix with (i,j)-th element $\frac{\partial^2 L}{\partial \dot{q}_i \partial \dot{q}_j}(q,\dot{q},u)$ is non-singular everywhere (which is generally the case for a mechanical system), then $p = (p_1,\dots,p_n)$ are independent functions, and one defines the *Hamiltonian* function as the Legendre transform of $L(q,\dot{q},u)$

$$(1.10) \qquad H(q,p,u) = \sum_{i=1}^{n} p_i \dot{q}_i - L(q,\dot{q},u),$$

where $\dot{q}$ and p are related by the equations (1.9). (Since by (1.9) the partial derivatives of the right-hand side of (1.10) with respect to $\dot{q}_i$ are zero, one immediately concludes that H does not depend on $\dot{q}$.) It is well-known that with (1.9) and (1.10) the Euler-Lagrange equations (1.5) transform into the *Hamiltonian* equations of motion

$$(1.11a) \qquad \dot{q}_i = \frac{\partial H}{\partial p_i}(q,p,u)$$

$$i = 1,\ldots,n.$$

$$(1.11b) \qquad \dot{p}_i = -\frac{\partial H}{\partial q_i}(q,p,u)$$

(In fact, (1.11b) follows from substituting (1.9) into (1.5), and (1.11a) follows by (1.10).) We call (1.11) a *Hamiltonian control system*. The main advantage of (1.11) in comparison with (1.5) is that (1.11) are explicit first-order differential equations and thus constitute a control system in state space form, with *state* (q,p) (in physics usually called the *phase*). Moreover the variables q and p are completely dual to each other; indeed it is well-known, see e.g. [A,AM], that there is an underlying geometric structure to equations (1.11), called the *symplectic* or canonical structure, (see also Section 2). The state space transformations which leave this structure invariant are called the *canonical* transformations. (Let us furthermore mention that the Hamiltonian formalism in physics does not only underly classical mechanics, but also statistical and quantum mechanics.)

As a particular case of (1.11) we note that if the Lagrangian is of the form $L_0(q,\dot{q}) + \sum_{j=1}^{m} u_j H_j(q)$ (as in (1.8)), then the Hamiltonian equals $H_0(q,p) - \sum_{j=1}^{m} u_j H_j(q)$, with $H_0(q,p)$ the Legendre transform of $L_0(q,\dot{q})$. More generally, a Hamiltonian control system with H(q,p,u) of the form

$$(1.12) \qquad H(q,p,u) = H_0(q,p) - \sum_{j=1}^{m} u_j H_j(q,p)$$

will be called an *affine* Hamiltonian control system with *internal* Hamiltonian H_0, and H_j being the *interaction* Hamiltonians.

Example 2. Consider a rigid two-link robot manipulator (double pendulum) with the relative joint angles q_1, q_2 being the configuration coordinates and as controls u_1, u_2 the torques at the joints (as delivered by actuators). The Hamiltonian is of the form $\frac{1}{2}\, p^T M^{-1}(q)p + V(q) - u_1 q_1 - u_2 q_2$, with M(q) the (positive-definite) inertia matrix and V(q) the potential energy (gravity). □

Example 3. Consider the system of Example 1 where $L = \frac{1}{2} m(\dot{q}+u)^2 - \frac{1}{2} kq^2$. We obtain $p = m(\dot{q}+u)$ and $H(q,p,u) = H_0(q,p) - up$, with $H_0(q,p) = \frac{1}{2m} p^2 + \frac{1}{2} kq^2$ the internal energy. Notice that even though L is quadratic in u, H is still of the form (1.12). □

Example 4. Consider k point masses m_i, with positions $q^i \in \mathbb{R}^3$, $i \in \underline{k}$, in their own gravitational field corresponding to the potential energy $V(q^1,..,q^k) = \sum_{i<j} m_i m_j / \|q^i - q^j\|$. Suppose the *positions* of the first ℓ masses ($\ell \le k$) can be controlled. We obtain the Hamiltonian control system

$$\dot{q}^i_j = \frac{\partial H}{\partial p^i_j}, \quad \dot{p}^i_j = -\frac{\partial H}{\partial q^i_j}, \qquad i = \ell+1,\dots,k, \qquad j = 1,2,3,$$

with $H(q^{\ell+1},..,q^k,p^{\ell+1},..,p^k,u^1,..,u^\ell) = V(u^1,..,u^\ell,q^{\ell+1},..,q^k) + \sum_{i=\ell+1}^{k} \frac{1}{2m_i} \|p^i\|^2$. □

We now come to the definition of the *natural outputs* of a Hamiltonian control system (1.11a,b). Indeed, we define the natural outputs associated to (1.11a,b) as

$$(1.11c) \qquad y_j = -\frac{\partial H}{\partial u_j}(q,p,u), \qquad j = 1,\dots,m.$$

(Equivalently for the Lagrangian control system (1.5), we define $y_j = \frac{\partial L}{\partial u_j}(q,\dot{q},u)$, $j \in \underline{m}$.) We note that if $H(q,p,u)$ is of the affine form (1.12), then the natural outputs are simply given as the interaction Hamiltonians $H_j(q,p)$.

There are several reasons for adopting this definition of natural outputs. First, in this way *duality* between inputs and outputs is induced. For instance, if $u_1,\dots,u_m$ are generalized forces, then $y_1,\dots,y_m$ will be the corresponding generalized configuration coordinates. (In the theory of flexible structures this is called the case of *co-located* sensors and actuators.) Indeed, a Hamiltonian system (1.11a,b,c) can be regarded as a conservative "mechanical m-port" with pairs of dual variables (u_j,y_j), $j \in \underline{m}$, since it forms a direct analogue of an electrical m-port, in which case the external behavior is specified by the evolution of m voltage/current pairs. Secondly a strong type of *symmetry* or *reciprocity* between the inputs and the natural outputs results, as will become clear in Section 3. We call the total system (1.11a,b,c) a *Hamiltonian input-output system*, or briefly a *Hamiltonian system*.

Example 2 (continued). The natural outputs are the relative angles q_1, q_2. If, on the other hand, the controls would be the horizontal and vertical force at the end-point of the manipulator, then the natural outputs are the Cartesian coordinates of the endpoint. □

Example 3 (continued). The natural output is the momentum p. □

Example 4 (continued). Natural outputs are the *forces* as experienced by the first ℓ point masses. □

Example 5 (Nonlinear electrical LC m-port, see [S3]). Let L be a nonlinear inductive n_1-port (a set of inductors connected to each other in some way satisfying Kirchhoff's laws, with n_1 external channels). Let i_j and φ_j be the current, respectively flux, corresponding to the j-th external channel. Similarly let C be a nonlinear capacitive n_2-port (only capacitors), and let v_j and q_j be the voltages, respectively charges, at the external channels. We interconnect the first n channels of L and C ($n \le n_1$, $n \le n_2$), i.e.

$$(1.13) \qquad i_j = -\dot{q}_j, \qquad v_j = \dot{\varphi}_j, \qquad j = 1,\dots,n.$$

Suppose that L can be parametrized by $(\varphi_1,\dots,\varphi_n, i_{n+1},\dots,i_{n_1})$. Then L is given by the constitutive relation

$$(1.14) \qquad i_j = \frac{\partial S}{\partial \varphi_j},\ j \in \underline{n}, \qquad \varphi_{n+k} = -\frac{\partial S}{\partial i_{n+k}},\ k \in \underline{n_1 - n}$$

with $S(\varphi_1,\dots,\varphi_n,i_{n+1},\dots,i_{n_1})$ the Legendre transform with respect to $\varphi_{n+1},\dots,\varphi_{n_1}$ of the magnetic energy [BM]. Analogously, suppose that C can be parametrized by $(q_1,\dots,q_n, v_{n+1},\dots,v_{n_2})$, then C is given as

$$(1.15) \qquad v_j = \frac{\partial T}{\partial q_j},\ j \in \underline{n}, \qquad q_{n+k} = -\frac{\partial T}{\partial v_{n+k}},\ k \in \underline{n_2 - n}$$

with $T(q_1,\dots,q_n,v_{n+1},\dots,v_{n_2})$ the Legendre transform with respect to $q_{n+1},\dots,q_{n_2}$ of the electric energy [BM]. Defining the Hamiltonian H as S + T we obtain

$$(1.16a) \qquad \begin{aligned} \dot{\varphi}_j &= \frac{\partial H}{\partial q_j} \\ \dot{q}_j &= -\frac{\partial H}{\partial \varphi_j} \end{aligned}, \qquad j = 1,\dots,n,$$

$$(1.16b)\quad \begin{aligned} \varphi_{n+k} &= -\frac{\partial H}{\partial i_{n+k}}, & k &= 1,\dots,n_1-n, \\ q_{n+k} &= -\frac{\partial H}{\partial v_{n+k}}, & k &= 1,\dots,n_2-n, \end{aligned}$$

which is a Hamiltonian system with state $(\varphi_1,\dots,\varphi_n,q_1,\dots,q_n)$, inputs $(i_{n+1},\dots,i_{n_1},v_{n+1},\dots,v_{n_2})$ and outputs $(\varphi_{n+1},\dots,\varphi_{n_1},q_{n+1},\dots,q_{n_2})$. □

Example 6 ([S3,S6]). Consider the (smooth and unrestricted) Bolza problem of minimizing with respect to $u(\cdot)$ the cost functional

$$(1.17a)\quad J(x_0,u(\cdot)) = K(x(T)) + \int_0^T L(x(t),u(t))dt,$$

(with x and u in $\mathbb{R}^n$, respectively $\mathbb{R}^m$), under the dynamical constraints

$$(1.17b)\quad \dot{x}(t) = f(x(t),u(t)), \qquad x(0) = x_0 .$$

In order to solve this *optimal control* problem the Maximum Principle tells us to consider the pseudo-Hamiltonian $H(x,p,u) := p^T f(x,u) - L(x,u)$, with p the *co-state*, and the Hamiltonian control system

$$\begin{aligned} &(1.18a) & \dot{x}_i &= \frac{\partial H}{\partial p_i}(x,p,u) & & x(0) = x_0 \\ & & & & i = 1,\dots,n, & \\ &(1.18b) & \dot{p}_i &= -\frac{\partial H}{\partial x_i}(x,p,u) & & p_i(T) = -\frac{\partial K}{\partial x_i}(x(T)) \end{aligned}$$

where $x(T)$ is the solution of (1.18a) - (1.17b) at time T.
A *necessary* condition for a control function $u^*(\cdot)$ on $[0,T]$ to be optimal is that for every $t \in [0,T]$, $H(x^*(t),p^*(t),u^*(t)) = \max_u H(x^*(t),p^*(t),u)$, where $x^*(\cdot),p^*(\cdot)$ is the solution of (1.18). So we are led to the following problem: Find for every (x,p) a u^* such that $H(x,p,u^*) = \max_u H(x,p,u)$. Since u belongs to $\mathbb{R}^m$ this yields the first-order conditions $\frac{\partial H}{\partial u_j}(x,p,u^*) = 0$, $j = 1,\dots,m$. Hence the Maximum Principle leads in a natural way to the *Hamiltonian system* given by (1.18a,b) together with the output equations

$$(1.18c)\quad y_j = -\frac{\partial H}{\partial u_j}(x,p,u), \qquad j = 1,\dots,m,$$

and a necessary condition for $u^*(\cdot)$ to be optimal is that the *outputs* y_j of this system are *zero* on $[0,T]$. □

Another major consequence of definition (1.11c) is the following *energy balance*. Assume that $H(q,p,u)$ is of the form (1.12) (see for the general case [B,S3]), then

$$\frac{dH_0}{dt} = \sum_{j=1}^{m} u_j \dot{y}_j . \tag{1.19}$$

Hence the increase of the internal energy H_0 is a function of the inputs and time-derivatives of the outputs; indeed the right-hand side of (1.19) is the total external work performed on the system by the controls $u_1, \dots, u_m$.
From a mathematical point of view equations (1.11c) and (1.19) suggest that the natural structure for the space of inputs and outputs in case of a Hamiltonian (1.12) is a *cotangent bundle* T^*Y, with Y an output manifold, so that $(u_1, \dots, u_m)$ is a *cotangent vector* ([S3], see also [AS]). For general Hamiltonians $H(q,p,u)$ the natural space of inputs and outputs is a *general* symplectic manifold with canonical coordinates $(y,u) = (y_1, \dots, y_m, u_1, \dots, u_m)$, see [S1,S3]. In this last case, since y and u are regarded at the same footing, one does not have to distinguish a priori between inputs and outputs (cf. Willems [W3,W4]); see for details [S1,S3].

As remarked before, the adopted definition of a Lagrangian or Hamiltonian system only covers *conservative* mechanical systems. (By (1.19) the internal energy H_0 is conserved for $u = 0$.) Although in practice mechanical systems always *do* possess inherent damping, especially for *weakly damped* mechanical systems the conservative idealization usually forms a natural starting point for analysis as well as for control purposes. Damping is often difficult to quantify, and the actual presence of unmodelled damping will in many cases only *improve* the characteristics of the controlled system, see also Section 4. Nevertheless it is unsatisfactory that we have not yet been able to incorporate dissipation on a conceptual level in our definition of a mechanical control system. (Although of course damping can be always taken into account as an additional feedback term.) A possible non-conservative extension of the definition of a Hamiltonian system, which retains as much as possible the reciprocal structure of the Euler-Lagrange or Hamiltonian equations, is suggested by adding to equations (1.2) a dissipation term in the following way

$$\frac{d}{dt}\left(\frac{\partial L}{\partial \dot{q}_i}\right) - \frac{\partial L}{\partial q_i} + \frac{\partial R}{\partial \dot{q}_i} = F_i^e, \qquad i = 1, \dots, n, \tag{1.20}$$

where $R(\dot{q})$ is the classical *Rayleigh's dissipation* function ([G,Wh]), modelling for example viscous damping.

2. GEOMETRIC DEFINITION

In this section we briefly sketch how Hamiltonian systems, in local canonical coordinates given as in (1.11a,b,c), can be defined in a global *coordinate free* way, using tools from *symplectic geometry* (see [AM,A]). Readers not familiar with symplectic geometry can skip this definition without loss of continuity. At the end of the section we will illustrate the definition for *linear* Hamiltonian systems.

We recall (see e.g.[A,AM]) that a two-form ω on a manifold M is a *symplectic* form if ω is *non-degenerate* (i.e., for any tangent vector X at $x \in M$ there is a tangent vector Y at x such that $\omega(x)(X,Y) \neq 0$), and *closed* ($d\omega = 0$). Then (M,ω) is called a *symplectic manifold*. By Darboux's theorem [A,AM] there exist local coordinates $(q,p) = (q_1,\ldots,q_n,p_1,\ldots,p_n)$ for M (necessarily dim M is even) such that $\omega = \sum_{i=1}^{n} dp_i \wedge dq_i$. Such coordinates are called *canonical*. A submanifold $L \subset (M,\omega)$ is called *Lagrangian* if ω restricted to L is zero, and furthermore L is *maximal* with respect to this property. It follows that $\dim L = \frac{1}{2}.\dim M = n$.

Now let (M,ω) be a 2n-dimensional symplectic manifold, denoting the *state space* (generalized configuration and momentum coordinates). The symplectic form ω on M *prolongs* to a symplectic form, denoted by $\dot{\omega}$, on the tangent bundle TM, cf. [Tu]; if (q,p) are local canonical coordinates then $\dot{\omega} = \sum_{i=1}^{n} (d\dot{p}_i \wedge dq_i + dp_i \wedge d\dot{q}_i)$, with $(q,p,\dot{q},\dot{p})$ being the natural coordinates for TM.

Furthermore, let (W,ω^e) be a 2m-dimensional symplectic manifold, denoting the space of *external variables** (inputs and outputs taken together, cf.[W3]). The product manifold $TM \times W$ is again a symplectic manifold, with product symplectic form $\omega \oplus (-\omega^e) =: \Omega$. (More precisely, let π_1, resp. π_2, be the projections from $TM \times W$ onto TM, resp. W, then $\Omega = \pi_1^* \dot{\omega} - \pi_2^* \omega^e$.)

Definition 2.1. [S1,S3] *Let $(TM \times W,\Omega)$ be as above. A Hamiltonian system with state space M, and space of external variables W, is defined by a submanifold $L \subset TM \times W$ having the following two properties:*

(2.1a) L can be parametrized by coordinates for M together with m coordinates for W.

(2.1b) L is a Lagrangian submanifold of $(TM \times W,\Omega)$.

---

*Apparently the use of the letter W for the space of external variables is not due to N. Wiener.

Let $(q,p) = (q_1,\dots,q_n,p_n,\dots,p_n)$ be local canonical coordinates for M. It follows from (2.1a) that we can find local coordinates $(y,u) = (y_1,\dots,y_m, u_1,\dots,u_m)$ for W with $\omega^e = \sum_{i=1}^{n} c_i \, du_i \wedge dy_i$, $c_i = \pm 1$ (such coordinates are called semi-canonical), such that L is parametrized by (q,p,u). By (2.1b) there exists locally a *generating function* ([A,AM,Wh]) $H(q,p,u)$ for L, implying that L as a submanifold of $TM \times W$ in its natural local coordinates $(q,p,\dot{q},\dot{p},y,u)$ is given by the equations

$$
\begin{aligned}
\dot{q}_i &= \frac{\partial H}{\partial p_i}(q,p,u) \\
& && i = 1,\dots,n, \\
\dot{p}_i &= -\frac{\partial H}{\partial q_i}(q,p,u) \\
y_j &= -c_j \frac{\partial H}{\partial u_j}(q,p,u) && j = 1,\dots,m,
\end{aligned}
\tag{2.2}
$$

thereby recovering (1.11a,b,c) if $c_j = 1$, $j \in \underline{m}$. (The case of alternating c_j corresponds to what in electrical network theory is called a *hybrid* representation; in a mechanical context this situation arises if, for example, the inputs consist of force variables *and* position variables while the outputs are the complementary position and force variables, see [S3].) We remark that in most mechanical systems the symplectic manifolds M and W appear as *cotangent bundles* with their natural symplectic forms [AM].
Definition 2.1 covers two extreme cases. First if M is absent (i.e. no dynamics), then $L \subset W$ defines a *static reciprocal* system (see [AM,p.412] for references). Alternatively if W is absent (no external variables) then $L \subset TM$ defines a locally *Hamiltonian vectorfield* on M ([Tu]).

Definition 2.1 specializes to linear systems as follows. A linear symplectic form on a linear space $M = \mathbb{R}^{2n}$ is simply a skew-symmetric non-degenerate bilinear form, and thus is given by a skew-symmetric invertible matrix J. By Darboux's theorem there exist *linear* canonical coordinates (q,p) for $\mathbb{R}^{2n}$, in which J takes the form $\begin{pmatrix} 0 & -I_n \\ I_n & 0 \end{pmatrix}$. The prolonged symplectic form $\dot{J}$ on $TM \simeq \mathbb{R}^{2n} \times \mathbb{R}^{2n}$ is given by the matrix $\begin{pmatrix} 0 & J \\ J & 0 \end{pmatrix}$. Let now $(M = \mathbb{R}^{2n}, J)$ and $(W = \mathbb{R}^{2m}, J^e)$ be linear symplectic spaces. A Lagrangian subspace L of $TM \times W$ is a (2n+m)-dimensional subspace of $\mathbb{R}^{2n} \times \mathbb{R}^{2n} \times \mathbb{R}^{2m}$, with the property that the symplectic form given by the matrix

$$
\dot{J} \oplus (-J^e) = \begin{bmatrix} 0 & J & 0 \\ J & 0 & 0 \\ 0 & 0 & -J^e \end{bmatrix}
\tag{2.3}
$$

is *zero* restricted to L, and a linear Hamiltonian system is a Lagrangian subspace of $\mathbb{R}^{2n} \times \mathbb{R}^{2n} \times \mathbb{R}^{2m}$, parametrized by the first 2n coordinates and m coordinates of $\mathbb{R}^{2m}$. It follows that we can take linear canonical coordinates $(y,u) = (y_1,\ldots,y_m,u_1,\ldots,u_m)$ for $\mathbb{R}^{2m}$, such that the Hamiltonian system is given as $\dot{x} = Ax + Bu$, $y = Cx$, satisfying $A^TJ + JA = 0$ (A is called a *Hamiltonian matrix*) and $B^TJ = C$.

3. REALIZATION AND SYNTHESIS

As remarked before, the definition of a Hamiltonian system (1.11) can be regarded as the conservative mechanical analogue of the definition of an electrical m-port. For linear electrical networks it has been quite well understood (using frequency-domain methods as in [Be], and subsequently state-space methods in e.g. [AV,W2]) which additional properties the transfer matrix of a linear system should have in order to be realizable as the driving-point impedance (or admittance) of an electrical m-port consisting of basic elements (resistors, capacitors,...) in some a priori specified class. Similar *synthesis* problems could be formulated for Hamiltonian systems, linear as well as nonlinear. For instance, it is of interest to know which (nonlinear) input-output behaviors can be synthesized by certain types of robotic manipulators.
Here we will not go into the synthesis problem itself (although some results on *linear* mechanical synthesis have been obtained in [B,S3]), but consider as a preamble the *Hamiltonian Realization* problem; i.e., we want to characterize those input-output behaviors which come from Hamiltonian systems (1.11). Two aspects of this problem can be distinguished: first to identify the special properties possessed by Hamiltonian systems, and secondly to single out amongst all (nonlinear) systems the intrinsically Hamiltonian ones. This last aspect is of some practical importance, since if we know a system to be Hamiltonian then this extra information may be used for controlling the system (see Section 4). As discussed in [S3,CS], a special case of the Hamiltonian Realization problem is the classical Inverse Problem in Mechanics, see [Sa],[T]. For the relationship with *quantization* of classical (control) systems we refer to [Sa,THC,S8].

A first result in the Hamiltonian Realization problem was obtained by Brockett & Rahimi [BR], showing that a minimal linear system $\dot{x} = Ax + Bu$, $y = Cx$, is Hamiltonian if and only if its transfer matrix $G(s) = C(Is-A)^{-1}B$ satisfies $G(s) = G^T(-s)$. Since $G^T(-s)$ is the transfer matrix of the *adjoint* system $\dot{z} = - A^T z - C^T u$, $y = B^T z$, this implies that a linear system is Hamiltonian iff it is equivalent to its adjoint system. (This property will be called *self-adjointness*.) Indeed the state space uniqueness theorem yields a skew-symmetric matrix J of full rank such that $A^T J + JA = 0$, $B^T J = C$, implying that the system is Hamiltonian with respect to the linear symplectic form defined by J.

This basic idèa was generalized in [CS] to nonlinear systems

$$(3.1) \qquad \begin{aligned} \dot{x} &= f(x,u), & x &\in M, \\ y &= h(x,u), & u &\in \mathbb{R}^m,\ y \in \mathbb{R}^m, \end{aligned}$$

with M a k-dimensional manifold. Along any solution curve $(\bar{u}(t),\bar{x}(t),\bar{y}(t))$ of (3.1), with $\bar{u}$, say, piecewise constant, the *variational system* is defined as the time-varying linear system

$$(3.2) \qquad \begin{aligned} \dot{v}(t) &= \frac{\partial f}{\partial x}(\bar{x}(t),\bar{u}(t))v(t) + \frac{\partial f}{\partial u}(\bar{x}(t),\bar{u}(t))\ u^v(t) \\ y^v(t) &= \frac{\partial h}{\partial x}(\bar{x}(t),\bar{u}(t))v(t) + \frac{\partial h}{\partial u}(\bar{x}(t),\bar{u}(t))\ u^v(t) \end{aligned} \qquad v \in \mathbb{R}^k,$$

with $u^v \in \mathbb{R}^m$, $y^v \in \mathbb{R}^m$ the variational inputs, resp. variational outputs. Furthermore the *adjoint variational system* is given as

$$(3.3) \qquad \begin{aligned} \dot{p}(t) &= -\left(\frac{\partial f}{\partial x}(\bar{x}(t),\bar{u}(t))\right)^T p(t) - \left(\frac{\partial h}{\partial x}(\bar{x}(t),\bar{u}(t))\right)^T u^a(t) \\ y^a(t) &= \left(\frac{\partial f}{\partial u}(\bar{x}(t),\bar{u}(t))\right)^T p(t) + \left(\frac{\partial h}{\partial u}(\bar{x}(t),\bar{u}(t))\right)^T u^a(t) \end{aligned} \qquad p \in \mathbb{R}^k,$$

with $u^a \in \mathbb{R}^m$, $y^a \in \mathbb{R}^m$ the adjoint variational inputs, resp. adjoint variational outputs. (For global coordinate-free considerations consult [CS].)

Theorem 3.1 [CS] (Self-adjointness condition) *A minimal, analytic and complete system (3.1) is Hamiltonian iff along any trajectory* $(\bar{u}(t),\bar{x}(t),\bar{y}(t))$, $t \geq 0$, *with* $\bar{x}(0)$ *fixed and* $\bar{u}$ *piecewise constant, the input-output maps of (3.2) and (3.3) for* $v(0) = 0 = p(0)$ *are equal, i.e. all variational systems are self-adjoint.*

Remark Furthermore, minimal Hamiltonian realizations are unique up to canonical transformations, while also the internal energy is uniquely determined (modulo constants) by the input-output behavior ([S3,BG,CS]).

Note that Theorem 3.1, as it stands, does not yield easily verifiable conditions for a system to be Hamiltonian, since in principle it requires the integration of the system. Up to now only in particular cases (including the Inverse Problem in Mechanics) the conditions of Theorem 3.1 have been recast into a more constructive form ([CL,Sa,T]).

A rather different line of characterizing Hamiltonian systems was pursued by Crouch and Irving [CI1,CI2] and Jakubczyk [J1,J2,J3], by considering the input-output map of (3.1) for $x(0) = x_0$, expanded in a Volterra series or generating power series. Very elegant algebraic conditions on the kernels of these series have been obtained, vastly generalizing the Brockett-Rahimi condition for linear systems interpreted as a condition on the impulse response matrix $H(t) = C\exp(At)B$, namely $H(t) = -H^T(-t)$.

A geometrically appealing characterization of Hamiltonian systems, solely in terms of their *variational* input-output behavior, was conjectured in [S3], and proved in modified form in [CS]. This characterization starts from the following observations. Let $(u(t,\epsilon),x(t,\epsilon),y(t,\epsilon))$, $|\epsilon|$ small, be a *family* of trajectories of (3.1) parametrized by ϵ, with $u(t,0) = \bar{u}(t)$, $x(t,0) = \bar{x}(t)$, $y(t,0) = \bar{u}(t)$. Then the variational trajectory $u^v(t) = \frac{\partial}{\partial\epsilon}u(t,0) =: \delta u(t)$, $v(t) = \frac{\partial}{\partial\epsilon}x(t,0) =: \delta x(t)$, $y^v(t) = \frac{\partial}{\partial\epsilon}y(t,0) =: \delta y(t)$ is a solution of the variational system (3.2), and moreover all solutions of (3.2) are obtained in this way. Now let $(\bar{u}(t),\bar{x}(t),\bar{y}(t))$ be any trajectory of a *Hamiltonian system*, and let $(\delta_i u,\delta_i x,\delta_i y)$, $i = 1,2$, be any two variational trajectories along $(\bar{u}(t),\bar{x}(t),\bar{y}(t))$. Then by the self-adjointness property we have in canonical coordinates for any t_1,t_2

$$(3.4)\qquad \left(\delta_1^T p(t_2)\delta_2 q(t_2)-\delta_2^T p(t_2)\delta_1 q(t_2)\right) - \left(\delta_1^T p(t_1)\delta_2 q(t_1)-\delta_2^T p(t_1)\delta_1 q(t_1)\right)$$
$$= \int_{t_1}^{t_2} \left(\delta_1^T u(t)\delta_2 y(t) - \delta_2^T u(t)\delta_1 y(t)\right)dt$$

where $\delta_i x = (\delta_i q,\delta_i p)$, $i = 1,2$. (This can be formulated globally using the symplectic forms ω on M and ω^e on $\mathbb{R}^m\times\mathbb{R}^m$, cf. Section 2.) Now let us consider variational trajectories $(\delta u,\delta x,\delta y)$ of $(\bar{u},\bar{x},\bar{y})$ satisfying the property that if support $(\delta u,\delta y)\subset[T_1,T_2]$ then $\delta x(T_1) = 0 = \delta x(T_2)$. This implies that for $t_i = T_i$, $i = 1,2$, the left-hand side of (3.4) is *zero*. Omitting the $(\bar{x},\delta x)$-part we will say that $(\delta u,\delta y)$ are *admissible variations of* $(\bar{u},\bar{y})$ *with compact support*. Fully *external* characterizations of admissible variations with compact support were given in [CS]. We obtain (see [CS] for details and other versions).

Theorem 3.2. [CS] (Variational condition) *Consider a minimal, analytic and complete system (3.1). If for any input-output trajectory* $(\bar{u},\bar{y})$, *with* $\bar{u}$ *piecewise constant, all possible pairs of admissible variations* $(\delta_i u,\delta_i y)$, $i = 1,2$, *with compact support of* $(\bar{u},\bar{y})$ *satisfy*

$$(3.5) \qquad \int_{-\infty}^{+\infty} \left(\delta_1^T u(t)\delta_2 y(t) - \delta_2^T u(t)\delta_1 y(t)\right)dt = 0$$

then every variational system is self-adjoint, and thus (Theorem 3.1) the system is Hamiltonian. Furthermore assume (3.1) is Hamiltonian. Let (Du,Dy) *be any time function from* $\mathbb{R}$ *to* $\mathbb{R}^m \times \mathbb{R}^m$ *with compact support. Assume that*

$$(3.6) \qquad \int_{-\infty}^{+\infty} \left(D^T u(t)\delta y(t) - \delta^T u(t)Dy(t)\right)dt = 0$$

for all admissible variations $(\delta u,\delta y)$ *with compact support of* $(\bar{u},\bar{y})$. *Then also* (Du,Dy) *is an admissible variation of* $(\bar{u},\bar{y})$ *with compact support.*

The above theorem implies that, at least formally, the input-output behavior of a Hamiltonian system (1.11) is geometrically characterized as a Lagrangian "submanifold" (cf. Section 2) of the "manifold" of *all* input-output behaviors of systems (3.1) modelled on the state space M, with the symplectic form suggested by the left-hand side of (3.5), and with the admissible variations of compact support being the "tangent vectors". For a more precise statement we refer to [CS]; in the linear case, however, the formulation can be easily given as follows. Consider a linear input-output system, for example given in matrix polynomial form

$$(3.7) \qquad D\left(\frac{d}{dt}\right)y(t) = N\left(\frac{d}{dt}\right)u(t) \qquad\qquad u,y \in \mathbb{R}^m$$

where the polynomial matrices D(s) and N(s) are such that $D^{-1}(s)N(s)$ is a strictly proper transfer matrix, and D(s) and N(s) are *left co-prime* (i.e., no pole-zero cancellations).

Define on the linear space $\mathcal{N}$ of all piecewise continuous functions $(u,y):\mathbb{R} \longrightarrow \mathbb{R}^m \times \mathbb{R}^m$ with compact support the linear (weak) symplectic form [AM]

$$(3.8) \qquad \Omega\left(\binom{u_1}{y_1},\binom{u_2}{y_2}\right) = \int_{-\infty}^{+\infty} \left(u_1^T(t)y_2(t) - u_2^T(t)y_1(t)\right)dt$$

Denote the linear space of all input-output trajectories (*the input-output behavior*) of (3.7) by $\mathcal{B}$, and define $\mathcal{B}_c := \mathcal{B} \cap \mathcal{N}$. Then Theorem 3.2 implies the

following: *A linear system (3.7) can be realized by a minimal linear Hamiltonian system if and only if* $\mathcal{B}_c$ *is a Lagrangian subspace of* $(\mathcal{N},\Omega)$.
The main drawback of all the above characterizations of Hamiltonian systems is that basically they are concerned with *input-output maps*, and so they do not fit very well into the general realization theory as developed, especially for linear systems, by Willems [W3,W4]. In particular Theorem 3.1 relies for the construction of the symplectic form on the state space M on the nonlinear state-space uniqueness theorem for initialized system, cf. [Su1]. At least for linear systems we will now show how we can obtain minimal Hamiltonian realizations *directly* from the input-output behavior. Let $w_1 = (u_1,y_1)$, $w_2 = (u_2,y_2)$ be elements of $\mathcal{B}_c$, with $\mathcal{B}_c$ a Lagrangian subspace of $(\mathcal{N},\Omega)$. Define the equivalence relation (see [W3,W4])

(3.9) $$w_1 \sim w_2 \iff \{w_1^-.w^+ \in \mathcal{B}_c \iff w_2^-.w^+ \in \mathcal{B}_c\}$$

(Here $w_1^-.w^+$ denotes the concatenation of the external signals w_1 and w in $t = 0$; w_1^- denotes the past of w_1, and w^+ the future of w.) By the assumption of co-primeness of D(s) and N(s) in (3.7) the state space of a minimal realization of (3.7) is isomorphic to the linear space [W3,W4]

(3.10) $$X = \mathcal{B}_c/\sim$$

Thus the equivalence class $[w]$ is the state at $t = 0$ corresponding to any input-output trajectory $\bar{w} \in [w]$. Denoting $x_1 = [(u_1,y_1)]$, $x_2 = [(u_2,y_2)]$, we define on X the function

(3.11) $$J(x_1,x_2) = \int_{-\infty}^{0} \left(u_1^T(t)y_2(t) - u_2^T(t)y_1(t)\right)dt$$

This is correctly defined. Indeed, let $x_1 = 0$ then $(u_1^-.0^+, y_1^-.0^+) \in \mathcal{B}_c$ and thus, since Ω is zero on $\mathcal{B}_c$,

(3.12) $$\int_{-\infty}^{0} \left(u_1^T(t)y_2(t) - u_2^T(t)y_1(t)\right)dt = -\int_{0}^{\infty} \left(0.y_2(t) - u_2^T(t).0\right)dt = 0$$

proving that J only depends on the equivalence classes x_1, x_2.
Clearly J is a bilinear, skew-symmetric form. Non-degeneracy of J is proved as follows. Suppose $\tilde{x} = [(\tilde{u},\tilde{y})]$ is such that for every $x = [(u,y)]$

(3.13) $$\int_{-\infty}^{0} \left(\tilde{u}^T(t)y(t) - u^T(t)\tilde{y}(t)\right)dt = 0$$

Then define $(\bar{u},\bar{y}) := (\tilde{u}^-.0^+, \tilde{y}^-.0^+)$. It follows that

$$(3.14)\qquad \int_{-\infty}^{+\infty} \left(\bar{u}^T(t)y(t) - u^T(t)\bar{y}(t)\right)dt = 0,$$

for all $(u,y) \in \mathcal{B}_c$. Since $\mathcal{B}_c$ is a Lagrangian subspace of $(\mathcal{N},\Omega)$ this implies that $(\bar{u},\bar{y}) \in \mathcal{B}_c$. Clearly $[(\bar{u},\bar{y})] = 0$, and hence $\tilde{x} = [(\tilde{u},\tilde{y})] = 0$, thus proving non-degeneracy of J.

In fact we did not only prove that J is a nondegenerate skew-symmetric bilinear form on X, but also

Proposition 3.3. *Let $\mathcal{B}_c$ be a Lagrangian subspace of $(\mathcal{N},\Omega)$. Define on $\mathcal{B}_c$ the bilinear form*

$$(3.15)\qquad \tilde{J}(w_1,w_2) = \int_{-\infty}^{0} \left(u_1^T(t)y_2(t) - u_2^T(t)y_1(t)\right)dt$$

for any $w_1 = (u_1,y_1), w_2 = (u_2,y_2) \in \mathcal{B}_c$. Then $w_1 \sim w_2$ if and only if $w_1 - w_2 \in \mathrm{Ker}\,\tilde{J}$. Hence $X = \mathcal{B}_c/\mathrm{Ker}\,\tilde{J}$.

Remark. Notice that the above results still make sense if X is not finite-dimensional.

From now on we write the bilinear form $J(x_1,x_2)$ defined by (3.11) in a basis for X as $x_1^TJx_2$, with J a skew-symmetric matrix of rank 2n (= dim X).

By taking concatenations at arbitrary times t_1,t_2, and by time-invariance, (3.11) implies that

$$x_1^T(t_2)Jx_2(t_2) - x_1^T(t_1)Jx_2(t_1) = \int_{t_1}^{t_2} \left(u_1^T(t)y_2(t) - u_2^T(t)y_1(t)\right)dt$$

or in differential form

$$(3.16)\qquad \frac{d}{dt}\, x_1^T(t)Jx_2(t) = u_1^T(t)y_2(t) - u_2^T(t)y_1(t)$$

By co-primeness of D(s) and N(s) in (3.7) it follows that $\mathcal{B}_c$ is generated by an observable and controllable system $\dot{x} = Ax + Bu$, $y = Cx$, $x \in X$. Insertion in (3.16) then immediately yields the identities $A^TJ + JA = 0$, $B^TJ = C$, which ensure that the realization is Hamiltonian. However the matrices A,B,C can be also directly obtained from the behavior $\mathcal{B}_c$. In fact, the internal energy for the system in state $x = [(u,y)]$ at $t = 0$ can be defined as

$$(3.17)\qquad H(x) = \int_{-\infty}^{0} u^T(t)\dot{y}(t)dt.$$

Similar arguments as employed above show that indeed H only depends on the equivalence class x. Furthermore it follows that H is quadratic in x, and so can be represented in a basis for X as

$$(3.18) \qquad H(x) = \frac{1}{2} x^T Qx$$

with Q a symmetric matrix. Then A is defined by the relation $JA = Q$. Furthermore $C:X \longrightarrow \mathbb{R}^m$ is simply defined by setting $C[(u,y)] = y(0)$ for $(u,y) \in \mathcal{B}_c$, and the definition of B follows from the relation $B^T J = C$.

4. CONTROL OF HAMILTONIAN SYSTEMS

The design of a controller for a complex (nonlinear) system ideally should be based on all knowledge available about the system. In particular, as argued already by several authors [B,C,S3,S1], in modern control theory more emphasis should be put on exploiting the special physical properties enjoyed by classes of systems. Indeed it is claimed that a further insight into the physics of a system can provide us with attractive and rewarding shortcuts both on the theoretical and computational level, and that a more physically motivated controller will tend to be more robust, as well as closer to engineering methodology such as dynamic analysis. This general philosophy will now be illustrated on the class of Hamiltonian systems, as dealt with before, in three examples. (For other uses of the Hamiltonian structure for control purposes we refer e.g. to [HS,M,NS,S4].)

The first example concerns *stabilization by feedback* of Hamiltonian systems. For clarity of exposition we will restrict ourselves to Hamiltonian systems (1.11), with H(q,p,u) of the form

$$(4.1) \qquad H(q,p,u) = \frac{1}{2} p^T G(q)p + V(q) - \sum_{j=1}^{m} u_j H_j(q), \qquad q,p \in \mathbb{R}^n$$

Here $\frac{1}{2}p^T G(q)p$ is the kinetic energy (thus $G(q) > 0$), $V(q)$ the potential energy, and the controls are the generalized forces corresponding to the configuration coordinates $y_j = H_j(q)$, $j \in \underline{m}$. Most mechanical systems are of this type. Equilibria for $u = 0$ are the points $(q_0, 0)$ in phase space satisfying $dV(q_0) = 0$. Suppose now that $V(q)$ has a strict local minimum in q_0. Then $H_0(q,p) = \frac{1}{2} p^T G(q)p + V(q)$ has a strict local minimum in $(q_0, 0)$. By conservation of energy for $u = 0$ (see (1.19)) $\frac{dH_0}{dt} = 0$, and thus H_0 is a Lyapunov function, implying that the uncontrolled system is critically (but never asymptotically [AM]) *stable*.

Physically it is evident that in order to *asymptotically* stabilize the system we should add damping. Indeed define

$$(4.2) \qquad u_j = -c_j\dot{y}_j, \qquad c_j > 0, \qquad j \in \underline{m},$$

with $y_j = H_j(q)$, $j \in \underline{m}$, the natural outputs. By (1.19) the controlled system satisfies

$$(4.3) \qquad \frac{dH_0}{dt} = -\sum_{j=1}^{m} c_j\dot{y}_j^2 \leq 0.$$

We now have the following application of LaSalle's invariance principle. Define the *Poisson bracket* of two functions $F(q,p)$ and $G(q,p)$ in local canonical coordinates as

$$(4.4) \qquad \{F,G\}(q,p) = \sum_{i=1}^{n}\left(\frac{\partial F}{\partial p_i}\frac{\partial G}{\partial q_i} - \frac{\partial G}{\partial p_i}\frac{\partial F}{\partial q_i}\right)(q,p)$$

and denote inductively $ad_F^0 G = G$, $ad_F^k G = \{F, ad_F^{k-1}G\}$, $k \geq 1$. (Recall, cf. [AM], that $\{F,G\}$ equals the time-derivative of the function G along the Hamiltonian vectorfield with Hamiltonian F.) Define for any (q,p) the linear space (co-distribution)

$$(4.5) \qquad P(q,p) = \text{span}\,\{dH_0(q,p),\ d(ad_{H_0}^k H_j)(q,p)\},\ j \in \underline{m},\ k \geq 0\}$$

Theorem 4.1. [S5]. *Consider the Hamiltonian system with Hamiltonian (4.1). Let* V *have a strict local minimum in* q_0, *and assume that* $dV(q) \neq 0$ *for* $q \neq q_0$ *in some neighborhood of* q_0. *Furthermore assume that dim* $P(q,p) = 2n$ *for all* (q,p), *with* $q \neq q_0$, *in some neighborhood of* $(q_0,0)$. *Then the feedback (4.2) locally asymptotically stabilizes the Hamiltonian system.*

(The condition on $P(q,p)$ is an *observability* condition, which ensures that the damping (4.2) spreads through the whole system ([Jo],[S5]).)
The main restriction of Theorem 4.1 is that V is assumed to have a (strict) local minimum in q_0. On the other hand, application of the proportional feedback $u_j = -k_j u_j + v_j$, $j \in \underline{m}$, with v_j the new controls, is easily seen to result in another Hamiltonian system with modified Hamiltonian

$$\bar{H}(q,p,v) = \tfrac{1}{2}p^T G(q)p + \bar{V}(q) - \sum_{j=1}^{m} v_j H_j(q)$$

where

$$(4.6) \qquad \bar{V}(q) = V(q) + \frac{1}{2}\sum_{j=1}^{m} k_j y_j^2, \qquad y_j = H_j(q), \qquad j \in \underline{m},$$

is the *modified potential energy*. The possibilities of "shaping" in this manner the potential energy $V(q)$ are indicated in

Proposition 4.2. *Let* $dV(q_0) = 0$ *and* $H_j(q_0) = 0$, $j \in \underline{m}$. *Assume that the* $n \times n$ *Hessian matrix* $\left[\frac{\partial^2 V}{\partial q_i \partial q_j}(q_0)\right]$ *is positive definite when restricted to the subspace* $\bigcap_{j=1}^{m} \ker \operatorname{grad} H_j(q_0)$. *Then there exists a feedback* $u_j = -k_j y_j$, $k_j \geq 0$, $j \in \underline{m}$, *such that* $\bar{V}(q)$ *defined by (4.6) satisfies the assumptions of Theorem 4.1, i.e.,* $\bar{V}$ *has a strict local minimum in* q_0 *with* $d\bar{V}(q) \neq 0$ *for all* q *around* q_0, *with* $q \neq q_0$.

Combination of Theorem 4.1 and Proposition 4.2 leads naturally to the consideration of controllers of PD-type

$$(4.7) \qquad u_j = -k_j y_j - c_j \dot{y}_j, \qquad k_j \geq 0, \quad c_j > 0, \quad j \in \underline{m}.$$

Let us now assume that the conditions of Theorem 4.1 and Proposition 4.2 are indeed satisfied, so that there exists an asymptotically stabilizing feedback (4.7). Since (4.7) mimics the addition of linear springs and dashpots it is immediately clear from physical considerations that the PD-controller (4.7) has very good robustness properties. Indeed the closed-loop system remains asymptotically stable for any perturbation of the matrix $G(q)$ in (4.1) to any other positive definite matrix, and for all perturbations of $V(q)$, as long as $\bar{V}(q)$ defined in (4.6) continues to have a strict local minimum in q_0. Also the robustness with respect to series perturbations (static perturbations on u) is very good; in particular k_j and c_j in (4.7) may become arbitrarily large without affecting the asymptotic stability, while $c_j > 0$ may be arbitrarily small and k_j only has to satisfy a lower bound (see e.g. [Gl] for some appropriate concepts in this context). Furthermore, cf. [Sl], the robustness properties of (4.7) with respect to *unmodelled*, but assumed to be *physically structured*, dynamics seem to be favorable. A main designing task is to tune the gain parameters k_j and c_j in (4.7) (or to take suitable nonlinear functions of y and $\dot{y}$), so as to achieve "optimal" transient behavior of the closed-loop system (such as the nonlinear analogue of *critical damping* for second order linear systems; see [K2] for a discussion in a robotics context).

Having very classical physical roots, the feedback scheme (4.7) has been used in various contexts (satellite control, flexible structures, etc.) for stabilization purposes ([ALMN],[Ba],[G],[Jo]). For point-to-point control of rigid robot manipulators its use has been recognized in [TA] (see also [Kl]); indeed, for $m = n$ and H_j, $j \in \underline{n}$, being independent, the conditions of Proposition 4.2 and Theorem 4.1 are trivially satisfied.

A second example where the use of the Hamiltonian structure provides some attractive shortcuts is in the computation of the *zeros* of a Hamiltonian system (1.11). For linear systems it is well-known that the location of the zeros is very important for controller design. In the nonlinear case the notion of zeros, or better *zero dynamics*, can be defined, extending the linear notion, as follows.

For clarity of exposition we only consider Hamiltonians $H(q,p,u)$ of the form (1.12). First we define the *clamped* or *constrained* dynamics of a Hamiltonian system (1.11) as that part of the system dynamics *compatible* with the constraints $y_1 = \ldots = y_m = 0$, where y_j, $j \in \underline{m}$, are the natural outputs (1.11c). Now define for any $i \in \underline{m}$ the integer ρ_i as the smallest integer ≥ 0 such that $\{H_j, ad_{H_0}^{\rho_i} H_i\} \neq 0$ for some $j \in \underline{m}$. Throughout we shall assume that $\rho_i < \infty$, $i \in \underline{m}$. Define then $A(q,p)$ as the $m \times m$ matrix with (r,s)-th element $\{H_s, ad_{H_0}^{\rho_r} H_r\}$, $r,s \in \underline{m}$. We have

Theorem 4.3. [S7,S9] *Consider a Hamiltonian system (1.11) with Hamiltonian of the form (1.12). Assume that*

$$\text{rank } A(q,p) = m, \qquad \textit{for every point } (q,p) \textit{ in}$$

(4.8)

$$N^* = \{(q,p) \in M \mid H_i(q,p) = ad_{H_0} H_i(q,p) = \ldots = ad_{H_0}^{\rho_i} H_i(q,p) = 0,\ i \in \underline{m}\}$$

Then N^* is, if non-empty, a symplectic submanifold of the phase space M of codimension $\sum_{i=1}^{m} (\rho_i + 1)$, and the clamped dynamics are given by the Hamiltonian vectorfield on N^*

(4.9)

$$\dot{\bar{q}} = \frac{\partial \bar{H}_0}{\partial \bar{p}}(\bar{q},\bar{p})$$

$$\dot{\bar{p}} = -\frac{\partial \bar{H}_0}{\partial \bar{q}}(\bar{q},\bar{p})$$

where $(\bar{q},\bar{p})$ are canonical coordinates for N^*, and $\bar{H}_0(\bar{q},\bar{p})$ is the restriction of the internal energy $H_0(q,p)$ to N^*.

Moreover, under the assumption (4.8) the clamped dynamics (4.9) can be also correctly called the *zero dynamics* [BI] of the Hamiltonian system [S7,S9]. (For *linear* systems the assumption (4.8) implies left-invertibility of the system.) We remark that for Hamiltonians of the form (4.1), assumption (4.8) is *automatically* satisfied, with $\rho_i = 1$, $i \in \underline{m}$ [S7]. For linear Hamiltonian systems we can dispense with assumption (4.8) ([S7]), and in [S9] it is conjectured that if assumption (4.8) is not satisfied then also in the nonlinear case still the zero dynamics will be given by a Hamiltonian vectorfield of the form (4.9).

Summarizing, the zero dynamics for a Hamiltonian system, at least under some extra assumptions, are given by a Hamiltonian vectorfield on $N^* \subset M$, with Hamiltonian $\bar{H}_0$ obtained by restricting H_0 to N^*. Since the *free dynamics* of a Hamiltonian system are simply given by the Hamiltonian vectorfield on M with Hamiltonian H_0, we immediately obtain strong relations between the "poles" and "zeros" of a Hamiltonian system. For example, if H_0 is positive definite (and thus the free dynamics are critically stable), then so is $\bar{H}_0$ (and thus the zero dynamics are critically stable). In particular, if H_0 is also of the form (4.1), then in the single-input case the poles and zeros of the (linearized) system are *interlacing* on the imaginary axis. Furthermore just by the fact that the free and zero dynamics are Hamiltonian it follows that neither of them can be asymptotically stable, and that the poles as well as the zeros of a linear or linearized nonlinear Hamiltonian system are located symmetrically with respect to the imaginary axis. For some control applications of these observations we refer to [HS,S7].

As a third example, we mention the study of *controllability* of a Hamiltonian system (1.11), e.g. with Hamiltonian (1.12). It is well-known that, at least in the analytic case, the controllability properties are determined by all (repeated) Lie brackets of the drift and the input vectorfields [Su2]. Now the Lie bracket of two Hamiltonian vectorfields with Hamiltonian F, respectively G, is again a Hamiltonian vectorfield, with Hamiltonian $\{F,G\}$ (the Poisson bracket of F and G). Hence all Lie brackets in this case are *Hamiltonian vectorfields*, and the controllability properties of the Hamiltonian system are determined by the (repeated) Poisson brackets of $H_0, H_1, .., H_m$. The theoretical implications of this observation are not clear at this moment; at least it follows that controllability is very much related to observability [S3]. Notice also the computational advantages of this observation. It implies that we do not have to go through the equations of motion; the knowledge of the Hamiltonian (1.12) suffices. (This is also valid for other problems, see e.g. the computation of the zero dynamics or the conditions of Theorem 4.1.)

5. CONCLUDING REMARKS

In the development of modern system and control theory over the last decades there has been a tendency to neglect the natural structures imposed by the physical character of the system. Instead the emphasis has been on *general* (state-space) systems. On the other hand, in *modelling* physical systems the underlying physical structure is often exploited in a crucial way (this is especially clear in the context of nonlinear mechanical systems, see also Section 1). Thus it seems that an appropriate system theoretic framework for modelling and representation issues should incorporate the relevant physical structure on a fundamental level. In Section 4 we have shown in some examples that also for *control* purposes it is advantageous to use the physical structure of the system under consideration in an explicit way, especially when dealing with complex (nonlinear, infinite-dimensional) systems. Already from the theory of dynamical systems (without inputs) it is well-known that one cannot hope for one *single* theory covering all nonlinear systems; instead one also has to concentrate on special subclasses, for instance Hamiltonian vectorfields. Furthermore, many of the constructions used in modern control theory do not have a clear physical interpretation when applied to systems with a specific physical structure. Apart from being unsatisfactory from a theoretical point of view this could also be a drawback in applications.

Summarizing, a lot of work remains to be done, both in the system theoretic approach to modelling of physical systems, as well as in the control of systems with physical structure (Hamiltonian systems, gradient systems, dissipative systems).

Finally, not only system and control theory can gain from a closer study of physics, but conversely, as already argued by Willems [W3,W1,W2], physics could benefit from the consideration of fundamental system theoretic concepts (like (minimal) state, controllability and observability, realization).

References

[A] V.I. Arnold, *Mathematical Methods of Classical Mechanics*, Springer, Berlin (1978) (translation of the 1974 Russian edition).

[ALMN] J.N. Aubrun, K. Lorell, T.S. Mast, J.E. Nelson, "Dynamic analysis of the actively controlled segmented mirror of the W.M. Keck ten-meter telescope", IEEE Contr.Syst.Mag., 7, no. 6 (1987).

[AM] R.A. Abraham & J.E. Marsden, *Foundations of Mechanics* (2nd edition), Benjamin/Cummings, Reading, Mass. (1978).

[AS] H. Abesser, J. Steigenberger, "On feedback transformations in hamiltonian control systems, Wiss. Z.TH Ilmenau 33 (1987), 33-42.

[AV] B.D.O. Anderson, S. Vongpanitlerd, *Network Analysis and Synthesis*, Prentice Hall, Englewood Cliffs, N.J. (1973).

[B] R.W. Brockett, "Control theory and analytical mechanics", in *Geometric Control Theory* (eds. C. Martin & R. Hermann), Vol VII of Lie groups: History, Frontiers and Applications, Math.Sci.Press, Brookline (1977), 1-46.

[Ba] M. Balas, "Direct velocity feedback control of large space structures", J.Guid.Contr., 4 (1979), 480-486.

[Be] V. Belevitch, *Classical Network Theory*, Holden-Day, San Francisco (1968).

[BG] J. Basto Gonçalves, "Realization theory for Hamiltonian systems", SIAM J.Contr.Opt., 25 (1987), 63-73.

[BI] C.I. Byrnes, A. Isidori, "A frequency domain philosophy for nonlinear systems, with applications to stabilization and adaptive control", Proc. 23rd IEEE Conf.Decision and Control (1984), 1569-1573.

[BM] R.K. Brayton, J.K. Moser, "A theory of nonlinear networks", Quart. Appl. Math., 22 (1964), Part I, pp. 1-33, Part II, pp. 81-104.

[BR] R.W. Brockett, A. Rahimi, "Lie algebras and linear differential equations", *Ordinary Differential Equations* (ed. L. Weiss), Academic, New York (1972).

[BS] A.G. Butkovskii, Yu.I.Samoilenko, "Control of quantum systems", I & II, Autom.Rem.Contr.40 (1979), 485-502, 629-645.

[C] P.E. Crouch, "Geometric structures in systems theory", Proc. IEE, 128, Pt.D (1981), 242-252.

[CI1] P.E. Crouch, M. Irving, "On finite Volterra series which admit Hamiltonian realizations", Math.Syst.Th., 17 (1984), 293-318.

[CI2] P.E. Crouch, M. Irving, "Dynamical realizations of homogeneous Hamiltonian systems", SIAM J.Contr.Opt., 24 (1986), 374-395.

[CL] P.E. Crouch, F. Lamnabhi-Lagarrigue, "State space realizations of nonlinear systems defined by input-output differential equations", *Analysis and Optimization of Systems* (eds. A. Bensoussan, J.L. Lions) Lect.Notes in Contr.Inf.Sc. 111, Springer, Berlin (1988), 138-149.

[CS] P.E. Crouch, A.J. van der Schaft, *Variational and Hamiltonian Control Systems*, Lect.Notes in Contr.Inf.Sc. 101, Springer, Berlin (1987).

[G] H. Goldstein, *Classical Mechanics*, Addison-Wesley, Reading, Mass. (1950).

[Ge] W.B. Gevarter, "Basic relations for control of flexible vehicles", AIAA Journal, 9 (1970).

[Gl] S.T. Glad, "Robustness of nonlinear state feedback - a survey", Automatica, 23 (1987), 425-435

[HS] H.J.C. Huijberts, A.J. van der Schaft, "Input-output decoupling with stability for Hamiltonian systems", University of Twente, Dept. Applied Mathematics, Memo 722 (1988), to appear in Math. of Control, Signals and Systems.

[J1] B. Jakubczyk, "Poisson structures and relations on vectorfields and their Hamiltonians", Bull.Pol.Ac.: Math, 34 (1986), 713-721.

[J2] B. Jakubczyk, "Existence of Hamiltonian realizations of nonlinear causal operators", Bull.Pol.Ac.: Math, 34 (1986), 737-747.

[J3] B. Jakubczyk, "Hamiltonian realizations of nonlinear systems", *Theory and Applications of Nonlinear Control Systems* (C.I. Byrnes, A. Lindquist, eds.), North-Holland, Amsterdam (1986), 261-271.

[Jo] E.A. Jonckheere, "Lagrangian theory for large scale systems", preprint, University of Southern California, Dept. of Electrical Engineering (1981).

[Kl] D.E. Koditschek, "Natural motion for robot arms", IEEE Proc. 23rd Conf. Decision and Control (1984), 733-735.

[K2] D.E. Koditschek, "Robot-control systems", *Encyclopedia of Artificial Intelligence* (Stuart Shapiro, ed.), Wiley (1987), 902-923.

[KM] P.S. Krishnaprasad, J.E. Marsden, "Hamiltonian structures and stability for rigid bodies with flexible attachments", Arch.Rat.Mech.Anal., 98 (1987), 71-93.

[M] R. Marino, "Hamiltonian techniques in control of robot arms and power systems", *Theory and Applications of Nonlinear Control Systems* (C.I. Byrnes, A. Lindquist, eds.), North-Holland, Amsterdam (1986), 65-73.

[MB] N.H. McClamroch, A.M. Bloch, "Control of constrained Hamiltonian systems and applications to control of constrained robots", *Dynamical Systems Approaches to Nonlinear Problems in Systems and Circuits* (F.M.A. Salam, M.L. Levi, eds.), SIAM, 1988, 394-403.

[NS] H. Nijmeijer, A.J. van der Schaft, "Input-output decoupling of Hamiltonian systems: The nonlinear case", *Analysis and Optimization* (J.L. Lions, A. Bensoussan, eds.), Lect.Not.Contr.Inf.Sci. 83, Springer, Berlin (1986), 300-313.

[S1] A.J. van der Schaft, "Hamiltonian dynamics with external forces and observations", Math.Syst.Th., 15 (1982), 145-168.

[S2] A.J. van der Schaft, "Symmetries, conservation laws and time-reversibility for Hamiltonian systems with external forces", J.Math.Phys., 24 (1983), 2095-2101.

[S3] A.J. van der Schaft, *System theoretic descriptions of physical systems,* CWI Tract 3, CWI, Amsterdam (1984).

[S4] A.J. van der Schaft, "Controlled invariance for Hamiltonian systems", Math.Syst.Th., 18 (1985), 257-291.

[S5] A.J. van der Schaft, "Stabilization of Hamiltonian systems", Nonl.An. Th.Math.Appl., 10 (1986), 1021-1035.

[S6] A.J. van der Schaft, "Optimal control and Hamiltonian input-output systems", *Algebraic and Geometric Methods in Nonlinear Control Theory* (M.Fliess, M. Hazewinkel, eds.), Reidel, Dordrecht (1986), 389-407.

[S7] A.J. van der Schaft, "On feedback control of Hamiltonian Systems", *Theory and Applications of Nonlinear Control Systems* (C.I. Byrnes, A. Lindquist, eds.), North-Holland, Amsterdam (1986), 273-290.

[S8] A.J. van der Schaft, "Hamiltonian and quantum mechanical control systems", 4th Int.Sem.*Mathematical Theory of Dynamical Systems and Microphysics* (A. Blaquiére, S. Diner, G. Lochak, eds.), CISM Courses and Lectures 294, Springer, Wien (1987), 277-296.

[S9] A.J. van der Schaft, "Hamiltonian control systems: decomposition and clamped dynamics", *Control Theory & Multibody Systems*, AMS Contemporary Mathematics, to appear (1989).

[Sa] R.M. Santilli, *Foundations of Theoretical Mechanics I,* Springer, New-York (1978).

[San] G. Sanchez de Alvarez, *Geometric Methods of Classical Mechanics applied to Control Theory*, Ph.D.Thesis, Dept.Mathematics, Univ. of California, Berkeley (1986).

[Sl] J.J.E. Slotine, "Putting physics in control - the example of robotics", IEEE Control Systems Magazine, December 1988, 12-18.

[Su1] H.J. Sussmann, "Existence and uniqueness of minimal realizations of nonlinear systems", Math. Syst. Th., 10 (1977), 263-284.

[Su2] H.J. Sussmann, "Lie brackets, real analyticity and geometric control", *Differential Geometric Control Theory* (R.W. Brockett, R.S. Millman, H.J. Sussmann, eds.), Birkhaüser, Boston (1983), 1-116.

[T] F. Takens, "Variational and conservative systems", Report ZW 7603, Univ. of Groningen (1976).

[TA] M. Takegaki, S. Arimoto, "A new feedback method for dynamic control of Manipulators", Trans. ASME, J. Dyn.Syst.Meas.Contr., 103 (1981), 119-125.

[THC] T.J. Tarn, G. Huang, J.W. Clark, "Modelling of quantum mechanical control systems", Math.Modelling, 1 (1980), 109-121.

[Tu] W.M. Tulczyjew, "Hamiltonian systems, Lagrangian systems and the Legendre transformation", Symp. Math. 14 (1974), 247-258.

[W1] J.C. Willems, "Dissipative dynamical systems", Part I & Part II, Arch.Rat.Mech.Anal., 45 (1972), 321-392.

[W2] J.C. Willems, "Realization of systems with internal passivity and symmetry constraints", J. Franklin Inst., 301 (1976), 605-621.

[W3] J.C. Willems, "System theoretic models for the analysis of physical systems", Ricerche di Automatica, 10 (1979), 71-106.

[W4] J.C. Willems, "From time-series to linear system - Part I. Finite dimensional linear time invariant systems", Automatica, 22 (1986), 561-580.

[Wh] E.T. Whittaker, *A treatise on the analytical dynamics of particles and rigid bodies*, 4th edition, Cambridge Univ.Press, Cambridge (1959).

The Impact of the Singular Value Decomposition in System Theory, Signal Processing, and Circuit Theory

J. Vandewalle, L. Vandenberghe, M. Moonen

ESAT (Electrical Engineering Department)
Katholieke Universiteit Leuven
Kardinaal Mercierlaan 94, 3030 Heverlee, Belgium

L. Vandenberghe and M. Moonen are supported by the N. F. W. O.
(Belgian National Fund of Scientific Research).

Abstract

System theory is a discipline which applies mathematical methods in order to provide a unified approach for many application areas. Such a unification is not only useful for communication between experts, but also in order to be able to carry over concepts, methods and software between areas. Also intellectually such a unified approach is mandatory since the same mathematical results are reused and the same derivations apply in the different application areas. The first aim of this contribution is to discover some common grounds in signal processing, circuit theory and some other engineering areas. Second the impact of singular value decomposition and its generalizations will be discussed. Third it will be shown how system theory can provide a unified approach to the problems and open up new avenues in these fields.

1 Introduction

Anniversaries are very good occasions to reflect about the broader scope of past activities and planned projects, certainly in times when most energy is dissipated in order to satisfy immediate needs. It is a pleasure for us to share these reflections with you at the occasion of Jan Willems' fiftieth anniversary. It is the aim of the contribution to extract from our past engineering research experience common system grounds and to point to some future orientations.

It is quite natural that disciplines like continents drift apart. However on the earth surface continents drift apart in one sense and converge in the other. Similarly system theory can bring together several remote disciplines and consolidate the mathematical framework.

This unification of system theory is important not only because it provides a platform for mathematically correct reasoning but also because it leads to parsimony of concepts, ideas, tools and software.

Indeed it is much easier to teach, understand, extend and apply the same concepts, methods and even software. Such a unified system approach provides an analytical link between on the one hand classes of systems and models and on the other hand classes of behavior (stability, bifurcation, sensitivity, ...). These unification efforts can be seen at the same level as the great unification efforts in mathematics with the advent of modern algebra. However from an engineering point of view design is more important than analysis. Since technical systems such as electrical, mechanical or computational or the combination of these systems (often called mechatronic systems) become more and more complex, a global system approach is mandatory.

In this paper we illustrate these ideas with certain concrete problems encountered during engineering research in mathematical modeling, circuit theory and complexity and we extract from these the common system grounds.

The vehicle for this tour is the singular value decomposition (SVD) which is an old concept from matrix theory (Autonne 1902 [30]) which was used extensively in numerical linear algebra since the sixties and which is now widely applied in signal processing, system theory and control under various generalizations.

This contribution is organized as follows. In Section 2 some general thoughts are given on the role of mathematical modeling in engineering and education. In Section 3 an overview of singular value decomposition and its generalizations is presented. In Section 4 SVD is used in the study of static systems. The identification of dynamic systems from input-output measurements with SVD is discussed in Section 5. In the last section the conclusions are presented. In the whole paper special emphasis is placed on the rank reduction properties that can be performed based on the spectrum of singular values. This is especially relevant in order to compare the complexity of models [3], in approximate modeling, model reduction and separation of signals and noise based on their strength.

2 On the role of mathematical modeling in engineering and education.

When dealing with complex analysis, design and control problems, engineers more and more turn to mathematical models and computer aids. In fact computers can only deal with mathematical models of physical or technical systems. Hence one should not forget that a computer simulation or a computer aided design is based on an underlying model. It is then the responsibility of the user or engineer to verify whether the underlying assumptions of the model (e.g. linearity, time invariance ...) are satisfied in reality. This can be a serious problem since certain types of behaviour can be undetected because a too restrictive class of models is used. So one can confuse in certain simulations [7,8] periodic behavior (with a large period), chaotic behavior and noise.

Of course there is always a trade-off in engineering and in other aplication areas between on the one hand the complexity of the model and on the other hand the accuracies of the model. The black box model as introduced by J. Willems [2,3] for a dynamical system is very adequate for many engineering applications like signal processing, circuit theory and identification.

Definition 1 *A dynamical system is defined by the time set $T \subseteq R$, the signal alphabet W and the behavior $B \in W^T$.*

Usually the time set T is R for the continuous time systems or the integer multiples τZ of the sampling interval τ for discrete time systems. This together with the alphabet (R,C or discrete) is known in advance to the engineer. The most relevant part for the engineer however is the behavior set B which is in fact the collection of all admissible signals for the external variables. In this general framework [2,3] the external variables should reflect the interaction of the dynamical system with the rest of the world. They can be inputs or outputs. In general it may not be clear which one is an input or an output. Although this is an unusual situation in control theory it is a very realistic approach in circuit theory, many measurement and process control situations.

Example : A diode is characterized by the time set R the alphabet $R \times R$ and the behavior

$$B = \{v(t), i(t) \mid v(t)i(t) = 0, v(t) \leq 0, i(t) \geq 0\} \tag{1}$$

This description of a diode corresponds with the traditional description of an ideal diode in circuit theory. On an unknown twoterminal element, called device under test, in a complex electronic circuit one can do electrical measurements and verify that the voltage $v(t)$ and the current $i(t)$ satisfy $v(t)i(t) = 0$ and $v(t) \leq 0, i(t) \geq 0$.
Thereby we assume that the only interaction with the outside world is via $v(t)$ and $i(t)$. When the diode can emit light (LED) the model should be modified. It is also not possible to call $v(t)$ or $i(t)$ an input or an output. In circuit theory one can deal with diodes and many other components like transistors, resistors, capacitors, inductances, transformers, ... in a similar way without specifying inputs or outputs [10]. Such a black box approach is useful, handy and appropriate in circuit theory and IC design. However for those who are involved in device physics and IC technology the diode is described by equations in quantum physics. Although this physical description is useful because it explains much more of the internal mechanisms like the temperature dependence and the dependence on the geometry, for system design and even VLSI design the black box approach or a slight refinement of it are quite sufficient. Also in control theory and signal processing mathematical models are quite often used because they show how the system interacts with the external world via external variables. Of course one should not use such models in order to obtain complete information on the internal mechanisms. However for the purpose of design (e.g. the buildup of a complex system with the black box) and for process control (e.g. quality control, ...) the black box model is adequate for engineers. Hence we can conclude that the definition of a dynamical system as given

in mathematical system theory, is quite natural for engineers involved in system design, signal processing and control.

It is at this level of abstraction that system theory provides a variety of concepts like time constants, transfer function, impulse response, state equations, transient, steady state, poles, convolution which are particularly suited to analyze, design and control linear systems. The fact that these concepts exist for any linear system implies that these can be applied in many individual engineering disciplines like on electrical, mechanical, hydraulic, thermodynamic systems and even on more global systems which encompass several individual disciplines like ecology, mechatronics, sensors, transducers and actuators. At the didactical level, system theory can be a general course for all engineering students. At the industrial level a unified systems approach to problems has many advantages. Since the problems often have aspects at the material, information and organization level, a global systems approach is mandatory. Even if there are only qualitative relations and not quantitative, they can be quite rewarding [29].
In a wider scope, a general systems approach can introduce to engineers a wealth of unifying concepts, e.g. convolution, chain fractions expansion, sensitivities, orthogonality, singular value decomposition, least squares, modulo calculus, recursion, fast Fourier transform, simulated annealing, stability, convergence, adaptivity, complexity theory, dynamic systems, average, standard deviation, yield, graph theory. In order to illustrate the convolution example, one can observe that convolution is used in the study of the sum of two random variables, the multiplication of polynomials, the multiplication of integers with a digital circuit, the relationship between input and output of a dynamical system .... Chain fractions expansion of rational numbers appear to be useful in gear boxes and electronic counters, whereas chain fractions expansions of ratios of polynomials are used in numerical analysis and filter design.

Quite often one system representation is more useful for a certain aspect than another. Hence the conversion from one representation to another is important. Many objects can be described in a geometric form (plots, block diagram, flow charts, ...) as well as in a linguistic form (data, formulas, programs, ...). For a given discrete time dynamical system for example one can find many representations by algebraic expressions like the transfer function, the impulse response .... Most often an engineer is more interested in a graphical or geometric representation. Students can train their understanding of the different representations and the relationships by varying some parameters and observing the differences between the plots as generated by a personal computer. It is not only in engineering but also in mathematics that one can observe an evolution towards the use of personal computers and of graphical representations. The publications of the Visual Mathematics Library [11] are quite interesting in this respect.

An important contribution of systems theory is that it provides links between reality and computer. In fact the same computer infrastructure and simulation tools can be used often for many different systems. The set of available tools should be considered in a very wide sense and should include audiovisual tools, computer tools (spreadsheet, symbolic manipulation e.g. MACSYMA), simulation tools, optimization tools. Quite often engineering students and engineers feel so confident about the simulations en-

vironment (workstation or PC and a user friendly softwarepackage), that they accept whatever results are obtained by the simulation. However computers need a mathematical model of reality for simulation and optimization and depending on the kind of model at hand one can observe certain phenomena. Hence engineers should be quite critical with respect to the results generated by the computer. Here again a solid mathematical understanding of the models, the phenomena as well as the numerical methods is important in order to make the correct decisions.
In this respect one can observe that the distinction between software and hardware becomes more and more artificial. Software is used to design hardware and hardware can help to accelerate the execution of software. Here again it is important that engineers have an integrated and broad view.

3 Singular value decomposition (SVD) and SVD based concepts and generalizations

In this section, the theorems stating the existence and the properties of singular value decomposition and its generalizations are presented. For a proof, software and computational requirements the reader is referred to the literature [44].

Theorem 1 *The* **singular value decomposition.** *If A is an $m \times n$ real matrix of rank r then there exist real orthogonal matrices*

$$U = [u_1\ u_2 \dots u_m], \quad V = [v_1\ v_2 \dots v_n] \tag{2}$$

such that

$$U^t \cdot A \cdot V = \begin{bmatrix} diag(\sigma_1, \sigma_2, \dots, \sigma_r) & 0 \\ 0 & 0 \end{bmatrix} = \Sigma \tag{3}$$

where

$$\sigma_1 \geq \sigma_2 \geq \dots \geq \sigma_r > \sigma_{r+1} = \dots = 0 \tag{4}$$

The σ_i are the singular values of A and the vectors u_i and v_i are respectively the i-th left and the i-th right singular vector.

The set $\{u_i, \sigma_i, v_i\}$ is called the i-th singular triplet. The singular vectors (triplets) corresponding to large (small) singular values are called large (small) singular vectors (triplets). The SVD of A

$$A = U\Sigma V^t$$

reveals a great deal about the structure of a matrix as evidenced by the following well known corollaries :

Corollary 1 *Let the SVD of A be given as in theorem 1 then*
(1). Rank property

$$r(A) = r \quad and \quad \begin{array}{l} N(A) = span\{v_{r+1}, \dots, v_n\} \\ R(A) = span\{u_1, \dots, u_r\} \end{array} \tag{5}$$

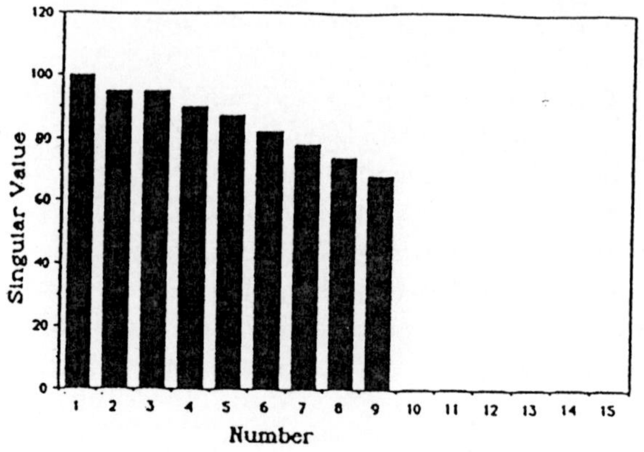

Figure 1: Typical singular spectrum of a 15 × 15 matrix.

(2). Dyadic decomposition

$$A = \sum_{i=1}^{r} u_i \cdot \sigma_i \cdot v_i^t \tag{6}$$

(3). Norms

$$\begin{aligned} \|A\|_F^2 &= \sigma_1^2 + \ldots + \sigma_r^2 \\ \|A\|_2 &= \sigma_1 \end{aligned} \tag{7}$$

(4). Rank k approximation. Define A_k by

$$A_k = \sum_{i=1}^{k} u_i \cdot \sigma_i \cdot v_i^t \quad with \quad k < r \tag{8}$$

then

$$\min_{r(B)=k} \|A - B\|_2 = \|A - A_k\|_2 = \sigma_{k+1} \tag{9}$$

$$\min_{r(B)=k} \|A - B\|_F^2 = \|A - A_k\|_F^2 = \sigma_{k+1}^2 + \cdots + \sigma_r^2 \tag{10}$$

This important result is the basis of many concepts and applications such as total linear least squares, data reduction, image enhancement, dynamical system realization theory and in all possible problems where the heart of the solution is the approximation, measured in 2-norm or Frobeniusnorm, of a matrix by one of a lower rank.

A typical singular value spectrum of a 15 × 15 matrix is depicted in Fig. 1. Such spectra typically occur in situations where the matrix A is generated with measurements of limited accuracy. Due to the limited accuracy of the measurements (range of 0.1) the matrix is of full rank (15). However with some small perturbations on the matrix A (2-norm in the range of $0.1\sqrt{15}$), the matrix can be approximated by a rank 9 matrix. In most systems, control and signal processing applications where such matrices occur, the lower rank approximation of the matrix is much more relevant than the data matrix A. In Figure 1 there is a clear difference between the large and relevant singular values $\sigma_1 \ldots \sigma_9$ and the small singular values, which are due to the inaccuracies in the measurements. Although in some applications the distinction may not be clear, it is usually appropriate to approximate a data matrix by making all singular values zero which are within the range of inaccuracies. This procedure is often essential in order to come up with meaningful and practical models for systems (see Section 4 and 5). The same arguments can be applied to the generalizations of the singular value decomposition (see below) which involve more than one matrix and which also produce a generalized singular spectrum.

The SVD provides an important tool in the generalization and characterization of important geometrical concepts. One of these is the notion of **angles between subspaces,** which is a generalization of the angle between two vectors.

Definition 2 *Let F and G be subspaces in R^m whose dimensions satisfy*

$$p = \dim(F) \geq \dim(G) = q \geq 1$$

The principal angles $\theta_1, \theta_2, \ldots, \theta_q \in [0, \pi/2]$ between F and G are defined recursively by

$$\cos(\theta_k) = \max_{u \in F} \max_{v \in G} u^t.v = u_k^t.v_k^t$$

subject to

$$\|u\| = \|v\| = 1;$$
$$u^t.u_i = 0 \quad i = 1, \ldots, k-1$$
$$v^t.v_i = 0 \quad i = 1, \ldots, k-1$$

The vectors u_i, v_i, $i = 1, \ldots, q$ are called the principal vectors of the subspace pair (F, G).

If the columns of P $(m \times p)$ and Q $(m \times q)$ define orthonormal bases for the subspaces F and G respectively, then it follows from the minimax characterization of singular values that :

$$\begin{aligned} [u_1, \ldots, u_p] &= P.Y \\ [v_1, \ldots, v_q] &= Q.Z \\ \cos\theta_k = \sigma_k & \qquad k = 1 \ldots q \end{aligned}$$

where Y, Z and the σ_k are given by the SVD of the ('generalized inner') product

$$P^t.Q = Y.diag(\sigma_1, \ldots, \sigma_q).Z^t$$

From this, it is not difficult to devise an algorithm to compute the intersection of subspaces that are for instance the column spaces of two given matrices A and B. This is precisely the idea behind the technique of canonical correlation, which appears to be very fruitful in the identification of linear dynamical state space models from noisy input-output measurements [25,26].

There are several ways to compute the canonical correlation structure of a matrix pair (A, B), roughly all possible ways of computing an orthonormal basis for the row spaces (e.g. two QR decompositions, two singular value decompositions) followed by an SVD of the generalized inner product of the orthonormal bases matrices. Another method is the computation of the right null space of the concatenated matrix

$$\begin{bmatrix} A \\ B \end{bmatrix}$$

However, it is expected that depending on the application at hand, one method could be preferable with respect to the others.

From now on we call the singular value decomposition the ordinary SVD or OSVD. A proposal [31] has been made to standardize the nomenclature for the generalizations of SVD. These are PSVD, QSVD, RSVD, SSVD and TSVD. We will briefly overview these different generalizations and explain in more detail the QSVD because it is needed in the next sections. However some other generalizations (like SSVD) have an important impact on control [34][36] and more applications of the generalizations are expected.

Theorem 2 *The* **quotient singular value decomposition** *(QSVD). If A is an $m \times n$ matrix with $n \geq m$ and B is an $m \times p$ matrix, then there exist orthogonal matrices Q_A $(n \times n)$ and Q_B $(p \times p)$ and an invertible X $(m \times m)$ such that :*

$$\begin{array}{llll} X^t \cdot A \cdot Q_A = D_A = diag(\alpha_i) & \alpha_i \geq 0 & i = 1, \ldots, m \\ X^t \cdot B \cdot Q_B = D_B = diag(\beta_i) & \beta_i \geq 0 & i = 1, \ldots, q = \min(m, p) \end{array} \tag{11}$$

where

$$\beta_1 \geq \beta_2 \geq \ldots \geq \beta_r \geq \beta_{r+1} = \ldots = \beta_q = 0 \quad r = rank(B) \tag{12}$$

Observe that the QSVD reduces to the OSVD in the case that $B = I_m$. The elements of the set

$$\sigma(A, B) = \{\alpha_1/\beta_1, \ldots, \alpha_r/\beta_r\} \tag{13}$$

are referred to as the quotient singular values of A and B. The quotient singular values corresponding to the $\beta_i = 0$ are infinite. They are considered to be equal and come first. For our purposes it is more convenient to order the diagonal elements in D_A and D_B according to decreasing quotient singular value rather than by (12).

The recent introduction of the fundamental concepts of **oriented energy** and **oriented signal-to-signal ratio** [21] has provided a rational framework in which both the estimation of ranks and subspaces can be formalized in a rigorous way. Moreover extremal directions of oriented energy and oriented signal-to-signal ratio can be calculated with OSVD and QSVD.

Let A and B contain measurement vector sequences (typically a number of consecutive sample vectors from m measurements channels). The columns of A and B are denoted by a_k, b_k.

Definition 3 *The oriented energy of the matrix A, measured in a direction q is defined as :*

$$E_q[A] = \sum_{k=1}^{n} (q^t a_k)^2 = \|q^t A\|^2 \tag{14}$$

Definition 4 *The oriented signal-to-signal ratio of the two vector sequences A and B in the direction q is defined as :*

$$E_q[A,B] = E_q[A]/E_q[B] \tag{15}$$

There are straightforward generalizations of these definitions to oriented energy and signal-to-signal ratios in subspaces Q^r. In [21] it is shown that the analysis tool for the oriented energy distribution of a matrix A is the singular value decomposition, while the analysis tool of the oriented signal-to-signal ratio of two vector sequences A and B is the quotient singular value decomposition of the matrix pair (A, B). These well understood matrix factorizations allow to characterize the directions of extremal oriented energy and oriented signal-to-signal ratio.

Theorem 3 *Extremal directions of oriented energy.*
Let A be an $m \times n$ matrix with OSVD $A = U\Sigma V^t$ where $\Sigma = diag\{\sigma_i\}$. Then each direction of extremal oriented energy of A is generated by a left singular vector u_i with extremal energy equal to the corresponding singular value squared σ_i^2.

Theorem 4 *Extremal directions of oriented signal-to-signal ratio.*
Let A $(m \times n)$ and B $(m \times p)$ be matrices with QSVD :

$$\begin{aligned} A &= X^{-t} \cdot D_a \cdot Q_A^t \quad D_a = diag\{\alpha_i\} \\ B &= X^{-t} \cdot D_b \cdot Q_B^t \quad D_b = diag\{\beta_i\} \end{aligned} \tag{16}$$

where the quotient singular values (possible infinite) are ordered such that $(\alpha_1/\beta_1) \geq (\alpha_2/\beta_2) \geq \ldots \geq 0$. Then each direction of extremal signal-to-signal ratio of A and B is generated by a column x_i of the matrix X and the corresponding extremal signal-to-signal ratio is the quotient singular value squared $(\alpha_i/\beta_i)^2$.

These two theorems are illustrated for two dimensions in figure 2. Observe that for the oriented energy the maximum and minimum correspond to the largest resp. smallest singular vectors while a saddle point would correspond to the intermediate singular vector. Observe that the extremal directions of oriented energy are orthogonal while this is not necessarily the case for the signal-to-signal ratio. Now one can proceed by investigating in which directions of the ambient space the vector signal in the matrix A can be best distinguished from the vector signal in the matrix B. This leads to the definition of maximal minimal and minimal maximal signal-to-signal ratios of two vector sequences.

Definition 5 *Maximal minimal and minimal maximal signal-to-signal ratio.*
The maximal minimal signal-to-signal ratio of two m-vector sequences contained in the matrices A and B over all possible r-dimensional subspaces $(r < m)$ is defined as :

$$MmR[A,B,r] = \max_{Q^r \subset R^m} \min_{q \in Q^r} E_q[A,B]$$

Similarly, the minimal maximal signal-to-signal ratio is defined as :

$$mMR[A,B,r] = \min_{Q^r \subset R^m} \max_{q \in Q^r} E_q[A,B]$$

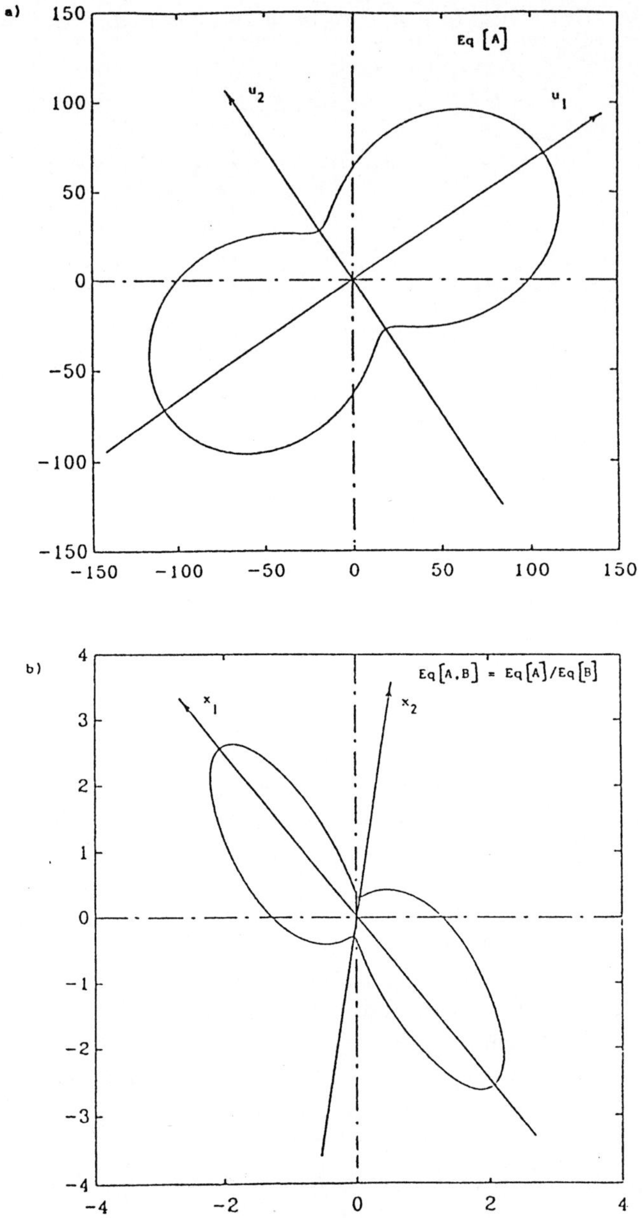

Figure 2: a) Oriented energy of a 2-vector sequence A and b) oriented signal-to-signal distribution of two 2-vector sequences A and B.

The idea behind these definitions is the following : for a given subspace Q^r of the m-dimensional ambient space ($r < m$) there is a certain direction $q \in Q^r$ for which the signal-to-signal ratio of the two vectorsequences A and B is minimal. This direction corresponds to the worst direction q in the sense that in this direction the energy of A is difficult to distinguish from the energy of B.

This worst case of course depends upon the precise choice of the subspace Q^r. Among all r-dimensional subspaces, at least one r-dimensional subspace has to exist where the worst case is better than all other worst cases. This subspace is the r-dimensional subspace of maximal minimal signal-to-signal ratio. It comes as no surprise that the QSVD allows to find this subspace : it is the r-dimensional subspace generated by the first r columns of X, when the quotient singular values are ordered as in theorem 4.

Hence, the concept of oriented signal-to-signal ratio and the QSVD allow to formalize all model identification approaches, in which

- the determination of a suitable rank r provides the complexity of the model.
- the model parameters follow from the corresponding subspace of maximal minimal signal-to-signal ratio.

Moreover, it can be shown that when the vector sequence B consists of an unobservable stochastic vector signal with known first and second order statistics (as is the case in most engineering applications), the QSVD solution corresponds precisely to the 'classical' Mahalanobis transformation that is commonly used in statistical estimators as a kind of prewhitening filter.

Still other extensions to the OSVD exist and these will now be discussed briefly. One unified generalization is the **Restricted Singular Value Decomposition,** introduced by Zha [41], see also De Moor and Golub [32]. The RSVD involves a matrix triplet (A, B, C) and allows to compute a lower rank approximation to the matrix A by subtracting a matrix which is restricted to have columns in the column space of B and rows in the row space of C. Given the matrices $A \in R^{m\times n}$, $B \in R^{m\times p}$ and $C \in R^{q\times n}$, the restricted singular values of the (A, B, C) are defined as

$$\sigma_{k+1}(A, B, C) = \min_{E\in R^{p\times q}} \{\|E\|_2 \mid \text{rank } (A + BEC) \leq k\} \quad k = 0 \ldots n-1 \tag{17}$$

If no such matrix E exists, σ_k is defined to be ∞, if $m < n$ and $m < k \leq n$, $\sigma_k = 0$. Thus the interpretation of the restricted singular values is very similar to the ordinary singular values (see eqs. (8) and (9) and fig.1).

A number of special cases of this RSVD are important in their own right :

The **OSVD** of A is obtained by choosing $B = I_m$ and $C = I_n$,

The **QSVD** of (A, B) is obtained obtained by taking $C = I_n$,

The **PSVD** (Product Singular Value Decomposition) of the pair (B^t, C) is the RSVD of (I_m, B, C). The PSVD was introduced by Fernando and Hammarling in 1987 [37].

A different type of restriction leads to the definition of the **Structured Singular Value**, introduced by Doyle in 1982 [34]. Consider a block partition of a matrix A as

$$A = \begin{bmatrix} A_{11} & \dots & A_{1q} \\ \dots & \dots & \dots \\ A_{p1} & \dots & A_{pq} \end{bmatrix}$$

and a matrix ΔA, partitioned in the same way as A, consisting of zero and nonzero blocks ΔA_{ij}, with possibly some constraints $\Delta A_{ij} = \Delta A_{kl}$.

Definition 6 *The structured singular value σ_{SSV} is defined as :*

$$\sigma_{SSV} = \min \|\Delta A\|_2 \text{ such that } rank(A + \Delta A) < \ rank\ A.$$

Applications are mainly in H_∞ control theory and some characterizations and algorithms are found in [36].

Finally, we mention the *Takagi Singular Value Decomposition* (TSVD) :

Theorem 5 *If A is a symmetric complex matrix, there exists a unitary U and a real nonnegative matrix $\Sigma = diag(\sigma_1, \dots, \sigma_n)$ such that $A = U\Sigma U^t$. The columns of U are an orthonormal set of eigenvectors for $A\bar{A}$ and the corresponding diagonal entries of Σ are the nonnegative square roots of the corresponding eigenvalues of $A\bar{A}$.*

This was introduced by Takagi in 1925 [39].

4 SVD for static systems

Static systems have no memory or state and are hence more easy to deal with. Here we will study the use of OSVD for static models for electric components and the use of both OSVD and QSVD for signal processing.

Quite often algebraic methods of physics or algebraic approaches for identification lead to mathematical models that are very sensitive to the data on which the model is based. This is a situation which is well known in numerical algebra [13] and in filter design. The approach that is followed in both cases is that one should investigate the collection of algebraically equivalent approaches and select in that class the one that is least sensitive to the variations in the data. Often these are orthogonal methods. When there exist many algebraically equivalent models then one can also best use that model that is least sensitive.

To make the point clear we consider a simple and hence a somewhat trivial example. Consider a two port resistor which is nothing but a parallel connection of the two ports. A straightforward implicit description is given by

$$\underbrace{\begin{bmatrix} 1 & -1 & 0 & 0 \\ 0 & 0 & 1 & 1 \end{bmatrix}}_{F} \begin{bmatrix} v_1 \\ v_2 \\ i_1 \\ i_2 \end{bmatrix} = \begin{bmatrix} 0 \\ 0 \end{bmatrix} \tag{18}$$

Observe that F constrains the admissible values of $v_1v_2i_1i_2$.

If voltage and current measuring devices are used which have the same precision (say 1 %) then one can easily check that any set of measured variables presents a solution to eq.(18) within the expected accuracy limits. By hook or by crook we obtain another implicit description with new matrix

$$\begin{bmatrix} 1 & -1 & 10^6 & 10^6 \\ 0 & 0 & 1 & 1 \end{bmatrix} \begin{bmatrix} v_1 \\ v_2 \\ i_1 \\ i_2 \end{bmatrix} = \begin{bmatrix} 0 \\ 0 \end{bmatrix} \tag{19}$$

A set of measured $v_1v_2i_1i_2$ will only satisfy eq. (19) with a reasonable precision when the current is measured very accurately. Algebraically equations (18) and (19) are however perfectly equivalent. So clearly it is desirable to dispose of some tools to compare algebraically equivalent multiports in less trivial cases. The basic tool is OSVD.

$$F = U\Sigma V \tag{20}$$

Let us now consider any implicit description of a resistive n-port [10]

$$F\begin{bmatrix} v \\ i \end{bmatrix} = 0 \tag{21}$$

where F is an $m \times 2n$ matrix. The constant offset has been dropped for simplicity. Such a description is very general and can arise from a physical device e.g. by a linearization, physical equations or from network equations. It may also be the specification of a multiport to be synthesized. In both situations the OSVD of F provides useful information concerning the implicit description. The action of F on the exact part and the inaccuracy of the data are easily analyzed with OSVD. First the orthonormal transformation V does not increase or decrease the sizes of both parts. Then the different components of the exact part and the inaccuracy are scaled differently with $\sigma_1, \sigma_2, \ldots \sigma_r, 0, 0$. Afterwards an orthonormal transformation does not affect the sizes again. The condition number is then $\kappa(F) = \sigma_1/\sigma_r$ which is the largest increase in error of the measured data. Clearly the best situation which can occur is that $\sigma_1 = \sigma_2 \ldots = \sigma_r$.

If the representation (21) is the exact specification for a design, the OSVD of F produces an algebraically equivalent but numerically most reliable description by

$$[A \;\; B]\begin{bmatrix} v \\ i \end{bmatrix} = 0 \tag{22}$$

where $[\ A\ B\ ]$ are the top r rows in $V = \begin{bmatrix} A & B \\ C & D \end{bmatrix}$. Indeed by multiplying (21) with U we obtain

$$\Sigma V\begin{bmatrix} v \\ i \end{bmatrix} = 0 \tag{23}$$

By scaling the equations with $\sigma_1^{-1}, \sigma_2^{-1}, \ldots \sigma_r^{-1}$ we obtain eq. (22).

If eq. (21) is an inaccurate model of a physical device this scaling is dangerous and one has to stick with (23). The OSVD then allows us to determine in a meaningful way the dimension of the n-port, which is by definition $2n$ minus the rank of the matrix F (22). A straightforward application of this definition on an inaccurate model of eq. (21) would generically produce a dimension which is $2n - m$, if F is an $m \times 2n$ matrix and $m \leq 2n$. In this way the inaccuracy of the parameters is used to generate restrictions on the admissible $v-i$ pairs, which is unacceptable. Rather the model should be modified by replacing by zero all singular values $\sigma_s, \sigma_{s+1}, \ldots \sigma_r$ such that σ_s/σ_1 is smaller than the accuracy in the parameters of F. This operation is justified by the fact that these modifications can be obtained while varying the parameters within the prescribed inaccuracy intervals.

Another problem is to classify the explicit representations for algebraic multiports and to generate if possible optimal explicit representations. To set the stage, classically one considers different representations for a resistive multiport (resistance, conductance, hybrid, transmission, scattering, ...). Remember that existence as well as measurement issues have stimulated the use of hybrid parameters for transistors, common and differential modes in op amps and scattering parameters in high frequency applications. Though the existence of these parametrizations is not always trivial, a more important issue is whether the explicit representation as a relationship between variables can blow up inaccuracies in the variables. In general using port coordinates x, y an explicit description is given [10] by

$$y = \Lambda x \quad \text{where} \quad \begin{bmatrix} y \\ x \end{bmatrix} = \Omega \begin{bmatrix} v \\ i \end{bmatrix}. \tag{24}$$

By choosing appropriate coordinate transformations Ω one can obtain the resistance, the conductance, any hybrid or scattering or transmission representation and a wealth of other explicit representations which have not received a name. The condition number of Λ is then the upper limit of the amplification in signal to noise ratio from x to y. If Λ is invertible this is also valid for the relation from y to x. Thus the different representations can be ranked according to their condition number. This information brings the whole set of existing parametrizations into perspective. Clearly the best situation is when the condition number is $\kappa(\Lambda) = 1$ or Λ is orthonormal up to a scaling factor. One can now wonder whether any multiport resistor has an optimal representation i.e. is there an orthonormal Ω such that $\kappa(\Lambda) = 1$. We start from the implicit representation of an n-dimensional n-port resistor i.e.

$$F \begin{bmatrix} v \\ i \end{bmatrix} = 0, \quad \text{rank} \quad = n \tag{25}$$

Applying the OSVD e.g. (20) of F, we obtain eq. (22) with $r = n$. Choosing the orthogonal coordinate transformation

$$\begin{bmatrix} y \\ x \end{bmatrix} = \underbrace{\begin{bmatrix} 1/\sqrt{2} & 1/\sqrt{2} \\ -1/\sqrt{2} & 1/\sqrt{2} \end{bmatrix} \begin{bmatrix} A & B \\ C & D \end{bmatrix}}_{\Omega} \begin{bmatrix} v \\ i \end{bmatrix}, \tag{26}$$

we can express the implicit description eq. (22) with the orthogonality of Ω as

$$\begin{aligned} [A \;\; B]\begin{bmatrix} v \\ i \end{bmatrix} &= [A \;\; B]\begin{bmatrix} A^T & C^T \\ B^T & D^T \end{bmatrix}\begin{bmatrix} 1/\sqrt{2} & -1/\sqrt{2} \\ 1/\sqrt{2} & -1/\sqrt{2} \end{bmatrix}\begin{bmatrix} y \\ x \end{bmatrix} \\ &= [1/\sqrt{2} \;\; -1/\sqrt{2}]\begin{bmatrix} y \\ x \end{bmatrix} = 0 \end{aligned} \tag{27}$$

The conclusion is then that by using the orthogonal transformation of eq. (26), which is obtained from the OSVD of eq. (20), an explicit description $y = x$ is obtained which is numerically optimal.

Of course the choice of the coordinates assumes already a certain description (25) or may be done approximately in advance but as long as it is close to that of eq. (26) the explicit description $y = \Lambda x$ will have a good condition number i.e. $\kappa(\Lambda)$ is close to 1. The point to be made here is that there is advantage to work and to compute with the $x - y$ coordinates rather than with others.

As a second application of SVD for static systems we consider the extraction of fetal ECG. The measurements are obtained from cutaneous electrodes placed at the heart and the abdomen of the mother. If there are p measurement channels (typically 6 to 8), the sampled data are stored in a $p \times q$ matrix M_{pq} where q denotes the number of consecutive samples that are processed. The p observed signals $m_i(t)$ (the rows of M_{pq}) are modeled as unknown linear combinations (modeled by a static $p \times r$ matrix T) of r source signals $s_j(t)$, corrupted by additive noise signals $n_i(t)$ with known (or experimentally verified) second order statistics. Hence the model has the well known factor-analysis-like structure :

$$M_{pq} = T_{pr} \cdot S_{rq} + N_{pq}$$

where the rows of S_{rq} are the source signals. The problem now consists of a rank decision to estimate r and of a subspace determination problem to determine the subspace generated by the columns of the matrix T, which are the so-called lead vectors. Since the second order statistics are assumed known, the conceptual framework of oriented signal-to-signal ratio could be applied. However, it has been verified [18,24] that for this specific application with an appropriate position of the electrodes, the subspace spanned by the lead vectors of the mother heart is three dimensional and orthogonal to the three-dimensional subspace generated by the lead vectors of the fetal heart transfer. Moreover, the source signals of mother heart and fetal heart are orthogonal vectors if considered over a sufficiently long time wherein the contribution of the mother heart is much stronger than of the fetal heart. For all these reasons, one single OSVD suffices to identify the subspace corresponding to the fetal ECG and by projecting the measurements on this subspace, the MECG can be eliminated almost completely. For more details on this separation based on the strength of the signals we refer to [18,24].

5 SVD for dynamic systems.

5.1 Total Linear Least Squares approach

In [1-3] a conceptual framework is developed in which the modeling problem is translated into an approximation context based upon the paradigm of low complexity and high accuracy models. The key concepts in this approach are the complexity of a model and the misfit between a model and the observations. Approximate modeling then consists of implementing the principle that either the desired optimal model is the least complex one in a given model class which approximates the observed data up to a preassigned tolerated misfit, or that it is the most accurate model within a preassigned tolerated complexity level. A particularly simple example is the *total linear least squares approach* [45,48] which essentially consists of fitting a linear subspace to a finite number of points. Consider an $m \times n$ matrix A ($n \gg m$) containing n measurements on a m-vector signal. Denote by a_i its i-th column. Let Q^r be an r-dimensional subspace of $\mathcal{R}^m$ then, the complexity is defined as :

$$c(r) : Q^r \rightarrow C = [0,1] : c(r) = \dim(Q^r)/m = r/m$$

This r-dimensional subspace Q^r can be considered as a lower rank approximation to the range of A, with a misfit defined as

$$\epsilon(A, Q^r) = \max_{q \perp Q^r}\left(\frac{\sqrt{\frac{1}{n}E_q[A]}}{\|q\|}\right)$$

where $E_q[A]$ is the oriented energy as defined in section 3. Then, we have the following theorem [3] :

Theorem 6 *Let $\frac{1}{\sqrt{n}}A = U.\Sigma.V^t$ be the SVD of the $m \times n$ matrix A of rank s ($s \leq m < n$) with singular values $\sigma_1 \geq \ldots \geq \sigma_s > 0$ and left singular vectors $u_i, i = 1, \ldots, m$. The unique optimal approximate model Q^r with complexity $c(Q^r) = \frac{r}{m}$ and misfit $\epsilon(A, Q^r) = \sigma_{r+1}$ is an r-dimensional subspace where :*

- *If c_{adm} is the maximal admissible complexity, then :*
  - *if $\text{int}[m.c_{\text{adm}}] = 0$, $r = 0$ and $Q^r = 0$.*
  - *if $\text{int}[m.c_{\text{adm}}] \geq s, r = s, Q^r = \text{span}_{\text{col}}[A]$*
  - *if $\sigma_k > \sigma_{\text{int}[m.c_{\text{adm}}]+1}, r = k, Q^r = S_U^k$*
- *If ϵ_{tol} is the maximal tolerated misfit, then :*
  - *if $\epsilon_{\text{tol}} \geq \sigma_1, r = 0$ and $Q^r = 0$*
  - *if $\epsilon_{\text{tol}} < \sigma_s, r = s, Q^r = \text{span}_{\text{col}}[A]$*
  - *if $\sigma_k > \epsilon_{\text{tol}} \geq \sigma_{k+1}, r = k$ and $Q^r = S_U^k$*

Proof: see [3]. □

In this framework of approximate modeling, the appropriate rank r is thus determined from either an a priori fixed admissible complexity or a maximal tolerable misfit. As for the identification of state space models for dynamic systems, it will now be shown how these concepts apply to the determination of a suitable model order. Both an OSVD approach for the white noise case, and a QSVD approach for the coloured noise case will be highlighted. Each time, the model order n follows from a $r = 2mi + n$ -law, where r is the appropriate rank choice, and m and i are certain constants yet to be defined.

5.2 OSVD-based system identification, the white noise case

For the time being, we consider time invariant linear, discrete time, multivariable systems with state space representation

$$\begin{aligned} x_{k+1} &= A.x_k + B.u_k \\ y_k &= C.x_k + D.u_k \end{aligned}$$

where u_k, y_k and x_k denote the input (m-vector), output (l-vector) and state vector at time k, the dimension of x_k being the minimal system order n. A, B, C and D are the unknown system matrices to be identified, making use only of recorded I/O-sequences $u_k, u_{k+1}, \ldots$ and $y_k, y_{k+1}, \ldots$

Let us first present the identification scheme. In [47] it was shown how a state vector sequence can be computed from I/O-measurements only, as follows. Let H_1 and H_2 be defined as

$$H_1 = \begin{bmatrix} u_k & u_{k+1} & \cdots & \cdots & \cdots & u_{k+j-1} \\ y_k & y_{k+1} & \cdots & \cdots & \cdots & y_{k+j-1} \\ u_{k+1} & u_{k+2} & \cdots & \cdots & \cdots & u_{k+j} \\ y_{k+1} & y_{k+2} & \cdots & \cdots & \cdots & y_{k+j} \\ \cdots & \cdots & \cdots & \cdots & \cdots & \cdots \\ u_{k+i-1} & u_{k+i} & \cdots & \cdots & \cdots & u_{k+j+i-2} \\ y_{k+i-1} & y_{k+i} & \cdots & \cdots & \cdots & y_{k+j+i-2} \end{bmatrix}$$

$$H_2 = \begin{bmatrix} u_{k+i} & u_{k+i+1} & \cdots & \cdots & \cdots & u_{k+i+j-1} \\ y_{k+i} & y_{k+i+1} & \cdots & \cdots & \cdots & y_{k+i+j-1} \\ u_{k+i+1} & u_{k+i+2} & \cdots & \cdots & \cdots & u_{k+i+j} \\ y_{k+i+1} & y_{k+i+2} & \cdots & \cdots & \cdots & y_{k+i+j} \\ \cdots & \cdots & \cdots & \cdots & \cdots & \cdots \\ u_{k+2i-1} & u_{k+2i} & \cdots & \cdots & \cdots & u_{k+2i+j-2} \\ y_{k+2i-1} & y_{k+2i} & \cdots & \cdots & \cdots & y_{k+2i+j-2} \end{bmatrix}$$

$$j \geq 2(m+l)i$$

and let the state vector sequence $\mathcal{X}$ be defined as

$$\mathcal{X} = [x_{k+i} \; x_{k+i+1} \; \cdots \; x_{k+i+j-1}]$$

then, under certain conditions (see [47])

$$\mathrm{span}_{\mathrm{row}}(\mathcal{X}) = \mathrm{span}_{\mathrm{row}}(H_1) \cap \mathrm{span}_{\mathrm{row}}(H_2)$$

so that any basis for this intersection constitutes a valid state vector sequence $\mathcal{X}$ with the basis vectors as the consecutive row vectors.

Once $\mathcal{X} = [x_{k+i}\ x_{k+i+1}\ \dots\ x_{k+i+j-1}]$ is known, the system matrices can be identified by solving an (overdetermined) set of linear equations:

$$\begin{bmatrix} x_{k+i+1} & \dots & x_{k+i+j-1} \\ y_{k+i} & \dots & y_{k+i+j-2} \end{bmatrix} = \begin{bmatrix} A & B \\ C & D \end{bmatrix} . \begin{bmatrix} x_{k+i} & \dots & x_{k+i+j-2} \\ u_{k+i} & \dots & u_{k+i+j-2} \end{bmatrix}$$

The above results constitute the heart of a two-step identification scheme. First a state vector sequence is realized as the intersection of the row spaces of two block Hankel matrices, constructed with I/O-data. Then the system matrices are obtained at once from the least squares solution of a set of linear equations.

Let us now discuss the computational details. The following derivation (which is slightly different from the one in [47]), shows how these computations can be carried out quite easily, resulting in a consistent (see below) **double OSVD identification algorithm.**

In a first step, the intersection of the row spaces spanned by H_1 and H_2, can be recovered from the OSVD of the concatenation

$$H = \begin{bmatrix} H_1 \\ H_2 \end{bmatrix}$$

$$\begin{aligned} H &= \begin{bmatrix} H_1 \\ H_2 \end{bmatrix} \\ &= U_H . S_H . V_H^t \\ &= \begin{bmatrix} U_{11} & U_{12} \\ U_{21} & U_{22} \end{bmatrix} . \begin{bmatrix} S_{11} & 0 \\ 0 & 0 \end{bmatrix} . V_H^t \end{aligned}$$

$$\begin{aligned} \dim(U_{11}) &= (mi + li) \times (2mi + n) \\ \dim(U_{12}) &= (mi + li) \times (2li - n) \\ \dim(U_{21}) &= (mi + li) \times (2mi + n) \\ \dim(U_{22}) &= (mi + li) \times (2li - n) \\ \dim(S_{11}) &= (2mi + n) \times (2mi + n) \end{aligned}$$

(see [47] for details). From

$$U_{12}^t . H_1 = -U_{22}^t . H_2$$

it follows that the row space of $U_{12}^t . H_1$ equals the required intersection. However, $U_{12}^t . H_1$ contains $2li - n$ row vectors , only n of which are linearly independent (dimension of the intersection). Thus, it remains to select n suitable combinations of these row vectors. As U_{12} and U_{22} form an orthogonal matrix, they can be decomposed as follows [44, p.22]

$$U_{12} = \left[\begin{array}{ccc} U_{12}^{(1)} & U_{12}^{(2)} & U_{12}^{(3)} \end{array} \right] . \left[\begin{array}{ccc} I_{(li-n)\times(li-n)} & & \\ & C_{n\times n} & \\ & & 0_{(li-n)\times(li-n)} \end{array} \right] .V_*^t$$

$$U_{22} = \left[\begin{array}{ccc} U_{22}^{(1)} & U_{22}^{(2)} & U_{22}^{(3)} \end{array} \right] . \left[\begin{array}{ccc} O_{(li-n)\times(li-n)} & & \\ & S_{n\times n} & \\ & & I_{(li-n)\times(li-n)} \end{array} \right] .V_*^t$$

$$\begin{aligned} C &= diag(c_1, \ldots, c_n) \\ S &= diag(s_1, \ldots, s_n) \\ I_{n\times n} &= C^2 + S^2 \end{aligned}$$

where $U_{12}^{(1)}$ then constitutes the $(li - n)$-dimensional orthogonal complement of H_1. Clearly, only $U_{12}^{(2)}$ delivers useful combinations for the computation of the intersection, and we can take

$$\mathcal{X} = U_{12}^{(2)^t}.H_1$$

The above expressions for U_{12} and U_{22} are in itself OSVD's of these matrices, and can be computed as such. It thus suffices to compute e.g. the OSVD of U_{12}. The computation of the required intersection then reduces to the computation of two successive OSVD's (for H and U_{12} respectively).

Up till now, we have assumed that the data were error-free. If there is some **inaccuracy on the measurement data** in H_1 and H_2, generically the row spaces of these matrices do not intersect, and all singular values in S_H are non-zero. Hence, one should approximate H by a matrix of lower rank by setting the smallest singular values equal to zero, in order to obtain the model that is least conflicting with the data. A suitable determination of this lower rank, is then to be carried out along the lines of the *total least squares approach* to approximate modeling, as was detailed in the previous section.

Note that the above derivation is nothing more than a **double OSVD approach** to computing the QSVD of the matrix pair (H_1, H_2), following from the constructive QSVD-proof in [48]. From this last remark, one might be tempted to immediately apply a one stage QSVD-procedure to the matrix pair. This latter method would however compute the exact intersection of the row spaces, which in the presence of noise turns out to be completely absent (generically). The outcome of applying such an algorithm would then be a zero dimensional intersection, as could be guessed beforehand. The difference between these methods turns out to be the intermediate rank decision after the first OSVD in the first approach (double OSVD), that fixes the dimension of the approximate intersection to be computed next. Although this (possibly difficult) intermediate rank decision has been a main motive for developing a one stage QSVD algorithm, for our purpose it is somehow inevitable.

In the second step, the system matrices are to be identified from a set of linear equations. Much like it was done in [47], it can straightforwardly be shown that the system matrices can be computed from the following reduced set as well (obtained after discarding the common orthogonal factor V_H)

$$\begin{bmatrix} U_{12}^{(2)^t}.U_H(m+l+1:(i+1)(m+l),1:2mi+n).S_{11} \\ U_H(mi+li+m+1:(m+l)(i+1),1:2mi+n).S_{11} \end{bmatrix}$$
$$= \begin{bmatrix} A & B \\ C & D \end{bmatrix} \begin{bmatrix} U_{12}^{(2)^t}.U_H(1:mi+li,1:2mi+n).S_{11} \\ U_H(mi+li+1:mi+li+m,1:2mi+n).S_{11} \end{bmatrix}$$

where $U_H(r:s,v:w)$ is a submatrix of U_H at the intersection of rows $r, r+1, \ldots, s$ and columns $v, v+1, \ldots, w$.

The identification procedure is proven to be **consistent** if the number of columns in H tends to infinity and if the input-output measurements are corrupted with additive white measurement noise, or in other words, if the columns in H are subject to independently and identically distributed errors with zero mean and common error covariance matrix equal to the identity matrix, up to a factor of proportionality. For that case, it can indeed be shown [43] that the left singular basis U_H can be computed consistently (as opposed to the singular values S_H and the right singular basis V_H). As the system matrices are next computed essentially from U_H only (see the above set of equations[1]), the model estimate is clearly consistent. The corresponding noise model is depicted in Figure 1.

5.3 QSVD-based system identification, the coloured noise case

Let us now proceed to the case where the I/O-data are corrupted by coloured noise. Assume that the columns in the concatenated matrix

$$H = \begin{bmatrix} H_1 \\ H_2 \end{bmatrix}$$

are subject to independently and identically distributed errors with zero mean and common error covariance matrix Δ up to a factor of proportionality, where

$$\Delta = R_\Delta . R_\Delta^t$$

is the Cholesky factorization of Δ (R_Δ lower triangular).

One can easily verify that the columns in the transformed matrix $R_\Delta^{-1}.H$ have an error covariance matrix equal to the identity matrix up to a factor of proportionality. One way of carrying out the identification would then consist in having the identification based on the SVD of $R_\Delta^{-1}.H$ (with a consistent computation of the left singular basis,

[1]The matrix S_H in this set imposes weights on the different equations. This does not influence the outcome if the set of equations can be solved exactly, which is particularly the case under the assumed conditions.

see the previous section) instead of H, and including some kind of a re-transformation with R_Δ in order to compensate for the first transformation with R_Δ^{-1}. The overall identification scheme would then deliver a consistent estimate. However, if R_Δ is singular or ill-conditioned, the matrix inverse R_Δ^{-1} should not be computed explicitly. Instead, one should make use of the **quotient singular value decomposition (QSVD)** of the matrix pair (H, R_Δ) (which in the non-singular case indeed reduces to the SVD of $R_\Delta^{-1}.H$). We can now show how a **double QSVD** identification scheme can be designed, analogously to the **double OSVD** scheme for the white noise case (the latter being a special case of the former where every single QSVD reduces to an OSVD, as can easily be verified).

The QSVD of (H, R_Δ) is defined as

$$\begin{aligned} X^t.H.Q_H &= \Sigma_H \\ X^t.R_\Delta.Q_{R_\Delta} &= \Sigma_{R_\Delta} \end{aligned}$$

where

$$\begin{aligned} \Sigma_H &= diag(\alpha_1, \ldots, \alpha_{2li+2mi}) \\ \Sigma_{R_\Delta} &= diag(\beta_1, \ldots, \beta_{2li+2mi}) \end{aligned}$$

$$\frac{\alpha_1}{\beta_1} > \frac{\alpha_2}{\beta_2} > \ldots > \frac{\alpha_{2li+2mi}}{\beta_{2li+2mi}}$$

Much like it was done for the white noise case, where the intersection of the row spaces of H_1 an H_2 was computed making use of the directions of minimal oriented signal energy for $H = \begin{bmatrix} H_1 \\ H_2 \end{bmatrix}$, viz. $\begin{bmatrix} U_{12} \\ U_{22} \end{bmatrix}$ (remember $U_{12}^t.H_1 = -U_{22}^t.H_2$) we can now compute the intersection making use of the directions of minimal oriented signal-to-noise ratio, viz. $\begin{bmatrix} X_{12} \\ X_{22} \end{bmatrix}$, to be defined next.

It is instructive to first consider the noise free case (error covariance proportional to Δ, but with a zero factor of proportionality), and then demonstrate that the derivations still hold if there is a non-zero error contribution.

If the data are noise free, then from the above QSVD definition, it follows that

$$\begin{aligned} H &= \begin{bmatrix} H_1 \\ H_2 \end{bmatrix} \\ &= X^{-t}.\Sigma_H.Q_H^t \\ &= \begin{bmatrix} X_{11} & X_{12} \\ X_{21} & X_{22} \end{bmatrix}^{-t} . \begin{bmatrix} \Sigma_{11} & 0 \\ 0 & 0 \end{bmatrix} .Q_H^t \end{aligned}$$

$$\begin{aligned} \dim(X_{11}) &= (mi + li) \times (2mi + n) \\ \dim(X_{12}) &= (mi + li) \times (2li - n) \\ \dim(X_{21}) &= (mi + li) \times (2mi + n) \\ \dim(X_{22}) &= (mi + li) \times (2li - n) \\ \dim(\Sigma_{11}) &= (2mi + n) \times (2mi + n) \end{aligned}$$

Again, from

$$X_{12}^t.H_1 = -X_{22}^t.H_2$$

it follows that the row space of $X_{12}^t.H_1$ equals the required intersection . As $X_{12}^t.H_1$ contains $2li - n$ row vectors , only n of which are linearly independent (dimension of the intersection), it remains to select n suitable combinations of these row vectors. Making use of a QSVD, one can easily show that

$$X_{12} = \left[\begin{array}{ccc} X_{12}^{(1)} & X_{12}^{(2)} & X_{12}^{(3)} \end{array}\right] . \left[\begin{array}{ccc} I_{(li-n)\times(li-n)} & & \\ & C_{n\times n} & \\ & & 0_{(li-n)\times(li-n)} \end{array}\right] .T_\star^t$$

$$X_{22} = \left[\begin{array}{ccc} X_{22}^{(1)} & X_{22}^{(2)} & X_{22}^{(3)} \end{array}\right] . \left[\begin{array}{ccc} O_{(li-n)\times(li-n)} & & \\ & S_{n\times n} & \\ & & I_{(li-n)\times(li-n)} \end{array}\right] .T_\star^t$$

$$\begin{aligned} C &= diag(c_1, \ldots, c_n) \\ S &= diag(s_1, \ldots, s_n) \\ I_{n\times n} &= C^2 + S^2 \end{aligned}$$

Clearly, only $X_{12}^{(2)}$ delivers useful combinations for the computation of the intersection, and we can take

$$\mathcal{X} = X_{12}^{(2)^t}.H_1$$

Note that in the white noise case, this last QSVD reduced to a CS-decomposition and could then be computed from a single SVD, resulting in an overall *double SVD scheme* for the computation of the intersection. In the general case, the computation of this intersection is carried out in a *double QSVD scheme*.

In the second step, the system matrices can be computed from the following reduced set of equations (obtained after discarding the common orthogonal factor Q_H)

$$\begin{aligned} &\left[\begin{array}{c} X_{12}^{(2)^t}.\mathcal{X}^{-t}(m+l+1:(i+1)(m+l), 1:2mi+n).\Sigma_{11} \\ \mathcal{X}^{-t}(mi+li+m+1:(m+l)(i+1), 1:2mi+n).\Sigma_{11} \end{array}\right] \\ &= \left[\begin{array}{cc} A & B \\ C & D \end{array}\right] \left[\begin{array}{c} X_{12}^{(2)^t}.\mathcal{X}^{-t}(1:mi+li, 1:2mi+n).\Sigma_{11} \\ \mathcal{X}^{-t}(mi+li+1:mi+li+m, 1:2mi+n).\Sigma_{11} \end{array}\right] \end{aligned}$$

where $\mathcal{X}^{-t}(r:s, v:w)$ is a submatrix of $\mathcal{X}^{-t}$ at the intersection of rows $r, r+1, \ldots, s$ and columns $v, v+1, \ldots, w$.

If there is some **inaccuracy on the measurement data** in H_1 and H_2, all quotient singular values in S_H are non-zero. Again, one should set the smallest quotient singular values equal to zero, in order to obtain the model that is least conflicting with the data and a suitable determination of this lower rank is to be carried out along the lines of the *total least squares approach* to approximate modeling.

It remains to show that the above identification scheme delivers **consistent** results if the number of columns in H tends to infinity, and if the columns in H are subject to independent and identically distributed errors with zero mean and common error covariance matrix equal to Δ, up to a factor of proportionality. For that case, it can again be shown [43] that the matrix X in the QSVD can be computed consistently. As the system matrices are next computed essentially from X only (the matrix Σ_H in the above set of equations again imposes weights that do not influence the solution in the considered case, see section 2), the model estimate is clearly consistent.

6 Conclusions.

System theory can provide a unifying framework for the study of many complex problems in and outside engineering.

Thereby it creates a unique platform for the exchange of concepts, tools, and software among different engineering disciplines. Hence it can stimulate the interaction and communication among specialist and lead to a holistic approach of problems. It also leads to a parsimony of concepts, tools and software which is very valuable in education.

The engineering approach to systems and models is usually not the deductive one from general to specific. On the contrary one usually starts with the simplest model or the specification at a high level and one gradually refines and further specifies the systems.

In this paper, it is claimed that the singular value decomposition and the quotient singular value decomposition have a great potential for system theory and signal processing in much the same way as the FFT had a great impact on digital signal processing in the seventies and eighties. Several applications were presented or referred to. The benefits of using the (quotient) singular value decomposition are most pronounced in those applications :

- where essentially rank decisions and the computation of the corresponding subspaces determine the complexity and parameters of the model
- where numerical reliability is of crucial importance and the potential loss of numerical accuracy is to be avoided.
- where a conceptual framework, such as the notion of oriented signal-to-signal ratio, may provide unrevealed additional insight, such as in factor-analysis-like problems.
- where the problem can be stated directly in terms of the (quotient) singular value decomposition, which leads immediately to a reliable and robust solution, such as in a canonical correlation analysis environment.

Moreover, in most engineering applications the number of measurements or the data acquisition poses only minor organisational problems (although the design of a measurement set up causes considerable efforts). The cost of the sensors however increases with higher accuracy and signal-to-noise requirements. In this environment the (quotient) singular value decomposition is the optimal bridge between limited measurement precision and robust modeling.

References

[1] Willems J.C., "Dissipative dynamical systems, part I General Theory; part II Linear systems with quadratic supply rates" Arch. Rational Mech. Analysis Vol. 45 pp. 321-351; pp. 352-393, 1972.

[2] Willems J.C., "System theoretic models for the analysis of physical systems", Richerche di Automatica,Vol. 10, no. 2, pp. 71-106, 1979.

[3] Willems J.C., "From times series to linear systems, parts I, II" Automatica, Vol. 21, 1986, pp. 561-580 and pp. 675-694, Vol. 23, 1987, pp. 87-115.

[4] Anderson B.D.O. and Vongpanitlerd S., "Network analysis and synthesis", Prentice Hall, Englewood Cliffs 1973.

[5] Willsky A.S., "Relationships between digital signal processing and control and estimation theory", Proc. IEEE, Vol. 66, no. 2, pp. 996-1027, 1978.

[6] "Matlab Manual", The Mathworks, Mass. 1985.

[7] Mees A.I. and Sparrow C.T., "Chaos", IEE Proc., Vol. 128, Pt. D, No. 5, Sept. 1981, pp. 201-205.

[8] Sugarman R. and Wallich P. "The limits to simulation." IEEE Spectrum, p. 36-41, April 1983.

[9] Chua L.O., Komuro M., and Matsumoto T., "The double scroll family Part I, II", IEEE Trans. on Circuits and Systems, Vol. CAS-33, Nov. 1986, pp. 1072-1097, pp. 1097-1118.

[10] Chua L.O., "Dynamic nonlinear networks : State-of-the-art", IEEE Trans. on Circuits and Systems, Vol. CAS-27, pp. 1014-1044, Nov. 1980.

[11] Abraham R. and Shaw C., "Dynamics - the geometry of behaviour, part 1,2,3" The visual mathematics library, Aerial Press, Santa Cruz, CA, 1985.

[12] Van Dooren P., "Numerical linear algebra : An increasing interest in linear system theory.", Proc. ECCTD The Hague, 1981, pp. 243-251.

[13] Staar J., Wemans M. and Vandewalle J., "Comparison of multivariable MBH realization algorithms in the presence of multiple poles, and noise disturbing the Markov sequence", in "Analysis and Optimization of Systems", ed. by A. Bensoussan and J.L. Lions, Springer Verlag, pp. 141-160, Berlin, 1980.

[14] Staar J. and Vandewalle J., "Numerical implications of the choice of a reference node in the nodal analysis of large circuits", Int. Journal of Circuit Theory and Applications, Vol. 9, pp. 488-492, 1981.

[15] Staar J. and Vandewalle J., "Singular value decomposition : A reliable tool in the algorithmic analysis of linear systems.", Journal A, Vol. 23, pp. 69-74, 1982.

[16] Vandewalle J. and Staar J., "Modelling of linear systems : critical examples, problems of numerically reliable approaches.", Proc. IEEE Int. Symp. on Circuits and Systems, ISCAS-82, Rome, pp. 915-918, 1982.

[17] Vandewalle J., Vanderschoot J. and De Moor B., "Source separation by adaptive singular value decomposition.", Proc. IEEE ISCAS Conf. Kyoto 5-7 June 1985, pp. 1351-1354.

[18] Vanderschoot J., Callaerts D., Sansen W., Vandewalle J., Vantrappen G., Janssens J., "Two methods for optimal MECG elimination and FECG detection from skin electrode signals.", IEEE Trans. on Biomedical Engineering, Vol. BME-34, no. 3, pp. 233-243, March 1987.

[19] De Moor B., Vandewalle J., "Non-conventional matrix calculus in the analysis of rank deficient Hankel matrices of finite dimensions.", System and Contr. Lett., Vol. 9, pp. 401-410, 1987.

[20] De Moor B., Vandewalle J., "An adaptive singular value decomposition algorithm based on generalized Chebyshev recursions." in : Mathematics in signal processing, T.S. Durrani, J.B. Abbiss, J.E. Hudson, R.N. Madan, J.G. McWirther and T.A. Moore (ed.), Clarendon Press-Oxford, 1987, pp. 607-635.

[21] De Moor B., Staar J. and Vandewalle, "Oriented energy and oriented signal-to-signal ratio concepts in the analysis of vector sequences ant time series.", in "SVD and Signal Processing" E. Deprettere (ed.), North Holland, 1988, pp. 209-232.

[22] Van Huffel S., Vandewalle J., "The total least squares technique : computation, properties and applications.", in "SVD and Signal Processing" E. Deprettere (ed.), North Holland, 1988, pp. 189-207.

[23] Vandewalle J., De Moor B., "A variety of applications of singular value decomposition in identification and signal processing." in "SVD and Signal Processing" E. Deprettere (ed.), North Holland, 1988, pp. 43-91.

[24] Callaerts D., Vandewalle J., Sansen W. and Moonen M., "On-line algorithm for signal separation based on SVD." in "SVD and Signal Processing" E. Deprettere (ed.), North Holland, pp. 269-276, 1988.

[25] De Moor B., Moonen M., Vandenberghe L. and Vandewalle J., "Identification of linear state space models with singular value decomposition using canonical correlation concepts.", in "SVD and Signal Processing" E. Deprettere (ed.), North Holland, pp. 161-169, 1988.

[26] De Moor B., Vandewalle J., Moonen M., Van Mieghem P. and Vandenberghe L., "A geometrical strategy for the identification of state space models of linear multivariable systems with singular value decomposition.", Preprints 8th IFAC/IFORS Symposium on identification and system parameter estimation, Beijing, August 27-31, 1988, pp. 700-704.

[27] Vandewalle J., "Trends in the need of mathematics for engineering and the impact on engineering education.", SEFI, Proc. 5th European Seminar on Mathematics in Engineering Education, Plymouth, March 23-26, 1988, pp. 94-105.

[28] De Moor B., Moonen M., Vandenberghe L. and Vandewalle J., "The application of the canonical correlation concept to the identification of linear state space models", in A. Bensousan, J.L. Lions, (Eds) Analysis and Optimization of Systems, Springer Verlag, Heidelberg, 1988, pp. 1103-1114.

[29] Van Belle H. en Van Brussel H., "Inleiding tot de systeemtheorie. Pleidooi voor een ruimere toepassing.", Het Ingenieursblad, no. 12, 1979.

[30] Autonne L., "Sur les groupes linéaires, réelles et orthogonaux", Bull. Soc. Math., France, Vol. 30, pp. 121-133, 1902.

[31] De Moor B., Golub G.H., "Generalized singular value decompositions : A proposal for a standardized nomenclature." Internal Report, Department of Computer Science, Stanford University, January 1989 (submitted for publication).

[32] De Moor B., Golub G.H., "The restricted singular value decomposition : properties and applications." Internal Report, Department of Computer Science, Stanford University, March 1989 (submitted for publication).

[33] Deprettere Ed. (Editor), "SVD and Signal Processing : Algorithms, Applications and Architectures", North Holland, 1988.

[34] Doyle J.C. "Analysis of feedback systems with structured uncertainties." Proc. IEE, Vol. 129, no. 6, pp. 242-250, Nov. 1982.

[35] Eckart G., Young G., "The approximation of one matrix by another of lower rank.", Psychometrika, 1 : 211-218, 1936.

[36] Fan M.K.H., Tits A.L., "Characterization and efficient computation of the structured singular value.", IEEE Trans. Automatic Control, Vol. AC-31, no. 8, August 1986, pp. 734-743.

[37] Fernando K.V., Hammarling S.J., "A Product Induced Singular Value Decomposition for two matrices and balanced realisation.", NAG Technical Report, TR8/87.

[38] Paige C.C., Saunders M.A., "Towards a generalized singular value decomposition.", SIAM J. Numer. Anal., 18, pp. 398-405, 1981.

[39] Takagi T., "On an algebraic problem related to an analytic theorem of Caratheodory and Fejer and on an allied theorem of Landau.", Japan. J. Math., 1, pp. 83-93, 1925.

[40] Van Loan C.F., "Generalizing the singular value decomposition." SIAM J. Numer. Anal., 13, pp. 76-83, 1976.

[41] Zha H., "Restricted SVD for matrix triplets and rank determination of matrices." Scientific Report 89-2, Berlin (submitted for publication).

[42] Damen A.A.H, Van den Hof P.M.J., Hajdasinski A.K., "Approximate realization based upon an alternative to the Hankel matrix : the Page matrix.", Systems and Control Letters, Vol.II, No. 4, pp. 202-208, 1982.

[43] De Moor B., "Mathematical Concepts and techniques for modelling static and dynamic systems", Doct. Diss., K.U.Leuven, 1988.

[44] Golub G.H. and Van Loan C.F., "Matrix computations.", North Oxford Academic Publishing Co., Johns Hopkins University Press, 1983.

[45] Golub G.H., Van Loan C.F., "An analysis of the total least squares problem.", SIAM J. Numer. Anal., Vol. 17, No. 6, pp. 883-893, 1980.

[46] Kung S.Y., "A new identification and model reduction algorithm via singular value decomposition.", Proc. 12th Asilomar Conf. on Circuits, Systems and Computers. Pacific Grove, pp. 705-714, 1978.

[47] Moonen M., De Moor B., Vandenberghe L., Vandewalle J., "On- and off-line identification of linear state space models", International Journal of Control, Vol. 49, No. 1, pp. 219-232, 1989.

[48] Van Huffel S., "Analysis of the total least squares problem and its use in parameter estimation.", Doct. Diss. K.U.Leuven, 1987.

[49] Van Huffel S., "Analysis and properties of the generalized total least squares problem $AX \approx B$ when some or all columns in A are subject to error.", SIAM J. Matrix Anal. Appl., to appear 1989.

[50] Zeiger H.P. and McEwen A.J., "Approximate linear realizations of given dimensions via Ho's algorithm.", IEEE Trans. Aut. Control, vol. AC-19, (pp. 153), 1974.

[51] Van den Hof P., "On residual-based parametrization and identification of multivariable systems", Doct. Diss. T.U.Eindhoven, 1989.

Stochastic Realization Problems

J. H. van Schuppen

Centre for Mathematics and Computer Science (CWI)
Kruislaan 413, 1098 SJ Amsterdam, the Netherlands

The stochastic realization problem asks for the existence and the classification of all stochastic systems for which the output process equals a given process in distribution or almost surely. This is a fundamental problem of system and control theory. The stochastic realization problem is of importance to modelling by stochastic systems in engineering, biology, economics etc. Several stochastic systems are mentioned for which the solution of the stochastic realization problem may be useful. As an example recent research on the stochastic realization problem for the Gaussian factor model and a Gaussian factor system is discussed.

This paper is dedicated to J. C. Willems on the occasion of his fiftieth birthday.

1. Introduction

The purpose of this paper is to introduce the reader to stochastic realization theory. This will be done by presentation of a verbal introduction, a survey of Gaussian stochastic realization theory, formulation of open stochastic realization problems, and a discussion of the stochastic realization problem for Gaussian factor models. This tutorial and survey-like paper is written for researchers in system and control theory, but may also be of interest to researchers dealing with mathematical models in engineering, biology, and economics.

The Kalman filter and stochastic control algorithms have proven to be very useful for those control and signal processing problems in which there is a considerable amount of noise in the observation processes. Examples of such problems are: minimum variance control of a paper machine, access control of communication systems, and prediction of water levels. The solution of stochastic control and filtering problems depends crucially on the availability of a model in the form of a stochastic system in state space form. There is thus a need for modelling and realization of noisy processes by stochastic systems. Stochastic realization theory addresses this modelling problem.

System and control theory is the subject within engineering and mathematics that deals with modelling and control problems for dynamic processes or phenomena. Such a phenomenon may initially be described by specifying the observation process or trajectories, which description will be termed the *external description.* For reasons of modelling and control it is often better to work with an *internal description.* The form of such an internal description depends on the properties of the observation process. For deterministic linear systems it may be a description in state space form. The state of such a system at any particular time contains all information from the past necessary to determine the future behavior of the state and output process. For stochastic systems the internal description is a stochastic system in state space form. Here the state is that amount of information that makes the past and the future of the observations and the state process conditionally independent. For a vector valued random variable one may consider the internal description of a Gaussian factor model, see section 5. For models of images and spatial phenomena in the form of random fields, other internal descriptions are needed.

The *realization problem* of system theory can then be formulated as how to determine an internal description of a model given an external description. Motivation for this problem comes from engineering, in particular from system identification and signal processing, from biology, and from econometrics. In these subject areas one may want to estimate parameters of the internal description from observations. The question should then be posed whether these parameters can be uniquely determined from the observations, that is whether they are identifiable. This question may be resolved by solution of the realization problem. First one must impose the condition that the model is minimal in some sense. The concept of minimality will depend on the class of internal descriptions. Secondly, there is in general no unique internal description for a phenomenon given an external description. The realization problem therefore also asks for a classification of all minimal internal descriptions that correspond to a given external description. Such internal descriptions may be called equivalent. Once the equivalence class has been determined one may choose a canonical form for it. From that point on standard techniques from system identification and statistics may be used to determine the internal description of the model. The part of system theory that deals with modelling questions is referred to as *realization theory.* It treats topics such as transformations between representations, parametrization of model classes, identifiability questions, and approximate modelling.

A brief description of this paper's content follows. Section 2 contains a verbal introduction to the modelling procedure of system theory. In section 3 a tutorial is presented on Gaussian stochastic realization theory. Several examples of stochastic systems for which the stochastic realization problem is open and relevant for engineering and economics, are mentioned in section 4. As an example the stochastic realization problem for the Gaussian factor analysis model is discussed in section 5, and for Gaussian factor systems or error-in-variables systems in section 6.

2. MODELLING AND SYSTEM THEORY

2.1. Introduction

As identified in the previous section there is a need for stochastic models of engineering and economic phenomena. The purpose of this section is to describe the modelling procedure of system and control theory. Particular attention will be devoted to modelling of economic processes.

2.2. The modelling procedure

It is assumed that data, possibly in the form of time series, are available for the modeller. It is well-recognized that useful data are easy to obtain in the technical sciences but hard to obtain in economics. One reason is that economics is not a laboratory science; experiments are often impossible or if possible cannot be repeated. Also data gathering is much more expensive in economics than in the technical sciences.

The objective of modelling is to obtain a model for a phenomenon that is realistic and of low complexity. A model is called *realistic* if its observed behaviour is in close agreement with the phenomenon. A measure of fit for this agreement has to be formulated. The term *low complexity* should be considered as in ordinary use. A mathematical definition of this term is very much model dependent. Models of high complexity are mathematically not well analyzable and computationally not feasible. The two modelling objectives mentioned are conflicting. Therefore a compromise or trade-off between these objectives is necessary.

The preferred modelling procedure consists of the following two steps:

- selection of a model class;
- selection of an element in the model class involving the above mentioned trade-off.

This procedure must be applied in an iterative fashion. If the selected element in the model class is not a realistic model then the model class may be adjusted. The two steps of this procedure will now be discussed separately.

2.3. Selection of a model class

In the selection of a model class one has to keep in mind the objectives of a realistic model and a model of low complexity. The selection procedure demands application of concepts and results both from the research area of the object to be modelled, and from system and control theory.

The formulation of realistic economic models is difficult for several reasons. One reason is that economic transactions involve multiple decisionmakers compared with a single decisionmaker in most engineering problems. The appropriate mathematical models are therefore game and team models and their dynamic counterparts. The status of dynamic game and team theory is not yet at a level at which a body of results is available for applications. A second reason, closely related to the first, is that a decisionmaker must also model the decisionmaking process of the other decisionmakers. This remark is well-known in the literature on stochastic dynamic games. The discussion about rational expectation also illustrates this point. A third reason is that the rules of the economic process change quickly compared with the periods over which economic data are available. Assumptions of time-invariance or stationarity are often unrealistic.

In system theory a formalism has been developed for the formulation of mathematical models of dynamic phenomena and for a modelling procedure. For a dynamic phenomenon in the form of a time series a preferred deterministic model is called a *dynamic system* in state space form. One distinguishes *inputs* and *outputs* of such a system, and a state process. The *state* of a dynamic system at any particular time is that amount of information that together with the future inputs completely determines the future outputs. The trajectories of the input, output and state process are the basic objects of a dynamical system. The reader is referred to [78] for material on linear systems.

Stochastic systems have proven to be useful models in several areas of engineering such as signal processing, communications and control. Within economics they are used for example in connection with portfolio theory. In stochastic system theory, probability theory is used as a mathematical model for uncertainty. A stochastic system is specified by a measure on the space of trajectories. This is a fundamental difference between deterministic and stochastic systems. For a stochastic system without inputs the state at any particular time makes the past and the future of the output and state processes conditionally independent. Despite the fact that a stochastic system is specified by a measure, the representation in terms of trajectories, for example by a stochastic differential equation, is crucial to the solution of control and filtering problems.

Why are stochastic models realistic in certain cases? Within economics reasons for this are that such modelling involves:

- aggregation over many decisionmakers;
- uncertainty over future actions of other decisionmakers;
- uncertainty in the measurement process, due to vague definitions and averaging.

Remark that the costs involved often prevent the gathering of full information. Therefore aggregation must be used. The variability of the data then suggests a stochastic model. This author is not optimistic about the applicability of stochastic models to economic phenomena. Reasons for this are the relatively short time series and the frequent change in structural relations.

Should one use a deterministic or a stochastic model class to model a certain phenomenon? What is needed is a criterion to decide whether for a specific phenomenon the class of deterministic systems or that of stochastic systems is the appropriate model class.

A crucial observation from system theory is that the *choice of model class* is all-important. Of course, a model must be realistic and of low complexity. But within these constraints there is left some freedom in the mathematical formulation of the model. Given this freedom it is advisable to choose a model class for which the motivating control problem is analytically tractable. An example of such a choice is the Gaussian system that leads to the Kalman filter. Filtering theory was formulated by N. Wiener and A. N. Kolmogorov for stationary Gaussian processes. R. E. Kalman restricted attention to a particular class of stationary Gaussian processes, those generated by linear stochastic systems driven by white noise. For this class of systems the solution of the filtering problem has proven to be straightforward. That this class may be extended to include non-stationary processes is then a useful corollary. How is this observation to be used in economic modelling? As suggested by R. E. Kalman, a detailed study must be made of economic models that are published in the literature to see whether changes in the mathematical formulation of these models are advantageous for the solution of control problems. The selection of the model class seems a creative process that involves knowledge of both the research area of the phenomenon to be modelled and of system theory.

For stochastic processes indexed by the real line the model class of stochastic systems seems an appropriate model. See section 3.1 for a definition of this concept. For a vector of random variables the model class of Gaussian factor models may be useful, see section 5. For random fields it is not yet clear what the appropriate model class should be.

Once the model class has been determined, the modelling procedure prescribes the solution of the stochastic realization problem. In section 3 this problem is formulated and the solution shown for the case of Gaussian processes.

2.4. Selection of an element in the model class

Given the data and the model class, the problem arises of how to select an element in the model class. As indicated earlier, the selection of a model is a trade-off between the objective of a realistic model and the objective of a model with low complexity. For deterministic dynamical systems results on the selection of an element in the model class are reported in [35, 79].

For stochastic systems a formalism for the selection of an element in the class of stochastic systems is described below. Consider first a measure of fit between the observations of the phenomenon and the external behaviour of a stochastic system. Recall that the observations consist of numbers while the external behaviour consists of a measure on the sample space of observation trajectories. The way to proceed is to use the observations, the numbers, to estimate the measure on the sample space of observation trajectories. In case this measure is Gaussian and the observation process is stationary it suffices to estimate the mean and covariance function of this measure.

One can define a measure of fit between the measure for the output trajectories estimated from observations and the measure associated with the external description of the system. Examples of such a measure are the Kullback-Leibler measure and the Hellinger measure; see section 3.7.

For stochastic systems one also needs a measure of complexity. A stochastic complexity measure introduced by J. Rissanen [60-64] seems the appropriate tool for this purpose. Stochastic complexity is based on A. N. Kolmogorov's complexity theory. Since this subject is well covered elsewhere the reader is referred to the indicated references.

The actual selection procedure given data, a model class, and measures of fit and complexity, consists then of a combination of analysis and numerical minimization. The details of this will not be discussed here.

3. GAUSSIAN STOCHASTIC REALIZATION

The purpose of this section is to present the modelling procedure for Gaussian processes. In this tutorial part of the paper results for the Gaussian stochastic realization problem are summarized. For a reference on the weak Gaussian stochastic realization problem see the book [24] and for a shorter introduction in the English language [23]. For a survey of the strong Gaussian stochastic realization problem see [47].

Notation

The following notation is used. $\mathbb{N}=\{0,1,2,\cdots\}$. $\mathbb{Z}_+=\{1,2,\cdots\}$. $\mathbb{Z}=\{\cdots,-1,0,1,\cdots\}$. $\mathbb{Z}_k=\{1,2,\cdots,k\}$. $\mathbb{R}$ denotes the set of real numbers, and $\mathbb{R}_+=[0,\infty)$. For a probability space (Ω,F,P) consisting of a set Ω, a σ-algebra F and a probability measure P, denote

$$L^+(F) = \{x:\Omega\rightarrow\mathbb{R}_+ \mid x \textit{ is a random variable measurable with respect to } F\}.$$

$x\in G(0,Q)$ denotes that the random variable x has a Gaussian distribution with mean zero and variance Q.

For a stochastic process $y:\Omega\times T\rightarrow\mathbb{R}^k$ the following notation is used for the σ-algebra's generated by the process $F_t^y=F_t^{y-}=\sigma(\{y(s),\forall s\leq t\})$ and $F_t^{y+}=\sigma(\{y(s),\forall s\geq t\})$.

DEFINITION 3.0.1. *The σ-algebra's F_1,F_2 are called* conditionally independent *given the σ-algebra G if*

$$E[z_1z_2\mid G] = E[z_1\mid G]E[z_2\mid G]$$

for all $z_i\in L^+(F_i)$. The notation

$$(F_1,F_2\mid G)\in CI$$

will be used to denote that F_1,F_2 are conditionally independent given G and CI will be called the conditional independence relation.

3.1. Stochastic systems and Gaussian systems

The purpose of this section is to define stochastic dynamic systems. Attention is restricted to discrete-time stochastic dynamic systems. Stochastic systems with inputs will not be considered here.

A motivation for the definition of a discrete time stochastic dynamic system follows. Consider the object that is usually called a stochastic system,

$$x_{t+1} = Ax_t + Mv_t,\ x_0, \tag{3.1.1}$$

$$y_t = Cx_t + Nv_t, \tag{3.1.2}$$

where $x_0:\Omega\to\mathbf{R}^n, x_0\in G(m_0,Q_0)$, $v:\Omega\times T\to\mathbf{R}^m$ is a Gaussian white noise process with $v_t\in G(0,V)$, F^{x_0}, F^v_∞ are independent σ-algebras, $A\in\mathbf{R}^{n\times n}$, $M\in\mathbf{R}^{n\times m}$, $C\in\mathbf{R}^{p\times n}$, $N\in\mathbf{R}^{p\times m}$, $x:\Omega\times T\to\mathbf{R}^n$ and $y:\Omega\times T\to\mathbf{R}^p$ defined by the above equations. It may be shown that this object is equivalent with the object specified by:

$$x_0\in G(m_0,Q_0); \tag{3.1.3}$$

$$E[\exp(iu^Tx_{t+1}+iw^Ty_t)\,|\,F_t^{x-}\vee F_{t-1}^{y-}] = \exp\left(i\begin{pmatrix}u\\w\end{pmatrix}^T\begin{pmatrix}Ax_t\\Cx_t\end{pmatrix}-\tfrac{1}{2}\begin{pmatrix}u\\w\end{pmatrix}^T S\begin{pmatrix}u\\w\end{pmatrix}\right), \tag{3.1.4}$$

for all $t\in T$ and some $S\in\mathbf{R}^{(n+p)\times(n+p)}$. Observe that the conditional characteristic function of (x_{t+1},y_t) given $(F_t^{x-}\vee F_{t-1}^{y-})$ depends only on the random variable x_t. It then follows that

$$E[\exp(iu^Tx_{t+1}+iw^Ty_t)\,|\,F_t^{x-}\vee F_{t-1}^{y-}] = E[\exp(iu^Tx_{t+1}+iw^Ty_t)\,|\,F^{x_t}] \tag{3.1.5}$$

for all $t\in T$. A stochastic dynamic system could now be defined as a state process x and an output process y such that for all $t\in T$ there is a map

$$x_t \mapsto \textit{distribution of } (x_{t+1}, y_t)$$

This definition may be found in [42; p. 5]. Below a different definition will be adopted. It may be shown that (3.1.5) is equivalent with the condition that for all $t\in T$

$$(F_t^{y+}\vee F_t^{x+}, F_{t-1}^{y-}\vee F_t^{x-}\,|\,F^{x_t})\in CI,$$

where $F_t^{x+}=\sigma(\{x_s,\forall s\geq t\})$, $F_t^{x-}=\sigma(\{x_s,\forall s\leq t\})$, and similar definitions for F_t^{y+}, F_t^{y-}. The property that the past and future of the state and output process are conditionally independent given the current state will be taken as the definition of a stochastic dynamic system.

DEFINITION 3.1.1. *A discrete-time* stochastic dynamic system *is a collection*

$$\sigma = \{\Omega, F, P, T, Y, B_Y, X, B_X, y, x\},$$

where

- $\{\Omega, F, P\}$ *is a complete probability space;*
- $T=\mathbf{Z}$, *to be called the* time index set;
- (Y, B_Y) *is a measurable space, to be called the* output space;
- (X, B_X) *is a measurable space, to be called the* state space;
- $y:\Omega\times T\to Y$ *is a stochastic process, to be called the* output process;
- $x:\Omega\times T\to X$ *is a stochastic process, to be called the* state process;

such that for all $t\in T$

$$(F_t^{y+} \vee F_t^{x+}, F_{t-1}^{y-} \vee F_t^{x-} \mid F^{x_t}) \in CI. \tag{3.1.6}$$

A stochastic dynamic system on $T \subset \mathbb{Z}$ is defined analogously. The class of stochastic systems is denoted by $S\Sigma$.

The above definition of a stochastic dynamic system is based on related concepts given in [48, 52, 72, 73].

From the definition of a stochastic dynamic system one obtains that the state process satisfies the condition

$$(F_t^{x+}, F_t^{x-} \mid F^{x_t}) \in CI$$

for all $t \in T$. This is equivalent with x being a Markov process. Markov processes are thus also stochastic dynamic systems, and the latter class thus contains the classical model of state processes.

The defining condition of a stochastic dynamic system is more or less symmetric with respect to time in the past and future of the state and output process. This is an advantage over the asymmetric formulation given in the representation (3.1.1) and (3.1.2).

The condition (3.1.6) is asymmetric with respect to the output process. This is a convention. A priori there are four possible conditions for a stochastic dynamic system which are listed below:

$$(F_t^{y+} \vee F_t^{x+}, F_t^{y-} \vee F_t^{x-} \mid F^{x_t}) \in CI \;\; \forall t \in T; \tag{3.1.7.1}$$

$$(F_{t+1}^{y+} \vee F_t^{x+}, F_{t-1}^{y-} \vee F_t^{x-} \mid F^{x_t}) \in CI \;\; \forall t \in T; \tag{3.1.7.2}$$

$$(F_t^{y+} \vee F_t^{x+}, F_{t-1}^{y-} \vee F_t^{x-} \mid F^{x_t}) \in CI \;\; \forall t \in T; \tag{3.1.7.3}$$

$$(F_{t+1}^{y+} \vee F_t^{x+}, F_t^{y-} \vee F_t^{x-} \mid F^{x_t}) \in CI \;\; \forall t \in T. \tag{3.1.7.4}$$

Condition (3.1.7.1) and a property of conditional expectation imply that

$$F^{y_t} \subset (F_t^{y+} \vee F_t^{y-}) \subset F^{x_t}$$

which fact is not compatible with the intuitive concept of state in that the output is in general not part of the state. Condition (3.1.7.2) is not suitable because it would allow examples that are counter-intuitive to the concept of state, see example 3.1.6. The conditions (3.1.7.3) and (3.1.7.4) thus remain, of which condition 3 has been chosen. This is a convention. Condition (3.1.7.4) results in the representation

$$x_{t+1} = Ax_t + Mv_t,$$

$$y_{t+1} = Cx_t + Nv_t,$$

which form is inconsistent with the system theoretic convention of (3.1.1 & 3.1.2). The option of taking condition (3.1.7.3) or (3.1.7.4) in the definition of a stochastic dynamic system is related to the option of considering Moore or Mealey machines in automata theory, see [50; I. A.2].

The definition of a stochastic system is formulated in terms of σ-algebras rather than in terms of stochastic processes. This is a geometric formulation in which emphasis is put on spaces and subspaces rather than on the variables or processes that generate those spaces.

DEFINITION 3.1.2. *Given a stochastic dynamic system*

$$\sigma = \{\Omega, F, P, T, Y, B_Y, X, B_X, y, x\} \in S\Sigma.$$

This system is called:

a. stationary *or* time-invariant *if* (x,y) *is a jointly stationary process;*

b. Gaussian *if* $Y=\mathbf{R}^p$, $X=\mathbf{R}^n$ *for certain* $p,n \in \mathbf{Z}_+$, $B_Y=B^p$ *and* $B_X=B^n$ *are Borel σ-algebras on Y respectively X, and if* (x,y) *is a jointly Gaussian process; by way of abbreviation, a Gaussian stochastic dynamic system will be called a* Gaussian system *and the class of such systems is denoted by* $GS\Sigma$;

c. finite *if* Y,X *are finite sets and* B_Y,B_X *are the σ-algebras on* Y,X *generated by all subsets; by way of abbreviation a finite stochastic dynamic system will be called a* finite stochastic system *and the class of such systems is denoted by* $FS\Sigma$.

PROPOSITION 3.1.3. *Consider a collection*

$$\{\Omega, F, P, T, Y, B_Y, X, B_X, y, x\}$$

as defined in 3.1.1 but without condition (3.1.6). The following statements are equivalent:

a. *for all* $t \in T$

$$(F_t^{y+} \vee F_t^{x+}, F_{t-1}^{y-} \vee F_t^{x-} \mid F^{x_t}) \in CI;$$

b. *for all* $t \in T$

$$(F^{y_t} \vee F^{x_{t+1}}, F_{t-1}^{y-} \vee F_t^{x-} \mid F^{x_t}) \in CI;$$

c. *for all* $t \in T$

$$(F_t^{y+} \vee F_t^{x+}, F^{y_{t-1}} \vee F^{x_{t-1}} \mid F^{x_t}) \in CI.$$

The following result is a useful sufficient condition for a stochastic dynamic system.

PROPOSITION 3.1.4. *Consider the collection*

$$\sigma = \{\Omega, F, P, T, Y, B_Y, X, B_X, y, x\}$$

as defined in 3.1.1 but without condition (3.1.6). If for all $t \in T$

1. $(F_t^{y+}, F_\infty^{x-} \vee F_{t-1}^{y-} \mid F^{x_t}) \in CI;$

2. $(F_t^{x+}, F_t^{x-} \vee F_{t-1}^{y-} \mid F^{x_t}) \in CI;$

then $\sigma \in S\Sigma$.

Below two examples of stochastic dynamic systems are presented.

EXAMPLE 3.1.5. Consider a Gaussian system representation

$$x_{t+1} = Ax_t + Mv_t, \tag{3.1.8}$$

$$y_t = Cx_t + Nv_t, \tag{3.1.9}$$

with the conventions given below (3.1.1 & 3.1.2). As indicated there this representation is equivalent with

$$E[\exp(iu^T x_{t+1} + iw^T y_t) \mid F_t^{x-} \vee F_{t-1}^{y-}]$$

$$= \exp(i \begin{bmatrix} u \\ w \end{bmatrix}^T \begin{bmatrix} Ax_t \\ Cx_t \end{bmatrix} - ½ \begin{bmatrix} u \\ w \end{bmatrix}^T S \begin{bmatrix} u \\ w \end{bmatrix}),$$

for all $t \in T$ and $x_0 \in G$. This and a property of conditional independence imply that

$$(F^{x_{t+1}} \vee F^{y_t}, F^{y-}_{t-1} \vee F^{x-}_t \mid F^{x_t}) \in CI, \ \forall t \in T,$$

and from 3.1.3 then follows that, with x,y specified by (3.1.8 & 3.1.9),

$$\sigma = \{\Omega, F, P, T, \mathbb{R}^p, B^p, \mathbb{R}^n, B^n, y, x\} \in S\Sigma.$$

From properties of Gaussian random variables follows that (x,y) is a jointly Gaussian process, hence σ is a Gaussian system or $\sigma \in GS\Sigma$. In the following (3.1.8 & 3.1.9) will be called a forward representation of a Gaussian system.

EXAMPLE 3.1.6. Let $v : \Omega \times T \to \mathbb{R}$ be a standard Gaussian white noise process. Define $y : \Omega \times T \to \mathbb{R}$, $x : \Omega \times T \to \mathbb{R}$ by

$$x_t = v_{t-1}, \ y_t = x_t + v_t = v_{t-1} + v_t.$$

Then the following hold.

a. For all $t \in T$ $(F^{y+}_{t+1}, F^{y+}_{t-1} \mid N) \in CI$, where $N \subset F$ is the trivial σ-algebra. Thus the process y is the output process of a stochastic dynamic system according to (3.1.7.2) with a trivial state space.

b. For all $t \in T$

$$E[\exp(iuy_t) \mid F^{y+}_{t-1}]$$

is nondeterministic, indicating that the process y has some kind of memory.

c.

$$(F^{y+}_t \vee F^{x+}_t, F^{y-}_{t-1} \vee F^{x-}_t \mid F^{x_t}) \in CI$$

for all $t \in T$, hence

$$\sigma = \{\Omega, F, P, T, Y, B, X, B, y, x\} \in GS\Sigma.$$

3.2. Forward and backward representations of Gaussian systems

The purpose of this subsection is to show that a Gaussian system has both a forward and a backward representation, and to derive relations between these representations.

PROPOSITION 3.2.1. *Let*

$$\sigma = \{\Omega, F, P, T, \mathbb{R}^p, B^p, \mathbb{R}^n, B^n, y, x\} \in GS\Sigma$$

be a Gaussian system. Assume that for all $t \in T$ $E[x_t]=0$, $E[y_t]=0$ *and that* $Q : T \to \mathbb{R}^{n \times n}, Q(t) = E[x_t x_t^T] > 0$.

a. *The Gaussian system has what will be called a* forward representation *given by*

$$x_{t+1} = A^f(t)x_t + Mv^f_t, \ x_0, \qquad (3.2.1)$$

$$y_t = C^f(t)x_t + Nv^f_t, \qquad (3.2.2)$$

where $v^f:\Omega\times T\rightarrow\mathbf{R}^{n+k}$ *is a Gaussian white noise process with intensity* V^f. *Given* σ *then*

$$A^f(t) = E[x_{t+1}x_t^T]Q(t)^{-1},$$

$$C^f(t) = E[y_t x_t^T]Q(t)^{-1},$$

$$V^f(t) = \begin{pmatrix} Q(t+1) & E[x_{t+1}y_t^T] \\ E[y_t x_{t+1}^T] & E[y_t y_t^T] \end{pmatrix} - \begin{pmatrix} A^f(t) \\ C^f(t) \end{pmatrix} Q(t)^{-1} \begin{pmatrix} (A^f(t))^T & (C^f(t))^T \end{pmatrix},$$

$$M = (I_n\ 0)\in\mathbf{R}^{n\times(n+p)},\ N = (0\ I_p)\in\mathbf{R}^{p\times(n+p)}.$$

Conversely, given a forward representation with A^f, C^f, V^f, M, N *functions and* x,y *defined by the above forward representation (3.2.1 & 3.2.2), then* σ *is a Gaussian system.*

b. *The given Gaussian system has also a* backward representation *given by*

$$x_{t-1} = A^b(t)x_t + Mv_t^b,\ x_0, \tag{3.2.3}$$

$$y_{t-1} = C^b(t)x_t + Nv_t^b, \tag{3.2.4}$$

where $v^b:\Omega\times T\rightarrow\mathbf{R}^{n+k}$ *is a Gaussian white noise process with intensity* V^b. *Given* σ

$$A^b(t) = E[x_{t-1}x_t^T]Q(t)^{-1}, \tag{3.2.5}$$

$$C^b(t) = E[y_{t-1}x_t^T]Q(t)^{-1}, \tag{3.2.6}$$

$$V^b(t) = \begin{pmatrix} Q(t-1) & E[x_{t-1}y_t^T] \\ E[y_t x_{t-1}^T] & E[y_t y_t^T] \end{pmatrix} - \begin{pmatrix} A^b(t) \\ C^b(t) \end{pmatrix} Q(t)^{-1} \begin{pmatrix} (A^b(t))^T & (C^b(t))^T \end{pmatrix}, \tag{3.2.7}$$

$$M = (I_n\ 0),\ N = (0\ I_p).$$

Conversely, given a backward representation with A^b, C^b, V^b, M, N *and* x,y *as defined by the above backward representation, then* σ *is a Gaussian system.*

c. *The relation between the forward and backward representation of a Gaussian system is given by*

$$A^f(t)Q(t) = Q(t+1)(A^b(t+1))^T, \tag{3.2.8}$$

$$C^b(t)Q(t) = C^f(t-1)Q(t-1)(A^f(t-1))^T + NV^f(t-1)M^T, \tag{3.2.9}$$

$$C^f(t)Q(t) = C^b(t+1)Q(t+1)(A^b(t+1))^T + NV^b(t+1)M^T. \tag{3.2.10}$$

d. *Assume that the given Gaussian system is stationary. Then* $A^f, C^f, V^f, A^b, C^b, V^b$, *do not depend explicitly on* $t\in T$ *and* $Q(t)=Q\in\mathbf{R}^{n\times n}$, $Q=Q^T>0$. *The relation between the forward and backward representation is then given by*

$$A^f = Q(A_b)^TQ^{-1}, \tag{3.2.11}$$

$$A^b = Q(A^f)^TQ^{-1}, \tag{3.2.12}$$

$$C^b = C^fQ(A^f)^TQ^{-1} + NV^fM^TQ^{-1} = C^fA^b + NV^fM^TQ^{-1}, \tag{3.2.13}$$

$$C^f = C^bQ(A^b)^TQ^{-1} + NV^bM^TQ^{-1} = C^bA^f + NV^bM^TQ^{-1}. \tag{3.2.14}$$

In the following the superscripts f and b will be omitted when it is clear from the context which representation is referred to.

3.3. Stochastic observability and stochastic reconstructibility

The theorem on the characterization of minimality of a stochastic realization makes use of the concepts of stochastic observability and stochastic reconstructibility. Below these concepts are introduced.

DEFINITION 3.3.1. *Consider a stochastic system*

$$\sigma = \{\Omega, F, P, T, \mathbb{R}^p, B^p, \mathbb{R}^n, B^n, y, x\} \in S\Sigma.$$

a. *This system is called* stochastically observable on the interval $\{t, t+1, \ldots, t+t_1\}$ *if the map*

$$x_t \mapsto E[\exp(i \sum_{s=0}^{t_1} u_s^T y_{t+s}) \mid F^{x_t}]$$

from x_t to the conditional characteristic function of $\{y_t, y_{t+1}, \ldots, y_{t+t_1}\}$ given x_t is injective on the support of x_t.

b. *Assume that the system σ is stationary. Then it is called* stochastically observable *if there exists a t, $t_1 \in T$, $0 < t_1 < \infty$, such that it is stochastically observable on the interval $\{t, t+1, \ldots, t+t_1\}$ as defined above. By stationarity this then holds for all $t \in T$.*

The interpretation of a stochastically observable stochastic system is that if one knows the conditional distribution of $\{y_t, y_{t+1}, \ldots, y_{t+t_1}\}$ given x_t, then one can uniquely determine the value of x_t. Note that the conditional distribution of $\{y_t, \ldots, y_{t+t_1}\}$ given x_t can in principle be determined from measurements.

PROPOSITION 3.3.2. *Consider the Gaussian system*

$$\sigma = \{\Omega, F, P, T, \mathbb{R}^p, B^p, \mathbb{R}^n, B^n, y, x\} \in GS\Sigma,$$

with forward representation

$$x_{t+1} = A(t)x_t + Mv_t,$$

$$y_t = C(t)x_t + Nv_t,$$

with $v_t \in G(0, V(t))$.

a. *The system σ is stochastically observable on $\{t, t+1, \ldots, t+t_1\}$ iff*

$$rank \begin{pmatrix} C(t) \\ C(t+1)\Phi(t+1,t) \\ \ldots \\ C(t+t_1)\Phi(t+t_1,t) \end{pmatrix} = n, \qquad (3.3.1)$$

iff

$$rank(\sum_{s=0}^{t_1} C(t+s)\Phi(t+s,t)\Phi(t+s,t)^T C(t+s)^T) = n.$$

b. *Assume that the system is stationary with forward representation*

$$x_{t+1} = Ax_t + Mv_t,$$

$$y_t = Cx_t + Nv_t,$$

with $v_t \in G(0, V)$. *Then this system is stochastically observable iff*

$$\text{rank} \begin{bmatrix} C \\ CA \\ \cdots \\ CA^{n-1} \end{bmatrix} = n. \tag{3.3.2}$$

DEFINITION 3.3.3. *Consider the stochastic system*

$$\sigma = \{\Omega, F, P, T, \mathbf{R}^p, B^p, \mathbf{R}^n, B^n, y, x\} \in S\Sigma.$$

a. *This system is called* stochastically reconstructible *on the interval* $\{t-1, t-2, \ldots, t-t_1\}$ *if the map*

$$x_t \mapsto E[\exp(i \sum_{s=1}^{t_1} u_s^T y_{t-s}) \mid F^{x_t}]$$

is injective on the support of x_t.

b. *Assume that the system is stationary. Then it is called* stochastically reconstructible *if there exist* $t, t_1 \in T$, $0 < t_1 < \infty$, *such that it it stochastically reconstructible on the interval* $\{t-1, \ldots, t-t_1\}$. *By stationarity this then holds for any* $t \in T$.

PROPOSITION 3.3.4. *Consider the Gaussian system*

$$\sigma = \{\Omega, F, P, T, \mathbf{R}^p, B^p, \mathbf{R}^n, B^n, y, x\} \in GS\Sigma$$

with backward representation

$$x_{t-1} = A(t)x_t + Mv_t,$$

$$y_{t-1} = C(t)x_t + Nv_t,$$

with $v_t \in G(0, V(t))$.

a. *The system* σ *is stochastically reconstructible on the interval* $\{t-1, t-2, \ldots, t-t_1\}$ *iff*

$$\text{rank} \begin{bmatrix} C(t) \\ C(t-1)\Phi(t-1,t) \\ \cdots \\ C(t-t_1)\Phi(t-t_1,t) \end{bmatrix} = n, \tag{3.3.3}$$

iff

$$\text{rank} \left(\sum_{s=1}^{t_1} C(t-s)\Phi(t-s,t)\Phi(t-s,t)^T C(t-s)^T \right) = n.$$

b. *Assume that the system* σ *is stationary with backward representation*

$$x_{t-1} = Ax_t + Mv_t,$$

$$y_{t-1} = Cx_t + Nv_t,$$

with $v_t \in G(0,V)$. *Then it is stochastically reconstructible iff*

$$\text{rank} \begin{pmatrix} C \\ CA \\ \cdots \\ CA^{n-1} \end{pmatrix} = n. \tag{3.3.4}$$

Note that the condition (3.3.2) is expressed in terms of the matrices (A,C) of the forward representation of the Gaussian system and the condition (3.3.4) is expressed in terms of the matrices (A,C) of the backward representation. See section 3.2 for the way the matrices of the forward and backward representation are related.

3.4. The weak Gaussian stochastic realization problem

Attention is again directed to the problem of modelling by a stochastic system. So, one is given a measure on the observed process that has been estimated from the data. One is asked to determine a stochastic system in the model class such that the measure restricted to the observation process equals the given measure.

PROBLEM 3.4.1. *The* weak Gaussian stochastic realization problem *for a stationary Gaussian process is, given a stationary Gaussian process on* $T=\mathbb{Z}$ *taking values in* $(\mathbb{R}^p, B^p)$ *having mean value function zero and covariance function* $W: T \rightarrow \mathbb{R}^{p \times p}$, *to solve the following subproblems.*

a. *Does there exist a stationary Gaussian system*

$$\sigma = \{\Omega, F, P, T, \mathbb{R}^p, B^p, \mathbb{R}^n, B^n, y, x\} \in GS\Sigma$$

such that the output process y of this system equals the given process in distribution. This means that these processes have the same family of finite dimensional distributions. Effectively this means that the covariance function of the output process must be equal to the given covariance function W because both processes are Gaussian. If such a system exists, then one calls σ a weak Gaussian stochastic realization *of the given process, or, if the context is known, a* stochastic realization.

b. *Classify all minimal stochastic realizations of the given process. A weak Gaussian stochastic realization is called* minimal *if the dimension of the state space is minimal. The following subproblems must be solved:*
   1. *characterize those stochastic realizations that are minimal;*
   2. *obtain the classification as such;*
   3. *indicate the relation between two minimal stochastic realizations;*
   4. *produce an algorithm that constructs all minimal weak Gaussian stochastic realizations of the given process.*

In problem 3.4.1 one is given a stationary Gaussian process with zero mean value function. Such a process is thus completely characterized by its covariance function. In part a. of this problem the question is whether the given process can be the output of a stationary Gaussian system. Because by definition such a Gaussian system has a finite-dimensional state space, not all stationary Gaussian processes can be the output process of a Gaussian system. The question should therefore be interpreted as to determine a necessary and sufficient condition on the given process, or its covariance function, such that it can be the output process of a Gaussian system.

In part b. of problem 3.4.1 a classification is asked for. This question arises because a stochastic realization, if it exists, is in general nonunique. This will be indicated below. The dimensions of the state space of two stochastic realizations may also be different in general. For system theoretic reasons, such as identifiability, one should restrict attention to those stochastic realizations for which the dimension of the state space is minimal. Such a realization is called *minimal.* In general minimal stochastic realizations are also nonunique. A classification of all minimal stochastic realizations is then useful for the solution of the identifiability question. The above defined problem is related to the problem of determining spectral factorizations of the spectral density of the given process.

Below a notation is used for the parameters of a time-invariant finite-dimensional linear system of the form

$$x(t+1) = Ax(t) + Bu(t),$$

$$y(t) = Cx(t) + Du(t),$$

with $U=\mathbb{R}^m$, $X=\mathbb{R}^n$, $Y=\mathbb{R}^p$, $u:T\rightarrow U$, $x:T\rightarrow X$, $y:T\rightarrow Y$. The notation is then

$$pls = \{p,n,m,A,B,C,D\} \in L\Sigma P.$$

In the formulation of theorem 3.4.2 use is made of the set $\mathbf{Q}_{\overline{pls}}$. The definition of this set is given in subsection 3.5.

THEOREM 3.4.2. *Consider the weak Gaussian stochastic realization problem for a stationary Gaussian process as posed in 3.4.1. Assume that* $\lim_{t\rightarrow\infty} W(t)=0$ *and that* $W(0)>0$.

a. *There exists a weak Gaussian stochastic realization of the given process iff there exists a* $pls=\{p,n,p,F,G,H,J\}\in L\Sigma P$ *with* $J=J^T$ *such that*

$$W(t) = \begin{cases} HF^{t-1}G, & \text{if } t>0, \\ 2J, & \text{if } t=0, \\ G^T(F^T)^{-t-1}H^T, & \text{if } t<0. \end{cases} \tag{3.4.1}$$

(a function having the form (3.4.1) will be called a discrete-time Bohl function; *the right hand side of (3.4.1) will be called a* covariance realization *of the covariance function W.)*
iff

$$\hat{W}(\lambda) = \sum_{t\in\mathbf{Z}} W(t)\lambda^{-|t|} \tag{3.4.2}$$

is a rational function. The dimension n in the covariance realization (3.4.1) is also called the McMillan degree *of the covariance function.*

b. *A weak Gaussian stochastic realization is minimal iff it is stochastically observable and stochastically reconstructible.*

c. *A minimal weak Gaussian stochastic realization is nonunique in two ways.*
   1. *If* $pgs_1=\{p,n,m,A,C,M,N,V\}\in GS\Sigma P$ *are the parameters of a forward representation of a minimal stochastic realization, and if* $S\in\mathbb{R}^{n\times n}$ *is nonsingular, then* $pgs_2=\{p,n,m,SAS^{-1},CS^{-1},SM,N,V\}\in GS\Sigma P$ *are also the parameters of a forward representation of a minimal stochastic realization.*
   2. *Fix the parameters of a minimal covariance realization as given in a. above,*

$pls = \{p,n,p,F,G,H,J\} \in L\Sigma P_{\min}$.

Denote the parameters of a forward representation of a minimal Gaussian stochastic realization by $\{p,n,A,C,V\}$ *and the set of such parameters by* $WGSRP_{\min}$. *Define the classification map*

$$c_{pls}: \mathbf{Q}_{\overline{pls}} \to WGSRP_{\min},\ c_{pls}(Q) = \{p,n,A,C,V\}, \tag{3.4.3}$$

by $A=F$, $C=H$,

$$V = \overline{V}(Q) = \begin{pmatrix} Q-FQF^T & G-FQH^T \\ G^T-HQF^T & 2J-HQH^T \end{pmatrix}.$$

Then, for fixed $pls \in L\Sigma P_{\min}$ *is* c_{pls} *a bijection. Thus all minimal weak Gaussian stochastic realizations are classified by the elements of* $\mathbf{Q}_{\overline{pls}}$.

d. *The stochastic realization algorithm as defined in 3.4.3 below is well defined and constructs all minimal weak Gaussian stochastic realizations.*

ALGORITHM 3.4.3. *The stochastic realization algorithm for weak Gaussian stochastic realizations of stationary Gaussian processes.*

Data: given a stationary Gaussian process with zero mean value function and covariance function $W:T \to \mathbb{R}^{p\times p}$. *Assume that the condition of 3.4.2.a. holds.*

1. *Determine a minimal covariance realization of W via a realization algorithm for time-invariant finite-dimensional linear systems, or* $pls = \{p,n,p,F,G,H,J\} \in L\Sigma P_{\min}$, *such that*

$$W(t) = \begin{cases} HF^{t-1}G, & \text{if } t>0, \\ 2J, & \text{if } t=0, \\ G^T(F^T)^{-t-1}H^T, & \text{if } t<0. \end{cases} \tag{3.4.4}$$

For algorithms for this step see books on linear system theory.

2. *Determine a* $Q \in \mathbf{Q}_{\overline{pls}}$, *or a* $Q \in \mathbb{R}^{n\times n}$ *satisfying* $Q=Q^T \geq 0$,

$$\begin{pmatrix} Q-FQF^T & G-FQH^T \\ G^T-HQF^T & 2J-HQH^T \end{pmatrix} \geq 0. \tag{3.4.5}$$

3. *Let*

$$A = F,\ C = H,\ M = (I_n\ 0) \in \mathbb{R}^{n\times(n+p)},\ N = (0\ I_p) \in \mathbb{R}^{p\times(n+p)},$$

$$V = \overline{V}(Q) = \begin{pmatrix} Q-FQF^T & G-FQH^T \\ G^T-HQF^T & 2J-HQH^T \end{pmatrix} \in \mathbb{R}^{(n+p)\times(n+p)},$$

construct a probability space by

$$\Omega = (\mathbb{R}^{(n+p)})^T,\ F = \Pi_T \otimes B^{(n+p)},\ v:\Omega\times T \to \mathbb{R}^{(n+p)},\ v(\omega,t) = \omega(t),\ P:F \to [0,1]$$

a probability measure such that v is a Gaussian white noise process with intensity V, $x:\Omega\times T\to\mathbb{R}^n$ $y:\Omega\times T\to\mathbb{R}^p$ *defined by*

$$x_{t+1} = Ax_t + Mv_t,\ x_{-\infty}=0, \tag{3.4.6}$$

$$y_t = Cx_t + Nv_t. \tag{3.4.7}$$

Then

$$\sigma = \{\Omega, F, P, T, \mathbb{R}^p, B^p, \mathbb{R}^n, B^n, y, x\} \in GS\Sigma \tag{3.4.8}$$

is a minimal weak Gaussian stochastic realization of the given process, meaning that the output process y is a Gaussian process with covariance function equal to the given covariance function W.

A mistake that is sometimes made is the following. Consider the following forward representation of a Gaussian system

$$x_{t+1} = Ax_t + Mv_t,$$

$$y_t = Cx_t + Nv_t,$$

with $v_t \in G(0, V)$. A statement is that if the pair of matrices $(A, MV^{1/2})$ is a reachable pair and if (A, C) is an observable pair, that then the stochastic realization described by the above system representation is a minimal realization of the output process. This statement is false as the following example shows.

EXAMPLE 3.4.4. Consider the Gaussian system

$$\sigma = \{\Omega, F, P, T, \mathbb{R}, B, \mathbb{R}, B, y, x\} \in GS\Sigma$$

with forward representation

$$x_{t+1} = ax_t + bv_t,$$

$$y_t = x_t + v_t,$$

with $v_t \in G(0, 1)$, $a \in (-1, +1)$, $a \neq 0$, $b = (a^2 - 1)/a$.

a. Then (a, b) is a reachable pair and $(a, 1)$ is an observable pair.
b. The system σ is a nonminimal realization of its output process.

It is possible to interpret certain stochastic realizations as a Kalman filter but this will not be done here. For a reference see [24].

The implication of the weak Gaussian stochastic realization problem for the identifiability question is illustrated by the following example.

EXAMPLE 3.4.5. Consider the time-invariant Gaussian system

$$\sigma = \{\Omega, F, P, T, \mathbb{R}, B, \mathbb{R}, B, y, x\} \in GS\Sigma$$

with forward representation

$$x_{t+1} = ax_t + (1\ 0)v_t, \tag{3.4.9}$$

$$y_t = cx_t + (0\ 1)v_t, \tag{3.4.10}$$

with $v_t \in G(0, V)$,

$$V = \begin{bmatrix} v_{11} & 0 \\ 0 & v_{22} \end{bmatrix}. \tag{3.4.11}$$

Consider the asymptotic Kalman filter for the Gaussian system (3.4.9 & 3.4.10)

$$\hat{x}_{t+1} = a\hat{x}_t + k(y_t - c\hat{x}_t), \tag{3.4.12}$$

$$\bar{v}_t = y_t - c\hat{x}_t, \tag{3.4.13}$$

in which $\bar{v}:\Omega \times T \rightarrow \mathbb{R}$ is a Gaussian white noise process with $\bar{v}_t \in G(0,r)$. This asymptotic Kalman filter may be rewritten as

$$\hat{x}_{t+1} = a\hat{x}_t + k\bar{v}_t = a\hat{x}_t + (1\ 0)v_1(t), \tag{3.4.14}$$

$$y_t = c\hat{x}_t + \bar{v}_t = c\hat{x}_t + (0\ 1)v_1(t), \tag{3.4.15}$$

in which $v_1:\Omega \times T \rightarrow \mathbb{R}^2$ is a Gaussian white noise process with $v_1(t) \in G(0, V_1)$,

$$V_1 = \begin{pmatrix} k^2 r & kr \\ kr & r \end{pmatrix} = \begin{pmatrix} k \\ 1 \end{pmatrix} r(k\ 1). \tag{3.4.16}$$

From these forward representations one deduces that (3.4.9 & 3.4.10) and (3.4.14 & 3.4.15) are both weak Gaussian stochastic realizations of the output process y. This may be verified by computing the covariance function of the output process. This example shows that one may not be able to uniquely determine the parameters of the noise process of a Gaussian system, here (3.4.11) and (3.4.16), from the covariance function of the output process. For results on the parametrization of Gaussian systems see [34].

Attention has also been devoted to the partial weak Gaussian stochastic realization problem in which one is not given a covariance function on all of $T = \mathbb{Z}$ but only on a finite time set, say $T = \{-t_1, -t_1+1, \ldots, -1, 0, 1, \ldots, t_1\}$. The motivation for this problem is that in practice one can estimate from a finite time series only the covariance function on a finite time set.

3.5. The dissipation matrix inequality

In subsection 3.4 it has been stated that the minimal weak Gaussian stochastic realizations are classified by the set $\mathbf{Q}_{\overline{pls}}$. In this section the set $\mathbf{Q}_{\overline{pls}}$ and its dual $\mathbf{Q}_{pls}$ will be considered. Throughout this section $J = J^T$. The results of this subsection may be found in [23, 24].

DEFINITION 3.5.1. Let $pls = \{p,n,p,F,G,H,J\} \in L\Sigma P$ with $J \geq 0$ and

$$\mathbf{Q}_{pls} = \{Q \in \mathbb{R}^{n \times n} \mid Q = Q^T \geq 0,\ V(Q) = \begin{pmatrix} Q - F^T Q F & H^T - F^T Q G \\ H - G^T Q F & 2J - G^T Q G \end{pmatrix} \geq 0\}, \tag{3.5.1}$$

and for $\overline{pls} = \{p,n,p,F^T,H^T,G^T,J\} \in L\Sigma P$

$$\mathbf{Q}_{\overline{pls}} = \{Q \in \mathbb{R}^{n \times n} \mid Q = Q^T \geq 0,\ \bar{V}(Q) = \begin{pmatrix} Q - FQF^T & G - FQH^T \\ G^T - HQF^T & 2J - HQH^T \end{pmatrix} \geq 0\}. \tag{3.5.2}$$

PROBLEM 3.5.2. GIVEN $pls \in L\Sigma P$ AND $\mathbf{Q}_{pls}$.

a. Classify all elements of $\mathbf{Q}_{pls}$.

b. Determine an algorithm that constructs all elements of $\mathbf{Q}_{pls}$.

PROPOSITION 3.5.3. *Consider* $pls = \{p,n,p,F,G,H,J\} \in L\Sigma P_{\min}$ *and* $\mathbf{Q}_{pls}$. *Assume that* $\mathbf{Q}_{pls} \neq \emptyset$, *and that* $J > 0$. *Then* $\mathbf{Q}_{pls}$ *is a convex, closed and bounded set, and there exists a* $Q^-, Q^+ \in \mathbf{Q}_{pls}$

such that for any $Q \in \mathbf{Q}_{pls}$, $Q^- \leq Q \leq Q^+$.

DEFINITION 3.5.4.

a. *The* regular part of $\mathbf{Q}_{pls}$ *is defined as*

$$\mathbf{Q}_{pls,r} = \{Q \in \mathbf{Q}_{pls} \mid 2J - G^T Q G > 0\}.$$

The set $\mathbf{Q}_{pls}$ *will be called* regular *if* $\mathbf{Q}_{pls} = \mathbf{Q}_{pls,r}$.

b. *For* $Q \in \mathbf{R}^{n \times n}$ *with* $Q = Q^T$ *and* $2J - G^T Q G > 0$ *define*

$$D(Q) = Q - F^T Q F - [H^T - F^T Q G][2J - G^T Q G]^{-1}[H^T - F^T Q G]^T. \tag{3.5.3}$$

c. *Correspondingly define*

$$\mathbf{Q}_{\overline{pls},r} = \{Q \in \mathbf{Q}_{\overline{pls}} \mid 2J - HQH^T > 0\},$$

$$\overline{D}(Q) = Q - FQF^T - [G - FQH^T][2J - HQH^T]^{-1}[G - FQH^T]^T, \tag{3.5.4}$$

and $\mathbf{Q}_{\overline{pls}}$ *is regular if* $\mathbf{Q}_{\overline{pls}} = \mathbf{Q}_{\overline{pls},r}$.

PROPOSITION 3.5.5. *Let* $pls = \{p,n,p,F,G,H,J\} \in L\Sigma P$. *Let* $Q \in \mathbf{R}^{n \times n}$, $Q = Q^T$.

a. *Assume that* $2J - G^T Q G > 0$, *and let*

$$T = \begin{bmatrix} I & 0 \\ -[2J - G^T Q G]^{-1}[H - G^T Q F] & I \end{bmatrix} \in \mathbf{R}^{(n+p) \times (n+p)}. \tag{3.5.5}$$

Then

$$\begin{bmatrix} D(Q) & 0 \\ 0 & 2J - G^T Q G \end{bmatrix} = T^T V(Q) T, \tag{3.5.6}$$

and

$$V(Q) = T^{-T} \begin{bmatrix} D(Q) & 0 \\ 0 & 2J - G^T Q G \end{bmatrix} T^{-1}, \tag{3.5.7}$$

where $V(Q)$ *is as defined in 3.5.1.*

b. *Assume that* $2J - G^T Q G > 0$. *Then* $V(Q) \geq 0$ *iff* $D(Q) \geq 0$. *Also* $V(Q) > 0$ *iff* $D(Q) > 0$. *In fact,* $rank\,(V(Q)) = rank\,(D(Q)) + p$.

c.

$$\mathbf{Q}_{pls,r} = \{Q \in \mathbf{R}^{n \times n} \mid Q = Q^T \geq 0,\ 2J - G^T Q G > 0,\ D(Q) \geq 0\}.$$

Notation for the boundary of $\mathbf{Q}_{pls}$ will be needed. The following notation will be used in the sequel,

$$\|Q\|_2 = \left[sup_{x \in \mathbf{R}^n, x \neq 0} \frac{x^T Q^T Q x}{x^T x} \right]^{1/2}, \tag{3.5.8}$$

$$B(Q, \epsilon) = \{S \in \mathbf{R}^{n \times n} \mid \|S - Q\|_2 \leq \epsilon\}. \tag{3.5.9}$$

DEFINITION 3.5.6. Let $pls \in L\Sigma P$ and consider $\mathbf{Q}_{pls}$. Define the *boundary* of $\mathbf{Q}_{pls}$ as the set

$$\partial\mathbf{Q}_{pls} = \{Q\in\mathbf{Q}_{pls} \mid \forall\epsilon\in\mathbf{R},\ \epsilon>0,\ \exists S\in B(Q,\epsilon) \text{ such that } S=S^T,\ S\neq Q,\ S\notin\mathbf{Q}_{pls}\},$$

and the *interior of* $\mathbf{Q}_{pls}$ as the set

$$int(\mathbf{Q}_{pls}) = \mathbf{Q}_{pls}\cap(\partial\mathbf{Q}_{pls})^c.$$

PROPOSITION 3.5.7. *Let* $pls=\{p,n,p,F,G,H,J\}\in L\Sigma P$.

a. $Q\in\partial\mathbf{Q}_{pls}$ *iff* $V(Q)$ *is singular.* $Q\in int(\mathbf{Q}_{pls})$ *iff* $V(Q)>0$.

b. *Assume that* $\mathbf{Q}_{pls}$ *is regular. Then* $Q\in\partial\mathbf{Q}_{pls}$ *iff* $D(Q)$ *is singular; and* $Q\in int(\mathbf{Q}_{pls})$ *iff* $D(Q)>0$.

DEFINITION 3.5.8. *Let* $pls=\{p,n,p,F,G,H,J\}\in L\Sigma P$ *and consider* $\mathbf{Q}_{pls}$.

a. *The set of* singular boundary points of $\mathbf{Q}_{pls}$ *is defined as*

$$\partial\mathbf{Q}_{pls,s} = \{Q\in\partial\mathbf{Q}_{pls} \mid rank(V(Q)) = rank(2J-G^TQG)\}.$$

b. *The set of* singular boundary points of the regular part of $\mathbf{Q}_{pls}$ *is defined as*

$$\partial\mathbf{Q}_{pls,r,s} = \{Q\in\mathbf{Q}_{pls,r}\cap\partial\mathbf{Q}_{pls} \mid rank(V(Q))=p\}.$$

THEOREM 3.5.9. *Let* $pls=\{p,n,p,F,G,H,J\}\in L\Sigma P_{\min}$. *Assume that* $\mathbf{Q}_{pls}\neq\emptyset$ *and that it is regular. Let*

$$F^{-} = F - G[2J-G^TQ^{-}G]^{-1}[H^T-FQ^{-}G]^T.$$

Then $Q^{-}+\Delta Q\in\mathbf{Q}_{pls}$ *and* $\Delta Q>0$ *iff*

1. $\Delta Q\in\mathbf{R}^{n\times n}$, $\Delta Q>0$;
2. $$(\Delta Q)^{-1} - F^{-}(\Delta Q)^{-1}(F^{-})^T - G[2J-G^TQ^{-}G]^{-1}G^T - S = 0, \quad (3.5.10)$$

   for some $S\in\mathbf{R}^{n\times n}$, $S=S^T\geq 0$;
3. $sp(F^{-})\subset C^{-}$.

3.6. The strong Gaussian stochastic realization problem

PROBLEM 3.6.1. *The* strong Gaussian stochastic realization problem *for a stationary Gaussian process is, given a probability space* (Ω,F,P), *a time index set* $T=\mathbf{Z}$ *and a stationary Gaussian process* $z:\Omega\times T\to\mathbf{R}^p$ *having zero mean value function and covariance function* $W:T\to\mathbf{R}^{p\times p}$, *to solve the following subproblems.*

a. *Does there exist a stationary Gaussian system*

$$\sigma = \{\Omega,F,P,T,\mathbf{R}^p,B^p,\mathbf{R}^n,B^n,y,x\}\in GS\Sigma$$

with forward representation

$$x_{t+1} = Ax_t + Mv_t,\ x_0,$$

$$y_t = Cx_t + Nv_t,$$

such that

1. $y_t = z_t$ *a.s. for all* $t \in T$;
2. $F^{x_t} \subset F^{y}_{\infty}$ *for all* $t \in T$.

If such a system exists then one calls σ *a* strong Gaussian stochastic realization *of the given process, or, if the context is known, a* stochastic realization.

b. *Classify all minimal stochastic realizations of the given process. A strong Gaussian stochastic realization is called* minimal *if the dimension of the state space is minimal.*

The difference between the weak and the strong Gaussian stochastic realization problems is that the given process and the output process of the Gaussian stochastic system are equal in the sense of the family of finite-dimensional distributions respectively equal in the sense of almost surely. For the strong Gaussian stochastic realization problem this requires that the stochastic system is constructed on the same probability space as the given process. Therefore the state process has to be constructed from the given process, and this explains condition 2 of problem 3.6.1.a.

For a survey of the strong Gaussian stochastic realization problem the reader is referred to the paper [47].

3.7. Pseudo-distances on the set of probability measures

The purpose of this subsection is define distances on the set of probability measures as a preparation for the approximate stochastic realization problem to be discussed in the next subsection.

DEFINITION 3.7.1. *Let X be a set. A* pseudo-distance *is a function* $d: X \times X \rightarrow \mathbf{R}$ *such that*

1. $d(x,y) \geqslant 0$ *for all* $x,y \in X$;
2. $d(x,y) = 0$ *iff* $x = y$.

If a pseudo-distance is not symmetric then one may construct its symmetrized version. A pseudo-distance need not satisfy the triangle inequality.

DEFINITION 3.7.2. *Let*

$$F_{2s} = \{f: \mathbf{R}_+ \rightarrow \mathbf{R} \mid f \in C^2,\ f(1)=0,\ \forall x \in (0,\infty),\ f''(x) > 0\}.$$

DEFINITION 3.7.3. *Given a measurable space* (Ω, F), *let*

$$\underline{P} = \{P: F \rightarrow \mathbf{R}_+ \mid P \text{ is a probability measure }\}.$$

For $f \in F_{2s}$ *define the* pseudo-distance $d_f: \underline{P} \times \underline{P} \rightarrow \mathbf{R}$ *on the set of probability measures* $\underline{P}$ *on* (Ω, F) *by*

$$d_f(P_1, P_2) = E_Q[f(\frac{r_1}{r_2}) r_2] = E_{P_2}[f(\frac{r_1}{r_2})]$$

where Q is a σ*-finite measure on* (Ω, F) *such that*

$$P_1 < Q \text{ with } \frac{dP_1}{dQ} = r_1,\ P_2 < Q \text{ with } \frac{dP_2}{dQ} = r_2.$$

The pseudo-distance d_f *is also called the* f-information measure, *the* f-entropy *or the* f-divergence.

A σ-finite measure Q as mentioned above always exists, for example $Q = P_1 + P_2$ will do. In case $(\Omega, F) = (\mathbf{R}, B)$ one may sometimes take Q to be Lebesgue measure. Because $r_2 > 0$ a.s. P_2 the above expression is well defined. The above definition has been given in [1].

PROPOSITION 3.7.4. *[1].*

a. *The function d_f defined in 3.7.3. is a pseudo-distance.*

b. *The pseudo-distance d_f does not depend on the choice of the σ-finite measure Q.*

DEFINITION 3.7.5. *The* Kullback-Leibler *pseudo-distance is defined as* $d_{f_1}:\underline{P}\times\underline{P}\to\mathbf{R}$ *with*

$$f_1:\mathbf{R}_+\to\mathbf{R},\ f_1(x)=\begin{cases} x\ln(x), & x>0,\\ 0, & x=0,\end{cases}$$

$$d_{f_1}(P_1,P_2)=E_{P_2}[f_1(\frac{r_1}{r_2})]=E_Q[f_1(\frac{r_1}{r_2})r_2]=E_Q[r_1\ln(\frac{r_1}{r_2})I_{(r_2>0)}].$$

DEFINITION 3.7.6. *The* Hellinger *pseudo-distance is defined as* $d_{f_2}:\underline{P}\times\underline{P}\to\mathbf{R}$ *with*

$$f_2:\mathbf{R}_+\to\mathbf{R},\ f_2(x)=(\sqrt{x}-1)^2,$$

$$d_{f_2}(P_1,P_2)=E_{P_2}[(\sqrt{r_1/r_2}-1)^2]=E_Q[(\sqrt{r_1}-\sqrt{r_2})^2].$$

The Hellinger pseudo-distance is symmetric.

Consider the set of functions on $T=\mathbf{Z}$ with values in $\mathbf{R}^k$. Let $\underline{P}$ be the set of Gaussian measures on this space that make the underlying process a stationary Gaussian process with zero mean value function. An expression for the Kullback-Leibler pseudo-distance on this set was derived in [43].

PROPOSITION 3.7.7. *Let P_1,P_2 be two probability measures on the set of functions defined on $T=\mathbf{Z}$ with values in $\mathbf{R}^k$. Assume that these measures are such that the underlying process is Gaussian, stationary, has zero mean value function, and covariance functions W_1,W_2 respectively. Moreover, assume that these covariance functions admit spectral densities $\hat{W}_1,\hat{W}_2$ respectively and that they satisfy condition C of [43]. Then the Kullback-Leibler pseudo-distance is given by the expression*

$$d_{KL}(P_1,P_2)=\frac{1}{4\pi}\int_{-\pi}^{\pi}[\,\mathrm{tr}(\hat{W}_1^{-1}(\lambda)[\hat{W}_2(\lambda)-\hat{W}_1(\lambda)])-\ln(\hat{W}_1^{-1}(\lambda)\hat{W}_2(\lambda))\,]d\lambda.$$

3.8. The approximate weak Gaussian stochastic realization problem

How to fit to data a model in the form of a Gaussian system? In engineering, in biology and in economics there are many modelling problems for which an answer to this question is useful. As indicated in section 2, from data one may estimate a measure on the set of observation trajectories. In case that one models the observations as a sample function of a Gaussian process, one may estimate its covariance function. Suppose further that one wants to model the observations as the output process of a stationary Gaussian system. Such a system has a finite-dimensional state space. In theorem 3.4.2 it has been shown that a covariance function has a stochastic realization as a Gaussian system only if it has a covariance realization as indicated or if it is rational. Now an arbitrary covariance function obtained from data may not correspond to such a covariance function. Therefore one has to resort to approximation.

The approximate stochastic realization problem is then to determine a stochastic system in a specified class such that the measure on the output process of this system approximates the measure on the same space determined from the data. Attention below will be restricted to the class of stationary Gaussian systems with dimension of the state space less or equal to $n \in \mathbb{Z}_+$. As a measure of fit the Kullback-Leibler pseudo-distance will be taken as mentioned in subsection 3.7. A measure of complexity will not be considered here; it may be based on stochastic complexity as indicated in section 2.

PROBLEM 3.8.1. Approximate weak Gaussian stochastic realization problem. *Let Y^T denote the set of time series defined on $T=\mathbb{Z}$ with values in $\mathbb{R}^p$, and let $P(Y^T)$ denote the set of probability measures on Y^T. Given is a Gaussian measure $P_0 \in P(Y^T)$ such that the underlying process corresponds to a stationary Gaussian process with zero mean function. Given is also an integer $n \in \mathbb{Z}_+$ and let $GS\Sigma(n)$ be the set of Gaussian systems with state space dimension $\leq n$. Solve the optimization problem*

$$\inf_{\sigma \in GS\Sigma(n)} d_{KL}(P_0, P(\sigma))$$

where d_{KL} is the Kullback-Leibler pseudo-distance on the set of probability measures on $P(Y^T)$, and $P(\sigma) \in P(Y^T)$ is the probability measure on Y^T associated with the Gaussian system $\sigma \in GS\Sigma(n)$.

As indicated in 3.7.7, if the pseudo-distance on the set of Gaussian measures is the Kullback-Leibler measure then the pseudo-distance may be expressed as a pseudo-distance on the set of covariance functions

$$d_{KL}(P_0, P(\sigma)) = d_1(W_0, W(\sigma))$$

where W_0 is the covariance function associated with the Gaussian measure P_0 and $W(\sigma)$ the covariance function associated with the Gaussian measure $P(\sigma)$. Note that the covariance function $W(\sigma)$ is a rational function with McMillan degree less or equal to n because it corresponds to a Gaussian system of state space dimension less or equal than n. The approximate weak Gaussian stochastic realization problem may therefore be considered as an approximation problem for a covariance function. In this problem the approximant $W(\sigma)$ has to be a rational function of McMillan degree at most n while the given covariance function W_0 may neither be rational nor of finite McMillan degree.

The approximate stochastic realization problem 3.8.1 is unsolved. Approaches along three different lines have been investigated.

Approach 1. Given any pseudo-distance d_1, problem 3.8.1 can be reformulated as an approximation problem for covariance functions with the criterion

$$d_1(W_0, W(\sigma))$$

where W_0 is the covariance function associated with the Gaussian measure P_0 and $W(\sigma)$ the covariance function associated with the Gaussian measure $P(\sigma)$ related to $\sigma \in GS\Sigma$.

PROBLEM 3.8.2. *Given a covariance function $W_0 : T \rightarrow \mathbb{R}^{p \times p}$ solve*

$$\inf_{\sigma \in GS\Sigma_n} d_1(W_0, W(\sigma)).$$

The pseudo-distance d_1 on the set of covariance functions may be taken to be the Hankel norm or the H-infinity norm. Possibly the L_2-norm is suitable.

The above problem may be rephrased as, given a not necessarily rational covariance function, to determine a rational covariance function that approximates the given covariance with respect to an approximation criterion. Note that a function is a covariance function iff it is anti-symmetric and a positive definite function.

It seems that a Hankel norm approximation of a covariance function is not itself a covariance function. The positive definiteness of a covariance function is therefore an essential constraint. References on this approach are [28, 29, 31, 38, 51, 65].

There is a related approach in which one first determines a spectral factor of the given covariance function and then a rational approximation of the spectral factor. This approach seems too restrictive to start with, although it may be the solution to some approximation criterion.

Of course, given any rational approximation of the covariance function one will still have to determine a state space realization for it.

Approach 2. By analogy with the approximate prediction problem for finite-dimensional Gaussian random variables, algorithms have been proposed for the approximate weak Gaussian stochastic realization problem.

ALGORITHM 3.8.3. LET BE GIVEN A COVARIANCE FUNCTION W_0.

1. Solve an approximate prediction problem. Fix $t \in T$. Let

$$y^+(t) = \begin{pmatrix} y_t \\ y_{t+1} \\ \cdots \\ y_{t+r} \end{pmatrix}, \quad y^-(t) = \begin{pmatrix} y_{t-1} \\ y_{t-2} \\ \cdots \\ y_{t-s} \end{pmatrix}.$$

The variance of the pair $(y^+(t), y^-(t))$ may be computed from the covariance function W_0. Let $n \in \mathbf{Z}_+$. Determine a matrix $S \in \mathbf{R}^{n \times s}$ such that with $x(t) = Sy^-(t)$ the following prediction criterion is minimized

$$inf_{S \in \mathbf{R}^{n\times s}}\ tr(\,E[(y^+(t) - E[y^+(t) \mid F^{x(t)}])(y^+(t) - E[y^+(t) \mid F^{x(t)}])^T]\,).$$

2. Determine a Gaussian system via regression by proceeding as follows,

$$\begin{pmatrix} x(t+1) \\ y(t) \end{pmatrix} = \begin{pmatrix} A \\ C \end{pmatrix} x(t) + v(t), \quad v(t) \in G(0, V),$$

where

$$\begin{pmatrix} A \\ C \end{pmatrix} = E[\begin{pmatrix} x(t+1) \\ y(t) \end{pmatrix} x(t)^T](E[x(t)x(t)^T])^{-1},$$

$$v(t) = \begin{pmatrix} x(t+1) \\ y(t) \end{pmatrix} - \begin{pmatrix} A \\ C \end{pmatrix} x(t).$$

Finally, replace the Gaussian process v with a Gaussian white noise process w with variance V.

The above algorithm in a somewhat different form appeared first in a paper of H. Akaike [3]. Other references are [11, 12, 44-46, 75, 76]. These papers differ mainly in the way they perform step 1 of the above algorithm. For canonical correlation analysis and the prediction problem see

[27, 57].

It is not clear in what sense the Gaussian system determined in step 2 of the above algorithm is a good approximation to the given Gaussian process. In other words, the approximation criterion, although inspired by the static approximate prediction problem, is never mentioned. The replacement of the process v by a Gaussian white noise process is also unmotivated.

Approach 3. Canonical correlation analysis for finite-dimensional Gaussian random variables has been generalized to infinite-dimensional Hilbert spaces in [36, 37, 49]. One has investigated approximate prediction problems for time series by canonical correlation analysis techniques. Approximation bounds have been derived [30]. It remains to be seen whether this approach is useful in practice.

Approach 4. Inspired by the above mentioned second approach to the approximate weak Gaussian stochastic realization problem yet another approach has been formulated. This approach has been worked out by M. Stöhr at the Centre for Mathematics and Computer Science. The following results up to the end of section 3 are due to M. Stöhr and are as of yet unpublished.

NOTATION 3.8.4. *Let $k_1, k_2, n \in \mathbf{Z}_+$, $k = k_1 + k_2$. Recall that $G(0,Q)$ denotes a Gaussian measure, say on $\mathbf{R}^k$, with zero mean and variance Q. For $Q \in \mathbf{R}^{k \times k}$ the decomposition*

$$Q = \begin{bmatrix} Q_{11} & Q_{12} \\ Q_{12}^T & Q_{22} \end{bmatrix}$$

will be used in which $Q_{11} \in \mathbf{R}^{k_1 \times k_1}$, $Q_{22} \in \mathbf{R}^{k_2 \times k_2}$, and $Q_{12} \in \mathbf{R}^{k_1 \times k_2}$. Let

$$\mathbf{Q}(n) = \{Q \in \mathbf{R}^{k \times k} \mid Q = Q^T \geqslant 0,\ rank(Q_{12}) \leqslant n\}.$$

PROBLEM 3.8.5. The static approximate weak Gaussian stochastic realization problem. *Given are $k_1, k_2, n \in \mathbf{Z}_+$, $k = k_1 + k_2$, and a Gaussian measure $G(0,Q_0)$ with $Q_0 = Q_0^T > 0$. Let d_{KL} be the Kullback-Leibler pseudo-distance on the set of Gaussian measures on $\mathbf{R}^k$. Solve*

$$inf_{G(0,Q_1),\, Q_1 \in \mathbf{Q}(n)}\ d_{KL}(G(0,Q_0), G(0,Q_1)).$$

One may interpret the above problem in the light of approach 2 indicated above. Associate the space $\mathbf{R}^{k_1}$ with the past of the observations, and the space $\mathbf{R}^{k_2}$ with the future of the observations. The Gaussian measure $G(0,Q_0)$ may then be associated with that derived from the data. In problem 3.8.5 one is asked to determine the measure $G(0,Q_1)$ with $Q_1 \in \mathbf{Q}(n)$. The latter condition implies that the dimension of the state space associated with $G(0,Q_1)$ is less or equal to n. Therefore the essential constraint on the dimension of the state space is taken care of.

PROPOSITION 3.8.6. *Consider problem 3.8.5. The Kullback-Leibler measure of two Gaussian measures $G(0,Q_0)$ and $G(0,Q_1)$ on $\mathbf{R}^k$ is given by the expression*

$$\begin{aligned} d_{KL}(G(0,Q_0), G(0,Q_1)) &= \tfrac{1}{2}[\ \mathrm{tr}(Q_1^{-1}Q_0) - \ln(\det(Q_1^{-1}Q_0)) - k\ ] \\ &= \tfrac{1}{2}[\ \sum_{i=1}^{k} (\lambda_i(Q_0,Q_1) - \ln(\lambda_i(Q_0,Q_1))) - k\ ], \end{aligned}$$

where $\{\lambda_i(Q_0,Q_1), i\in\mathbf{Z}_k\}$ *are the generalized eigenvalues of* Q_0 *with respect to* Q_1*, here defined as the zeroes of* $\det(Q_1\lambda - Q_0)=0$.

It can be shown that the generalized eigenvalues are real and satisfy $\lambda_i(Q_0,Q_1)\geqslant 0$, for $i\in\mathbf{Z}_k$.

NOTATION 3.8.7. *For* $Q_0\in\mathbf{R}^{k\times k}$, $Q_0=Q_0^T>0$, $n\in\mathbf{Z}_+$ *let*

$$\Lambda(Q_0,n) = \left\{\begin{array}{c}\lambda\in\mathbf{R}^k_+ \mid \exists\, Q\in\mathbf{Q}(n) \text{ such that generalized eigenvalues} \\ \text{of } Q_0 \text{ with respect to } Q \text{ are } \{\lambda_1,\ldots,\lambda_k\}\end{array}\right\}.$$

and for $\lambda\in\mathbf{R}^k_+$ *let*

$$\mathbf{Q}_s(Q_0,n,\lambda) = \left\{\begin{array}{c} Q\in\mathbf{Q}(n)\mid \text{generalized eigenvalues} \\ \text{of } Q_0 \text{ with respect to } Q \text{ are } \{\lambda_1,\ldots,\lambda_k\}\end{array}\right\}.$$

$$f:\mathbf{R}^k_+\to\mathbf{R}_+,\; f(\lambda) = \tfrac{1}{2}\Big[\sum_{i=1}^{k}(\lambda_i - \ln(\lambda_i)) - k\Big].$$

It may be shown that the function f is convex. There are results on the structure of the matrices in the set $\mathbf{Q}_s(Q_0,n,\lambda)$.

PROBLEM 3.8.8. *Consider problem 3.8.5 and the notation 3.8.7. Solve*

$$\inf_{\lambda\in\Lambda(Q_0,n)} f(\lambda).$$

Suppose that there exists a $\lambda^*\in\Lambda(Q_0,n)$ such that

$$f(\lambda^*) = \inf_{\lambda\in\Lambda(Q_0,n)} f(\lambda).$$

The solution set of problem 3.8.5 is then given by $\mathbf{Q}_s(Q_0,n,\lambda^*)$. Note that problem 3.8.8 is the infimization of a convex function over the set $\Lambda(Q_0,n)$. The latter set is a cone. It is conjectured that it is a polyhedral cone. It may be shown that the optimal solution of problem 3.8.8 is such that $\sum_{i=1}^{k}\lambda_i = k$. This property simplifies the function f. If this constraint is taken into account then the set $\Lambda(Q_0,n)$ is reduced to a shifted simplex. It is not yet known whether problem 3.8.8 admits an explicit expression as solution or whether one has to resort to numerical minimization.

The hope is that the solution of problem 3.8.5 provides information on the solution of the approximate weak Gaussian stochastic realization problem 3.8.1.

4. SPECIFIC OPEN STOCHASTIC REALIZATION PROBLEMS

The purpose of this section is to present several stochastic systems and processes for which the solution to the stochastic realization problem may be useful for engineering, economics etc. The presentation of these models is brief. The tutorial and survey-like character of this paper may make it useful to mention these models.

Gaussian systems

The approximate weak Gaussian stochastic realization problem, as described in subsection 3.8, is unsolved. For Gaussian systems there are unsolved problems for specific subclasses of systems that may be of interest to specific application areas. Some of these problems and models are described below.

The co-integration and the error correction model. As a model for economic processes that move about an equilibrium, C.W.J. Granger [32] has proposed a model that is known as the *co-integration model.*

The components of a vector valued process $y:\Omega \times \mathbf{Z} \to \mathbf{R}^k$ are said to be *co-integrated of order* 1,1 if

1. after differencing once ($\nabla y(t) = y(t) - y(t-1)$) the resulting process has a stationary invertible AutoRegressive-Moving-Average (ARMA) representation without deterministic component;
2. there exists a vector $\alpha \in \mathbf{R}^k$, $\alpha \neq 0$, such that $z(t) = \alpha^T y(t)$ has again a stationary invertible ARMA representation without deterministic component.

The interpretation of this model is that the economic process that is modelled consists of a trend and stationary fluctations, but is such that a linear combination of the process is stationary. The linear combination should be associated with some difference of economic processes, say income minus consumption. According to the model this difference fluctuates around some equilibrium value and it may be considered as forced towards this equilibrium by economic forces. A generalization of this model has been proposed, see [22]. That paper also reports on the suitability of the co-integration model for economic processes.

A vector valued proces $y:\Omega \times T \to \mathbf{R}^k$ is said to have an *error correction representation,* see [22], if it can be expressed as:

$$A(B)(1-B)y(t) = -\gamma z(t-1) + u(t)$$

in which u is a stationary process representing a disturbance, $A(.)$ is a matrix polynomial with $A(0) = I$, B is the delay operator defined by $By(t) = y(t-1)$, there exists a $\alpha \in \mathbf{R}^k$ such that $z(t) = \alpha^T y(t)$ and $\gamma \in \mathbf{R}^k$, $\gamma \neq 0$.

The interpretation of an error correction model is that the disequilibrium of one period, $z(t-1)$, is used to determine the economic process in the next period.

For recent work on the co-integration and error correction model see a special issue of *Journal of Economic Dynamics and Control* that is opened by the special editor M. Aoki with the paper [8]. In that issue there is another paper by M. Aoki [9] in which he shows that the co-integration model may be obtained from a Gaussian system representation under a condition on the poles of the system. In that approach a co-integration vector is not assumed, nor are assumptions needed on trends or periods.

An approach to the stochastic realization problem for the co-integration model and the error correction model may be based on stochastic realization theory for a particular class of Gaussian systems.

Gaussian systems with inputs. A time-invariant Gaussian system with inputs has a forward representation of the form

$$x(t+1) = Ax(t) + Bu(t) + Mv(t),$$

$$y(t) = Cx(t) + Du(t) + Nv(t),$$

where $u : \Omega \times T \to \mathbb{R}^m$ is an input process, and $v : \Omega \times T \to \mathbb{R}^k$ is a Gaussian white noise process. Such systems are used in stochastic control. The stochastic realization problem for this class of systems has not yet been treated. It is motivated by stochastic control theory. An unsolved question is whether such a stochastic system is a minimal realization of the measure on the observation processes of output y and input u. The conditions for minimality should be related to the solvability conditions of the linear-quadratic-Gaussian stochastic control problem.

For this class of systems one has also to investigate the stochastic realization problem associated with the solution to the linear-exponential-quadratic-Gaussian stochastic control problem [14, 77]. This solution is related to recent results in H-infinity theory.

The Gaussian factor model

This model and the associated stochastic realization problem are discussed in section 5 of this paper.

Factor systems

These systems and the associated stochastic realization problem are discussed in section 6.

Positive stochastic linear systems

A stochastic system in which the state and observations process take values in the vector space $\mathbb{R}_+$ will be called *a positive stochastic system.* The gamma distribution is an example of a probability distribution on $\mathbb{R}_+$. Such systems may be appropriate stochastic models in economics, biology, and communication systems where the state variables are economic quantities, concentrations etc. Examples from biology may be found in [56]. Several examples of such systems follow.

Portfolio models. A portfolio model is a dynamic model for the growth of assets such as shares, bonds and money in savings accounts. After the fall of share prices in October 1987 there is a renewed interest in portfolio models.

A stochastic portfolio model may be specified by

$$dp(t) = ap(t)dt + p(t)dv(t), p(0),$$

where $p : \Omega \times T \to \mathbb{R}$ represents the price of the asset, $a \in \mathbb{R}$ represents a growth trend and $v : \Omega \times T \to \mathbb{R}$ represents random fluctuations. More refined models can be defined to account for control of buying and selling, and for switch-over costs. A realistic portfolio model would require a realistic macro economic model for short-term and long-term economic growth, preferably on an international scale.

The portfolio model should be seen as a special case of a growth model. In addition, growth models that exhibit saturation should be investigated in connection with market saturation effects.

The realization problem for the stochastic portfolio model would have to deal with questions as whether the trends and variances of these models can be determined from observed prices. This problem becomes more interesting if, for example, the price of a share is related to development of the markets in which the company is active, to its management structure, and to long-term growth of the economy.

The Gale model and a Leontieff system. For production planning of firms a model proposed by D. Gale is used. For references on this model see the book by V. I. Arkin and I. V. Evstigneev [10]. The classical Leontieff model is a matrix relation between inputs and outputs of an economic unit. A dynamic version of this model has been proposed, it will be called a *Leontieff system.*

The Gale model is specified by

$$z(t) = \begin{bmatrix} x(t-1) \\ y(t) \end{bmatrix}, \quad x,y : T \to \mathbf{R}^n_+ \tag{4.1}$$

satisfying

$$z(t) \in Q(t), \tag{4.2}$$

$$y(t) \geq x(t), \tag{4.3}$$

where $Q(t) \in \mathbf{R}^{2n}_+$ is a convex set. Here $x(t-1)$ is called the *input*, and $y(t)$ the *output* in period $(t-1, t]$, and $z(t)$ the *technological process* at time $t \in T$. Condition (4.2) is a technological feasibility condition; condition (4.3) implies that the input at any time step cannot exceed the output of the previous step. A parametric form of this model is given in subsection 1.1.8 of [10].

There is also a stochastic version of the Gale model, see the subsections 2.4.1 and 2.4.7 of [10].

Optimal control problems for the Gale model are treated in [10]. The results are maximum principles and turnpike theorems.

Finite stochastic systems

In section 3 a finite stochastic system has been defined. It consists of an output process taking values in a finite set and a finite-state Markov process. The stochastic realization problem for this class of systems is then to classify all minimal stochastic systems such that the output process of such a system equals a given process either in distribution or almost surely. The motivation of this problem comes from the use of finite stochastic systems as models for communication or computers systems. For such technical problems, stochastic models with discrete variables arise naturally or are useful approximate models. The stochastic realization problem was formulated in 1957 in a paper by Blackwell and Koopmans [15]. During the 1960's several publications appeared that provide a necessary and sufficient condition for the existence of a finite stochastic realization. For references see [52]. Unsolved questions are the characterization of minimality of the state space and the classification of all minimal stochastic realizations. The main bottleneck is currently the characterization of the minimality of the state space. This question leads to a basic problem for positive linear algebra, that is, linear algebra over $\mathbf{R}_+$.

Counting process systems

An example of a counting process system is a continuous-time stochastic system of which the output process is a counting process with stationary increments and in which the intensity process of the counting process is a finite-state Markov process. The stochastic realization problem for this class of systems is unsolved.

The motivation for this stochastic realization problem comes from the use of counting process models in communication, queueing theory, computer science, and biology. The observation process may often be taken as a counting process with stationary increments.

The above mentioned class of stochastic systems has been investigated in [68, 69]. The question of characterizing the minimal size of the state space is closely related to the same question for the finite stochastic realization problem.

Gaussian random fields

For this class of stochastic objects new mathematical models are needed.

5. Factor analysis

In this section the stochastic realization problem for the Gaussian factor analysis model will be formulated and analyzed.

The factor analysis model was proposed early this century. For references on the factor analysis model see [7, 74]. Factor analysis is used as a quantitative model in sociology and psychology. R. Frisch has suggested the factor analysis model as a way to determine relations among random variables [25]. R. E. Kalman has emphasized this model and formulated the associated stochastic realization problem [39-41]. Since then several researchers have considered the stochastic realization problem for this model class. This problem is still unsolved. Below one finds a problem formulation, questions, partial results and conjectures for this stochastic realization problem. For recent publications on this problem see the special issue of *J. of Econometrics* that is opened by the paper [2].

Problem formulation

From economic data that exhibit variability one may estimate a covariance. Suppose that this data vector may be modelled by a Gaussian random variable. Effectively one is thus given a Gaussian measure, say on $\mathbb{R}^k$. The initial problem may then be stated as: how to represent this measure such that the dependencies between the components of the vector are exhibited? The factor analysis model will be used to describe these dependencies.

Definition 5.1. *A* Gaussian factor analysis model *or a* Gaussian factor model *is defined by the specification*

$$y = Hx + w, \tag{5.1}$$

or

$$y_i = H_i x + w_i, \quad i = 1, \ldots, k, \tag{5.2}$$

where $x:\Omega \to \mathbb{R}^n$, $x \in G(0, Q_x)$ *is called the* factor, $w:\Omega \to \mathbb{R}^k$, $w \in G(0, Q_w)$ *is called the* noise, $y:\Omega \to \mathbb{R}^k$, $y \in G(0, Q_y)$ *is called the* observation vector, $H \in \mathbb{R}^{k \times n}$ *is called the* matrix of factor loadings, Q_w *is a diagonal matrix, and* (x, w) *are independent random variables.*

The interpretation of the Gaussian factor analysis model (5.2) is that each component of the observation vector consists of a systematic part $H_i x$ and a noise part w_i. Observe that the condition that Q_w is diagonal is equivalent to the condition that $(w_1, \ldots, w_k)$ are independent random variables. A generalization of the above definition may be given to the case in which Q_w is block diagonal. The Gaussian factor model in rudimentary form goes back to [67]. The Gaussian factor analysis model is equivalent to the *confluence analysis model* introduced by R. Frisch [25]. In this model the representation of the observation vector is specified by

$$y = u + w, \quad Au = 0,$$

in which $A \in \mathbb{R}^{(k-n) \times k}$ u, w are independent random variables, and Q_w is a diagonal matrix. For other references on this approach see the publications of O. Reiersøl [58, 59].

The Gaussian factor analysis model, or, equivalently, the confluence analysis model, has been suggested as an alternative to regression analysis. Strong pleas for this approach are the introduction of the book by R. Frisch [25], and the papers of R. E. Kalman [39-41]. Within economic and statistical literature the questions regarding regression and factor models have been recognized, see for example [7, 66, 70, 80].

PROBLEM 5.2. *The* weak stochastic realization problem for a Gaussian factor model *is given a Gaussian measure* $G(0,Q)$ *on* $\mathbb{R}^k$ *to solve the following subproblems.*

a. *Determine a Gaussian factor model, say*

$$y = Hx + w,$$

such that the measure of y equals the given measure or

$$y \in G(0,Q_y) = G(0,Q).$$

If such a Gaussian factor model exists then it is called a weak stochastic realization *of the given measure.*

b. *Determine the minimal dimension* $n^*(Q)$ *of the factor x in a weak stochastic realization of the given measure* $G(0,Q)$. *Call a weak stochastic realization* minimal *if the dimension of the factor systems equals* $n^*(Q)$.

c. *Classify all minimal weak stochastic realizations of the given measure.*

Part a. of problem 5.2 is equivalent to: determine $(n, Q_x, Q_w, H) \in \mathbb{N} \times \mathbb{R}^{n \times n} \times \mathbb{R}^{k \times k} \times \mathbb{R}^{k \times n}$ such that

$$Q = HQ_xH^T + Q_w,$$

where $Q_x = Q_x^T \geqslant 0$, $Q_w = Q_w^T \geqslant 0$, and Q_w is diagonal. Part a of the above problem is trivial, the hard parts of the problem are b and c.

Corresponding to problem 5.2 there is a *strong stochastic relization problem for a Gaussian factor model.* In this problem one is given a probability space (Ω, F, P) and a Gaussian distributed random variable $z \in G(0,Q)$. The problem is then to construct a Gaussian factor model

$$y = Hx + w$$

on the given probability space such that

$$z = y \ \ a.\,s.$$

and to classify all minimal models of this type. This problem has been defined in [54], where a generalization of the Gaussian factor model for Hilbert spaces is introduced. The strong stochastic realization problem will not be discussed in detail here.

What is the main characteristic of the Gaussian factor model? To answer this question one has to introduce the following concept.

DEFINITION 5.3. *The σ-algebra's* $F_1, F_2, \ldots, F_m$ *are called* conditionally independent *given the σ-algebra G if*

$$E[z_1 \cdots z_m \mid G] = E[z_1 \mid G] \cdots E[z_m \mid G]$$

for all $z_i \in L^+(F_i)$. *The notation*

$$(F_1, F_2, .., F_m \mid G) \in CI$$

will be used to denote that $F_1, \ldots, F_m$ *are conditionally independent given G and CI will be called the* multivariate conditional independence relation.

The following elementary result then establishes the relation between the Gaussian factor model and the conditional independence relation.

PROPOSITION 5.4. *Let* $y_i : \Omega \to \mathbf{R}$, $i = 1, 2, ..., k$, $x : \Omega \to \mathbf{R}^n$. *The following statements are equivalent:*

a. *The random variables* $(y_1, .., y_k, x)$ *are jointly Gaussian with zero mean and satisfy*

$$(F^{y_1}, ..., F^{y_k} \mid F^x) \in CI.$$

b. *The random variables* y, x *satisfy the conditions of the Gaussian factor analysis model of 5.1 with the representation*

$$y = Hx + w.$$

The conditional independence property of a Gaussian factor model is now seen to be its main characteristic. It will be called the *factor property* of a Gaussian factor model. It allows extensions to non-Gaussian random variables. Such extensions have been considered in the literature, see for references [74]. The factor property is a generalization of the concept of state for a stochastic system. In such a system the future of the state and output process on one hand, and the past of the state and output process on the other hand are conditionally independent given the present state. The analogy is such that the state corresponds to the factor and the output process to the observation vector of the factor model. The factor property or the conditional independence property occurs in many mathematical models in widely different application areas.

Below the stochastic realization problem 5.2 will be discussed, first in terms of the external description and then in terms of the internal description.

The stochastic realization problem in terms of the external description.
In this subsection one is assumed to be given a Gaussian measure $G(0, Q_y)$. The weak stochastic realization problem for a Gaussian factor model specializes in this case to the following question.

QUESTION 5.5. *Given a Gaussian measure* $G(0, Q_y)$.

a. *What is the minimal dimension* $n^*(Q_y)$ *of the factor in a stochastic realization of* $G(0, Q)$?

b. *What is the classification of all minimal stochastic realizations of* $G(0, Q)$, *or all decompositions of the form*

$$Q_y = Q_1 + Q_w$$

in which $Q_1 = Q_1^T \geq 0$, $Q_w = Q_w^T \geq 0$ *is diagonal and* $\text{rank}(Q_1) = n^*(Q_y)$.

NOTATION 5.6.

a. *If* $Q \in \mathbf{R}^{k \times k}$ *then*

$$D(Q) \in \mathbf{R}^{k \times k}$$

is a diagonal matrix with on the diagonal the elements of the diagonal of the matrix Q.

b. *If* $Q \in \mathbb{R}^{k \times k}$ *then the matrix* $OD(Q) \in \mathbb{R}^{k \times k}$, *called the* off-diagonal part of Q, *is defined by*

$$OD(Q)_{ii} = 0,\ OD(Q)_{i,j} = Q_{i,j},\ \text{for all } i,j \in \mathbb{Z}_k,\ i \neq j.$$

c.

$$\mathbf{Q}(Q_y,k,n) = \left\{ \begin{array}{l} (Q_1,Q_w) \in \mathbb{R}^{k \times k} \times \mathbb{R}^{k \times k} \mid Q_1 = Q_1^T \geqslant 0,\ \operatorname{rank}(Q_1) = n, \\ Q_w = Q_w^T \geqslant 0,\ Q_w \text{ diagonal},\ Q_y = Q_1 + Q_w \end{array} \right\}$$

d.

$$n^*(Q_y) = \min\{n \in \mathbb{N} \mid \exists\ (Q_1,Q_w) \in \mathbf{Q}(Q_y,k,n)\}$$

It turns out to be useful to work with a standard form for the variance matrix, a canonical form.

DEFINITION 5.7. *One says that the matrices* $Q_1,Q_2 \in \mathbb{R}^{k \times k}$, *that are assumed to be strictly positive definite, are* equivalent *if there exists a diagonal matrix* $D \in (0,\infty)^{k \times k}$ *such that*

$$Q_1 = DQ_2D.$$

A canonical form with respect to this equivalence relation is then such that $D(Q) = I$. An investigation should be made of another equivalence relation defined as in 5.7 in which negative elements are also admitted on the diagonal.

Question 5.5.a is still unsolved. Characterizations of $n^*(Q_y)$ are known in the two extreme cases of $n^*(Q_y) = 1$ and $n^*(Q_y) = k-1$. These results are stated below. The characterization for $n^*(Q_y) = 1$ may go back to C. Spearman and co-workers. The formulation given here is from [13].

THEOREM 5.8. [13]. *Given* $Q_y \in \mathbb{R}^{k \times k}$, $Q_y = Q_y^T > 0$. *Assume that* $k \geqslant 4$, $Q_y \in (0,\infty)^{k \times k}$, *and that* Q_y *is irreducible. Then* $n^*(Q_y) = 1$ *iff*

$$\left\{ \begin{array}{l} q_{il}q_{jm} - q_{im}q_{jl} = 0,\ q_{il}q_{ji} - q_{ii}q_{jl} \leqslant 0, \\ \forall\ i,j,l,m \in \mathbb{Z}_k,\ l \neq m, j \neq l, j \neq m, i \neq j, i \neq l, i \neq m. \end{array} \right.$$

THEOREM 5.9. [13, 39, 58]. *Given* $Q_y \in \mathbb{R}^{k \times k}$, $Q_y = Q_y^T > 0$. *Then* $n^*(Q_y) = k-1$ *iff* Q_y^{-1} *has strictly positive elements, possibly after sign changes of rows and corresponding columns.*

What are the generic values of $n^*(Q_y)$? Below are stated the main results from a study by J. P. Dufour [20] on this question.

DEFINITION 5.10. *Let*

$$\mathbb{S}_k^+ = \{Q \in \mathbb{R}_+^{k \times k} \mid Q = Q^T\}.$$

Note that the condition of positive definiteness is not imposed in the definition of the set $\mathbb{S}_k^+$. In the following the Euclidean topology is used on the vector space $\mathbb{R}^n$.

THEOREM 5.11. [20].

a. *There exists an open and dense subset* $\mathbb{S} \subset \mathbb{S}_k^+$ *such that for all* $Q_y \in \mathbb{S}$

$$n^*(Q_y) \geq \tfrac{1}{2}(2k+1-\sqrt{1+8k}).$$

This inequality is known as the Ledermann bound.

b. *Let* $Q \in \mathbb{S}$. *For every* Q_1 *in a sufficiently small neighberhood of* Q *in* $\mathbb{S}$ *the relation*

$$n^*(Q) = n^*(Q_1)$$

holds.

c. *For any integer* p *such that*

$$\tfrac{1}{2}(2k+1-\sqrt{1+8k}) \leq p \leq k-1$$

there exists a $Q \in \mathbb{S}$ *such that* $n^*(Q)=p$.

By way of illustration there follow characterizations on the value of $n^*(Q_y)$ for variance matrices $Q_y \in \mathbb{R}^{k \times k}$ with several low values of k.

PROPOSITION 5.12. *Let* $Q_y \in \mathbb{R}^{3\times3}$, $Q_y = Q_y^T > 0$, $D(Q_y) = I$.

a. $n^*(Q_y)=0$ *iff* Q_y *is diagonal.*

b. $n^*(Q_y)=1$ *iff one of the following cases applies.*
Case 1. If $q_{12}>0$, $q_{13}>0$, $q_{23}>0$ *and*

$$\frac{q_{12}q_{13}}{q_{23}}, \frac{q_{12}q_{23}}{q_{13}}, \frac{q_{13}q_{23}}{q_{12}} \in [0,1].$$

Case 2. If $q_{12}>0$, $q_{13}=0$, $q_{23}=0$.
Other cases are derived from the above by permutations of signs and indices.

c. $n^*(Q_y)=2$ *iff otherwise.*

For the special case in which $Q_y \in \mathbb{C}^{4\times4}$ and $n^*(Q_y)=1$ a characterization is given in [6].

PROPOSITION 5.13. *Let* $Q_y \in (0,\infty)^{4\times4}$. *Then* $n^*(Q_y)=1$ *iff, up to a permutation of indices,*

1. $c = \dfrac{q_{12}q_{13}}{q_{23}} = \dfrac{q_{12}q_{14}}{q_{24}} = \dfrac{q_{13}q_{14}}{q_{34}} \in (0,1]$;

2. $c \geq q_{12}^2$, $c \geq q_{13}^2$, $c \geq q_{14}^2$.

Classification. In this subsubsection the classification question 5.5.b will be discussed. Thus, given $Q_y \in \mathbb{R}^{k\times k}$, the question is to classify all decompositions of the form

$$Q_y = Q_1 + Q_w$$

in which rank $(Q_1) = n^*(Q_y)$. Geometry seems the appropriate tool for this classification, in particular polyhedral cones and convex analysis. For an approach along these lines see [19]. Below another approach is indicated that combines analysis and geometry.

Remark that in the decomposition

$$Q_y = Q_1 + Q_w = HQ_xH^T + Q_w$$

the off-diagonal elements of Q_1 are equal to the off-diagonal elements of Q_y. Moreover, by convention $D(Q_y)=I$. Hence the set $\mathbf{Q}(Q_y,k,n^*(Q_y))$ may be classified by the diagonal of Q_1.

PROPOSITION 5.14. *Let*

$$\mathbf{D}(Q_y,k,n) = \left\{ D \in \mathbb{R}^{k \times k} \mid D \text{ diagonal, } -OD(Q_y) \leqslant D \leqslant I, \text{ rank } (D + OD(Q_y)) = n \right\},$$

$$f:\mathbf{D}(Q_y,k,n) \rightarrow \mathbf{Q}(Q_y,k,n), \quad f(D) = (D + OD(Q_y), I - D).$$

Then f is a bijection.

Remark that the set $\mathbf{D}(Q_y,k,n)$ without the rank condition is a closed convex set. From 5.14 and some linear algebra one obtains the following result on the classification.

THEOREM 5.15. *Let $Q_y \in \mathbb{R}^{k \times k}$, $Q_y = Q_y^T > 0$, $D(Q_y)=I$,*

$$\mathbf{D}_1(Q_y,k,n) = \left\{ \begin{array}{l} D_1 \in \mathbb{R}^{n \times n} \mid D_1 \text{ diagonal, } 0 < D_1 \leqslant I, \\[4pt] \exists \text{ permutation matrix } P \text{ such that if } PQ_yP^T = \begin{bmatrix} A & B \\ B^T & C \end{bmatrix}, \\[10pt] \text{then } D_1 + OD(A) > 0, \ D_2 := B^T[D_1 + OD(A)]^{-1}B - OD(C) \\ \text{is diagonal and satisfies } 0 \leqslant D_2 \leqslant I \end{array} \right\},$$

$$g:\mathbf{D}_1(Q_y,k,n) \rightarrow \mathbf{Q}(Q_y,k,n),$$

$$g(D_1) = (P^T \begin{bmatrix} D_1 + OD(A) & B \\ B^T & D_2 + OD(C) \end{bmatrix} P, \ P^T \begin{bmatrix} I - D_1 & 0 \\ 0 & I - D_2 \end{bmatrix} P).$$

Then:

a. *g is well defined;*
b. *g is surjective;*
c. *The diagonal matrix*

$$\begin{bmatrix} D_1 & 0 \\ 0 & D_2 \end{bmatrix}$$

  is unique up to a permutation.

The proof of the above theorem is elementary with the aid of the following lemma.

LEMMA 5.16. *Let $k,n \in \mathbb{Z}_+$, $k > n$, $A \in \mathbb{R}^{n \times n}$, $B \in \mathbb{R}^{n \times (k-n)}$, $C \in \mathbb{R}^{(k-n) \times (k-n)}$, $A = A^T$, $C = C^T$, rank $= n$,*

$$Q = \begin{bmatrix} A & B \\ B^T & C \end{bmatrix} \in \mathbb{R}^{k \times k}, \quad T = \begin{bmatrix} A^{-1/2} & -A^{-1}B \\ 0 & I \end{bmatrix} \in \mathbb{R}^{k \times k}.$$

a. *Then*

$$T^T Q T = \begin{bmatrix} I & 0 \\ 0 & C - B^T A^{-1} B \end{bmatrix}.$$

b. rank $(T)=k$.
c. rank $(Q)=n$ *iff* $C-B^TA^{-1}B=0$.
d. $Q\geqslant 0$ *iff* $C-B^TA^{-1}B\geqslant 0$.

The study of the classification along the lines sketched above must proceed by an investigation of the following relations for the diagonal matrix $D_1\in\mathbb{R}^{n\times n}$:

$$D_1+OD(A)>0,$$

$$D_2:=B^T[D_1+OD(A)]^{-1}B-OD(C),\ \ 0\leqslant D_2\leqslant I,\ \ D_2 \text{ is diagonal.}$$

For the cases $n^*(Q_y)=k-1$ and $n^*(Q_y)=1$ theorem 5.15 directly yields explicit classifications. The classifications of three low-dimensional examples are listed.

PROPOSITION 5.17. *For the case* $k=2$, $Q_y\in\mathbb{R}^{2\times 2}$, $n^*(Q_y)=1$ *with*

$$Q_y=\begin{bmatrix}1 & q\\ q & 1\end{bmatrix},\quad q\neq 0,$$

the classification, in the notation of 5.15, is given by

$$\mathbb{D}(Q_y,2,1)=\left\{d_1\in\mathbb{R}_+ \mid q^2\leqslant d_1\leqslant 1\right\}$$

and

$$g(d_1)=\left(\begin{bmatrix}d_1 & q\\ q & q^2/d_1\end{bmatrix},\begin{bmatrix}1-d_1 & 0\\ 0 & 1-q^2/d_1\end{bmatrix}\right).$$

PROPOSITION 5.18. *For the case* $k=3$, $Q_y\in(0,\infty)^{3\times 3}$, *and* $n^*(Q_y)=2$ *the classification according to 5.15 is given by*

$$\mathbb{D}_1(Q_y,3,2)=\left\{\begin{array}{c}\begin{bmatrix}d_1 & o\\ 0 & d_2\end{bmatrix}\in\mathbb{R}^{2\times 2}\ \mid\ d_1,d_2\in[0,1],\ d_1d_2-q_{12}^2\neq 0,\\ \dfrac{d_1q_{23}^2+d_2q_{13}^2-2q_{12}q_{13}q_{23}}{d_1d_2-q_{12}^2}\in[0,1]\\ \text{and conditions obtained by permutation of indices}\end{array}\right\}.$$

PROPOSITION 5.19. *For the case* $k=3$, $Q_y\in(0,\infty)^{3\times 3}$, $n^*(Q_y)=1$, *the decomposition is unique with*

$$Q_1=\begin{bmatrix}\dfrac{q_{12}q_{13}}{q_{23}} & q_{12} & q_{13}\\ q_{12} & \dfrac{q_{12}q_{23}}{q_{13}} & q_{23}\\ q_{13} & q_{23} & \dfrac{q_{13}q_{23}}{q_{12}}\end{bmatrix}.$$

PROPOSITION 5.20. *For the case* $k=4$, $Q_y\in(0,\infty)^{4\times4}$, $n^*(Q_y)=2$ *the classification according to 5.15 is given by*

$$\mathbf{D}_1(Q_y,2)=\left\{\begin{array}{c}\begin{bmatrix} d_1 & 0\\ 0 & d_2\end{bmatrix} \mid d_1,d_2\in[0,1],\ d_1d_2-q_{12}^2\neq0,\\ q_{34}=d_2q_{13}q_{14}+d_1q_{23}q_{24}-q_{12}q_{14}q_{23}-q_{12}q_{13}q_{24},\\ d_2q_{13}^2+d_1q_{23}^2-2q_{12}q_{13}q_{23}\in[0,1],\\ d_1q_{24}^2+d_2q_{14}^2-2q_{12}q_{14}q_{24}\in[0,1],\\ \text{and conditions obtained by permuting the indices}\end{array}\right\}.$$

The stochastic realization problem in terms of the internal description

The specification of the Gaussian factor model as given in 5.1 will be called the *internal description*. It is called internal because the specification is in terms of the matrices (H,Q_x,Q_w) rather than in terms of Q_y. The questions for the internal desciption require one definition.

DEFINITION 5.21. *The Gaussian factor model with representation*

$$y = Hx + w$$

is called minimal *if* $n=n^*(Q_y)$ *in which* $x:\Omega\to\mathbb{R}^n$, $Q_x>0$ *and*

$$Q_y = HQ_xH^T + Q_w.$$

Introduce the convention $Q_x=I$. The weak stochastic realization problem for a Gaussian factor model specializes in this case to the following question.

QUESTION 5.22.

a. *Which conditions on the matrices* (H,Q_x,Q_w) *are equivalent with minimality of the Gaussian factor model?*

b. *How are two minimal Gaussian factor models related?*

The above questions are still open. The minimality question 5.22.a seems most interesting because its answer will involve a new system theoretic concept like stochastic observability. To hint at what may be needed a special case is considered.

Consider a special Gaussian factor model of the form

$$\begin{bmatrix} y_1\\ y_2\end{bmatrix} = \begin{bmatrix} H_1\\ H_2\end{bmatrix}x + w$$

in which the variance Q_w is required to be block-diagonal, in particular it consists of two blocks only

$$Q_w = \begin{bmatrix} Q_{w_1} & 0\\ 0 & Q_{w_2}\end{bmatrix}.$$

One says that this Gaussian factor model is *stochastically observable* if the map

$$x \mapsto E[\exp(iu^T y_1) \mid F^x]$$

is injective on the support of x. Similarly one says that the Gaussian factor model is *stochastically reconstructible* if the map

$$x \mapsto E[\exp(iu^T y_2) \mid F^x]$$

is injective on the support of x. It may then be proven that the Gaussian factor model is minimal iff it is stochastically observable and stochastically reconstructible iff rank $(H_1)=n=$ rank (H_2) [71].

Let's return to question 5.22.a, when is a Gaussian factor model minimal in case Q_w is restricted to be diagonal. The following conjecture comes to mind first: A Gaussian factor model is minimal iff the map

$$x \mapsto E[\exp(iuy_i) \mid F^x], \text{ for } i \in \mathbb{Z}_k,$$

is injective on the support of x for all $i \in \mathbb{Z}_k$. This conjecture is false, because the effective dimension n of x may be larger than 1. Even if $n=1$ it is false, see 5.23 below. The special case of $k=3$ and $n=2$ mentioned in 5.24 shows that the equivalent condition for minimality of a Gaussian factor system needs more thinking. The minimality characterizations for the following special cases may be helpful in formulating conjectures for the general result.

PROPOSITION 5.23. *Consider a Gaussian factor model*

$$y = hx + w$$

with $k \geq 2$, $n=1$, $h \in \mathbb{R}^k$. *Then this model is minimal iff*

$$\exists\, i,j \in \mathbb{Z}_k,\ i \neq j, \text{ such that } h_i \neq 0 \text{ and } h_j \neq 0.$$

PROOF. The Gaussian factor model with $n=1$ is minimal iff the dimension of the factor cannot be reduced. This is true iff $n^* > 0$ or iff Q_y is non-diagonal. Note that $OD(Q_y)=OD(hh^T)$. □

PROPOSITION 5.24. *Consider the Gaussian factor model of 5.1 with* $k=3$, $n=2$,

$$H = \begin{pmatrix} h_1^T \\ h_2^T \\ h_3^T \end{pmatrix} \in \mathbb{R}^{3\times 2},\ D(Q_y)=I.$$

Assume that $h_1^T h_2 > 0$, $h_1^T h_3 > 0$, $h_2^T h_3 > 0$. *Then this Gaussian factor model is minimal iff one of the following conditions is satisfied:*

1. $\dfrac{(h_1^T h_2)(h_1^T h_3)}{(h_2^T h_3)} \notin [0,1]$,

2. $\dfrac{(h_1^T h_2)(h_2^T h_3)}{(h_1^T h_3)} \notin [0,1]$,

3. $\dfrac{(h_1^T h_3)(h_2^T h_3)}{(h_1^T h_2)} \notin [0,1]$.

PROOF. This follows from 5.12. □

Classification of internal description
The motivating question here is whether the internal description of a Gaussian factor model is uniquely determined by the variance of the observation vector. In general such a model is not unique. This question is related to question 5.5.b. For the classification of the internal description of factor analysis models with block-diagonal structure see [53]. To structure the discussion a definition is introduced.

DEFINITION 5.25. *Two Gaussian factor models*

$$y = Hx + w$$

and

$$\overline{y} = \overline{H}\overline{x} + \overline{w}$$

are called equivalent *if*

$$HQ_xH^T + Q_w = \overline{H}Q_{\overline{x}}\overline{H}^T + \overline{Q}_w.$$

Note that the two Gaussian factor models of 5.25 that are defined to be equivalent both have the same variance matrix Q_y, since

$$Q_y = H_1Q_{x_1}H_1^T + Q_{w_1} = H_2Q_{x_2}H_2^T + Q_{w_2}.$$

Therefore they cannot be distinguished given Q_y. It is well-known that if (n,H,Q_x,Q_w) are the parameters of a Gaussian factor system and if $S \in \mathbb{R}^{n\times n}$ is an orthogonal matrix ($SS^T=I$), the two Gaussian factor models specified by (n,H,Q_x,Q_w) and (n,HS,S^TQ_xS,Q_w) are equivalent. However, there may be other ways in which two Gaussian factor models are equivalent.

In applications of Gaussian factor analysis it has been recognized that there may be many equivalent models. To reduce the class of equivalent models practitioners fix certain elements of the matrix of factor loadings, based on prior knowledge about the observation vector or arbitrarily.

The question now is, given a Gaussian factor model, to describe the equivalence class of all Gaussian factor models that are equivalent with the given one. This question is still open.

6. GAUSSIAN FACTOR SYSTEMS

The purpose of this section is to formulate the concept of a Gaussian factor system and to survey the preliminary results of the stochastic realization problem for this class of systems.

A motivation for the study of this class of systems is the stochastic realization problem for Gaussian systems with inputs. One would like to know whether it is possible to determine from an observed vector-valued process which components are inputs and which are outputs of a Gaussian system. Another motivation for the study of this class of systems is the exploration of the extension of Gaussian factor models to dynamic systems.

DEFINITION 6.1. *A* Gaussian factor system, *in discrete time, is an object specified by the equations*

$$x(t+1) = Ax(t) + Bu(t),$$
$$y(t) = [Cx(t) + Du(t)] + w(t)$$

or

$$y(t) = \sum_{s \in T} H(t-s)u(s) + w(t)$$

where $u:\Omega \times T \rightarrow \mathbf{R}^p$ *is a stationary Gaussian process called the* factor process, $w:\Omega \times T \rightarrow \mathbf{R}^k$ *is a stationary Gaussian process called the* noise process, $y:\Omega \times T \rightarrow \mathbf{R}^k$ *is called the* observed process, $u, w_1, \ldots, w_k$ *are independent processes, the spectral densities of* $u, w_1, \ldots, w_k$ *are rational functions, and the Fourier transform of the transfer function* H *is rational and causal.*

A Gaussian factor system is said to have the factor property if the processes $u, w_1, \ldots, w_k$ are independent processes. This condition can also be rephrased in terms of conditional independence but this will not be done here. Note that the processes $w_1, \ldots, w_k$ need not be white noise processes.

Concepts similar to that of a Gaussian factor system have been introduced in the literature. An elementary version of a Gaussian factor system with H a constant matrix is introduced in [58]. In [26] a Gaussian factor system is defined without the rationality and causality conditions. In [21] one can find the definition 6.1 and a generalization. In [54] a generalization of 6.1 is presented in which the spectral density of the process w is not diagonal but block-diagonal and in which the transfer function H not be causal. The term *dynamic errors-in-variables systems* is used instead of Gaussian factor system in the publications of B. D. O. Anderson and M. Deistler [4-6, 16, 17]. An interpretation of this term follows.

Consider a deterministic finite-dimensional linear system in impulse response representation

$$\hat{y}(t) = \sum_{s \in T} H(t-s)\hat{u}(s).$$

Suppose that the variables of input $\hat{u}$ and output $\hat{y}$ of this system are observed with errors or noise, say by

$$u(t) = \hat{u}(t) + w_1(t), \; y(t) = \hat{y}(t) + w_2(t),$$

in which w_1, w_2 are independent Gaussian white noise processes. Combining these expressions one obtains

$$\begin{pmatrix} u(t) \\ y(t) \end{pmatrix} = \begin{pmatrix} I\delta(t-s) \\ \sum_{s \in T} H(t-s) \end{pmatrix} u(s) + \begin{pmatrix} w_1(t) \\ w_2(t) \end{pmatrix},$$

which is a Gaussian factor system except for the fact that the spectral density of the noise is not a diagonal function but block-diagonal with two blocks. The interpretation of the above defined system of which the variables are observed with error, illustrates the term errors-in-variables model.

Problem 6.2. *The* weak stochastic realization problem for a Gaussian factor system *is to solve the following subproblems. Assume given a stationary Gaussian process with zero mean function and covariance function* Q *or spectral density* $\hat{Q}$.

a. *Find conditions under which there exists a Gaussian factor system*

$$y(t) = \sum_{s \in T} H(t-s)u(s) + w(t)$$

such that the spectral density of y equals the given spectral density, or

$$\hat{Q} = \hat{Q}_y = \hat{H}\hat{Q}_u\overline{\hat{H}}^T + \hat{Q}_w.$$

If such a Gaussian factor system exists then it is called a weak stochastic realization *of the given process.*

b. *Classify all minimal weak stochastic realizations of the given process. A weak stochastic realization is called* minimal *if* rank $(\hat{H}\hat{Q}_u\overline{\hat{H}}^T)$ *is minimal.*

A difficulty with the above defined problem is the definition of minimality. In addition to the concept defined in 6.2, which is minimality of the dimension of the factor process u, one could also consider minimality of the degree of $\hat{H}\hat{Q}_u\overline{\hat{H}}^T$. From a viewpoint of linear system theory the latter concept would be preferable. Possibly a mixture of both the dimension of the factor process and the degree has to be considered. Because of this difficulty the author of this paper is not yet convinced that a Gaussian factor system is a suitable model for economic and engineering practice. However, what may be of interest is the special case in which the spectral density of the noise is block-diagonal with two blocks.

The weak stochastic realization problem for Gaussian factor systems is unsolved. Only for low-dimensional cases have results been published. For the case of an observed process with two components see [4, 18, 33] and for the case with three components see [6, 18]. A discussion of the problem may be found in [17]. Questions of identifiability and problems of parameter estimation for Gaussian factor systems have been discussed in [21, 26].

A strong version of the weak stochastic realization problem of 6.2 has been proposed in [54]; see also [55]. The case in which the spectral density $\hat{Q}_w$ of the noise consists of two diagonal blocks has been treated there.

ACKNOWLEDGEMENTS

The author acknowledges J. C. Willems for his inspiring conceptual approach to system and control theory. For the material on factor analysis and factor systems the author has benefited from discussions with L. Baratchart, M. Deistler, R. E. Kalman, and G. Picci.

REFERENCES

1. N. L. AGGARWAL (1974). Sur l'information de Fisher. *Théories de l'Information*, Lecture Notes in Mathematics 398, Springer-Verlag, Berlin, 111-117.
2. D. J. AIGNER and M. DEISTLER (1989). Latent variables models - Editor's introduction. *J. Econometrics 41*, 1-3.
3. H. AKAIKE (1976). Canonical correlation analysis of time series and the use of an information criterion. R. K. MEHRA, D. G. LAINIOTIS (eds.). *System identification - Advances and case studies*, Academic Press, New York, 27-96.
4. B. D. O. ANDERSON (1985). Identification of scalar errors-in-variables models with dynamics. *Automatica J.-IFAC 21*, 709-716.
5. B. D. O. ANDERSON and M. DEISTLER (1984). Identifiability in dynamic errors-in-variables models. *J. Time Series Anal. 5*, 1-13.
6. B. D. O. ANDERSON and M. DEISTLER (1987). Dynamic errors-in-variables systems with three

variables. *Automatica J.-IFAC 23*, 611-616.

7. T. W. Anderson and H. Rubin (1956). Statistical inference in factor analysis. J. Neyman (ed.). *Proc. Third Berkeley Symposium on Mathematical Statistics and Probability, Volume V*, University of California Press, Berkeley, 111-150.
8. M. Aoki (1988). Nonstationarity, cointegration, and error correction in economic modeling. *J. Econ. Dyn. Control 12*, 199-201.
9. M. Aoki (1988). On alternative state space representations of time series models. *J. Econ. Dyn. Control 12*, 595-607.
10. V. I. Arkin and I. V. Evstigneev (1987). *Stochastic models of control and economic dynamics*, Academic Press, New York.
11. K. S. Arun, D. V. Bhaskar Rao, and S. Y. Kung (1983). A new predictive efficiency criterion for approximate stochastic realization. *1983 Conference on Decision and Control*, IEEE, 1353-1355.
12. K. S. Arun and S. Y. Kung (1984). State space modeling and approximate realization methods for ARMA spectral estimation. *IEEE International Conference on Systems, Man and Cybernetics*, IEEE.
13. P. A. Bekker and J. De Leeuw (1987). The rank of reduced dispersion matrices. *Psychometrika 52*, 125-135.
14. A. Bensoussan and J. H. Van Schuppen (1985). Optimal control of partially observable stochastic systems with an exponential-of-integral performance index. *SIAM J. Control Optim. 23*, 599-613.
15. D. Blackwell and L. Koopmans (1957). On the identifiability problem for functions of finite Markov chains. *Ann. Math. Statist. 28*, 1011-1015.
16. M. Deistler (1986). Linear errors-in-variables models. S. Bittanti (ed.). *Time series and linear systems*, Lecture Notes in Control and Information Sciences, Springer-Verlag, Berlin, 37-67.
17. M. Deistler and B. D. O. Anderson (1988). Linear dynamic errors in variables models: Some structure theory. A. Bensoussan, J. L. Lions (eds.). *Analysis and optimization of systems*, Lecture Notes in Control and Information Sciences 111, Springer-Verlag, Berlin, 873-883.
18. M. Deistler and B. D. O. Anderson (1989). Linear dynamic errors-in-variables models - Some structure theory. *J. Econometrics 41*, 39-63.
19. B. De Moor (1988). *Mathematical concepts and techniques for modelling of static and dynamic systems*, Thesis, Katholieke Universiteit Leuven, Leuven.
20. J. P. Dufour (1983). *Resultats generiques en analyse factorielle*, Departement de Mathématiques, U. S. T. L., Montpellier.
21. R. Engle and M. Watson (1981). A one-factor multivariate time series model of metropolitan wage rates. *J. Amer. Statist. Assoc. 76*, 774-781.
22. R. F. Engle and C. W. J. Granger (1987). Co-integration and error correction: Representation estimation, and testing. *Econometrica 55*, 251-276.
23. P. Faurre (1976). Stochastic realization algorithms. R. K. Mehra, D. G. Lainiotis (eds.). *System Identification - Advances and Case Studies*, Academic Press, New York, 1-25.
24. P. Faurre, M. Clerget, and F. Germain (1979). *Opérateurs rationnels positifs*, Dunod, Paris.
25. R. Frisch (1934). *Statistical confluence analysis by means of complete regression systems*, Publ. no. 5, University of Oslo Economic Institute, Oslo.

26. J. F. GEWEKE and K. J. SINGLETON (1981). Maximum likelihood 'confirmatory' factor analysis of econometric time series. *Int. Economic Rev. 22*, 37-54.

27. R. D. GITTENS (1985). *Canonical analysis - A review with applications in ecology*, Springer-Verlag, Berlin.

28. K. GLOVER and E. JONCKHEERE (1986). A comparison of two Hankel-norm methods for approximating spectra. C. I. BYRNES, A. LINDQUIST (eds.). *Modelling, Identification and Robust Control*, Elsevier Science Publishers B. V. (North-Holland), Amsterdam, 297-306.

29. A. GOMBANI and M. PAVON (1985). On the Hankel-norm approximation of linear stochastic systems. *Systems & Control Lett. 5*, 283-288.

30. A. GOMBANI and M. PAVON (1986). On approximate recursive prediction of stationary stochastic processes. *Stochastics 17*, 125-143.

31. A. GOMBANI, M. PAVON, and B. COPPO (1986). On Hankel-norm approximation of stationary increment processes. C. I. BYRNES, A. LINDQUIST (eds.). *Modelling, Identification and Robust Control*, Elsevier Science Publishers B. V., Amsterdam, 307-323.

32. C. W. J. GRANGER (1981). Some properties of time series data and their use in econometric model specification. *J. Econometrics*, 121-130.

33. M. GREEN and B. D. O. ANDERSON (1986). Identification of multivariable errors in variable models with dynamics. *IEEE Trans. Automatic Control 31*, 467-471.

34. E. J. HANNAN and M. DEISTLER (1988). *The statistical theory of linear systems*, John Wiley & Sons, New York.

35. C. HEIJ (1988). *Deterministic identification of dynamical systems*, Thesis, University of Groningen, Groningen.

36. N. P. JEWELL and P. BLOOMFIELD (1983). Canonical correlations of past and future for time series: Definitions and theory. *Ann. Statist. 11*, 837-847.

37. N. P. JEWELL, P. BLOOMFIELD, and F. C. BARTMANN (1983). Canonical correlations of past and future for time series: Bounds and computation. *Ann. Statist. 11*, 848-855.

38. E. A. JONCKHEERE and J. W. HELTON (1985). Power spectrum reduction by optimal Hankel norm approximation of the phase of the outer spectral factor. *IEEE Trans. Automatic Control 30*, 1192-1201.

39. R. E. KALMAN (1982). System identification from noisy data. A. R. BEDNAREK, L. CESARI (eds.). *Dynamical Systems II*, Academic Press, New York, 135-164.

40. R. E. KALMAN (1982). Identification from real data. M. HAZEWINKEL, A. H. G. RINNOOY KAN (eds.). *Current developments in the interface: Economics, Econometrics, Mathematics*, D. Reidel Publishing Company, Dordrecht, 161-196.

41. R. E. KALMAN (1983). Identifiability and modeling in econometrics. P. R. KRISHNAIAH (ed.). *Developments in Statistics* 4, Academic Press, New York, 97-136.

42. R. E. KALMAN, P. L. FALB, and M. A. ARBIB (1969). *Topics in mathematical system theory*, McGraw-Hill Book Co., New York.

43. D. KAZAKOS and P. PAPANTONI-KAZAKOS (1980). Spectral distance measures between Gaussian processes. *IEEE Trans. Automatic Control 25*, 950-959.

44. S. Y. KUNG and K. S. ARUN (1983). Approximate realization methods for ARMA spectral estimation. *1983 IEEE International Symposium on Circuits and Systems*, IEEE, 105-109.

45. W. E. LARIMORE, S. MAHMOUD, and R. K. MEHRA (1984). Multivariate adaptive model algorithmic control. *Proceedings 23rd Conference on Decision and Control*, IEEE, 675-680.

46. W. E. LARRIMORE (1983). System identification, reduced-order filtering and modeling via canonical variate analysis. *Proceedings 1983 American Control Conference*, 445-451.

47. A. LINDQUIST and G. PICCI (1985). Realization theory for multivariate stationary Gaussian processes. *SIAM J. Control Optim. 23*, 809-857.

48. A. LINDQUIST, G. PICCI, and G. RUCKEBUSCH (1979). On minimal splitting subspaces and Markovian representations. *Math. Systems Th. 12*, 271-279.

49. M. PAVON (1984). Canonical correlations of past inputs and future outputs for linear stochastic systems. *Syst. Control Lett. 4*, 209-215.

50. A. PAZ (1971). *Introduction to probabilistic automata*, Academic Press, New York.

51. V.V. PELLAR and S.V. KHRUSHCHEV (1982). Hankel operators, best approximations, and stationary Gaussian processes. *Russian Math. Surveys 37*, 61-144.

52. G. PICCI (1978). On the internal structure of finite-state stochastic processes. *Proc. of a U. S.-Italy Seminar*, Lecture Notes in Economics and Mathematical Systems 162, Springer-Verlag, Berlin, 288-304.

53. G. PICCI (1989). Parametrization of factor analysis models. *J. Econometrics 41*, 17-38.

54. G. PICCI and S. PINZONI (1986). Factor analysis models for stationary stochastic processes. A. BENSOUSSAN, J. L. LIONS (eds.). *Analysis and Optimization of Systems*, Lecture Notes in Control and Information Sciences 83, Springer-Verlag, Berlin, 412-424.

55. G. PICCI and S. PINZONI (1986). Dynamic factor-analysis models for stationary processes. *IMA J. Math. Control and Info. 3*, 185-210.

56. P. PURDUE (1979). Stochastic compartmental models: A review of the mathematical theory with ecological applications. J. H. MATIS, B.C. PATTEN, G. C. WHITE (eds.). *Compartmental analysis of ecosystem models*, International Co-operative Publishing House, Fairland, MD, 223-260.

57. C. R. RAO (1964). The use and interpretation of principal component analysis in applied research. *Sankhya, Series A 26*, 329-358.

58. O. REIERSØL (1941). Confluence analysis by means of lag moments and other methods of confluence analysis. *Econometrica 9*, 1-24.

59. O. REIERSØL (1945). Confluence analysis by means of instrumental sets of variables. *Arkiv für Mathematik, Astronomi och Fysik 32A*.

60. J. RISSANEN (1978). Modeling by shortest data description. *Automatica J. IFAC 14*, 465-471.

61. J. RISSANEN (1983). A universal prior for integers and estimation by minimum description length. *Ann. Statist. 11*, 416-431.

62. J. RISSANEN (1985). *Modeling by the minimum description lenghth principle - A survey*, Preprint.

63. J. RISSANEN (1986). Stochastic complexity and modeling. *Ann. Statist. 14*, 1080-1100.

64. J. RISSANEN (1986). Predictive and nonpredictive minimum description length principles. S. BITTANTI (ed.). *Time series and linear systems*, Lecture Notes in Control and Information Sciences 86, Springer-Verlag, Berlin, 115-140.

65. J. RUCKEBUSCH (1978). *Sur l'approximation rationnelle des filtres*, Centre de Mathématiques Appliquées, Ecole Polytechnique, Palaiseau.

66. P. A. SAMUELSON (1979). A note on alternative regressions. J. E. STIGLITZ (ed.). *The collected scientific papers of Paul A. Samuelson, Fifth Printing*, M. I. T. Press, Cambridge, 694-697.

67. C. A. SPEARMAN (1904). General intelligence, objectively determined and measured. *Amer. J. Psych. 15*, 201-293.

68. P. J. C. SPREIJ (1986). *Selfexciting counting process systems with finite state space*, Report OS-R8613, Centre for Mathematics and Computer Science, Amsterdam.

69. P. J. C. SPREIJ (1987). *Counting process systems - Identification and stochastic realization*, Doctoral Thesis, University of Twente, Enschede.

70. J. H. STEIGER (1979). Factor indeterminacy in the 1930's and the 1970's some interesting parallels. *Psychometrika 44*, 157-167.

71. C. VAN PUTTEN and J. H. VAN SCHUPPEN (1983). The weak and strong Gaussian probabilistic realization problem. *J. Multivariate Anal. 13*, 118-137.

72. J. H. VAN SCHUPPEN (1979). Stochastic filtering theory: A discussion of concepts, methods and results. M. KOHLMANN, W. VOGEL (eds.). *Stochastic control theory and stochastic differential systems*, Lecture Notes in Control and Information Sciences 16, Springer-Verlag, Berlin, 209-226.

73. J. H. VAN SCHUPPEN (1982). The strong finite stochastic realization problem - Preliminary results. A. BENSOUSSAN, J. L. LIONS (eds.). *Analysis and optimization of systems*, Lecture Notes in Control and Information Sciences 44, Springer-Verlag, Berlin, 179-190.

74. J. H. VAN SCHUPPEN (1986). Stochastic realization problems motivated by econometric modeling. C. I. BYRNES, A. LINDQUIST (eds.). *Modelling, Identification and Robust Control*, Elsevier Science Publishers B. V. (North-Holland), Amsterdam, 259-275.

75. E. VERRIEST (1986). Projection Techniques for model reduction. C. I. BYRNES, A. LINDQUIST (eds.). *Modelling, Identification and Robust Control*, Elsevier Science Publishers B. V. (North-Holland), Amsterdam, 381-396.

76. J. V. WHITE (1983). Stochastic state-space models from empirical data. *Proceedings 1983 Conference on Acoustics, Speechs and Signal Processing*, IEEE, 243-246.

77. P. WHITTLE (1981). Risk-sensitive linear/quadratic/Gaussian control. *Adv. Appl. Prob. 13*, 764-777.

78. J. C. WILLEMS (1986). From time series to linear systems - Part I. Finite dimensional linear time invariant systems. *Automatica J. IFAC 22*, 561-580.

79. J. C. WILLEMS (1987). From time series to linear systems - Part III. Approximate modelling. *Automatica J. IFAC 23*, 87-115.

80. E. B. WILSON (1929). Review of 'Crossroads in the mind of man: A study of differentiable mental abilities' by T. L. Kelley. *J. Gen. Psychology 2*, 153-169.

Robust Stabilization of Uncertain Dynamic Systems

J. L. Willems

Engineering Faculty, University of Gent
Gent, Belgium

1. Introductory Remarks

It gives me a strange feeling to write a contribution for my twin brother's fiftieth birthday. It is almost like writing a paper for my own birthday! Only after accepting with great pleasure the invitation to contribute I began to realize how difficult the task was going to be. On the one hand it is not possible to write something purely professional for one's twin brother, because the relationship is not primarily professional, certainly not! On the other hand I felt it would not be fit to write a contribution with too personal a flavor in a book on "Mathematical System Theory". The book is indeed meant to be a scientific work for the mathematical system theory community.

It took me a long time to reach a conclusion. I finally decided to organize my contribution as follows: after some comments on our professional relationship, I will give a survey on robust stabilization of stochastic dynamical systems, a topic we worked on and published about together. This topic indeed embraces a number of concepts which are very dear to Jan: the Riccati equation, (almost) invariant subspaces, high gain feedback. It is like mentioning three of his middle names!

As is quite normal for twin brothers, we attented the same elementary school and the same high school. It is less common that after finishing high school both of us decided to study engineering at the university. It is probably even much less common that we both chose electromechanical engineering as our major field. We thus graduated together in electromechanical engineering from the Engineering Faculty of the University of Gent, Gent, Belgium, in July 1963. During our final year at the university, probably thanks to our father's advice, we both applied for a fellowship for graduate study in the United States, and were successful. At the end of the summer of 1963 we left for the U.S.; at that time our ways separated. Indeed, because we obtained graduate fellowships from different foundations, we traveled to the U.S. on different transatlantic ships; Jan made the voyage aboard the Queen Mary, I traveled on the Rotterdam. Since most people in the control field associate both of us with M.I.T. and with Roger Brockett in particular, it may surprise many that from that time onwards we never worked at the same institute at the same time. Indeed I studied at M.I.T. for obtaining my Master's degree during the academic year 1963-1964, and then returned to Belgium. Jan studied at the University of Rhode Island from 1963 to 1965, and came to M.I.T. in 1965 to work towards the Ph. D. degree.

Both of us were fortunate to work with Roger Brockett at M.I.T.; he supervised my Master's thesis and stimulated my interest in control theory in general, and in stability theory in particular. The same happened to Jan two years later. Because of our association with Roger Brockett we worked on related problems. This gave us the opportunity to do some joint research which led to a number of joint papers. The exchange of ideas took place during family visits and by letters, between comments on the progress of the children in school and on the health of some aunt or uncle. We almost succeeded to be in the same research group in 1970, when Jan was on the Faculty at M.I.T. and I went on a leave of absence from the University of Gent to work as a postdoctoral fellow with Roger Brockett. However by that time Roger Brockett had left M.I.T. and had joined the Division of Engineering and Applied Physics at Harvard.

Our first two joint articles were discussion notes [1,2] on stability criteria for nonlinear feedback systems; even Jan began with a couple of rather modest contributions! Thanks to our 'Lyapunov function education' at M.I.T we did some joint work on the generation of Lyapunov functions for the analysis of transient power system stability [3]. Our larger piece of joint research work was concerned with a problem of robust stabilization, namely the robust feedback control of systems with stochastic coefficients. The stochastic coefficients are modeled by white noise processes. This leads to the consideration of systems with state-dependent and/or control-dependent noise [4, 5]. In the first paper [4] we combined the ideas of Jan's earlier work on pole assignment of deterministic linear systems and on the algebraic Riccati equation, with my previous research results on the mean square stability of stochastic systems with state-dependent stochastic elements. In the second paper [5] we described further results applying the concept of almost (A,B)-invariant subspaces Jan developed. This research topic is discussed in the sequel of this contribution. The discussion concentrates on the development of the model, on the motivation of the approach, and on the interpretation of the results. For the technical details the reader is referred to the original papers, cited in the bibliography.

2. System Modeling

A large fraction of the research on control systems in the last quarter of a century is concerned with the linear control system, described by the set of ordinary differential equations

$$dx(t)/dt = Ax(t) + Bu(t) \tag{1}$$

where x denotes the state vector and u the control input vector, with n and m components respectively. A and B are constant matrices of appropriate dimension. In particular very complete results exist on the possibility of stabilizing system (1) or arbitrarily assigning its eigenvalues by means of state feedback control strategies [6, 7].

It is obvious that (1) is only an approximate representation of the real physical system. It is therefore interesting to investigate whether results obtained for (1) still hold for 'neighboring' systems, in other words whether properties, such as the stabilizability or the pole assignability property mentioned above, are robust with respect to perturbations of the system model. Robustness of control systems has been a very active research area in recent years [8]. It is indeed an important aspect of a design study; the model used for the analysis of a physical system is only an approximate representation since some phenomena are unknown or neglected and hence not included in the model.

The exact model of the physical system would in most cases require including nonlinear and time-varying phenomena as well as taking into account phenomena which are governed by external effects. The latter effects are unpredictable from the point of view of the system itself. The model perturbations can be modeled in different ways; for a discussion of this aspect the reader is referred to the extensive literature on robust control system design.

The model considered in the present contribution along the line of previous papers [4, 5] includes perturbations described by stochastic 'white noise' coefficients; a rigorous mathematical description requires the consideration of Itô differential equations. For simplicity of the analysis and the discussion in this paper only perturbations of the plant matrix A in (1) are considered, but no perturbations of the input matrix B. This leads to the equation

$$dx = (Ax+Bu)dt + \Sigma_i \sigma_i F_i x \, dW_i \tag{2}$$

where the last term ($i = 1, \ldots, M$) denotes the disturbances. The stochastic processes are assumed to be zero-mean uncorrelated stationary normalized standard Wiener processes. Thus

$$E(dW_i) = 0$$

$$E(dW_i^2) = dt$$

$$E(dW_i dW_j) = 0 \qquad \text{for } i \neq j$$

where E denotes the expected value. The factors σ_i indicate the intensities of the disturbances. Intuitively (2) should be regarded as the differential equation with stochastic parameters

$$dx(t)/dt = [A + \Sigma_i f_i(t) F_i] x(t) + Bu(t) \tag{3}$$

where the processes $f_i(t)$ are white noise stochastic processes.

The stabilizability problem considered in this paper is the question whether there exists a state feedback control law of the form

$$u = Kx \tag{4}$$

such that the controlled system

$$dx = (A+BK)x\,dt + \Sigma_i \sigma_i F_i x\,dW_i \tag{5}$$

is stable.

More explicitly, the following robustness issues are discussed in the present contribution:

Robust stabilizability. Find conditions such that a feedback control exists which stabilizes (5) for a given range of the noise intensities σ_i.

Robust stabilizability for all noise intensities. Find conditions such that for any range of the noise intensities a feedback control exists which stabilizes (5) for that range of noise intensities σ_i.

Perfect robust stabilizability. Find conditions such that there exists a feedback control which stabilizes (5) for all noise intensities.

The stability property considered is mean square asymptotic stability:

Definition 1

System (2) is said to be mean square asymptotically stable if for all initial states $x(0)$ the second moment matrix $E[x(t)x^T(t)]$ tends to zero as t tends to infinity.

The explicit definitions of the robust stabilizability properties mentioned above, are hence as follows.

Definition 2

System (2) is said to be robustly stabilizable for a set of noise intensities $(s_1, s_2,\ldots,s_M)$ if there exists a feedback control (4) such that (5) is mean square asymptotically stable for the noise intensities satisfying

$$\sigma_i \le s_i \qquad (i = 1,\ldots,M)$$

Definition 3

System (2) is said to be robustly stabilizable for all noise intensities if it is robustly stabilizable for all bounds $s_1, \ldots, s_M$.

Definition 4

System (2) is said to be perfectly robustly stabilizable if there exists a feedback control (4) such that (5) is mean square asymptotically stable for all noise intensities.

The property expressed by Definition 3 is weaker than the property expressed by Definition 4 in that the feedback matrix K may depend on the bounds s_i; some entries of K may hence increase without bound as the bounds s_i tend to infinity.

The analysis of robust stabilizability uses properties of the algebraic Riccati equation [9]. Perfect robust stabilizability can be characterized by means of (A,B)-invariant subspaces [10, 11]. Robust stabilizability for all noise intensities is closely related to the concept of almost invariant (A,B)-invariant subspaces and high gain feedback [10, 11]. Hence by discussing this problem area, the importance and relevance of many of Jan Willems' research results can be illustrated.

3. Preliminary results

For the linear deterministic system it is well known that asymptotic stability of the uncontrolled system (1) can be related to properties of the Lyapunov equation

$$A^TP + PA = -Q \tag{6}$$

with the superscript T denoting matrix transposition. On the other hand feedback stabilizability of the controlled system (1) is connected to the Riccati equation

$$A^TP + PA - PBR^{-1}B^TP = -Q \tag{7}$$

Indeed asymptotic stability of the uncontrolled system (1) requires that (6) has a positive definite solution P for a given positive definite matrix Q. Stabilizability by state feedback requires the existence of a positive definite solution P of (7) for given positive definite matrices Q and R. In both cases it can be shown that, if the condition is satisfied for some positive definite matrix Q, and in (7) for some positive definite matrix R, then it is also satisfied for all positive matrices Q and R. Moreover the stabilizing feedback control of (1) is derived from the positive definite solution of (7):

$$u = -R^{-1}B^{T}Px \tag{8}$$

Similarly, mean square asymptotic stability of the uncontrolled system (2) can be analysed by means of a linear matrix equation which is very close to the Lyapunov equation, and mean square stabilizability by means of an equation which is close to the matrix Riccati equation. From the equation for the second moments of the state variables of the uncontrolled system (2), we derive the condition that mean square asymptotic stability of (2) without control is implied by the existence of a positive definite solution P of the equation [12]

$$A^{T}P + PA + \Sigma_{i}\, \sigma_{i}^{2}F_{i}^{T}PF_{i} = -Q \tag{9}$$

for a positive definite matrix Q. The stabilizability of (2) by means of state feedback is equivalent to the existence of a positive definite solution P of the equation

$$A^{T}P + PA - PBR^{-1}B^{T}P + \Sigma_{i}\, \sigma_{i}^{2}F_{i}^{T}PF_{i} = -Q \tag{10}$$

for some positive definite matrices Q and R. Also here, a suitable feedback control realizing mean square asymptotic stability of the closed loop system is given by (8), with P the said solution of (10). An immediate consequence of the above discussion is that if the uncontrolled system (2) is mean square asymptotically stable for a set of noise intensities, it has the same property for all smaller noise intensities. The same is true for stabilizability.

It is immediately clear that (9) can be associated with a Lyapunov equation and (10) with a Riccati equation if

$$Q - \Sigma_{i}\, \sigma_{i}^{2}\, F_{i}^{T}PF_{i} > 0 \tag{11}$$

where P is the solution of (9) or (10) respectively. This remark points out how the analysis of (9) and (10) can be carried out by means of techniques available for the Lyapunov or Riccati equations for linear deterministic systems. This is further elaborated in the next sections.

By means of A-invariant subspaces a criterion can be derived for mean square asymptotic stability of the uncontrolled stochastic system (2) for all noise intensities σ_i. This result is used in the next sections to derive robust stabilizability conditions. Let the subspaces W_j be defined by the following recursive algorithm:

$$W_o := \{0\}$$
$$\ldots$$
$$W_j := \langle \cap_i F_i^{-1} W_{j-1} \mid A \rangle \qquad (12)$$
$$\ldots$$

where $\langle F|A \rangle$ denotes the maximal A-invariant subspace contained in the subspace F. Note that W_j is a subspace of W_{j+1}, hence the subspaces W_j are nested. This shows that the recursive algorithm leads to a limiting subspace, denoted by W^*, in a finite number of steps.

Criterion 1

The uncontrolled system (2) is mean square asymptotically stable for all noise intensities if and only if the matrix A is Hurwitz and W^* is the n-dimensional space.

Remarks

(i) The conditions of the above criterion are also equivalent [5] to the solvability of the matrix Lie algebra generated by the set of matrices A, F_1, ... , F_M, the nilpotency of the matrices F_i and the Hurwitz character of A.

(ii) The above condition is also equivalent to the possibility of simultaneously block-triangularizing the matrices A, F_1, ... , F_M, where the blocks on the diagonal are zero submatrices for the matrices F and Hurwitz submatrices for the matrix A. This form of the system representation clearly shows why the system is mean square asymptotically stable for all noise intensities [5].

Example 1

An example of a system which is mean square asymptotically stable for all noise intensities is

$$\begin{bmatrix} dx_1 \\ dx_2 \end{bmatrix} = \begin{bmatrix} \alpha & 0 \\ \delta & \beta \end{bmatrix} \begin{bmatrix} x_1 \\ x_2 \end{bmatrix} dt + \sigma \begin{bmatrix} 0 \\ 1 \end{bmatrix} [1 \quad 0] \begin{bmatrix} x_1 \\ x_2 \end{bmatrix} dW$$

where α and β are negative constants. Intuitively the reason is that the perturbation only affects the off-diagonal element δ, but not the diagonal elements α and β (the eigenvalues of the system matrix).

4. Robust stabilizability

The discussion of the previous section already points out how robustness criteria can be obtained. The feedback control (8), with P the positive definite solution of (7), yields a closed loop system which is mean square asymptotically stable for all noise intensities satisfying (11). However this only yields sufficient conditions; different choices of the matrix Q and R may lead to more or less conservative conditions. From the properties of the solutions of the algebraic Riccati equation it is clear that a smaller R yields less conservative conditions. It is therefore interesting to express R as βR and to compute the limiting value of P for β decreasing to zero. Also better bounds can be found by choosing

$$Q = Q_1 + \alpha Q_2$$

where the matrix Q_2 is positive definite, the matrix Q_1 is positive semi-definite, α is positive, and

$$Ker(Q_1) = \cap_i Ker(F_i) \qquad (13)$$

with Ker denoting the kernel or null space of a matrix. P is computed for the limiting case for α decreasing to zero.

This algorithm leads to a necessary and sufficient condition, and hence to the exact maximum allowable noise intensity, in the following special case:

* M = 1,

* F_1 has rank one: $F_1 = b_1c_1$ where b_1 is a column vector and c_1 a row vector,
* the system has a single input: $B = b$, where b is a column vector.

Then the system equation has the following form

$$dx = (Ax+bu)dt + \sigma\, b_1c_1\, dW \tag{14}$$

The above discussion leads to the necessary and sufficient conditions for stabilizability:

(i) (A,b) is stabilizable

$$\text{(ii)}\quad \sigma^2 b_1{}^T P_o b_1 < 1 \tag{15}$$

Here P_o is the limiting value, as the positive constant β tends to zero, of the unique positive semi-definite solution of

$$A^TP + PA - (1/\beta)Pbb^TP = c_1{}^Tc_1$$

which is such that $A-(1/\beta)bb^TP$ is a Hurwitz matrix. Another interpretation of $b_1{}^TP_ob_1$ is

$$b_1{}^TP_ob_1 = \inf_k \int_0^\infty (c_1x)^2\, dt \tag{16}$$

where the output c_1x in the integral is to be computed along the solutions of the <u>deterministic</u> system equation

$$dx(t)/dt = Ax(t) + bu(t) \tag{17}$$

with $x(0) = b_1$ and with state feedback control $u = kx$. The infimum in the above expression is to be taken over the set of stabilizing state feedback control strategies.

If $b_1{}^TP_ob_1$ is zero, then the system is robustly stabilizable for all noise intensities. This special case shows clearly the difference between <u>perfect robust stabilizability</u> on the one hand, and <u>robust stabilizability for all noise intensities</u> on the other hand. Both properties require

$$b_1{}^T P_o b_1 = 0 \tag{18}$$

Moreover for perfect robust stabilizability it is required that the infimum at the right side of (16) actually be a minimum, i.e. that it be realizable by some feedback control. If however the infimum corresponds to the limiting value for increasing values of the feedback gain (high gain feedback), then system (14) is not perfectly robustly stabilizable, but only robustly stabilizable for all noise intensities.

5. Perfect robust stabilizability

The results of Section 3 can be used to generate conditions for perfect robust stabilizability, that is for the existence of a feedback matrix such that the closed loop system satisfies the conditions of Criterion 1. The stabilizability criterion is that there should exist a feedback matrix K such that the system with plant matrix A+BK, instead of A, should satisfy the conditions of Criterion 1. This can be converted to an explicit condition on the system data by means of the concept of (A,B)-invariant subspaces and stabilizability subspaces. An (A,B)-invariant subspace is such that for any two states in it the system state can be transferred by an appropriate input from the one to the other state without leaving the subspace; the input can be chosen to be an open loop or a feedback control. For a stabilizability subspace it is moreover required that the feedback control stabilizes the system. These concepts were introduced in the control system literature by Wonham [7]; an early discussion was also given by Basile and Marro [13]. Many developments are due to Jan Willems [11]. To formulate an explicit criterion the following formal definitions are needed:

Definition 5

Let S be a subspace of R^n. Then a subspace V is (A,B)-invariant if there exists a matrix K such that V is (A+BK)-invariant, i.e. such that $(A+BK)V \subset V$. Let V(S) denote the set of (A,B)-invariant subspaces in S, and V*(S) the largest (A,B)-invariant subspace contained in S. $V_g(S)$ denotes the largest (A,B)-invariant subspace in S with the additional restriction that the matrix A+BK is Hurwitz:

$$V_g(S) := \sup\{V \in V(S) \mid (A+BK)V \subset V \text{ and } \sigma(A+BK) \subset C_g \text{ for some } K\} \tag{19}$$

with $\sigma(M)$ denoting the spectrum of the matrix M, and C_g denoting the left half complex plane

$$C_g := \{s \in C \mid Re(s) < 0\}$$

It can be proved that these concepts are well defined, and that there exist straightforward algorithms to compute them.

Definition 6

The recursive algorithm

$$V_{g,o} := \{0\}$$
$$\dots$$
$$V_{g,j} := V_g(\cap_i F_i^{-1} V_{g,j-1})$$
$$\dots$$

defines a limiting subspace V_g^* in a finite number of steps.

Criterion 2

System (2) is perfectly robustly stabilizable if and only if V_g^* is the n-dimensional vector space.

Remarks

(i) It can readily be proved that

$$V_g^* = R^n$$

is equivalent to

$$V_{g,k} \supset \cap_i \text{ im } F_i \tag{20}$$

for some k, where the sum denotes the direct sum of vector spaces and im denotes the image or range space of a matrix.

(ii) Suppose there is only one stochastic element and the associated matrix F_1 has rank one, such that the system is described by equation (14). It can be shown that for this case V_g^* is obtained in at most two steps. Hence the condition of Criterion 2 is equivalent to

$V_g(\mathrm{Ker}\ c_1) \supset \mathrm{im}\ b_1$ (21)

This implies the existence of a feedback matrix K such that A+BK is a Hurwitz matrix and $c(Is-A-BK)^{-1}b$ vanishes identically. It is equivalent to the condition of disturbance decoupling [14] with stability from the disturbance input im b_1 to the output c_1x. This can be expressed explicitly in terms of the transfer functions associated with the original system model: the necessary and sufficient condition for the perfect robust stabilizability of (14) is that the ratio of transfer functions

$$c_1(Is-A)^{-1}b_1/c_1(Is-A)^{-1}b$$

is strictly proper and has only poles with negative real parts, after cancellation of common factors. Note the relation of this condition with the (stronger) property that the system is minimum phase with respect to the output c_1x.

Example 2

The system with stochastic perturbations

$$\begin{bmatrix} dx_1 \\ dx_2 \end{bmatrix} = \begin{bmatrix} 0 & 1 \\ 0 & -1 \end{bmatrix} \begin{bmatrix} x_1 \\ x_2 \end{bmatrix} dt + \begin{bmatrix} 0 \\ 1 \end{bmatrix} dt + \sigma \begin{bmatrix} 2 \\ -1 \end{bmatrix} [1 \quad 2] \begin{bmatrix} x_1 \\ x_2 \end{bmatrix} dW$$

can be perfectly robustly stabilized by means of the feedback control

$$u = -\alpha x_1-(2\alpha-0.5)x_2$$

where α is an arbitrary positive constant. With this feedback the controlled system is mean square asymptotically stable for any intensity σ of the noise. The constant α can be used for pole assignment purposes. The system indeed satisfies the above criterion since

$$c_1(Is-A)^{-1}b_1/c_1(Is-A)b = 1/(2s+1)$$

6. Robust stabilizability for all noise intensities

In this section the question is considered to what extent the criterion of the previous section can be relaxed if only stabilizability of the system is required for all σ_i. This means that for any σ_i a stabilizing feedback matrix K should exist, such that

$$dM/dt = (A+BK)^T M + M(A+BK) + \Sigma_i \sigma_i^2 F_i M F_i^T \qquad (22)$$

is asymptotically stable in the cone of nonnegative definite (n x n) matrices. Since the matrix K may depend on the noise intensities σ_i, some elements of K may go to infinity as the noise intensities increase without bound. Then there does not exist a single feedback matrix which stabilizes (4) in the mean square for all noise intensities and the system is hence not perfectly robustly stabilizable.

It has been shown [5] that for stabilizability for all noise intensities a very elegant relaxation of the conditions of Criterion 2 can be obtained by means of the clever concept of almost (A,B)-invariant subspaces introduced by Jan Willems [11]. An almost (A,B)-invariant subspace is such that for any two states in the subspace there exists an input such that the system state is transferred from the one to the other while the trajectory remains arbitrarily close to the subspace. However, if the maximal distance becomes smaller and smaller, the required input may become larger and larger and tend to an impulsive input or an infinite gain feedback control. Let the subspaces $\underline{V}_{g,j}$ be defined in the same way as the subspaces $V_{g,j}$ in the previous section, but with the open left half complex plane C_g replaced by the closed left half plane

$$\underline{C}_g := \{s \in C \mid Re(s) \le 0\}$$

Let $R_b^*(S)$ denote the largest L_1-almost (A,B)-invariant subspace in S; this means that $R_b^*(S)$ is the largest subspace of the state space in which any two states can be transferred to one another while keeping the L_1-norm of the distance of the state trajectory to S arbitrarily small. For a formal definition the reader is referred to the literature [11]. Then we obtain

Criterion 3

System (2) is robustly stabilizable for all noise intensities if (A,B) is stabilizable and

$$\underline{V}_g(\cap_i F_i V_g^*) + R_b^*(\cap_i F_i^{-1} V_g^*) \supset \Sigma_i \text{ im } F_i$$

where V_g^* has the same meaning as in the previous section.

Remark

For the special case of system (14) the above criterion yields a necessary and sufficient condition

$$\underline{V}_g(\text{Ker } c_1) + R_b^*(\text{Ker } c_1) \supset \text{ im } b_1 \tag{23}$$

This can elegantly be expressed as a frequency domain criterion. Indeed (23) is equivalent to the condition that the system (A,b) is stabilizable and that the ratio of transfer functions

$$c_1(Is-A)^{-1}b_1/c_1(Is-A)^{-1}b$$

has no poles with positive real parts, after cancellation of common factors. The condition is equivalent to the property of almost disturbance decoupling [11] with stability from the input im b_1 to the output c_1x.

Example 3

The system with stochastic perturbations

$$\begin{bmatrix} dx_1 \\ dx_2 \end{bmatrix} = \begin{bmatrix} 0 & 1 \\ 0 & -1 \end{bmatrix} \begin{bmatrix} x_1 \\ x_2 \end{bmatrix} dt + \begin{bmatrix} 0 \\ 1 \end{bmatrix} dt + \sigma \begin{bmatrix} 2 \\ -1 \end{bmatrix} [a \quad 2] \begin{bmatrix} x_1 \\ x_2 \end{bmatrix} dW$$

can be robustly stabilized by means of state feedback for all noise intensities if a is nonnegative. It is readily seen that the system is not perfectly robustly stabilizable if $a \neq 1$. Indeed one of the conditions such that

$$u = -\alpha x - \beta x$$

stabilizes the system in the mean square sense, is

$$3\beta > 4(a-1)^2\sigma^2$$

This shows that, except for the case where a equals 1, the feedback gain must increase without bound with the noise intensity. This is in agreement with the frequency domain condition. We obtain

$$c_1(Is-A)^{-1}b_1/c_1(Is-A)^{-1}b = [2(a-1)s+a]/(2s+a)$$

The pole of this rational function is non-positive if a is nonnegative; then the stochastic system is robustly stabilizable for all noise intensities. The function is strictly proper only if a is equal to 1; only in that case is the stochastic system perfectly robustly stabilizable.

7. Further Remarks

1. A similar analysis can be performed on robust stabilizability of discrete-time systems with stochastic parameters

   $$x(t+1) = Ax(t) + Bu(t) + \Sigma_i \sigma_i f_i(t) F_i x(t)$$

   where the processes $f_i(t)$ are zero mean white noise discrete-time processes. The analysis of discrete-time systems with white noise is technically more straightforward than for continuous-time systems. Some results and examples are discussed in a previous paper [5].

2. It was shown that perfect robust stabilizability corresponds to the possibility of simultaneously block triangularizing the plant matrix A+BK of the closed loop system and the matrices F_i, such that the diagonal blocks of A+BK are Hurwitz and those of F_i are zero matrices. This readily shows that in this case perfect robustness is also valid for nonlinear time-varying perturbations:

$$dx(t)/dt = Ax(t) + Bu(t) + \Sigma_i f_i(x(t),t)F_i x(t) \tag{24}$$

It can be shown that some (minor) changes are required to make the criteria of Section 5 valid for such systems with nonlinear time-varying perturbations.

3. The condition for perfect robust stabilizability remains valid if the stochastic system (3) is considered with non-white stochastic processes, and also if other moment stability properties are considered.

8. Conclusion

In this contribution a review has been given of some joint research work with Jan Willems on the robust control of uncertain dynamic systems. Conditions on the admissible levels of the perturbations have been derived which do not destroy stability, stabilizability, or stabilization. It was shown that the analysis relies heavily on earlier original research results developed by Jan Willems, such as results on the qualitative properties of the algebraic Riccati equation and the very interesting concept of almost (A,B)-invariant or almost controllability subspaces.

References

[1] Willems, J.L. and J.C. Willems, "A stability criterion for a nonlinear nonautonomous system", Proceedings of the IEEE (Letters), vol. 56, p. 244-245, 1968.

[2] Willems, J.L. and J.C. Willems, "Untersuchung der Stabilität nichtlinearer Regelungssysteme im Frequenz-bereich", Messen, Steuern, Regeln, vol. 11, pp. 114-116, 1968.

[3] Willems, J.L. and J.C. Willems, "The application of Lyapunov methods to the computation of transient stability regions for multimachine power systems", IEEE Transactions on Power Apparatus and Systems, vol. PAS-89, pp. 795-801, 1970.

[4] Willems, J.L. and J.C. Willems, "Feedback stabilizability for stochastic systems with state and control dependent noise", Automatica, vol. 12, pp. 277-283, 1976.

[5] Willems, J.L. and J.C. Willems, "Robust stabilization of uncertain systems", **SIAM Journal of Control and Optimization**, vol. 21, pp. 352-374, 1983.

[6] Kwakernaak, H. and R. Sivan, **Linear Optimal Control Systems**, Wiley-Interscience, New York, 1972.

[7] Wonham, W.M., **Linear Multivariable Control: A Geometric Approach**, 2nd Ed., Springer Verlag, New York, 1972.

[8] Curtain, R.F. (ed.), **Modelling, Robustness and Sensitivity Reduction in Control Systems**, Proceedings of a NATO Advanced Research Workshop, Springer-Verlag, Berlin, 1987.

[9] Willems, J.C., "Least squares optimal control and the algebraic Riccati equation", **IEEE Transactions on Automatic Control**, vol. AC-16, pp. 621-634, 1971.

[10] Willems, J.C., "Almost A(modB)-invariant subspaces", **Astérisque**, vol. 75-76, pp. 239-248, 1980.

[11] Willems, J.C., "Almost invariant subspaces: an approach to high gain feedback design - Part I: almost controlled invariant subspaces", **IEEE Transactions on Automatic Control**, vol. AC-26, pp. 235-252, 1981.

[12] Willems, J.L., " Mean square stability criteria for stochastic systems", **Problems of Control and Information Theory**, vol. 2, pp. 199-217, 1973.

[13] Basile, G. and G. Marro, "Controlled and conditioned invariant subspaces in linear system theory", **Journal of Optimization Theory and Applications**, vol. 3, pp. 306-315, 1969.

[14] Willems, J.C. and C. Commault, "Disturbance decoupling by measurement feedback with stability or pole placement", **SIAM Journal on Control and Optimization**, vol. 19, pp. 490-504, 1981.

Acknowledgment

The author gratefully acknowledges partial research support from the Belgian Fund for Scientific Research (F. K. F. O. Grant).

On the Control of Discrete-Event Systems

W. M. Wonham

Systems Control Group, Dept. of Electrical Engineering
University of Toronto
Toronto, Ontario, Canada M5S 1A4

INTRODUCTION

A discrete-event system (DES) is a dynamic system whose behavior is characterized by the abrupt occurrence, at possibly unknown irregular intervals, of physical events. For example, an event may correspond to the arrival or departure of a customer in a queue, the breakdown or restoration to service of a machine, or the transmission or reception of a message packet. Thus DES arise in service and logistic systems, manufacturing, and communications, as well as in many other domains such as vehicular traffic, and robot and process control at the level of task coordination.

Abstractly the distinguishing features of DES are that they are discrete (in time and state space), asynchronous (event- rather than clock-driven), nondeterministic (generative and capable of internal choices), and modular (composed of quasi-independent component DES down to some level of primitives). In addition DES may be equipped with various means of control and intercommunication, notably for the enablement/disablement of selected controllable events and the signaling of observable events from one module to another. Control and communication are to be coordinated so that the flow of events within the system takes place in accordance with designer specifications.

The increasing complexity of man-made DES made possible by computer technology underlines the need for formal theories of and systematic design approaches to DES control. Standard control theory and design, though finding application to DES at the level of small-system optimization [Ho87, Co85], is linear-space-based and is ill-equipped to address the larger structural issues, which cannot be framed in a linear setting; while modelling approaches originating in computer programming theory (e.g. [Hr85]) neither capture the crucial properties of DES relating to their degree of controllability and observability nor formalize the issues of control and communication architecture.

A control paradigm adapted to the DES area was introduced in [RW82] and has since been actively developed by those authors, their coworkers, and others (see [RW89] for an extensive bibliography). We refer to this framework as RW. While based on the broad concepts of feedback control and communication, RW incorporates the distinguishing features of DES itemized above, via constructs from automata, formal language and formal logic. While unspecific as to applications, RW has already been exploited by workers in communication protocols [Ci88], database management [La87], and flexible manufacturing [Ma86].

In this paper we provide a summary overview of RW, referring the reader for most technical details to the literature.

REPRESENTATION OF DES IN RW

In RW a DES is represented either by a formal language or, more concretely, by the generator of a formal language. For the application of control-theoretic techniques it is convenient to take as the generator a state transition structure (automaton). A typical primitive example, that we shall call **MACH**, is the 'machine' displayed in Fig. 1. The three states are labelled I ('idle'), W ('working') and D ('broken down'); the corresponding transitions, or *events*, then have the obvious interpretations. The events are labelled by symbols from an event *alphabet*, in this case the set $\sigma=\{\alpha,\beta,\lambda,\mu\}$. In the absence of any control, **MACH** may be thought of as spontaneously generating strings of symbols $\sigma \in \Sigma$ in accordance with the graph, starting from I as the initial state. Depending on the purposes of analysis one may consider all possible infinite strings so generated, or focus attention only on all possible strings of finite length. For simplicity we restrict attention for now to the finite strings. In standard notation the set of finite strings formed from Σ is denoted by Σ^*; the strings (or *words*) that can be generated by **MACH** thus make up a subset of Σ^* that we call the *closed behavior* L(**MACH**). It may be useful to select from L(**MACH**) those words that correspond to completed cycles of the type $\alpha\beta$ or $\alpha\lambda\mu$, or finite sequences of completed cycles: this may be done by *marking* suitable states of the transition graph (not to be confused with 'marking' of Petri nets). In this example I is marked and the resulting subset of *marked* strings in L(**MACH**) is called the *marked behavior* of **MACH**, denoted by L_m(**MACH**).

A feature of RW is that the foregoing more-or-less standard description is augmented by a control function as follows. We select a subset $\Sigma_c \subseteq \Sigma$ of events to be *controllable*; the complementary subset Σ_u is *uncontrollable*. In **MACH** the controllable subset is $\Sigma_c = \{\alpha, \mu\}$ and is distinguished in Fig. 1 by a 'tick' on the event arrow. Controllable events have the interpretation that they can be *disabled* (prevented from occurring) or *enabled* (allowed but not forced to occur) by some control agent, for the moment unspecified; while an uncontrollable event can never be directly prevented from occurring if the DES happens to be at the appropriate state, hence is always enabled. Which events are declared to be controllable is a matter of modelling; for instance in **MACH** it may be not unreasonable to assume that the transition $I \rightarrow W$ (initiation of a work cycle) can be disabled, and that $D \rightarrow I$ can be disabled (by witholding servicing in the case of breakdown), but that $W \rightarrow I$ (successful completion of a work cycle) or $W \rightarrow D$ (breakdown) will occur uncontrollably, in accordance with underlying physical mechanisms which the controller is unable to access directly. In this scenario control by an external agency is *permissive* in the sense that no event is 'forced', except possibly by disabling all the alternatives and thus forcing by default. Extensions of the model to accommodate forced events explicitly have been proposed ([GR87], [BH88]) but will not be considered here.

Several constructions exist for combining primitive DES like **MACH** into more elaborate structures of the same type: the simplest is *shuffle*, which creates a product structure DES from components over disjoint alphabets, and models the situation in which the generating actions of the components proceed independently from and asynchronously with one another. More generally one can bring in the *synchronous product*, allowing *a priori* synchronization of events having common labels in distinct components, and the *concurrency product* (cf. [LiW88a]), which allows for the possibility of unsynchronized (i.e. unforced) simultaneity of events in distinct components.

Finally, there is no compelling requirement that the state sets be finite; nor is it necessary to consider the models simply as "raw" transition structures. For instance the algebraic structure of vector addition systems (as in Petri nets) can be adjoined to the foregoing model and exploited to advantage in situations where the system state, or one of its factors in a product structure, can be modelled on the nonnegative integers, as for the occupancy number of I, W or D in a group of machines, or the content of a buffer (cf. [LiW88b]).

CONTROLLABLE LANGUAGES AND CENTRALIZED SUPERVISION OF DES

The type of control problem for which the foregoing modelling approach is natural is that of manipulating the controllable events, in the light of system past history, in such a way that the closed and marked behaviors actually generated by the DES under control satisfy designer specifications. Let **G** denote the DES to be controlled, **S** denote the controller (its style of representation is not important at the moment) and L(S/G) resp. L_m(S/G) the closed resp. marked behavior of '**G** under control of **S**'. Let A, E be sublanguages of L_m(G) corresponding respectively to 'minimal acceptable' and 'maximum permissible' marked behavior. Then a possible specification on **S** is that $A \subseteq L_m(\mathrm{S/G}) \subseteq E$. The questions now are whether any such **S** actually exist; if so, whether or not some notion of optimality can be attached to make the selection; and finally whether or not the whole approach can be made constructive in principle and computationally feasible in practice.

To examine these problems we note first that a physically realizable controller can do no more than map strings $s \in L(\mathrm{G})$ to subsets Σ' of controllable events, the interpretation being that only events $\sigma \in \Sigma' \cup \Sigma_u$ are candidates for the event immediately following the generation of s; namely events in $\Sigma' \subseteq \Sigma_c$ are enabled, while events in $\Sigma_c - \Sigma'$ are disabled. Under this constraint, the next event σ (if one is possible) is generated in accordance with the transition structure of **G**, and the process is repeated. What sublanguages of L_m(G) can be generated by a controller, or *supervisor*, acting in this manner? To answer this question we need a concept of controllability. For any language $K \subseteq \Sigma^*$, denote by $\bar{K}$ the *prefix-closure* of K, namely K together with all the prefixes (initial segments), including the empty prefix, of strings in K. Then K is *controllable* provided

$$\bar{K}\Sigma_u \cap L(\mathrm{G}) \subseteq \bar{K};$$

namely the next occurrence of an uncontrollable event in **G** can never cause a string already in $\bar{K}$ to exit from $\bar{K}$. (Here $\bar{K}\Sigma_u$ denotes the set of strings of the form $s\sigma$, with $s \in \bar{K}$ and $\sigma \in \Sigma_u$). Controllability can be thought of as "invariance with respect to the occurrence of uncontrollable events". The answer to our question is now immediate: $K = L_m(\mathrm{S/G})$ for some supervisor **S** if and only if $K \subseteq L_m(\mathrm{G})$ and K is controllable.

Examination of the controllability condition reveals two properties that can be used to settle on a definition of optimality. The first is that the empty language is controllable: setting $K=\emptyset$ we get that $\bar{K}=\emptyset$, hence $\bar{K}\Sigma_u=\emptyset$ and the claim is proved. The second is that the condition is closed under arbitrary unions, a claim that is easily verified from the convenient fact that the prefix-closure of an arbitrary union of languages is the union of the prefix-closures. Thus if $E \subseteq L_m(\mathrm{G})$ and $\mathrm{C}(E)$ is the collection of controllable sublanguages of E, then $\mathrm{C}(E) \neq \emptyset$ since $\emptyset \in \mathrm{C}(E)$ and so, taking the union of members of $\mathrm{C}(E)$, we find that the *supremal element*

$$K_{\text{sup}} := \sup \mathbf{C}(E)$$

exists and belongs to $\mathbf{C}(E)$. Since K_{sup} is controllable and belongs to E, it is the natural candidate for the 'optimal' solution of the problem $L_m(\mathbf{S}/\mathbf{G}) \subseteq E$; a supervisor S that implements, or 'synthesizes' K_{sup}, is *maximally permissive* with respect to the constraint E. Finally we have an abstract solution to our supervisory existence problem: it is solvable if and only if

$$K_{\text{sup}} \supseteq A$$

It can be shown that K_{sup} can be characterized as the largest fixpoint of a certain mapping on $\mathbf{Pwr}(\Sigma^*)$, the set of sublanguages of Σ^*. In general this provides an approach to the explicit computation of K_{sup} by successive approximation. In the regular case (when all the given languages are representable by finite state generators) this computation converges to K_{sup} in a finite number of steps of worst case order $\|E\| \cdot \|L_m(\mathbf{G})\|$, where $\|\cdot\|$ denotes Nerode index (state size) of the indicated language. In practice the convergence is much faster.

We have now shown that satisfactory solutions are at hand to the problems raised at the beginning of this section, at least in the regular case. Software that implements the approach is available; examples and computational details can be found in [Wo88]. We now turn to architectural issues.

MODULAR SUPERVISION OF DES

As indicated in the Introduction, DES are often built up from modular elements, and so it makes sense to carry over the idea of modularity to control itself (cf. [WR88]). Very often a control task will consist of several specialized subtasks: for instance the group of machines making up a work cell may be subject to one control specification in respect to the prevention of overflow and underflow of buffers, and another that establishes priorities of repair when one or more machines are down. Controllers dealing with specialized subtasks may often be designed rather easily from the subtask specifications; and these subcontrollers can then be run concurrently to implement a modular solution of the original problem. In addition to being more easily synthesized, such a *modular supervisor* should ideally be more readily modified, updated and maintained. For example, if one subtask is changed, then it should only be necessary to redesign the corresponding subcontroller: in other words, the overall modular supervisor should exhibit greater flexibility than its 'monolithic' counterpart.

Unfortunately these advantages are not always to be gained without a price. The fact that the individual control modules are simpler implies that their control action must be based on a partial or 'local' version of the global system state; in linguistic terms, a subcontroller processes only a *projection* of the behavior of the DES to be controlled. A consequence of this relative insularity may be that different subsupervisors, acting quasi-independently on the basis of local information, come into conflict at the 'global' level, and the overall system thereby exhibits *blocking* (inability to complete the global task) or even *deadlock* (inability to continue operation). Thus a fundamental issue that always arises in the presence of modularity is how to guarantee the nonblocking property of the final synthesis.

We can focus more sharply on the blocking issue through the following definition: languages L_1 and L_2 are *nonconflicting* if

$$\overline{L_1 \cap L_2} = \bar{L}_1 \cap \bar{L}_2 ;$$

namely any string that is both a prefix of L_1 and a prefix of L_2 can be completed to a common word of L_1 and L_2. It can be shown that two individually nonblocking subsupervisors S_1 and S_2 acting concurrently will yield a nonblocking *conjunction* $S_1 \wedge S_2$ just in case the individual languages $L_m(S_1/G)$ and $L_m(S/G)$ are nonconflicting. Furthermore, if $E_1, E_2 \subseteq L_m(G)$ and if sup $C(E_1)$, sup $C(E_2)$ are nonconflicting, then

$$\sup C(E_1 \cap E_2) = \sup C(E_1) \cap \sup C(E_2)$$

While these results are more-or-less immediate consequences of the definitions, they lead to direct computational procedures for validating any proposed modular design in respect to nonblocking and optimality, as well as to analytical methods of inferring these desirable properties from an examination of specific modular structure in special cases.

The reduction in complexity gained by exploiting modularity can be dramatic. As a simple example consider three 'machines' of the type of **MACH** together with a buffer, arranged as in Fig. 2. The buffer, of capacity 3, serves as output (sink) for **MACH1** and **MACH2** and as input (source) for **MACH3**. The specifications are (i) the buffer must not overflow or underflow, (ii) **MACH1** and **MACH2** are repaired in order of breakdown, and (iii) **MACH3** has priority of repair over **MACH1** and **MACH2**. As the DES to be controlled we take the shuffle **BIGMACH** of **MACH1, MACH2** and **MACH3**, consisting of 27 states and 108 transitions (written (27,108)). Expressing the specifications as languages and combining these into their intersection, we obtain the 'monolithic' specification as a generator **BIGSPEC** (32,248). The optimal 'monolithic' supervisor is computed as an automaton **BIGSUP** (96,302), evidently a rather cumbersome structure to implement directly. By contrast the same behavioral result can be obtained by inspection using the conjunction of 4 subsupervisors based rather directly on the given individual specifications; the largest of these modular components has only 4 states. The details may be found in [Wo88].

HIERARCHICAL SUPERVISION OF DES

Hierarchical structure is a familiar feature of the control of complex dynamic systems, where the controlled system may be thought of as executing some overall high-level task. It may be described generally as a division of control action and the concomitant information processing according to scope. Commonly, the scope of a control action is defined by the extent of its temporal horizon, or by the depth of its logical dependence in a task decomposition. Generally speaking, the broader the temporal horizon of a control and its associated subtask, or the deeper its logical dependence on other controls and subtasks, the higher it is said to reside in the hierarchy. Frequently the two features of broad temporal horizon and deep logical dependency are found together.

Hierarchical structure in the control of DES can be investigated in RW by means of a mild extension of the framework already introduced. While different approaches to hierarchical control might be adopted even within this restricted framework, the theory to be summarized in this section does capture the basic feature of scope already mentioned, and casts some light on an issue that we call *hierarchical consistency*. Our account follows [ZW88].

In outline our setup will be the following. Consider a two-level hierarchy consisting of a low-level plant $\mathbf{G}_{lo}$ and controller $\mathbf{C}_{lo}$, along with a high-level plant $\mathbf{G}_{hi}$ and controller $\mathbf{C}_{hi}$. These are coupled as shown in Fig. 3. Our viewpoint is that $\mathbf{G}_{lo}$ is the actual plant to be controlled in the real world by $\mathbf{C}_{lo}$, the *operator*; while $\mathbf{G}_{hi}$ is an abstract, simplified model of $\mathbf{G}_{lo}$ that is employed for decision-making in an ideal world by $\mathbf{C}_{hi}$, the *manager*. The model $\mathbf{G}_{hi}$ is refreshed or updated every so often via the information channel (or mapping) labelled $\mathbf{Inf}_{lohi}$ (information low-to-high) to $\mathbf{G}_{hi}$ from $\mathbf{G}_{lo}$. Alternatively one can interpret $\mathbf{Inf}_{lohi}$ as carrying information sent up by the operator $\mathbf{C}_{lo}$ to the manager $\mathbf{C}_{hi}$: in our model the formal result will be the same. Another information channel, $\mathbf{Inf}_{lo}$ (low-level information), provides conventional feedback from $\mathbf{G}_{lo}$ to its controller $\mathbf{C}_{lo}$, which in turn applies conventional control to $\mathbf{G}_{lo}$ via the control channel labelled $\mathbf{Con}_{lo}$ (low-level control). Returning to the high level, we consider that $\mathbf{G}_{hi}$ is endowed with control structure, according to which it makes sense for $\mathbf{C}_{hi}$ to attempt to exercise control over the behavior of $\mathbf{G}_{hi}$ via the control channel $\mathbf{Con}_{hi}$ (high-level control), on the basis of feedback received from $\mathbf{G}_{hi}$ via the information channel $\mathbf{Inf}_{hi}$ (high-level information). In actuality, the control exercised by $\mathbf{C}_{hi}$ in this way is only 'virtual', in that the behavior of $\mathbf{G}_{hi}$ is determined entirely by the behavior of $\mathbf{G}_{lo}$, through the updating process mediated by $\mathbf{Inf}_{lohi}$. The structure is, however, completed by the command channel $\mathbf{Com}_{hilo}$ linking $\mathbf{C}_{hi}$ to $\mathbf{C}_{lo}$. The function of $\mathbf{Com}_{hilo}$ is to convey the manager's high-level control signals as commands to the operator $\mathbf{C}_{lo}$, which must translate these commands into corresponding low-level signals which will actuate $\mathbf{G}_{lo}$ via $\mathbf{Con}_{lo}$. State changes in $\mathbf{G}_{lo}$ will eventually be conveyed in summary form to $\mathbf{G}_{hi}$ via

$\mathbf{Inf_{lohi}}$. $\mathbf{G_{hi}}$ is updated accordingly, and then provides appropriate feedback to $\mathbf{C_{hi}}$ via $\mathbf{Inf_{hi}}$. In this way the hierarchical loop is closed. The forward path sequence $\mathbf{Com_{hilo}}$; $\mathbf{Con_{lo}}$ is conventionally designated "command & control", while the feedback path sequence $\mathbf{Inf_{lohi}}$; $\mathbf{Inf_{hi}}$ will be referred to as "report & advise".

As a metaphor, one might think of the command center of a complex system (e.g. electric power distribution system,...) as the site of the high-level plant model $\mathbf{G_{hi}}$, where a high-level decision-maker or manager $\mathbf{C_{hi}}$ is in command. The external real world and those operators coping with it are embodied in $\mathbf{G_{lo}}$, $\mathbf{C_{lo}}$.

The questions addressed by the theory concern the relationship between the behavior required, or expected, by the manager $\mathbf{C_{hi}}$ of the high-level model $\mathbf{G_{hi}}$, and the actual behavior implemented by the operator $\mathbf{C_{lo}}$ in $\mathbf{G_{lo}}$ in the manner described, when $\mathbf{G_{lo}}$ and $\mathbf{Inf_{lohi}}$ are given at the start. It will turn out that *hierarchical consistency* between these behaviors imposes rather stringent requirements on $\mathbf{Inf_{lohi}}$ and that, in general, it is necessary to refine the information conveyed by this channel before consistent hierarchical control structure can be achieved. This result accords with the intuition that for high-level control the information sent up by the operator to the manager must be timely, and sufficiently detailed for various critical low-level situations to be distinguished.

As usual we model $\mathbf{G_{lo}}$ as the generator of a language $L_{lo} := L(\mathbf{G_{lo}}) \subseteq \Sigma^*$, with the partition $\Sigma = \Sigma_c \cup \Sigma_u$ as before. For $M \subseteq L(\mathbf{G})$ we use the abbreviated notation

$$M^{\uparrow} := \sup \mathrm{C}(M)$$

Now let T be a new alphabet of 'significant event labels'. T may be thought of as the events perceived by the manager which will enter into the description of the high-level plant model $\mathbf{G_{hi}}$, of which the derivation will follow in a moment. First, to model the information channel $\mathbf{Inf_{lohi}}$ we postulate a map $\eta : L_{lo} \rightarrow T^*$ with the properties

$$\eta(\varepsilon) = \varepsilon \quad (\varepsilon \text{ denotes the empty string over any alphabet})$$

$$\eta(s\sigma) = \begin{cases} \text{either } \eta(s) \\ \text{or } \eta(s)\tau, \text{ some } \tau \in T \end{cases}$$

for $s \in \Sigma^*$, $\sigma \in \Sigma$. Thus η is *causal* in the sense that it is prefix-preserving: if $s \le s'$ then $\eta(s) \le \eta(s')$ (here $\le$ means "is a prefix of"). Intuitively η can be used to signal the occurrence of events that depend in some fashion on the past history of the behavior of $\mathbf{G_{lo}}$: for instance η might produce a fresh symbol τ' whenever $\mathbf{G_{lo}}$ has just generated a nonzero multiple of 10 of some distinguished symbol σ', but 'remain silent' otherwise. It is convenient to combine η with $\mathbf{G_{lo}}$ in a unified description. This may be done in standard fashion by replacing the pair $(\mathbf{G_{lo}}, \eta)$ by a Moore generator, $\mathbf{G_{lo,new}}$ say, having state-outputs over the alphabet $T_o := T \cup \{\tau_o\}$, where τ_o is a new symbol ($\notin T$) interpreted as the 'silent output symbol'. The abstract construction of $\mathbf{G}_{\mathbf{lo,new}}$ from $\mathbf{G}_{\mathbf{lo}}$ and η is a routine exercise that we omit; and we now rename $\mathbf{G}_{\mathbf{lo,new}}$ as simply $\mathbf{G}_{\mathbf{lo}}$. The states of $\mathbf{G}_{\mathbf{lo}}$ are either *vocal* (state-output in T) or *silent* (output τ_o); a *silent path* in $\mathbf{G}_{\mathbf{lo}}$ is a path in the transition graph of $\mathbf{G}_{\mathbf{lo}}$ that starts at a vocal state and whose subsequent states are all silent.

At this stage we temporarily define $\mathbf{G}_{\mathbf{hi}}$. For this we note that, in the absence of any control action, $\mathbf{G}_{\mathbf{lo}}$ generates the uncontrolled language L_{lo} (unchanged from before). For now, $\mathbf{G}_{\mathbf{hi}}$ will be taken as the canonical recognizer for the image of L_{lo} under η:

$$L(\mathbf{G}_{\mathbf{hi}}) = \eta(L_{lo}) \subseteq T^*$$

and we write $L(\mathbf{G}_{\mathbf{hi}}) =: L_{hi}$. As yet, however, the event label alphabet T of $\mathbf{G}_{\mathbf{hi}}$ needn't admit any natural partition into controllable and uncontrollable subalphabets; that is, $\mathbf{G}_{\mathbf{hi}}$ needn't admit any natural control structure.

This defect can be remedied by refining the state structure of $\mathbf{G}_{\mathbf{lo}}$ and by splitting the elements of T into 'siblings'. Thus each $\tau \in T$ is split into a pair (τ_c, τ_u); in the extended state structure $\mathbf{G}_{\mathbf{lo,ext}}$, say, a state-output τ at a vocal state q is replaced by either τ_c or τ_u according as q can or cannot be made unreachable along all silent paths in $\mathbf{G}_{\mathbf{lo,ext}}$ leading to q, by disablement of suitable $\sigma \in \Sigma_c$. It can be shown that the state size of $\mathbf{G}_{\mathbf{lo,ext}}$ is at most double that of $\mathbf{G}_{\mathbf{lo}}$; in random examples the factor is usually much less. Now T is replaced by T_{ext}, say, in this way, and T_o by $T_{o,ext} := T_{ext} \cup \{\tau_o\}$. The corresponding extended map η_{ext} will induce an extended high-level model $\mathbf{G}_{\mathbf{hi,ext}}$ over T_{ext}. Since each element in T_{ext} is now unambiguously controllable or uncontrollable, the structure $\mathbf{G}_{\mathbf{lo,ext}}$ is said to be *output-control-consistent* (OCC). $\mathbf{G}_{\mathbf{hi,ext}}$ will be a DES having standard RW control structure, to which the usual methods of Sects. 2 and 3 can be applied. Henceforth we assume that this extension procedure has been carried out (it has been implemented in the regular case), and drop the subscript 'ext'.

Despite the fact that $\mathbf{G}_{\mathbf{lo}}$ is now OCC, it needn't be true that the controllability property is mapped in either direction between $\mathbf{G}_{\mathbf{lo}}$ and $\mathbf{G}_{\mathbf{hi}}$: that is, $K_{lo} \subseteq L(\mathbf{G}_{\mathbf{lo}})$ may be controllable with respect to $\mathbf{G}_{\mathbf{lo}}$, yet $K_{hi} := \eta(K_{lo}) \subseteq L(\mathbf{G}_{\mathbf{hi}})$ not controllable with respect to $\mathbf{G}_{\mathbf{hi}}$; conversely K_{hi} controllable does not imply that $K_{lo} := \eta^{-1}(K_{hi})$ is controllable.*

The property that $\mathbf{G}_{\mathbf{lo}}$ is OCC is exploited as follows. High-level supervisory control is determined by a selection of high-level controllable events to be disabled, on the basis of high-level past history. That is, $\mathbf{C}_{\mathbf{hi}}$ is defined by a map

$$\gamma_{hi} : L_{hi} \times T \to \{0,1\}$$

such that $\gamma_{hi}(t,\tau)=1$ for all $t \in L_{hi}$ and $\tau \in T_u$. As usual, if $\gamma_{hi}(t,\tau)=0$ the event labelled τ is said

*This remark corrects an inessential but annoying error in [ZW88], where Proposition 3.1 should be deleted.

to be disabled; otherwise τ is enabled; of course, only controllable events ($\tau \in T_c$) can be disabled. The result of applying this control on the generating action of $\mathbf{G_{hi}}$ would amount to the construction of a suitable supervisor over T as input alphabet. However, in the hierarchical control loop direct implementation of $\mathbf{C_{hi}}$ is replaced by command & control: the action of $\mathbf{C_{hi}}$ on $\mathbf{G_{hi}}$ must be mediated via $\mathbf{Com_{hilo}}$ and $\mathbf{Con_{lo}}$ as already described. With γ_{hi} given, it turns out to be possible to construct a corresponding low-level disabled event map

$$\gamma_{lo} : L_{lo} \times \Sigma \rightarrow \{0,1\}$$

that matches the command and control structural constraint, by which the operator can execute a command of the form "disable τ" received from the manager. Now suppose that a nonempty closed specification language $E_{hi} \subseteq L_{hi}$ is established by the manager. It may be assumed that E_{hi} is controllable; otherwise the manager simply replaces E_{hi} by $E_{hi}^{\uparrow}$. Next γ_{hi} is determined in such a way that the corresponding high-level controlled language, $L(\gamma_{hi}, \mathbf{G_{hi}})$, say, is E_{hi} (or would be E_{hi} if direct control of $\mathbf{G_{hi}}$ by $\mathbf{C_{hi}}$ in the sense of Sect. 1 were possible). Define E_{lo} as the preimage in L_{lo} of E_{hi} under the map η corresponding to $\mathbf{Inf_{lohi}}$:

$$E_{lo} := \eta^{-1}(E_{hi}) \subseteq L_{lo}$$

In general, as we know, E_{lo} is not controllable. Let γ_{lo} be determined from E_{lo} as described above. The main consequence of output-control-consistency is that by use of γ_{lo} the closed-loop language $L(\gamma_{lo}, \mathbf{G_{lo}})$ synthesized in $\mathbf{G_{lo}}$ by command & control is as large as possible subject to the constraint E_{lo} just defined:

$$L(\gamma_{lo}, \mathbf{G_{lo}}) = E_{lo}^{\uparrow}$$

Obviously the transmitted high-level behavior will satisfy the required specification constraint:

$$\eta(L(\gamma_{lo}, \mathbf{G_{lo}})) \subseteq E_{hi}$$

but in general the inclusion will be proper. That is, while the 'expectation' of the high-level controller $\mathbf{C_{hi}}$ on using the control γ_{hi} might ideally be the synthesis in $\mathbf{G_{hi}}$ of the controllable behavior E_{hi}, only a proper subset of this behavior can in general actually be realized. The reason is simply that a call by $\mathbf{C_{hi}}$ for the disablement of some high-level event $\tau \in T_c$ may require $\mathbf{C_{lo}}$ (the control γ_{lo}), as an undesired side effect, to disable paths in $\mathbf{G_{lo}}$ that lead directly to outputs other than τ. However this result is the best that can be achieved under the current assumptions about $\mathbf{G_{lo}}$. The main result above will be called *low-level hierarchical consistency*. Intuitively it guarantees that the updated behavior of $\mathbf{G_{hi}}$ will always satisfy the high-level specification constraint, and that the 'real' low-level behavior in $\mathbf{G_{lo}}$ is as large as possible subject to this constraint. Nevertheless, the situation from the manager's viewpoint is still unsatisfactory: the high-level behavior he expects may be larger than what the operator of $\mathbf{G}_{lo}$ can optimally report.

The desirable situation would be that, whenever E_{hi} is controllable, then

$$\eta((\eta^{-1}(E_{hi}))^{\uparrow}) = E_{hi}$$

The foregoing property will be called *high-level hierarchical consistency*. In that case, the command and control process defined for E_{hi} will actually synthesize E_{hi} in $\mathbf{G}_{hl}$. Achieving this property in general requires a further refinement of the transition structure of $\mathbf{G}_{lo}$: in other words, the possibly costly step of enhancing the information transmitted by $\mathbf{G}_{lo}$ to $\mathbf{G}_{hl}$. Suffice it to say here that the appropriate construction can be carried out effectively (at least in the regular case), resulting in the property for the refined version of $\mathbf{G}_{lo}$ that it is now *strictly* output-control-consistent (SOCC). That is, with $\mathbf{G}_{lo}$ now SOCC, high-level hierarchical consistency is achieved for arbitrary high-level specification languages.

Two conclusions that may be drawn from this rather involved discussion are that, first, RW supports a plausible hierarchical control architecture; but secondly, the design of consistent hierarchical supervisory controls can demand quite refined consideration of low-level system structure and of the definition of high-level significant events.

The theory will be illustrated by developing a high-level hierarchical supervisor for Transfer Line, consisting of two machines **M1**, **M2** plus a test unit **TU**, linked by buffers **B1**, **B2** in the sequence: **M1**, **B1**, **M2**, **B2**, **TU** (Fig. 4). State transition diagrams of **M1**, **M2**, and **TU** are displayed in Fig. 5.

TU either "passes" or "fails" each processed workpiece, signaling its decision with events 60, 80 respectively. In case of "pass test", the workpiece is sent to the system output (event 62); in case of "fail test", it is returned to **B1** (event 82) for reprocessing by **M2**. There is no limit on the number of failure/reprocess cycles a given workpiece may undergo.

For ease of display we consider only the simplest case, where **B1** and **B2** each has capacity 1. Initially an optimal low-level supervisor is designed by any of the methods of previous sections, to ensure that neither of the buffers is subject to overflow or underflow. In detail, let

PL = shuffle(M1,M2,TU);

and let **B1SP**, **B2SP** be the buffer specification generators (Fig. 6)*. Then we set **BSP** = **meet(B1SP,B2SP)**, and

* Boldface function names refer to computing procedures of the software package TCT, available from the author.

PLSUP = supcon(PL,BSP)

as displayed in Fig. 7. With **PLSUP** as the starting point for the development of hierarchical structure, we must first assign the "significant" events to be signaled to the "manager". Let us assume that the manager is interested only in the events corresponding to "taking a fresh workpiece" (low-level event 1, signaled as high-level event $\tau 1$, say), and to "pass test" (low-level event 60, signaled as $\tau 2$) or "fail test" (low-level event 80, signaled as $\tau 3$). If too many failures occur the manager intends to take remedial action, which will start by disabling the failure/reprocess cycle. To this end the uncontrollable event 80 is now replaced in the low-level structure by a new controllable event 81. Furthermore, the meaning of the signaled events $\tau 1$, $\tau 2$, $\tau 3$ must be unambiguous, so a transition entering state 1 like [8,62,1] must not be confused with the "significant" transition [0,1,1]; namely a new state (say, 12) must be introduced, transition [8,62,1] replaced by [8,62,12], and a new transition [12,2,2] inserted. The final Moore structure, **GLO**, is displayed in Fig. 8. Here the vocal [state, output] pairs are [1,τ1], [8,τ1], [7,τ2] and [6,τ3].

We are now ready to carry out the procedures of the theory. By inspection of Fig. 8, it is clear that each of $\tau 1$, $\tau 2$, $\tau 3$ is unambiguously controllable, that is, **GLO** is already output-control-consistent. The corresponding high-level model **GHI** is displayed in Fig. 9.

However, for the manager to disable $\tau 2$ will require the operator to disable low-level event 5, which in turn disables the high-level event $\tau 3$ as an undesired side effect; thus **GLO** is not strictly-output-control-consistent (SOCC). To improve matters it is enough to vocalize the low-level state 5 with a new high-level output $\tau 4$, signaling the new "significant" event that "TU takes a workpiece". This step incidentally converts the status of $\tau 2$ from controllable to uncontrollable. With this the construction of a SOCC model, say **CGLO**, from **GLO** is complete (Fig. 10). The corresponding high-level model **CGHI** is displayed in Fig. 11, where $\tau 1$, $\tau 2$, $\tau 3$, $\tau 4$ have been coded respectively as 11, 20, 31, 41.

The simple model **CGHI** can be supervised by the manager to achieve his objective of "quality control". A possible high-level specification might be: "If two consecutive test failures (31) occur, allow TU to operate just once more, then shut down the system"; this is modeled by **HISP** as displayed (Fig. 12). The resulting supervisor

CGHISUP = supcon(CGHI,HISP)

is shown in Fig. 13. On termination of **CGHISUP** at state 7, it can be easily verified that **CGLO** will have halted at its marker state 0.

OTHER DEVELOPMENTS

The foregoing sections suffice to convey the flavor of RW control theory for DES. A number of important topics not touched on here have been discussed in the literature. Decentralized control based explicitly on local models of the global DES is investigated in [LnW88a], while supervision based on partial observations (i.e. observation of a subset of the event alphabet) is considered in [LnW88b]. Both points of view are combined in a concept of coordination explored in [LnW88c].

We comment briefly on an extension of the theory to supervisor synthesis subject to infinite-string specifications. While the framework of languages in Σ^* employed in previous sections may be adequate for supervisor synthesis subject to most practical 'safety' specifications, nothing good is ever *guaranteed* to happen. One way to address such *liveness* issues is to bring in languages with infinite strings -- so-called Σ^ω-languages. In this framework an event can be required to occur "eventually", without specifically stating when. While the theory becomes more technical, it is hoped that the final results will be natural and simply expressible. In addition the new framework ought to provide a semantics for the use of temporal logic as a convenient specification language. Appropriate definitions of controllability and nonblocking in the Σ^ω setting have been provided [TW87, TW88]. The supervisory synthesis problem has the same formal appearance as in Sect. 1, except that Σ^* is replaced by Σ^ω. Under technical conditions, a unique optimal solution will exist. An effective solution is possible at least when the languages A, E and $L(\mathrm{G})$ have representations as finite automata (generators) over infinite strings, a situation that can be formalized in terms of so-called Muller automata. In the solvable case, a finite, nonblocking supervisor that solves the synthesis problem can be effectively constructed.

Finally we mention a generalization of RW that may well be a promising approach to real-time control of DES, although as yet no formal synthesis methods have been developed for the systematic computation of 'optimal' supervisors. *Extended state machines* (ESMs) in the sense of [OW87], building on [MP83] and [Hr85], model DES as extended transition structures, involving a structured state space defined as the product of an automaton-like state set for *activity variables* and a state space (e.g. $\mathbf{Z}^n$) of conventional type for (e.g. numerical) *process variables*. Transitions are structured to include not only a transition label but also a boolean guard, program step (variable assignment or synchronous communication), and lower and upper time bounds. The closed loop system becomes a suitably defined concurrency product of a clock ESM, plant ESMs and controller ESMs. These component ESMs interact via shared and communicating transitions. Semantically, ESMs are interpreted as generators of *trajectories*, namely infinite sequences of states and transitions. Trajectories are established by initialization, followed by consistency with the guards, variable assignments and time bounds of the transitions. Specification of the behavior of a controlled system of this kind can be carried out in a version of temporal logic that includes the real-time feature; specifications may include properties of safety, priority and real-time liveness. Important current research problems in this area revolve around such issues as supervisor verification by effective decision procedures, criteria for supervisor existence, and the distributed control problems of modular design.

CONCLUSIONS

In this paper we have provided an overview of one trend among others in the development of a control theory for discrete-event systems. In view of the relatively long history of prior approaches to discrete-event control design (notably discrete-event system simulation, and analysis via Petri nets, starting in the 1960s; and investigations via queueing theory and its variants, including perturbation analysis, from the early 1970s) it is perhaps surprising that attempts to evolve a synthetic, control-theoretic overview of the problem area, especially in its qualitative, logical aspects, have been both few in number and recent in appearance. In any case, it can fairly be said that control of DES is now an established branch of control theory.

The current studies of control of DES in its qualitative aspects highlight the thesis that control science is defined in terms of problems and concepts, not in terms of techniques. In general control science may be described as the study of how information and dynamics are brought into purposeful interaction. Stimulated by the demands of technology and by developments in computer science, control science has entered a new phase, where discreteness, modularity and communication are fundamental. Alongside the traditional mathematics of control theory like differential equations and operator theory, new techniques are entering the field from automaton theory, formal language and formal logic; while developments in computer programming methodology, as for instance abstract data structures and the object-oriented paradigm, may strongly influence the way this new mathematics (new in control theory) will be put to work. For both researchers and educators in the control field, the challenges are plentiful.

REFERENCES

[BH88] Y. Brave, M. Heymann. Formulation and control of real time discrete event processes. Proc. 27th IEEE Conf. on Decision and Control, IEEE Control

Systems Society, New York, Dec. 1988, pp. 1131-1132.

[Ci88] R. Cieslak, C. Desclaux, A. Fawaz, P. Varaiya. Supervisory control of discrete event processes with partial observations. IEEE Trans. Aut. Control 33 (3), 1988, pp. 249-260.

[Co85] G. Cohen, D. Dubois, J.P. Quadrat, M. Viot. A linear-system-theoretic view of discrete-event processes and its use for performance evaluation in manufacturing. IEEE Trans. Aut. Control AC-30 (3), 1985, pp. 210-220.

[GR87] C.H. Golaszewski, P.J. Ramadge. Control of discrete event processes with forced events. Proc. 26th IEEE Conf. on Decision and Control, IEEE Control Systems Society, New York, Dec. 1987, pp. 247-251.

[Ho87] Y.C. Ho. Perturbation analysis explained. Proc. 26th IEEE Conf. on Decision and Control, IEEE Control Systems Society, New York, Dec. 1987, pp. 243-246.

[Hr85] C.A.R. Hoare. Communicating Sequential Processes. Prentice-Hall, Englewood Cliffs, 1985.

[La87] S. Lafortune. Modeling and analysis of transaction execution in database systems. Report CRL-TR-06-87, Computing Research Laboratory, The University of Michigan, Ann Arbor, Aug. 1987.

[LiW88a] Y. Li, W.M. Wonham. On supervisory control of real-time discrete event systems. Information Sciences 46 (2), 1988, pp. 159-183.

[LiW88b] Y. Li, W.M. Wonham. A state-variable approach to the modeling and control of discrete-event systems. Proc. 26th Annual Allerton Conference on Communication, Control, and Computing, University of Illinois, 1988, pp. 1140-1149.

[LnW88a] F. Lin, W.M. Wonham. Decentralized supervisory control of discrete-event systems. Information Sciences 44 (2), 1988, pp. 199-224.

[LnW88b] F. Lin, W.M. Wonham. On observability of discrete-event systems. Information Sciences 44 (2), 1988, pp. 173-198.

[LnW88c] F. Lin, W.M. Wonham. Decentralized control and coordination of discrete-event systems. Proc. 27th IEEE Conference on Decision and Control, IEEE Control Systems Society, New York, Dec. 1988, pp. 1125-1130.

[Ma86] O. Maimon, G. Tadmor. Efficient low-level control of flexible manufacturing systems. MIT LIDS Rpt. No. LIDS-P-1571, Cambridge, MA., 1986.

[MP83] Z. Manna, A. Pnueli. Verification of concurrent programs: A temporal proof system. Foundations of Computer Science IV, Mathematics Center Tracts, Amsterdam 1983, pp. 163-225.

[OW87] J.S. Ostroff, W.M. Wonham. Modelling, specifying and verifying real-time embedded computer systems. Proc. Eighth Real-Time Systems Symposium, IEEE Computer Society, New York, Dec. 1987, pp. 124-132.

[RW82] P.J. Ramadge, W.M. Wonham. Supervision of discrete-event processes. Proc. 21st IEEE Conf. on Decision and Control, IEEE Control Systems Society, New York, Dec. 1982, pp. 1228-1229.

[RW89] P.J. Ramadge, W.M. Wonham. Control of discrete-event systems. Proc. IEEE, Special Issue on Discrete Event Dynamic Systems, 77(1), Jan. 1989, pp. 81-98.

[TW87] J.G. Thistle, W.M. Wonham. Supervisory control with infinite-string specifications. Proc. Twenty-Fifth Annual Allerton Conference on Communication, Control and Computing, University of Illinois, 1987, vol. 1, pp. 327-334.

[TW88] J.G. Thistle, W.M. Wonham. On the synthesis of supervisors subject to ω-language specifications. Proc. 1988 Conference on Information Sciences and Systems, Dept. of Electrical Engineering, Princeton University, 1988, pp. 440-444.

[Wo88] W.M. Wonham. A control theory for discrete-event systems. In M.J. Denham, A.J. Laub (Eds.), Advanced Computing Concepts and Techniques in Control Engineering, NATO ASI Series, vol. F47, Springer-Verlag, Berlin, 1988; pp. 129-169.

[WR88] W.M. Wonham, P.J. Ramadge. Modular supervisory control of discrete event systems. Maths. of Control, Signals & Systems 1 (1), 1988, pp. 13-30.

[ZW88] H. Zhong, W.M. Wonham. On hierarchical control of discrete-event systems. Proc. 1988 Conference on Information Sciences and Systems, Dept. of Electrical Engineering, Princeton University, 1988, pp. 64-70.

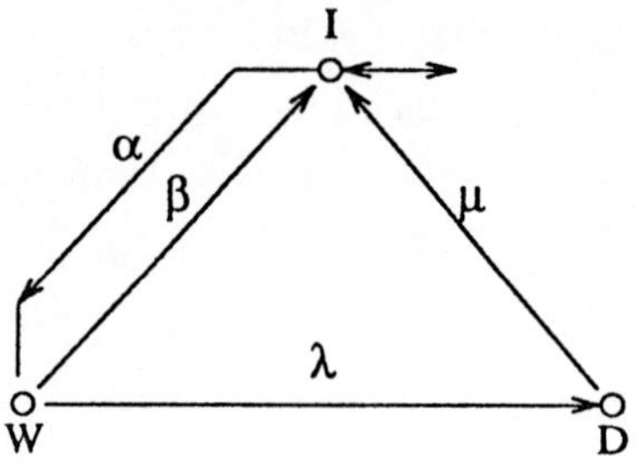

Fig. 1
'Machine' **MACH**

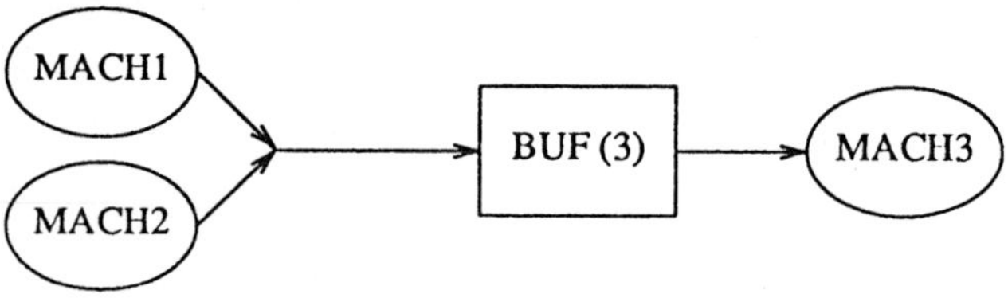

Fig. 2
'Factory'

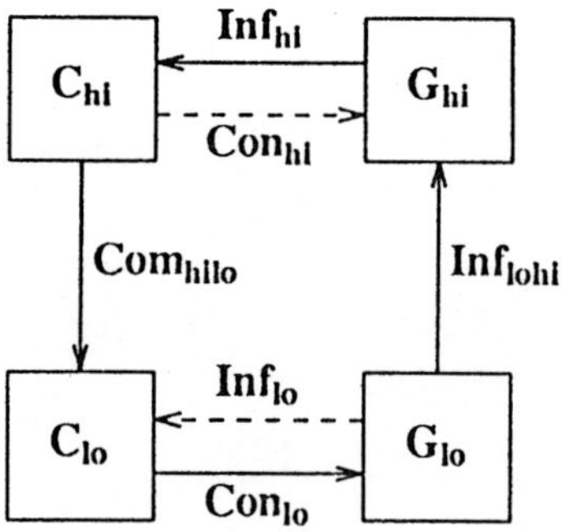

Fig. 3
Two-Level Hierarchy

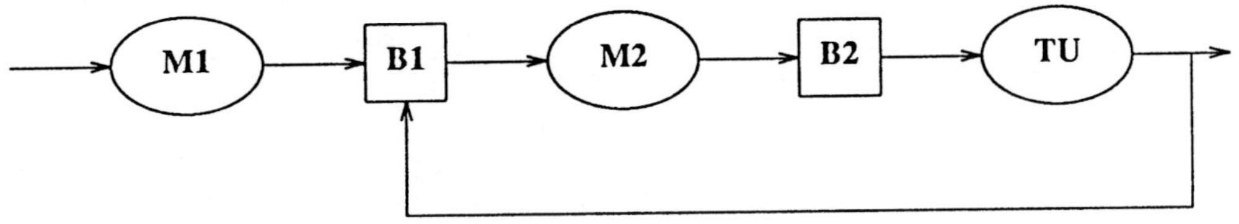

Fig. 4
Transfer Line

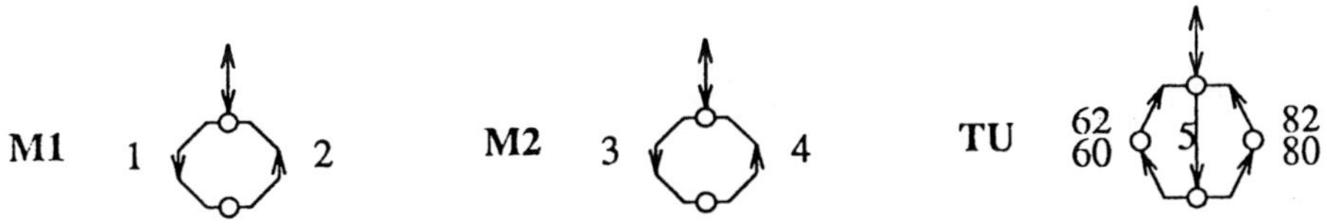

Fig. 5
State Diagrams - Transfer Line

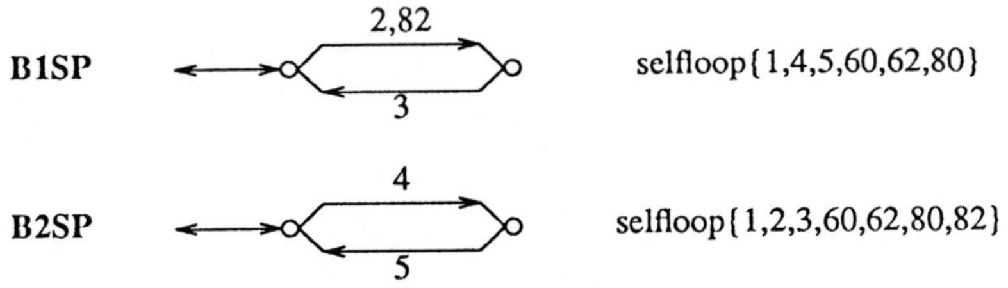

Fig. 6

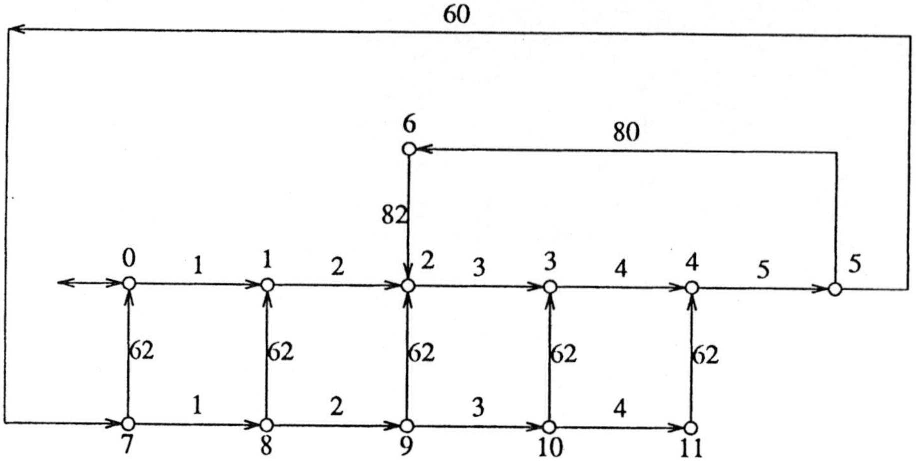

Fig. 7
PLSUP

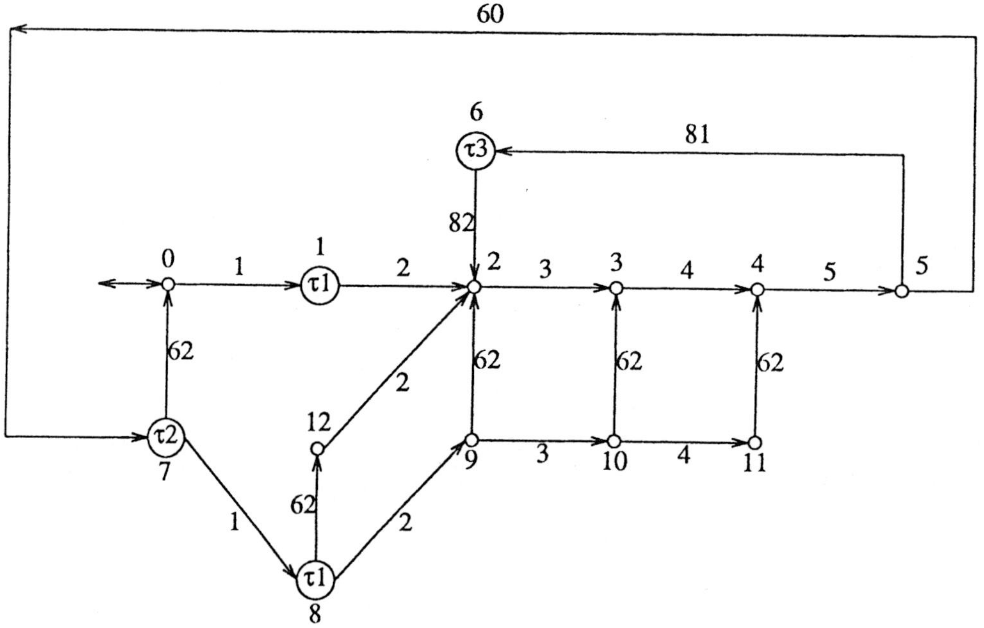

Fig. 8
GLO

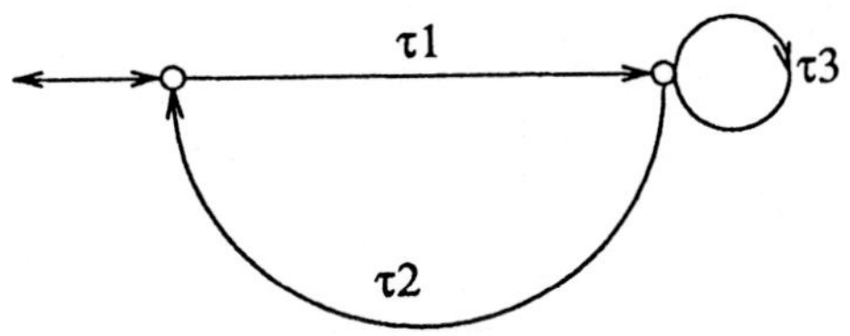

Fig. 9
GHI

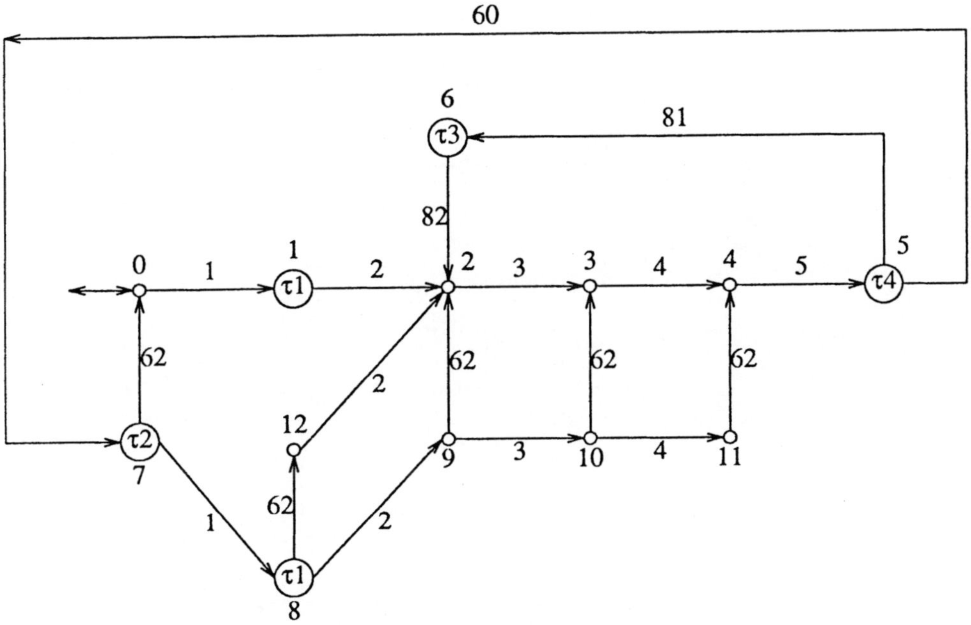

Fig. 10
CGLO

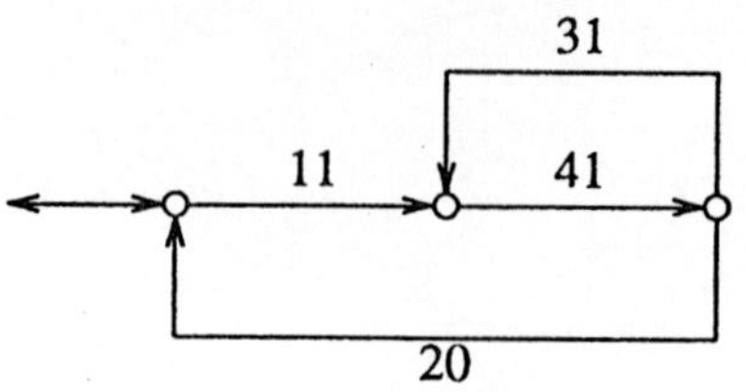

Fig. 11
CGHI

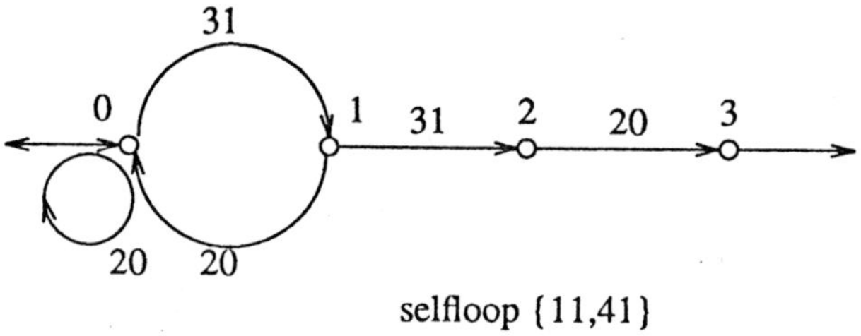

Fig. 12
HISP

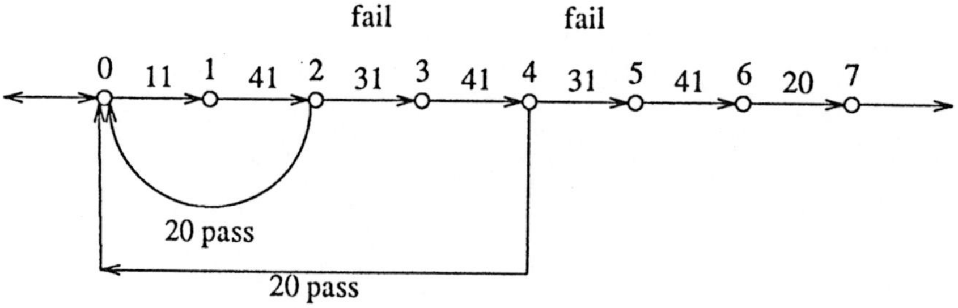

Fig. 13
CGHISUP

Lecture Notes in Control and Information Sciences

Edited by M. Thoma and A. Wyner

Lecture Notes in Control and Information Sciences

Edited by M. Thoma and A. Wyner

Lecture Notes in Control and Information Sciences

Edited by M. Thoma and A. Wyner

Vol. 117: K.J. Hunt
Stochastic Optimal Control Theory
with Application in Self-Tuning Control
X, 308 pages, 1989.

Vol. 118: L. Dai
Singular Control Systems
IX, 332 pages, 1989

Vol. 119: T. Başar, P. Bernhard
Differential Games and Applications
VII, 201 pages, 1989

Vol. 120: L. Trave, A. Titli, A. M. Tarras
Large Scale Systems:
Decentralization, Structure Constraints
and Fixed Modes
XIV, 384 pages, 1989

Vol. 121: A. Blaquière (Editor)
Modeling and Control of Systems
in Engineering, Quantum Mechanics,
Economics and Biosciences
Proceedings of the Bellman Continuum
Workshop 1988, June 13-14, Sophia Antipolis, France
XXVI, 519 pages, 1989

Vol. 122: J. Descusse, M. Fliess, A. Isidori,
D. Leborgne (Eds.)
New Trends in Nonlinear Control Theory
Proceedings of an International
Conference on Nonlinear Systems,
Nantes, France, June 13-17, 1988
VIII, 528 pages, 1989

Vol. 123: C. W. de Silva, A. G. J. MacFarlane
Knowledge-Based Control with
Application to Robots
X, 196 pages, 1989

Vol. 124: A. A. Bahnasawi, M. S. Mahmoud
Control of Partially-Known
Dynamical Systems
XI, 228 pages, 1989

Vol. 125: J. Simon (Ed.)
Control of Boundaries and Stabilization
Proceedings of the IFIP WG 7.2 Conference
Clermont Ferrand, France, June 20-23, 1988
IX, 266 pages, 1989

Vol. 126: N. Christopeit, K. Helmes
M. Kohlmann (Eds.)
Stochastic Differential Systems
Proceedings of the 4th Bad Honnef Conference
June 20-24, 1988
IX, 342 pages, 1989

Vol. 127: C. Heij
Deterministic Identification
of Dynamical Systems
VI, 292 pages, 1989

Vol. 128: G. Einarsson, T. Ericson,
I. Ingemarsson, R. Johannesson,
K. Zigangirov, C.-E. Sundberg
Topics in Coding Theory
VII, 176 pages, 1989

Vol. 129: W. A. Porter, S. C. Kak (Eds.)
Advances in Communications and
Signal Processing,
Proceedings of an International Conferenc
Baton Rouge, Louisiana, October 19-21, 19
VI, 376 pages, 1989.

Vol. 130: W. A. Porter, S. C. Kak,
J. L. Aravena (Eds.)
Advances in Computing and Control
Proceedings of an International Conference
Baton Rouge, Louisiana, October 19-21, 19
VI, 367 pages, 1989

Vol. 131: S. M. Joshi
Control of Large Flexible Space Structures
IX, 196 pages, 1989.

Vol. 132: W.-Y. Ng
Interactive Multi-Objective Programming
as a Framework for Computer-Aided Contr
System Design
XV, 182 pages, 1989.

Vol. 133: R. P. Leland
Stochastic Models for Laser Propagation
in Atmospheric Turbulence
VII, 145 pages, 1989.

Vol. 134: X. J. Zhang
Auxiliary Signal Design in Fault
Detection and Diagnosis
XII, 213 pages, 1989

Vol. 135: H. Nijmeijer, J. M. Schumacher (Eds
Three Decades of Mathematical System Th
A Collection of Surveys at the Occasion of
50th Birthday of Jan C. Willems
VI, 562 pages, 1989